Rotating Electric Machinery and Transformer Technology

Third Edition

Rotating Electric Machinery and Transformer Technology

Donald V. Richardson, MME, PE
Emeritus
Electrical Engineering Technology
Waterbury State Technical College

Arthur J. Caisse, Jr.
Assistant Professor
Electrical Engineering Technology
Waterbury State Technical College

A Reston Book
Prentice-Hall, Inc.
Englewood Cliffs, New Jersey 07632

Library of Congress Cataloging-in-Publication Data
Richardson, Donald V.
    Rotating electric machinery and transformer tech-
nology.

    "A Reston book."
    Bibliography: p.
    Includes index.
    1. Electric machinery. 2. Electric transformers.
I. Caisse, Arthur J. (date). II. Title.
TK2182.R5 1987   621.31′042   86-17084
ISBN 0-8359-6747-6

Editorial/production supervision and
    interior design: Linda Zuk, WordCrafters Editorial Services, Inc.
Cover design: Lundgren Graphics, Ltd.
Manufacturing buyer: Carol Bystrom

Printed in the United States of America

10 9 8 7 6 5 4 3 2 1

ISBN 0-8359-6747-6   025

Prentice-Hall International (UK) Limited, *London*
Prentice-Hall of Australia Pty. Limited, *Sydney*
Prentice-Hall Canada Inc., *Toronto*
Prentice-Hall Hispanoamericana, S.A., *Mexico*
Prentice-Hall of India Private Limited, *New Delhi*
Prentice-Hall of Japan, Inc., *Tokyo*
Prentice-Hall of Southeast Asia Pte. Ltd., *Singapore*
Editora Prentice-Hall do Brasil, Ltda., *Rio de Janeiro*

Contents

Preface

The years since World War II have seen two major forces develop in engineering education. On one hand the traditional four-year engineering courses are continuing to become more generalized and analytical and, at the same time, less equipment-oriented. This results in competent practicing engineering graduates who cannot work comfortably with actual apparatus—even though they are trained in the design analysis of equipment to be developed. On the other hand, technically-oriented new courses of study have appeared around the country in the community colleges, technical colleges, and private two-year institutions. These courses of study are designed to turn out a student who can both analyze and work with actual apparatus.

Many excellent texts exist that cater to the needs of the new type of four-year college student as far as the engineering study of transient conditions in rotating electric machinery. However, prior to this book, few really suitable texts have been available that cater to the needs of the two-year technical college student or to the four-year Bachelor of Engineering Technology student. This text is specifically prepared to enable the technical college student and the practicing engineer to recognize, understand, analyze, specify, connect, control, and satisfactorily apply the various existing types of electric motors and generators.

Though prices of oil fluctuate, the trend toward increased cost of electrical energy continues. Thus, there is sound reason for this book's emphasis on the efficiency of all machine processes. Strong emphasis is placed on the production and measurement of horsepower or its watt equivalent and the isolation of the accompanying losses. The intended result is that the reader who is trained by these methods will be competent and comfortable with the requirements to analyze or test the power required by a specific machine. The reader will also be able to determine the accompanying losses and specify the most effective machine for the job. He or she will be competent to connect the motor, generator, or transformer with its correct controls and safety devices. Finally, he or she will be able to test the installation with a wide variety of methods in order to realistically use the facilities on hand in a specific laboratory or industrial situation.

As the basis of electrical machine understanding progresses, experiments that illuminate and clarify the principles under discussion are referred to. This enables a school with almost any degree of electric machinery laboratory equipment to actually have its students perform the most relevant possible experiments.

The experimental methods and computational procedures described enable the practicing engineer and the maintenance technician to perform useful development work without specialized apparatus such as dynamometers. Efficiency calculations can be performed readily on machine tools, conveyors, and other installed and working equipment.

It is not expected that students in any one school will perform all of the course study or the related experiments, since more than an academic year's work is included. Rather, the material to be covered should be related to the experiments that use the highest level of equipment available and that most nearly fill the portion of the academic year to be used. The *Solutions Manual* offers suggested material to include or delete to match available time.

The manufacturers of electric motors have heeded the need for greater efficiency by changing their motor design procedures to maximize efficiency rather than minimize size and cost. This has resulted in a widespread trend toward the use of capacitor-run motors in fractional horsepower sizes. This is not a new type of motor (it has been known for years), but it is being used in an area of application that was not previously considered.

Internationally agreed standarization of motor frame sizes in new, logically progressing steps has, unfortunately, lagged. The world has not yet completely accepted System International, or SI, units of measure. It was hoped that by this time International Standards Organization (ISO)-sponsored physical sizes would find widespread agreement and application. Also, it was hoped that the use of SI units would be more fully accepted and that, as a result, the use of conventional English units could be reduced or dropped in this new edition. Neither of these desirable and hoped-for changes has come to pass. Therefore, this third edition continues the parallel and equal emphasis on English units for the regions that will not easily change over, and SI units for the regions that have already made the change. The latest stages of ISO-sponsored, International Electrotechnical Commission (IEC)-suggested standard sizes are presented and explained as well. Unfortunately, the world does not progress as rapidly as we might wish; but progress it does, and this text has endeavored to keep pace with that progress.

Increasing use is now being made of solid-state motor controls. The vast majority of installed motors continue to use the traditional electromagnetic relays and contactors, but newly-designed installations are turning more and more frequently to solid-state controls. Since the use of diodes, transistors, SCRs, DIACs, and TRIACs has moved from the laboratory to the commercially recognized product, it is necessary that the emerging technician and engineer understand their applications in motor control. The major changes in this edition are devoted to solid-state control technology as applied to motors.

At the same time, the new methods for efficiency determination and calibration of motors that have been presented from the first edition have found commercial recognition.

The determination of armature circuit losses under dynamic conditions with what we call Forgue's method has been successfully applied to diesel electric locomotive traction motors in the 900 horsepower range. When combined with rotational loss determinations as described herein, the operating efficiency of a motor under locomotive conditions has been measured without the need for a dynamometer installation.

These new tests have been performed at the Transportation Test Center in Pueblo, Colorado, under the auspices of the Association of American Railroads. The senior author (Richardson) wishes to thank Dr. Conan Furber of the AAR who foresaw the need for specific component efficiency tests; Mr. Joseph J. Schmidt, transportation consultant, who found these methods and recommended them; and Mr. Raymond P. Washburn, Senior Electrification Engineer of the TTC, who directed the test installation and its operation.

It has been the author's experience that the student or practicing engineer who is experienced in this material is well prepared either to go on to a four year engineering degree or to take a responsible position in industry. He is able to make use of motors, generators, transformers, and controls on a confident basis and without requiring help from some intermediary to make the needed installations.

In the preparation of any textbook, certain major areas of aid and decision stand out. Professor Joseph Aidala of Queensboro Community College gave very significant advice in the initial preparation. The freely given permission to use the laboratory and other facilities at the Waterbury State Technical College was of great advantage. The manuscript was edited, proofread, and typed in all editions by the senior author's wife, Betty, whose partnership in time and effort has fully matched his own. The senior author's late father, John W. Richardson, who was a retired mechanical engineer, proofread the entire manuscript and checked all worked example solutions.

Since the senior author is retired from active teaching, and also to assure modern information in the field of controls, a coauthor, Arthur J. Caisse, Jr., has been brought aboard to assure the continuity and periodic updating of this text in the controls area.

The coauthor wishes to thank his wife Ana, daughters Kelly and Coleen, mother Monica, and especially his father, Arthur Sr., for their encouragement and support during the writing of this book.

D.V.R.
A.J.C., Jr.

Rotating Electric Machinery and Transformer Technology

1

Introduction

Some major electrical developments that took place nearly a century ago started a chain of interlocking developments that have permanently changed our lives. Thomas Alva Edison experimentally developed a useful electric light. He immediately foresaw that widespread and effective use of his light would require a practical source and distribution of large amounts of electric power. His development work on steam-engine-driven direct-current (dc) generators enabled the commercial expansion of electric lighting. Figure 1-1 shows a very early dc generator.

Overlapping the early use of electric lighting was the development by Sprague and others of practical dc motors and motor speed controls. This enabled the building of what would now be called urban electric railroad systems. These early trolley cars enabled mass commuting to work and business toward the end of the last century. An early trolley car using dc motors is shown in Fig. 1-2.

1-1 CHANGE OF MAJOR USE TO ALTERNATING CURRENT

The widespread distribution of direct-current electric power has virtually disappeared. The single-unit street-surface electric rail car, as distinguished from the multiple-unit subway car, is about gone. Alternating-current electric distribution became nearly universal by the second half of the twentieth century.

The advantage of alternating current over direct current in power distribution is based on the ease of changing the ac voltage level. High-voltage power lines carry their energy with a correspondingly lower current. From the very basic Ohm's law, the voltage drop in a resistance is related to the current and the

1

Figure 1-1. Edison "Long Legged Mary Ann" dynamo, first commercial generator aboard S.S. *Columbia*, 1879. (*Collections of Greenfield Village and the Henry Ford Museum, Dearborn, Michigan*)

Figure 1-2. Typical urban trolley car. (*Courtesy The Trolley Museum at East Haven, Connecticut*)

Figure 1-3. Modern 345 kV distribution line.

resistance. Since the resistance of a power line, or any conductor, is related to its cross-sectional area, a larger-sized wire will have less resistance. On the other hand, a long-distance line will cost a prohibitive amount if it is made of a large cross section. A modern high-voltage ac distribution line is shown in Fig. 1-3.

By using very high voltage in power distribution lines, the required current can be kept low. This in turn means lower losses in the lines. The problem with dc power is that it can only be generated at a relatively few thousand volts owing to the limitations of the generator's commutator and brushes. Once generated, the voltage cannot be raised to a higher level by convenient means as far as high power level has been concerned. This limitation is being overcome in experimental installations, but it was a very real limit in the past.

Alternating current can be and is generated at 18 000 volts (V) and higher. Figure 1-4 is a typical high-voltage synchronous alternator. Then by use of transformers it can be raised to levels of many hundreds of thousands of volts. One million volt levels seem within practical reach in the near future. This means that enormous power can be carried long distances at low current levels. A large high voltage step-up transformer is shown in Fig. 1-5.

On a smaller scale, voltage levels can be raised or lowered locally in distribution transformers. The ac motors can then be designed and built for any of a number of standardized voltage levels. This ability to conveniently change voltage

Figure 1-4. Typical 150 000 kVA synchronous alternator.

level in a simple, efficient, and long-life device such as the transformer is not available in direct current. A local distribution transformer appears in Fig. 1-6.

Direct-current electric motors now move the vast majority of all railroad trains, but the power is usually developed from an on-board diesel-driven generator system. The dc electric railway that is supplied by central station power is used in only a small number of heavy traffic urban commuter railroads over relatively short distances. Figure 1-7 shows a typical modern locomotive traction motor.

It would appear then that the first developments have passed their cycle of usefulness. A bit of reflection, however, shows that the first uses were vitally necessary steps so that the following developments could have a basis. Charles Steinmetz was the focal driving force behind ac development. He originally understood that alternating current was the practical form of electricity for long-distance power transmission.

1-2 EARLY USE OF MOTORS

Once Edison and his competitors had made electric power available, it became practical to develop the first electric-motor-driven machines and appliances. Most people have not thought about their own dependence upon the enormous use of electric power until it is temporarily interrupted. Unless they pay the bills directly, many people have not thought about the cost of electric power or the importance of using it efficiently.

The importance of electric motors has spread beyond rail transportation until there are more uses than can be listed in any practical space. The ability to start, stop, and even to reverse direction made electric motors a natural power source

Figure 1-5. Typical large high-voltage step-up transformer and associated switching structure.

for machine tools. By the mid-1920s the conversion of machine tool design was virtually complete. After that, even older machines were individually motorized by conversion apparatus. Many older machines are still serving productively today after motorizing conversions.

The difficult and even dangerous task of starting an automobile was taken over by a special version of a dc series motor in about 1915. By 1925, even the lowest-cost automobiles were originally equipped with electric starters. A representative automobile starting motor is shown in Fig. 1-8. The design conversion of aircraft starting began in larger engines late in World War I. This was finally completed when even the small light plane engine was routinely started with an electric motor after World War II.

Similar conversions of starting means have taken place in outboard motor design. Even the larger domestic lawn mower is usually electrically started today.

All the reciprocating engine starting systems require a power storage medium, a starter motor, and a generator to return the power to the storage battery. Even

Figure 1-6. Typical local distribution transformer.

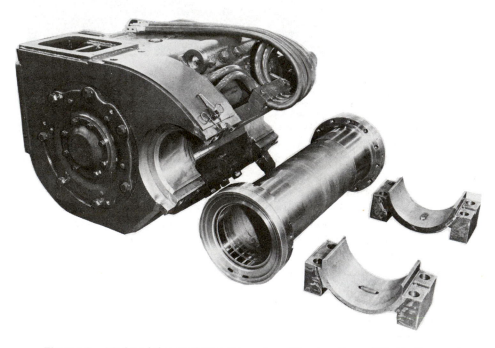

Figure 1-7. Modern dc locomotive traction motor. (*Courtesy General Electric Company*)

Figure 1-8. Automobile starting motor. (*Courtesy Delco-Remy Division, General Motors Corporation*)

in as widely produced item as an automobile, this system is complex and not usually understood.

1-3 ELECTRIC POWER DISTRIBUTION TO HOMES

Household lighting began before the turn of the century in urban areas and neighboring smaller towns.   The rural electrification efforts of the U.S. government made domestic electric power widely available to the most isolated areas by the 1930s.

Very soon after electric power became available for lighting, the start of the electric appliance industry began.   Many appliance applications, such as toasters and flatirons, are based upon resistive heating elements.   However, many more require the use of motor power.

The earliest uses were as labor-saving devices, such as clothes washing machines and water system pumps.   This was soon followed by practical electric-motor-driven refrigerators.   Motor-driven refrigeration is now used in almost all household units in the United States.   Offshoots of these mechanisms have led to the household food storage freezer and to air-conditioning units.   Most food retail stores now have extensive storage and packaged food selection areas that are refrigerated by small to medium motor-driven units.

The use of tall buildings only became practical with the development of safe, flexible passenger elevators.   These have been developed in both ac and dc versions.   Heating, cooling, ventilating, and water service for toilets and fire safety all depend upon electric-motor-driven packaged pump units.

The enormous proliferation of small single-phase motor use is discussed in

Figure 1-9. Group of single-phase appliance motors. (*Courtesy General Electric Company*)

Sect. 20-7. The degree of use of electric motors in the American household usually surprises even the student who has spent a full term studying motors and generators and their methods of control. Most people will guess that the number of motors in their home is less than 10. In reality, 30 to 40 single-phase motors of many different types and sizes seems to be closer to average. Typical single-phase appliance motors are shown in Fig. 1-9.

1-4 ELECTRIC MOTORS NOT WIDELY UNDERSTOOD

It seems incredible that there has been so little education attention paid to the design, application, and control of electric motors and generators. Most people are less able to care for the ills of their motors than they are able to service their far more complicated automobiles. The result is that in small sizes far more motors are junked than are repaired. The usual *repair* for a household appliance is a replacement of the motor or even the whole appliance. The usual *problem* is worn brushes in a universal motor or a failed starting switch in an induction motor.

Many graduate engineers, even electrical engineers, are unable to specify and develop a simple motor application and its associated control.

This textbook is conceived and designed to close this knowledge gap. The major principles are explored and quantified in a step-by-step manner. The basic principles are brought together so that a motor or generator can be understood. The procedures that can be applied for operation and control of all kinds of motors are studied.

Control, in the sense used in this book, means to turn on, to accelerate, to retard, to reverse, or to stop, all in a safe, reliable manner. This text does not treat with transient conditions as they affect the stability and response rate of a modern servomechanism control. Transient control study requires a mathematical background that is beyond the normal two-year college student.

1-5 PREREQUISITE REQUIREMENTS

The analytical approach used in this text is almost entirely algebra. Although most associate degree technical students study mathematics through differential and integral calculus, their real understanding remains keyed to simple algebraic processes. As in any subject, there are basic prerequisites. In the case of this course, in addition to algebra, it is necessary that the student have completed a full term of basic electrical circuits, both alternating and direct current.

It should be possible with this text to understand various dc and ac machines and how they are selected and controlled. It should be possible to follow and understand a typical motor control, such as is used on a machine tool. It should also be possible to correctly apply the many types of motors that are available commercially.

1-6 EFFICIENT USE OF ENERGY

The increasing cost and scarcity of energy is now widely known if not appreciated and understood. Because of energy cost and the moral obligation to use energy wisely, a strong emphasis has to be placed on the efficiency of various types of motors and generators. Definitions of power and efficiency are carefully developed. Energy is the ability to perform work. Efficient use of energy means to use the least energy possible to perform a given task. This sometimes involves the ethics of the task itself, but this is left to other courses to explore whether we should continue to lavishly motorize tasks that can be routinely done by hand or that should not be done at all.

There are many measures of the prosperity, output, or standard of living of a nation as a whole. One basic measure is the average power used by each worker in the performance of his job. In construction work, much of this power is now from diesel reciprocating engines installed in the various construction machines. It is easy to see that more powerful bulldozers or other construction machines can move more earth. Not quite so obvious, but equally true, is the fact that a more powerful machine can shape metal at a higher speed.

At the present, almost all production machinery for wood, metal, or other manufactured items is electric-motor powered. More machine power can produce more goods. More available manufactured goods per worker man-hour of labor

Figure 1-10. High-efficiency machine tool motor. (*Courtesy Gould Inc., Century Electric Division*)

require the application of more machine power per worker. As a result, one index of productivity that seems well correlated to a nation's material standard of living is the available tool power per worker. The steady rise of living standards in the United States and other developed nations is directly tied to the manufacturing use and household appliance use of electric motor power.

Future increases in living standards in the developing nations of the world will depend upon wise and selected increases in available power per worker. Much of the work of the world will continue to be made easier by the reciprocating engine. But whenever electric power can be made available, all sorts of tasks can be flexibly and efficiently performed by use of electric motors. Small motors perform at 50 to 70 percent efficiency. Typical machine sizes perform at 80 percent efficiency and more. This kind of efficiency cannot be approached by any reciprocating or turbine engine. A high-efficiency ac motor having many uses is shown in Fig. 1-10.

1-7 SYSTEM INTERNATIONAL METRIC UNITS

The electrical units of volts, amperes, ohms, and watts and resistance, inductance, or capacitance have long been internationally standardized. On the other hand, the physical dimensions have equally long been divided into units that were originated in England or units that were based on the French meter and gram. This division of measurement standards between English units and metric units is increasingly working to the disadvantage of the United States and its now badly obsolete English-based dimensions.

In the last few years the differences between various metric uses are being overcome. Out of these efforts has come the rationalized meter–kilogram–second (RMKS) system and finally the System International or SI system. As a result, the early centimeter–gram–second system (cgs) is as obsolete as our English system. It is only a matter of a few years before the whole world, including the United States, will legally use SI measurements. Many major nations such as Canada have switched over from English units to SI. Most U.S. corporations that sell significant manufactured products overseas have made, or will very shortly make, the change.

This text therefore works in both English and SI units. All problems involving physical dimensions, speeds, and powers are worked in both systems. It is felt that English-dimensioned tools and equipment will remain in use in decreasing quantities for the working lifetime of today's technical student. On the other hand, most new-product development that a student may be involved with after his graduation will make use of at least partial SI unit work. It therefore seems reasonable that a new technician or technically oriented engineer will need working competence in both systems. The scientist or scientifically based engineer may well make the transition in one stroke as long as he is not involved in existing equipment.

System International covers units of measurement and specifies correct abbreviations of these units. SI specifies usage of multiples and submultiples by increments of 10^3 or 10^{-3} (1000 or 0.001). Another allied usage in numerical work is that no comma should be used for grouping in numbers. This means that the number 123,456 is correctly shown as 123 456. The reason is that in many nations of the world the comma symbol is used as a decimal point. This text uses the space in place of the comma in all numerical work in either English or System International.

A course curriculum may be followed entirely in SI units when using this text if the school has really made the unit changeover decision. No paging back and forth will be necessary. Alternatively, if it is felt that curriculum change must still be delayed, English units may be used exclusively, again without special effort to hunt or to avoid.

1-8 FUTURE OF ELECTRIC MACHINES

Any prognostication of the future use of any device is foolish, and usually misses widely. Either the device does not catch on and quietly disappears or the ultimate uses exceed the developer's wildest dreams. However, certain long-term trends seem apparent, and some future constraints need to be allowed.

Electrical machines are energy-conversion devices. Almost all uses of electricity require conversion from electrical to mechanical energy, or the other way around. Basic energy is released by chemical, mechanical, or nuclear processes. Conversion to the final form of use should be as efficient as possible for both economic and humanitarian reasons. Since electric machines, when properly applied, are efficient and versatile, the broad future seems assured.

Any widespread new use of electric automobiles will probably use motors that are recognizable as related to today's types. High-speed urban commuting railroads may ultimately adopt the linear induction motor. This type is at present a laboratory curiosity that is moving into the automatically controlled machine field. Since it is really an unwrapped induction motor, its analysis and development seem to be dependent upon research and economic comparison study.

New power sources such as solar heating and even solar cooling will require electrically operated pumps and controls. Direct development of electricity from solar cells or hydrocarbon fuel cells will provide electrical energy that will largely be used by various types of electric machines.

New appliance uses are certain to appear and receive powerful commercial promotion. While the electric motor in most of its types is from one to three generations old, its applications are increasing rapidly. Pump packages for solar home-heating systems have only recently appeared on the market, and they make use readily of the more efficient types of existing motors.

Domestic small-scale waste-water filtration, partial purification, and use by recirculation would seem to be a requirement in the future. The systems do not exist now in smaller than factory or urban sizes, but when reduced and packaged for home use in water shortage areas, the use of motor drives seems logical.

Heat pumps for household heating have long been used experimentally. These reverse refrigeration units have a broad future and will need motors and controls.

Wind-driven or tidal-driven generators are another field that may well move from the laboratory curiosity to widely accepted practice. The conversion of these essentially mechanical forms of energy to conveniently distributed electrical energy will require differently proportioned but otherwise familiar generators.

If strip mining of coal or recovery of oil shale or tar sands is to be widely practiced, the processes will require electric motors at many stages. The increased awareness of the need for clean air will mean that many remote operations that are now performed by reciprocating engines with costly liquid or gaseous fuel will be converted to electric-motor drives.

Electric energy generation, distribution, and use will certainly continue.

2

Electromechanical Energy Conversion

Some thought about the subject will reveal that electricity is not often wanted for itself but rather for its ability to serve as an intermediary. Electrical energy is usually not available as such. It must be produced by various other means. In addition, electrical energy is not often wanted as such, but is usually converted at its application point for whatever purpose is needed or desired. There are, of course, exceptions to both of these statements.

2-1 ELECTRICITY IN THE MIDDLE

Electrical energy does have one overriding advantage in that it can be transmitted and controlled more easily than most forms of energy. There is no present competitor for long-distance transmission of energy unless it be petroleum and natural gas in pipeline distribution systems. Here again, the piped petroleum product usually needs some conversion. Natural gas may be piped underground through branching networks out to the individual dwelling unit, where it may be directly burned if heat for cooking or space heating is desired. We soon run out of direct energy-transmission systems.

Electricity is most often generated in large installations, transformed to appropriate voltage in units called *transformers*, and carried overland for various distances. Then it is usually retransformed to lower voltages and finally converted to whatever form is desired.

This book is concerned with the important devices that receive mechanical input energy and convert it to electrical energy, and those devices that receive electrical energy and convert it to appropriate mechanical force or motion. These broad fields of energy conversion are handled at the mechanical-to-electrical end by *generators* and at the electrical-to-mechanical end by *motors*.

A major reason that these devices are so very widely used is that they are relatively efficient and controllable; so much so, that our whole industrialized civilization is inseparably involved with their use.

2-2 SCARCITY OF USEFUL ENERGY-CONVERSION DEVICES

In motors and generators the means of coupling between mechanical and electrical energy is through appropriately located and controlled magnetic fields. There are a few known and commercially practical electromechanical energy conversion phenomena that govern the function of devices other than motors and generators. A few of these are as follows:

(1) Electrostatic forces between the plates of a capacitor, which require very high voltages for comparatively small forces.

(2) Piezoelectric transducer effects by which a crystal is deformed and voltages are produced. Here the motion is exceedingly small even though the forces may be very substantial.

(3) Magnetostrictive effects by which certain magnetic materials change their dimensions minutely under the influence of a magnetic field. Here as in item 2 the motion is small, even microscopic, but the force may be large.

These three phenomena are useful in instrumentation, communication, and control. They are at least substantially reversible in that they work both ways. However, they do not seem suited to handling either high power or continuous motion.

There are other useful energy conversion processes:

(4) Chemical-to-electrical energy, as in batteries. Here some cycles are reversible in that the energy can be put back in and stored. The processes, so far, are inefficient or heavy in relation to their power, or both. Intensive development is underway here.

(5) Thermal energy may be directly converted to electrical energy by thermocouples, but the voltage is low so the principal use is in measurement of temperature.

(6) The whole field of energy release through chemical combustion of gaseous, liquid, or solid hydrocarbon fuels is of enormous importance. These phenomena are not reversible in that we cannot usually put in energy and receive back the original fuel. To create a fuel, at the present time, takes about 15 times the energy that the combustion of that fuel will release. The study of the release of chemical energy in the form of heat and the conversion of the heat energy by the use of an intermediate fluid into mechanical energy is the science of *thermodynamics*. This field is normally studied in a mechanical, chemical, or physics curriculum and is apart from electricity.

(7) In the last generation, the release of heat energy from the fission of heavy

elements such as uranium and plutonium has made the whole new important science of atomic energy. There is promise that still more energetic energy release cycles may be perfected and controlled through nuclear fusion. Some practical means may be found whereby the fusion process may produce electricity directly without intervening thermodynamic cycles. As of now, this is still in the future, although badly needed as soon as possible.

The whole of science and engineering cries out for new and practical energy-conversion processes. The steadily worsening fuel energy availability coupled with the steadily increasing local, national, and world need for usable energy reinforces the need for new fruitful research. Whatever new sources of energy or energy-conversion processes become available, it is likely that at least a large portion of it will ultimately be used in electric motors to perform mechanical work.

2-3 FARADAY'S BASIC DISCOVERY

In 1831, the scientist Michael Faraday discovered that when a mechanical motion was introduced to move a closed circuit conductor in a magnetic field, an electrical voltage was somehow produced that caused a current to flow. Some scholars have gone so far as to say that this was the most important scientific discovery of all time! Even if this is not accepted, it has proved to be one of the primary discoveries of scientific history. All motors and generators depend directly on the applied use of this discovery.

Coupled with Faraday's discovery was the already known fact that magnetism could be generated by an electric current passing through a coil. It was known that like magnetic poles repelled each other and unlike poles attracted each other. Together, these phenomena, ruled by other electrical laws, have made motors and generators possible.

2-4 RELATED FUNDAMENTAL LAWS

All rotating electric machines as well as the related special transducers operate on the same principles and obey the same few fundamental laws. We will isolate and identify these key relationships and credit them to their discoverers:

(1) *Faraday's law of induction* was given quantity relation by Neumann and space relation by Fleming. Lenz further clarified transient relationships in magnetic induction.

(2) *Kirchhoff's voltage and current laws* relating to electric circuits. These enable us to understand and predict relationships in series and parallel arrangements of machine coils and load circuits.

(3) *Ampere's circuital law of the magnetic field*, which enables us to understand and design the magnetic portion of machines.

(4) *Ohm's great fundamental law relating voltage, current, and resistance*, which enables motor and generator coils to be proportioned.

(5) *Biot–Savart's law of force on a conductor in a magnetic field*, which determines output forces produced by a motor and input driving forces required by a generator.

(6) *Watt's determination of the relation between force, work, time, and power*, which quantified the horsepower term. Even though James Watt did not work electrically, his concepts were recognized when the electrical unit of power was named the watt.

2-5 FARADAY'S LAW OF INDUCTION

Up to the time of Faraday's discovery, the only way to develop a continuing electric current had been by chemical-to-electrical energy conversion in a battery. Faraday's great contribution was to show that a mechanical motion of a conductor in a magnetic field would also develop the same electricity. There must have been considerable discussion and thought as to whether it was the same force or not. However, as far as Faraday (or anyone since) could tell, it was the same force and it affected instruments in the same manner. He called it an "induced" voltage because it was developed without contact between the magnet and the conductor. His device was named the Faraday disc and has since been labeled as the *homopolar generator*. This device was simply a conducting disc that was rotated by a hand crank while part of its periphery was inserted between the poles of a magnet. We shall discuss this type of machine later because it has some very interesting properties. The Faraday disc is *not* typical of rotating machines, even though it started the whole field of development. We shall then revert to a manifestation of the same phenomena that can lead to what are now considered as "practical" voltages and currents. Remember that Faraday got there first and was not concerned with practical voltages, since no family of devices was awaiting a commercial birth that was dependent on his generator. They all came later.

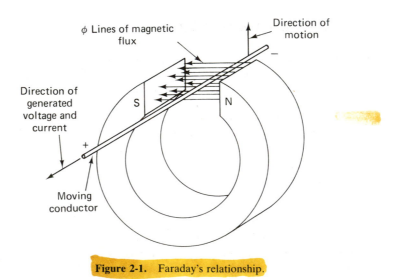

Figure 2-1. Faraday's relationship.

A voltage is developed if a conductor is moved through a magnetic field as shown in Fig. 2-1. *Faraday's law* then can be stated as follows:

> *The voltage induced in a turn or coil of a conductor is proportional to the rate of change of lines of force that pass through the coil.*

Lines of magnetic force are said to *link* a coil of wire when they pass through the coil like the links of a chain. Voltage is induced when the number of linked lines of magnetic force changes, either to more or less.

2-6 NEUMANN'S FORMULATION

Faraday's disc was widely demonstrated and studied as a scientific curiosity, but 14 years after Faraday's discovery Neumann made another breakthrough. Neumann was able to quantity Faraday's work with the following formula:

$$E_{av} = \frac{\Phi}{t} \text{ ab Volts} \quad \text{or} \quad E_{av} = \frac{\Phi}{t} \times 10^{-8} \text{ V} \qquad \text{Eq. (2-1}_E)$$

where the voltage developed was seen to be directly proportional to the *rate of change* of the linked lines of force, or a more rapid change of linkage will increase the generated voltage.

Equation (2-1$_E$) comes from fundamental quantities. Its basis is developed below:

E_{av} = average rectified volts generated per single turn of winding wire so the units are $\dfrac{\text{volts}}{\text{turn}}$

Φ = lines of force (maxwells) that are cut or linked by the single turn when moving from a zero flux to a maximum flux position. In a two-pole machine this takes place in one quarter of a revolution $\dfrac{\text{lines}}{\text{turn}} \Big/ \dfrac{\text{rev}}{4}$ or $\dfrac{\text{lines}}{\text{turn}} \dfrac{4}{\text{rev}}$

Note: Turn is number of loops in a fixed coil. Revolution is mechanical revolving.

t = time in seconds in which Φ lines are linked by the turn of wire as it moves through the magnetic field. Since the maximum change of linkage requires the same quarter-revolution as above, the units are seconds/rev/4. Inverting, we have $\dfrac{\text{rev}}{\text{sec } 4}$

Note: This term is inverted when used and this is permissible.

10^{-8} = relation between the old centimeter–gram–second (cgs) absolute units and modern practical units. In the original relationship, Neumann found $E_{av} = \Phi/t$ ab volts, but there are 10^8 absolute volts (ab volts) per practical volt. Therefore, a single turn must link 10^8 lines in 1 second to induce 1 volt. This becomes 10^8 lines per second per volt or $\dfrac{10^8 \text{ lines}}{\text{sec volt}}$ or, inverting, $\dfrac{10^{-8} \text{ sec volts}}{\text{lines}}$

Note and understand when working with units and stating (some unit per some other unit) that this means two distinct things: (1) *per* means "divided by" or, in this case, 10^8 lines divided by seconds and also divided by volts or (10^8 lines/sec volts) since the second "per" also refers to the original quantity; (2) per also means "equals." When working with *units*, the numerator *equals* the denominator. Since one equals the other, a unit expression of this sort is then algebraically equal to 1/1. This can be inverted without changing its value so that the inversion above is permissible. Another way of seeing this relation is taken from a normal table of equivalents: 1 meter = 39.37 inches. This may be expressed as (1 m/39.37 in.) or (39.37 in./1 m). Both expressions are true. Knowledge of and confidence in these elusive concepts will greatly simplify unit conversion operations. Gathering units together,

$$E_{av} = \frac{\Phi \text{ lines} \times \text{rev}}{\text{turn rev}} \times \frac{\text{rev}}{t \text{ sec} \times} \times \frac{10^{-8} \text{ sec volts}}{\text{lines}}$$

Canceling where applicable,

$$\boxed{E_{av} = \frac{\Phi}{t} \times 10^{-8} \frac{\text{volts}}{\text{turn}}} \qquad \text{Eq. (2-1_E)}$$

This equation holds for cgs units, or for present-day English units since Φ is in *lines*. When the ϕ term becomes webers, as in modern RMKS or in System International units, then, since there are 10^8 lines/weber by definition,

$$\left\langle E_{av} = \frac{\phi}{t} \frac{\text{volts}}{\text{turn}} \right\rangle \qquad \text{Eq. (2-1_{SI})}$$

Note that when using *lines* we use Φ for flux and with webers we use ϕ.

Neumann's original formula, which was quantized from Faraday's work, relates to the voltage generated in a single loop or turn of wire that is moved through a complete linkage of a magnetic field, or Φ lines in t sec. In a practical rotating machine construction of minimum complexity, this takes place four times in a revolution. This is because maximum flux linkage to zero flux linkage requires 90° of motion in a two-pole machine. The rather unorthodox mixture of constants in the dimensional analysis is followed in this explanation because it relates practical machines to construction of the formulas that are usually used.

In Eqs. (2-1) no physical size was assumed for the poles that contained Φ lines, ϕ webers, or the turn of conductor that was swept through the lines of force.

B = flux density in lines per square inch, which is simply total flux Φ divided by pole face area in square inches: $\dfrac{\text{lines}}{\text{in. (in.)}}$

or β' = flux density in lines per square centimeter: $\dfrac{\text{lines}}{\text{cm (cm)}}$

or β = flux density in webers per square meter: $\dfrac{\text{webers}}{\text{m (m)}}$

L = length of an individual conductor that is *within the field* of B density: inches. Within the field refers to the portion of the conductor or winding that intersects magnetic lines and ignores the remainder of a coil that is devoted to practical interconnection to terminals.

or l' = length of an individual conductor that is *within the field* of β' density: cm

or l = length of conductor in field of β density: meters

V = translational or peripheral velocity of the conductor in inches per second: $\dfrac{\text{in.}}{\text{sec}}$

or v' = same in centimeters per second: $\dfrac{\text{cm}}{\text{sec}}$

or v = same in meters per second: $\dfrac{\text{m}}{\text{sec}}$

10^{-8} = same basis as in Eq. (2-1$_\text{E}$) or $\dfrac{10^{-8} \text{ sec volts}}{\text{lines}}$

This holds for English or cgs: $\dfrac{10^{-8} \text{ sec volts}}{\text{lines}}$

The constant does not appear in SI but the *labels* are required.

e_inst = volts generated at any instant of time when the other conditions hold.

Gathering terms together again,

$$e_\text{inst} = \frac{B \text{ lines}}{\text{in. (in.)}} \times L \text{ (in.)} \times V\frac{\text{in.}}{\text{sec}} \times \frac{10^{-8} \text{ sec V}}{\text{lines}}$$

Canceling where indicated yields $\boxed{e_\text{inst} = BLV \times 10^{-8} \text{ V}}$ Eq. (2-2$_\text{E}$)

or: $$e_\text{inst} = \frac{\beta' \text{ lines}}{\text{cm (cm)}} \times l' \text{ (cm)} \times \frac{v' \text{ cm}}{\text{sec}} \times \frac{10^{-8} \text{ sec V}}{\text{lines}}$$

This produces $\boxed{e_\text{inst} = \beta' l' v' \times 10^{-8} \text{ V}}$ Eq. (2-2$_\text{cgs}$)

or: $$e_\text{inst} = \frac{\beta \text{ webers}}{\text{m (m)}} \times l \text{ (m)} \times \frac{\text{vm}}{\text{sec}} \times \frac{\text{sec V}}{\text{weber}}$$

And, in SI, $\langle e_\text{inst} = \beta l v \text{ V} \rangle$ Eq. (2-2$_\text{SI}$)

Example 2-1. If a single conductor is arranged so that 13.5 in. of its length passes through a uniform magnetic field of 52 300 lines/in.² and moves at the rate of 55 in. in 1 sec, determine the generated or induced voltage during any instant. This is a linear or translational velocity.

As in any problem the first step is to determine the most applicable of available formulas. Next, determine if all units fit; if they do not, convert data until they do.

Solution: In this situation Eq. (2-1) does not fit, but Eq. (2-2$_E$) does.

$$e_{inst} = BLV \times 10^{-8} \text{ V, from Eq. (2-2}_E)$$

$$= 52\ 300 \ \frac{\text{lines}}{\text{in. (in.)}} \times 13.5\ \text{in.} \times 55\ \frac{\text{in.}}{\text{sec}} \times 10^{-8} \ \frac{\text{sec volts}}{\text{lines}}$$

$$= 52\ 300 \times 13.5 \times 55 \times 10^{-8} \text{ V}$$

$$= 0.3883 \text{ V}$$

Example 2-2. If a single conductor is arranged so that 34.29 cm of its length passes through a uniform magnetic field of 8106 lines/cm² and moves at the rate of 139.7 cm in 1 sec, determine the generated or induced voltage at any instant. Here all the units are metrically related and Eq. (2-2$_{cgs}$) holds.

Solution:

$$e_{inst} = \beta' l' v' \times 10^{-8} \text{ V, from Eq. (2-2}_{cgs})$$

$$= 8106 \ \frac{\text{lines}}{\text{cm (cm)}} \times 34.29\ \text{cm} \times 139.7\ \frac{\text{cm}}{\text{sec}} \times 10^{-8} \ \frac{\text{sec V}}{\text{lines}}$$

$$= 8106 \times 34.29 \times 139.7 \times 10^{-8} \text{ V}$$

$$= 0.3883 \text{ V}$$

In this case the same absolute flux density, conductor length, and velocity were shown in cgs units and, being the same quantities, they produced the same results. If the magnetic flux density had been stated in (webers/meter²), the length in meters, and the velocity in meters per second, as will soon be conventional under the new SI units, the same equation would hold except that the 10^{-8} quantity would disappear. This form of the equation has been developed as Eq. (2-2$_{SI}$).

Example 2-3.

$$e_{inst} = \beta l v \text{ V, from Eq. (2-2}_{SI})$$

Solution:

$$e_{inst} = 0.8106 \ \frac{\text{weber}}{\text{met (met)}} \times 0.3429\ \text{meters} \times 1.397\ \frac{\text{meters}}{\text{sec}}$$

$$\times 1 \ \frac{\text{sec V}}{\text{weber}}$$

$$= 0.3883 \text{ V}$$

One may well expect during the years to come that mixed units will be encountered as often as not during the transition to SI. For example, in the same simple problem relation as shown in Examples 2-1, 2-2, and 2-3, the quantities could conceivably be very mixed because of the experimental instruments that may be on hand.

Example 2-4. If a single conductor is arranged so that 0.3429 m of its length passes through a uniform magnetic field of 8106 lines/cm² and moves at the rate of 275 ft/min, determine the generated or induced voltage during any instant.

Obviously, none of the variations of Eq. (2-2) fits directly here. A choice must be made and the other units converted to fit the situation. If we choose the modern situation and work in SI, 0.3429 m is correct as it is given. 8106 lines/cm² must be converted to webers/meter² and 275 ft/min must be converted to meters/second.

Solution: From the unit conversion chart at the end of the chapter, under *flux density* find (1 line/cm²) = (10^{-4} weber/meter²). Therefore,

$$\frac{8106 \text{ lines} \times 10^{-4}}{\text{cm}^2} = 0.8106 \frac{\text{Wb}}{\text{m}^2}$$

Next,

$$275 \frac{\cancel{\text{ft}}}{\cancel{\text{min}}} \times \frac{12 \cancel{\text{ in.}}}{\cancel{\text{ft}}} \times 0.0254 \frac{\text{m}}{\cancel{\text{in.}}} \times \frac{1 \cancel{\text{min}}}{60 \text{ sec}}$$

$$= \frac{275 \times 12 \times 0.0254}{60} = 1.397 \frac{\text{meters}}{\text{sec}}$$

Then, from Eq. (2-2$_{\text{SI}}$), $e_{\text{inst}} = \beta l v$ V

$$= 0.8106 \times 0.3429 \times 1.397$$
$$= 0.3883 \text{ V}$$

This example has been performed from four different sets of given data, but all using what are the same absolute values, and therefore producing the same result. This has been done to show that it really does not matter what units are used providing they are used consistently. System International or SI will be really the easiest when we can overcome our lack of familiarity. In general, this text will use both practical English units and System International and will minimize the use of cgs units.

Equation (2-3) is simply Eq. (2-2) with different velocity units, where now

$e_{\text{inst}} = \frac{1}{5}BLV \times 10^{-8}$ V; there is no System International equivalent of this equation due to more rational units

V = *feet per minute* this time instead of inches per second and the constant $\frac{1}{5}$ relates inches per second to feet per minute in the following manner:

$$e_{\text{inst}} = \frac{\overset{1}{\cancel{12 \text{ in.}}}}{1 \cancel{\text{ft}}} \times \frac{1 \cancel{\text{min}}}{\underset{5}{\cancel{60 \text{ sec}}}} \times \frac{B \cancel{\text{lines}}}{\cancel{\text{in.}} \text{ (in.)}} \times L \cancel{\text{in.}} \times V \frac{\cancel{\text{ft}}}{\cancel{\text{min}}} \times 10^{-8} \frac{\cancel{\text{sec}} \text{ V}}{\cancel{\text{lines}}}$$

Canceling,

$$e_{inst} = \tfrac{1}{5}BLV \times 10^{-8} \text{ V}$$

Eq. (2-3$_E$)

Use Eq. (2-1), (2-2), or (2-3) according to the units that are supplied or desired, but *use care*.

Equation (2-4) is a special case of Eq. (2-2) or Eq. (2-3), where the conductor does not move orthogonally with reference to the magnetic field. If the angle θ refers to the angle of motion of the velocity V (or v) in reference to the magnetic flux B (or β), then the new equation becomes, for example, $e = \tfrac{1}{5}BLV \sin \theta \times 10^{-8}$ V, but no units are changed since $\sin \theta$ is a dimensionless multiplier here. These forms are useful in the development and analysis of special transducers. They become

from Eq. (2-2$_E$)

$$e_{inst} = BLV \sin \theta \times 10^{-8} \text{ V}$$

Eq. (2-4$_E$)

from Eq. (2-2$_{SI}$)

$$e_{inst} = \beta l v \sin \theta \text{ V}$$

Eq. (2-4$_{SI}$)

Equation (2-5) derives from Eq. (2-1) by changing the time functions so that E_{av}(volts/turn) or (volts/coil) $= 4\Phi Ns \times 10^{-8}$ V.

$$t = \text{same quantity as in Eq. (2-1),} \quad \text{or} \quad \frac{\text{sec } 4}{\text{rev}}$$

$$s = \text{revolutions in 1 sec,} \quad \text{or} \quad \frac{\text{rev}}{\text{sec}}$$

Note that s is really the reciprocal of t except for the constant multiplier. Using s instead of t as in Eq. (2-1),

$$E_{av} \frac{\text{volts}}{\text{turn}} = s \frac{\text{rev}}{\text{sec}} \times \Phi \frac{\text{lines } 4}{\text{turn rev}} \times 10^{-8} \frac{\text{sec volts}}{\text{lines}}$$

$$= 4s\Phi \times 10^{-8} \frac{\text{volts}}{\text{turn}}$$

which is easily modified to volts per coil by multiplying by N, which is

$$\frac{\text{turns}}{\text{coil}}$$

Thus Eq. (2-5$_E$) becomes

$$E_{av/coil} = 4\Phi Ns \times 10^{-8} \frac{\text{volts}}{\text{coil}}$$

Eq. (2-5$_E$)

Again, when using SI units, the Φ becomes ϕ and the 10^{-8} constant disappears. If the velocity is left in revolutions per second, which is convenient, the equation becomes $E_{av/coil} = 4\phi Ns$(volts/coil). However, in true SI units rotational velocity is measured in ω radians per second, and there are 2π radians per revolution. As a result,

$$\omega\frac{\text{rad}}{\text{sec}} \times \frac{\text{rev}}{2\pi\ \text{rad}} \times \frac{\phi\ \text{webers}\ 4}{\text{turn rev}} \times \frac{\text{sec volts}}{\text{webers}} \times \frac{N\ \text{turns}}{\text{coil}}$$

This becomes

$$E_{\text{av/coil}} = \frac{4\omega\phi N}{2\pi}\frac{\text{volts}}{\text{coil}}$$ Eq. (2-5$_{\text{SI}}$)

or, rearranged,

$$E_{\text{av/coil}} = 0.636\ 62\ \phi N\omega\ \frac{\text{volts}}{\text{coil}}$$ Eq. (2-5$_{\text{SI}}$)

2-7 FLEMING'S RELATIONSHIPS

In electrical work, there are many right-hand rules and/or left-hand rules because they are convenient ways of remembering interrelationships when polarities and directions make a great difference. One of the first of this type is due to Fleming, who related Faraday's work as follows: If the magnetic field is considered as stationary in space, the conductor then is considered as moving orthogonally across it. The right hand is extended with the thumb, index finger, and second finger extended at right angles to each other so that they are orthogonally arranged. With this arrangement, the magnetic field is represented by the index finger with the field represented as traveling from the north to the south pole in the direction that the finger points. If the thumb then is considered as pointing in the direction of the motion of the conductor, the middle finger points in the direction that conventional current will flow. See Fig. 2-2 for this arrangement.

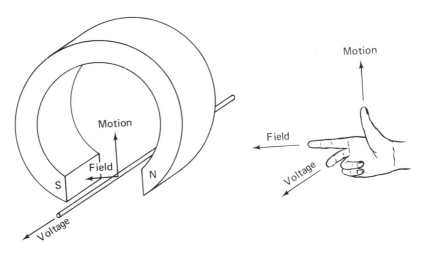

Figure 2-2. Fleming's right-hand rule.

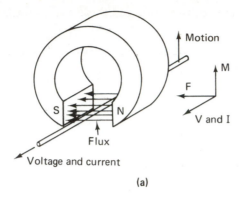

Motion

M

F

V and I

S   N

Flux

Voltage and current

(a)

If the direction of the induced voltage and its resulting current is as shown, following Faraday and Fleming,

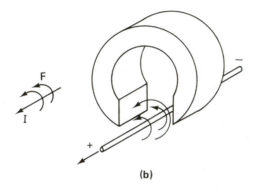

F

I

−

+

(b)

then a magnetic field proportional to the induced current will surround the conductor, in accordance with Ampere's laws.

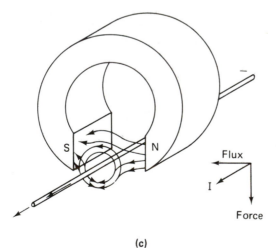

S   N

Flux

I

Force

−

(c)

As a result the combination of the two magnetic fields will *oppose* the motion of the conductor because the fields are in the same direction above the conductor in the diagram and therefore oppose each other. At the same time the fields are in opposite directions below the conductor and therefore attract each other.

Figure 2-3. Magnetic force on a conductor.

Remember that electron flow, which is opposite to conventional current, was unknown when this rule was developed. Benjamin Franklin had arbitrarily identified current flow and polarity a century before electrons were discovered. It is an irony of history that he was wrong on what was a 50–50 choice, but there was no chance whatever that he could have found a rational basis for his choice prior to the later discovery of the electron. He merely organized a positive and negative relationship on an arbitrary basis so that the scientific world could correspond and publish in an agreed fashion. All credit to him.

We shall find later that the left hand can be used in the same way to describe motor action.

2-8 LENZ'S LAW

Still relating to Faraday's pioneering work, Heinrich Lenz stated in 1833, only two years after Faraday's basic discovery, a relationship that turns out to be basic in electromechanical energy transformation:

> In all cases of electromagnetic induction, an induced voltage will cause a current to flow in a closed circuit in such a direction that the magnetic field which is caused by that current will oppose the change that produced the current.

This relationship is really a form of stating the basic fact of the conservation of energy. It is basic to the operation of inductances, transformers, motors, and generators.

What is being stated here as far as motors and generators are concerned is illustrated in Fig. 2-3.

In addition, the field caused by the conductor current is directly proportional to the current. Therefore, the forces caused by these opposing fields are directly related to the conductor current. This force in a generator or a motor is known as motor action. It is important to remember, and we will emphasize it again and again, that this action takes place at the same time as the generator action. Motor and generator action is present at the same time in a rotating electric machine, whether it is acting as a motor or as a generator.

2-9 DIRECT-CURRENT GENERATOR AND MOTOR MECHANICAL CONSTRUCTION

Up to this point we have been concerned with voltages that are generated in single turns of wire or in simple arrangements of multiturn coils of wire. In a practical working machine a mechanical arrangement must be used to support a relatively large number of coils in an arrangement that will allow the conductors of the coils to be moved in and out of a strong magnetic field. Many arrangements have been tried historically, but relatively few have reached and held a production status. The problems involved are interrelated and, as is frequently the case, are sometimes mutually conflicting.

A dense or concentrated magnetic field is necessary to reach a working voltage in a generator with coils of as few turns as possible. The fewer turns allows a large wire diameter in the limited available space. This usually means a flux density B (or β) sufficiently high that it nearly saturates the magnetic field structure. This, in turn, means a relatively large and heavy magnetic structure of specially chosen magnetic alloys. In a practical sense, this means that the bulk of the volume and weight of a motor or a generator is composed of a magnetic structure which is arranged so that the winding coils can be controllably passed into and out of the magnetic flux.

It is worth consideration that there are only a few types of motion that are continuous or cyclically continuous. For example, a straight-line motion may be used to bring a conductor into, through, and beyond a magnetic field. Unless that motion is made to oscillate in a cyclic fashion, it is a "one-shot" proposition. Similarly, an angular motion may be used to pass a coil into, through, and beyond a magnetic field, but unless it is an oscillatory motion it is again a "one-shot" situation. The exception that gives continuous smooth motion is rotary motion. We take it for granted, but rotary motion has no counterpart in nature. It is a man-conceived, mechanically developed way of doing things.

To build an effective magnetic structure that will allow rotary motion, a motor is configured around a cylindrical stack of magnetic alloy discs that are pressed or keyed to a shaft. This rotating magnetic structure or armature is then pierced or slotted to allow room for the coils that are mounted on it.

The magnetic poles that the conductors are passed by during the rotary motion must then face inward toward the cylindrical surface of the rotating structure. They are, in fact, curved on their inner face to conform, at least in part, to the rotating shape. Since a magnetic line of force must be a closed loop, there must be a return path between the outer ends of the magnetic poles. This function is taken by the heavy alloy structure of the main frame. A thoughtful inspection will show that far more material is provided on the exterior of a motor or generator than is needed for structural purposes alone. Conventional names have been developed for these various parts, as shown in Fig. 2-4.

This configuration is still incomplete until end structures or *end bells* are provided, which serve to support the *bearings*, which in turn allow the required rotary motion and yet confine that motion such that no mechanical contact takes place between the armature structure and the field poles. In addition, the bearings are required to confine the endways or axial position of the armature as a whole. Furthermore, the bearings must support any forces due to belts, gears, couplings, or other direct-mounted driving or driven mechanisms.

The magnetic field is usually provided by a set of multiturn *field coils* that surround the field poles in the space between the armature and the main frame. These coils are of various types depending on where their required current comes from. For the present time we will simply consider that enough turns and current will be provided so enough ampere turns exist to produce the required flux in the total magnetic circuit. The relationships that determine the flux will be discussed in Chapter 3.

The armature windings that are contained or mounted in the slots in the armature magnetic structure are proportioned according to the relations in Eq.

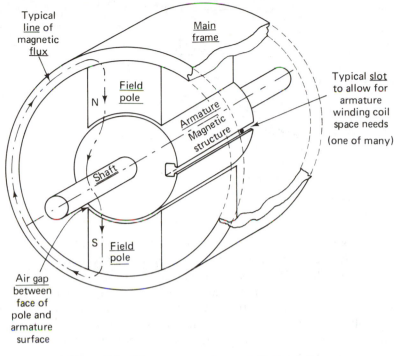

Typical
line of
magnetic
flux

Main
frame

Field
pole

N

Typical slot
to allow for
armature
winding coil
space needs

(one of many)

Armature
Magnetic
structure

Shaft

S   Field
pole

Air gap
between
face of
pole and
armature
surface

Figure 2-4.   Motor or generator magnetic structure.

(2-6) particularly.   However many coils and turns per coil are provided, there must be sufficient space to accommodate them.   It then follows that, if because of a high current which requires a large wire cross section or a high voltage which requires many turns, the physical size of the armature is determined by the dimensions of the windings it must contain and the total pole flux within it.

These windings are interconnected in a few practical configurations, but they always comprise a closed loop or loops of interconnected coils.

The voltages that are generated in an individual conductor actually vary in polarity as first a north pole and then a south pole is passed in conformity with Fleming's right-hand rule (see Fig. 2-5).

At the moment shown in Fig. 2-5, each conductor of the single-turn coil is moving at right angles to the magnetic field.   In this case, for an elementary or discussion generator, it can be assumed that the whole gap between the north and south magnetic poles is filled with a straight uniform field.   In that case, Fleming's rule will show that the right-hand conductor is generating an electromotive force that causes the current to flow into the page.   With the left-hand conductor, the opposite holds.   These two opposite-direction voltages are actually connected in series by the back coil connection at the rear and by any external circuit that is connected across the ends of the coil labeled + and −.

It is good to note in this and many other like diagrams used for illustrative purposes that no such configuration actually exists, because the coil has no support and the magnetic field is across a very substantial pole-to-pole gap in open air.

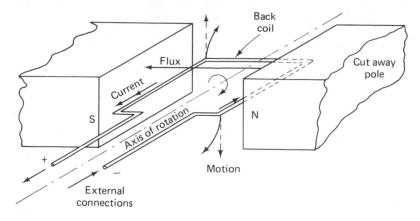

Figure 2-5. Elementary generator.

This type of diagram is used because the coils that are actually used cannot be readily seen. They are buried in the armature magnetic structure. If Fig. 2-6a is consulted, it can be seen that after 90° of rotation the single-turn coil is now moving parallel to the assumed uniform magnetic field. At this position the coil is momentarily not cutting any magnetic linkages and therefore is not generating any voltage. On the next 90° of motion the coil reverses the situation, and the induced voltage is of the opposite polarity from the original situation in Fig. 2-5.

Figure 2-6b shows an idealized or elementary generator case of a uniform field flowing from pole to pole. In this situation the voltage in any one turn will vary sinusoidally. This generation of an alternating current is true in a practical machine also, even though the wave form is not usually a sine wave, since the magnetic field does not flow straight across from a remote pole to a remote pole.

The rectification of this unwanted ac voltage is conventionally performed by the *commutator*. This is simply a rotating mechanical switch composed of insulated segments connected to the ends of the coils. Fixed brushes are arranged to contact the commutator segments, as shown in Fig. 2-7. A study of this illustration will show that the bottom brush in the view is always positive in polarity, since it connects to whichever commutator segment is positive at the moment owing to its position

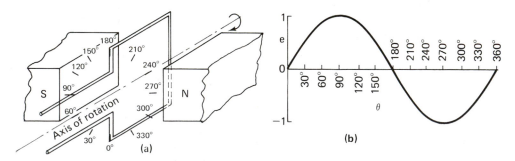

Figure 2-6. (a) Instantaneous angular positions; (b) generated voltage versus angular position.

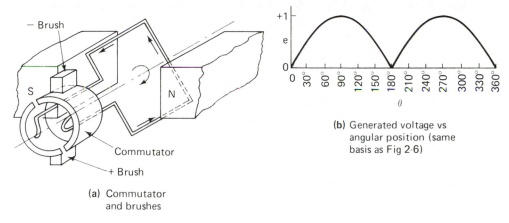

(a) Commutator and brushes

(b) Generated voltage vs angular position (same basis as Fig 2-6)

Figure 2-7. Generated voltage versus angular position (same basis as Fig. 2-6).

and motion in the magnetic field. A little visualization using your hand arranged according to Fleming's rule is necessary for this situation to be clearly understood.

The result of a commutated single-coil elementary generator is then a full-wave rectified alternating current. In a practical dc machine, the magnetic field direction is *not* straight across from field pole to field pole, but radially inward or radially outward at the surface of the armature owing to the magnetic properties of the armature structure. Furthermore, the field will be relatively uniform in strength after an initial entering change from no field to full field as seen by the coil. The result is that the generated voltage per coil is more accurately represented by the flat-top wave form in Fig. 2-8.

When more coils are added and spaced uniformly around the armature, as is always the case in a practical machine, there are always a number of coils generating voltage. These voltages are additive owing to the internal series connections of the coils. The result is a uniform direct current with only small voltage

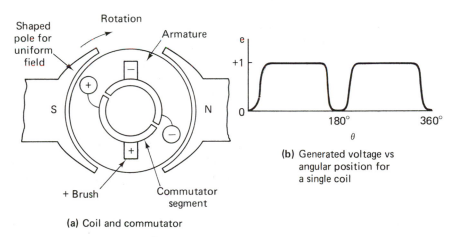

(a) Coil and commutator

(b) Generated voltage vs angular position for a single coil

Figure 2-8. Generated voltage versus angular position for a single coil.

variations as coils are switched in and out. It is usual in a dc machine to have the field poles so proportioned that about 70 percent of the outer surface of the armature is covered by and therefore influenced by the field poles. This is the practical maximum pole area without pole-to-pole flux leakage. For an actual magnetic circuit, it is necessary that there be at least two field poles and that the poles exist by integer numbers of pairs. Thus, a motor or generator is identified as "two pole," "four pole," "six pole," and so on.

Either by having multipoles or by direct result of the winding configuration, there will be some parallel paths in a dc armature. The actual winding configurations are discussed in Chapter 3.

2-10 PRACTICAL GENERATOR VOLTAGE DEVELOPMENT

Equation (2-6) is Eq. (2-5) modified by generator design factors of poles, parallel paths, total turns, etc., and with the velocity factor changed from s to S or rev/sec to rev/min.

Φ = the total flux per pole $\dfrac{\text{lines 4}}{\text{turn rev}}$

Z = total conductors in the armature (note that this is now involving all turns and all coils); Z is then $\dfrac{\text{cond}}{\text{arm}}$

$\frac{1}{2}$ = constant due to the necessity of keeping the constants balanced since Eq. (2-6) is derived from Eq. (2-5) and that equation was derived from volts per turn and volts per coil. This was traceable to Eq. (2-1), which was volts per turn, and there are *always two conductors per turn*, one out and one back, or $\dfrac{1 \text{ turn}}{2 \text{ cond}}$

S = (rev/min), which needs a term to correct it to (rev/sec), which is used in Eq. (2-5), so $\dfrac{1 \text{ min}}{60 \text{ sec}}$

P = number of poles in the field of the machine involved; however, previous formulas have been derived for a two-pole condition since poles must exist in pairs in a symmetrical machine $\dfrac{P \text{ pole pairs}}{2 \text{ gen}}$

a = number of parallel paths in the armature winding, but its *reciprocal* is used to keep units correct $\dfrac{\text{arm}}{a \text{ path}}$

Note: The terms *pole pairs*, *path*, and for that matter, *turn*, *conductor*, *arm*, and *gen* are not units but *labels* used in developing constants. They may be conveniently left in if they cancel or if they are used to develop the term, such as volts per *gen*. But where not used they are correctly left out.

10^{-8} = as before, or $\dfrac{\text{sec volts}}{\text{lines}}$

Gathering terms and canceling, we then have

$$E\frac{\text{volts}}{\text{gen}} = \frac{\Phi \cancel{\text{ lines }} 4}{\cancel{\text{turn}}\cdot\cancel{\text{rev}}} \times \frac{Z \cancel{\text{ cond}}}{\cancel{\text{arm}}} \times \frac{1 \cancel{\text{ turn}}}{2 \cancel{\text{ cond}}} \times \frac{S \cancel{\text{ rev}}}{\text{min}}$$

$$\times \frac{1 \cancel{\text{ min}}}{60 \cancel{\text{ sec}}} \times \frac{P}{\cancel{2}\text{ gen}} \times \frac{\cancel{\text{arm}}}{a} \times \frac{10^{-8} \cancel{\text{ sec}}\text{ volts}}{\cancel{\text{lines}}}$$

$$\boxed{E\frac{\text{volts}}{\text{gen}} = \frac{\Phi ZSP}{60a} \times 10^{-8} \frac{\text{volts}}{\text{gen}}} \qquad \text{Eq. (2-6}_\text{E}\text{)}$$

Note, if a term is not introduced into a formula it does not then exist after cancellation, thus

(1) Without *turn*, Eq. (2-1$_\text{E}$) becomes $E = \dfrac{\Phi}{t} \times 10^{-8}$ V.

(2) Without *coil*, Eq. (2-5$_\text{E}$) becomes $E = 4\Phi Ns \times 10^{-8}$ V.

(3) Without *gen*, Eq. (2-6$_\text{E}$) becomes $E = \dfrac{\Phi ZSP}{60a} \times 10^{-8}$ V.

Equation (2-6$_\text{E}$) contains many terms that are constants by virtue of the units used or the physical construction of a motor or a generator. These are $(ZP/60a) \times 10^{-8}$ and all of these may be expressed as a single constant K. Once a design is fixed, the generated voltage may be conveniently expressed as

$$\boxed{E\frac{\text{volts}}{\text{gen}} = K\Phi S \text{ volts/gen}} \qquad \text{Eq. (2-7}_\text{E}\text{)}$$

Following similar reasoning but using the SI version of Eq. (2-5) to develop an equivalent Eq. (2-6), we then have

$\Phi =$ flux in webers and carries the same added labels as Φ has used; this is flux per pole $\dfrac{\text{webers}}{\text{turn rev}} 4$

$Z =$ same as the English units, $\dfrac{\text{cond}}{\text{arm}}$

$\frac{1}{2} =$ same value, $\dfrac{1 \text{ turn}}{2 \text{ cond}}$

$\omega =$ radians per second; but ϕ carries a label of revolutions, which must be satisfied; thus, $\dfrac{\omega \text{ rad}}{\text{sec}} \times \dfrac{\text{rev}}{2\pi \text{ rad}}$

$P =$ same as the English, $\dfrac{P \text{ pole pairs}}{2 \text{ gen}}$

$a =$ the same, $\dfrac{\text{arm}}{a \text{ path}}$

The 10^{-8} constant does not appear, but we still need its labels with webers:

$$\frac{\text{sec volts}}{\text{weber}}$$

Gathering and canceling, we then have

$$E\,\frac{\text{volts}}{\text{gen}} = \frac{\phi\,\cancel{\text{weber}}\,\cancel{4}}{\cancel{\text{turn}}\,\cancel{\text{rev}}} \times \frac{Z\,\cancel{\text{cond}}}{\cancel{\text{arm}}} \times \frac{1\,\cancel{\text{turn}}}{\cancel{2}\,\cancel{\text{cond}}} \times \frac{\omega\,\cancel{\text{rad}}}{\cancel{\text{sec}}}$$

$$\times \frac{\cancel{\text{rev}}}{2\pi\,\cancel{\text{rad}}} \times \frac{P}{\cancel{2}\,\text{gen}} \times \frac{\cancel{\text{arm}}}{a} \times \frac{\cancel{\text{sec}}\,\text{volt}}{\cancel{\text{weber}}}$$

$$\left\langle\; E\,\frac{\text{volts}}{\text{gen}} = \frac{\phi Z \omega P}{2\pi a}\,\frac{\text{volts}}{\text{gen}} \;\right\rangle \qquad \text{Eq. (2-6}_{\text{SI}})$$

The same comments hold on the SI-related equations; if a term or label is not introduced, it does not appear in the final equation. Here we simply dropped *pole pairs* and *path* because they had served their purpose in developing the constants but are not units.

Equation (2-6$_{\text{SI}}$) may be expressed in simplified fashion much as Eq. (2-6$_{\text{E}}$). Here the constant is $(ZP/2\pi a) = k$.

Thus,

$$\left\langle\; E_{\text{volts/gen}} = k\phi\omega\,\frac{\text{volts}}{\text{gen}} \;\right\rangle \qquad \text{Eq. (2.7}_{\text{SI}})$$

Note that K and k are *not* the same even if Z and P were the same. Be careful when using *any* constant K so that the correct units are achieved.

One may expect to see formulas in many various fashions where the letters used are not standardized. For example, some authors use RPM instead of S, particularly in torque-horsepower formulas. We have chosen to use N as number of turns. We might have used T, for example. There is nothing sacred about this process. The real problem here is to separate correctly the velocity terms, where

ω = angular velocity in radians per second
s = angular velocity in revolutions per second
S = angular velocity or rotational speed in revolutions per minute
t = seconds per one-quarter revolution or simply seconds
V = inches per second or feet per minute linear velocity
v' = centimeters per second linear velocity
v = meters per second linear velocity

The magnetic flux terms are similary confusing and must be treated with due care.

These various formulas are then seen to be directly traceable one to another.

They can therefore be used as may be convenient, depending on whether individual conductors, turns, or multiturn coils are involved. If a length or velocity term is involved, the English or SI metric related terms must be used as good sense dictates. Chapter 3 will help explain the problems of numbers of poles and parallel paths, and the like.

Example 2-5. A two-pole dc generator is built with nine turns per coil, 48 coils, and two parallel paths. The field excitation is adjusted to produce 0.496 × 10^6 lines/pole (0.004 96 Wb/pole) and is rotated at 1750 rpm (183.2 rad/sec). Determine (a) volts per conductor; (b) volts per turn; (c) volts per coil; and (d) total generated voltage.

Note that the various parts of the problem are all related in that they look at successively larger elements of the armature circuit of the generator. All parts can be derived from Eq. (2-1), but other formulas are more convenient after part a. In this stage we have total flux per pole. So far, no size dimensions are given so we cannot develop B or β.

Solution:
(a) Using Eq. (2-1$_E$),

$$E_{av} = \frac{\Phi}{t} \times 10^{-8} \frac{\text{volts}}{\text{turn}}$$

Note that Φ is given as 0.496 × 10^6 lines/pole, but time t must be derived, since for an actual machine of two poles t is the time for one-quarter revolution, so seconds must be in the numerator of the time relation.

$$t = \frac{1}{1750} \frac{\text{min}}{\text{rev}} \times \frac{60 \text{ sec}}{\text{min}} \times \frac{\text{rev}}{4} = 0.008 \ 57 \frac{\text{sec}}{\text{rev}} 4$$

After finding the t term it is used inverted in the denominator.

Admittedly, this is an awkward procedure, but the formula is a generalized case for a single pass through a magnetic field.

$$E_{av} = \frac{0.496 \times 10^6}{0.008 \ 57} \times 10^{-8} \text{ V} = \frac{0.004 \ 96}{0.008 \ 57} = E_{av} = 0.579 \frac{\text{volts}}{\text{turn}}$$

Here and in the remaining parts of Example 2-5, if the flux units are used as webers (0.004 96 Wb), the 10^{-8} adjusting constant at the end of the equation is not applicable since there are 10^8 lines/weber (or 10^{-8} webers/line).

Since there are always *two* conductors per *turn*, one out and one back, the voltage per conductor is 0.579/2 = 0.289:

$$E_{av} = 0.289 \frac{\text{volts}}{\text{conductor}}$$

(b) From above,

$$E_{av} = 0.579 \frac{\text{volts}}{\text{turn}}$$

This can be verified by Eq. (2-5$_E$) if we consider a turn as a one-turn coil.

$$E_{av/coil} = 4\Phi Ns \times 10^{-8} \frac{\text{volts}}{\text{coil}}$$

$$s = \frac{\text{rev}}{\text{sec}}, \quad \text{so} \quad 1750 \frac{\text{rev}}{\text{min}} \times \frac{\text{min}}{60 \text{ sec}} = 29.17 \frac{\text{rev}}{\text{sec}}$$

$$E_{av/coil} = 4 \times 0.496 \times 10^6 \times 1 \times 29.17 \times 10^{-8} \frac{\text{volts}}{\text{coil}}$$

For a *single* turn

$$E_{av/turn} = 0.579 \frac{\text{volts}}{\text{turn}}$$

(c) The voltage for a nine-turn coil is simply part b above $\times N$:

$$E_{av/coil} = 4 \times 0.496 \times 10^6 \times 9 \times 29.17 \times 10^{-8} \frac{\text{volts}}{\text{coil}}$$

$$= 5.209 \text{ V}$$

There are two ways to solve for the total generator voltage, one using the coil voltage multiplied by the number of coils in series, the other using the overall Eq. (2-6$_E$).

(d) Since there are 48 coils in two parallel paths, there must be 24 coils connected in series in each path.

$$E_{gen} = 5.209 \times 24 = 125 \text{ V}$$

Normally, Eq. (2-6$_E$) would be used unless other voltages at various subdivisions were desired.

$$E \frac{\text{volts}}{\text{gen}} = \frac{\Phi ZSP}{60a} \times 10^{-8} \frac{\text{volts}}{\text{gen}}$$

All quantities are given in the statement of the example except Z, the total number of active conductors in the armature.

$$Z = 9 \times 48 \times 2 = 864 \text{ conductors}$$

Note that there are *always* two working conductors per turn of an armature coil in any modern winding type.

$$E \frac{\text{volts}}{\text{gen}} = \frac{0.496 \times 10^6 \times 864 \times 1750 \times 2 \times 10^{-8}}{60 \times 2} \frac{\text{volts}}{\text{gen}}$$

$$= 125 \text{ V}$$

Also note, in the case of a two-pole two-parallel-path armature, the multiplier two (2) appears three times since it is also a component of Z conductors. This is not true on more poles than two or even on a two-pole unit with multiple winding paths where quantity a might be four or six or more. Each of these quantities must be inspected individually for each problem. This will be further explored in Chapter 3.

The generator data developed above is from an actual example, with the construction data taken from a Hampden Engineering Corp. REM-1 school laboratory machine dc unit. Various other facets of this unit, which is a quite typical multiwinding dc machine, will be discussed in later examples. Students may wish to make similar calculations from the data of their specific school laboratory equipment.

This problem may also be followed by a sequence of SI-related equations. Since the absolute values of the data are the same, the results should be the same.

Solution:
(a) Using Eq. (2-1$_{SI}$),

$$E_{av} = \frac{\phi}{t} \frac{volts}{turn}$$

and the same derived t as before, where $t = 0.008\ 57$ (sec 4/turn) (read as second per quarter turn),

$$E_{av} = \frac{0.004\ 96}{0.008\ 57}$$

$$= 0.579 \frac{volts}{turn}$$

This is again the answer to part b. The same logic will produce the answer to part a by dividing volts per turn by 2 to get volts per conductor.

$$\frac{0.579}{2} = E_{av} = 0.289 \frac{volts}{cond}$$

(b) Equation (2-5$_{SI}$) will produce the same volts per turn if we again consider the coil as one turn per coil, so that $N = 1$ for this part.

$$E_{av/coil} = 0.636\ 62\ \phi N \omega \frac{volts}{coil}$$

$$= 0.636\ 62(0.004\ 96)1(183.2)$$

$$= 0.578 \frac{volts}{coil}$$

Note that *coil* is really *turn* since $N = 1$ here.
(c) Again the voltage for a nine-turn coil is simply part b times $N = 9$, using the same equation.

$$E_{av/coil} = 0.636\ 62(0.004\ 96)9(183.2)$$

$$= 5.206 \frac{volts}{coil}$$

(d) Using Eq. (2-6$_{SI}$) for total voltage,

$$E\frac{\text{volts}}{\text{gen}} = \frac{\phi Z\omega P}{2\pi a}\frac{\text{volts}}{\text{gen}}$$

$$= \frac{0.004\ 96(864)183.2(2)}{2\pi 2}$$

$$= 124.9\ \frac{\text{volts}}{\text{gen}}$$

2-11 BIOT-SAVART FORCE RELATIONSHIP

The law of the force on a conductor when in a magnetic field, which quantifies the effect in Lenz's law, is named after Biot and Savart. This law relates the magnetic flux per unit area B (or β), the length of the conductor that is immersed in the magnetic field L (or l), and the current I, which together produce a force in whichever units are consistent.

The basic law was proportioned in the old cgs or centimeter–gram–second units, as they became known. With this unit system we use the prime (′) designations. Originally, absolute amperes were used where 1 ab ampere = 10 amperes.

If we use modern ampere units, then starting from the Biot–Savart relation,

$$F = \frac{\beta' I l'}{10}\ \text{dynes}$$

where

β' = lines per square centimeter, $\dfrac{\text{lines}}{\text{cm (cm)}}$

I = current in the conductor, amp

l' = conductor length in the field, cm

β' = Φ lines divided by area A in cm²; a line is a unit of magnetic force reduced from the relationship

$$1\ \text{line magnetic flux} = \frac{1\ \text{dyne of force at 1 cm distance}}{10\ \text{A in conductor}}$$

The 10 is from 10 A = 1 ab ampere again, so that lines equal $\dfrac{\text{dyne cm}}{\text{amp 10}}$

Also, β' flux per unit area $= \dfrac{\Phi\ \text{flux}}{A\ \text{area}}$.

In cgs units area is in square centimeters, cm (cm). Gathering units and canceling, we have

$$f' = \frac{\Phi \text{ dyne } \cancel{\text{cm}} \; I \; \cancel{\text{amp}} \; l' \; \cancel{\text{cm}}}{\cancel{\text{amp}} \; 10 \; A \; \cancel{\text{cm}} \; (\cancel{\text{cm}})} = \frac{\Phi I l'}{A \; 10}$$

$$\boxed{f' = \frac{\beta' \; I l'}{10} \text{ dynes}} \qquad\qquad \text{Eq. } (2\text{-}8_{\text{cgs}})$$

By this process we have established the original rational basis so that we can modify the units to the "practical" English units or to the new SI units. Note that SI closely follows the previous rationalized meter–kilogram–second or RMKS system, except that a more rigorous use of units times 10^3, 10^6, or 10^{-3}, 10^{-6}, etc., is used and deci, centi, etc., are not used.

In the English system of units, Eq. (2-8) usually appears as

$$F = \frac{BIL \times 10^{-7}}{1.13} \quad \text{or} \quad F = \frac{BIL}{0.113 \times 10^8} \text{ lb}$$

The constants $10^{-7}/1.13$ or $1/0.113 \times 10^8$ or 0.8849×10^{-7}, or even other versions, which are of course equivalent to one another, come from

$$1 \text{ line } = \frac{\text{dyne cm}}{\text{amp } 10}$$

which can be corrected to English units:

$$1 \text{ line } = \frac{\cancel{\text{dyne}} \; (0.22481 \times 10^{-5} \text{ lb}) \cancel{\text{cm}} \; (\text{in.})}{\text{amp } 10 \qquad \cancel{\text{dyne}} \qquad 2.54 \cancel{\text{cm}}}$$

which is

$$\frac{1 \text{ lb in. } \times 10^{-7}}{\text{amp} \times 1.13}$$

$B = \Phi$ lines/A in.^{2}; then

$$B = \frac{\text{lb } \cancel{\text{in.}} \times 10^{-7}}{\text{amp (in.) } \cancel{\text{in.}} \times 1.13}$$

I = current in the conductor in amperes
L = length of the conductor in the field in inches

Gathering the terms and canceling we have

$$F = \frac{B \text{ lb}}{\cancel{\text{amperes in.}}} \times I \cancel{\text{amperes}} \times L \cancel{\text{in.}} \times \frac{10^{-7}}{1.13}$$

$$\boxed{F = \frac{BIL}{1.13} \times 10^{-7} \text{ lb}} \qquad\qquad \text{Eq. } (2.8_{\text{E}})$$

one piece of wire

In SI, Eq. (2-8) becomes $F = \beta Il$ newtons, where 1 Wb = 1 newton meter/ ampere, and 1 Wb = 10^8 lines by definition. In this case the weber is the unit for flux, ϕ, and meters^2 is the unit for area A. Then

$$\beta = \frac{\phi \text{ newton meter}}{\text{ampere } A \text{ meter}^2} \quad \text{so} \quad \frac{\text{newton}}{\text{amp meter}}$$

Again, I is the current in the conductor in ampere and l is the length of the conductor, this time in meters.

Then gathering terms and canceling

$$f = \frac{\beta \text{ newton}}{\text{amperes meter}} \times I \text{ amperes} \times l \text{ meter}$$

$$f = \beta Il \text{ newton} \qquad\qquad \text{Eq. (2-8}_{\text{SI}}\text{)}$$

Again we can work in any of these unit systems. Practical motor and generator work in the United States has mostly been performed in English units in the larger sizes and cgs units for instrument sizes. Both of these will inevitably change to SI units, and competence should be achieved in all units at this time. The only real difficulty will be in visualizing the size of the basic units, since flux density quantities, for example, cover a range of 10^4 to 1 or 10 000 to 1 for the *same* flux.

The use of the Biot–Savart law is not confined to a motor but is essential in both a motor and a generator. Any conductor that is moving across a magnetic field and at the same time carrying a current will exert a force on its confining structure. This force is termed *motor action*. The arrangement of the conductors used in a motor or generator to produce their generator action results in the conductors being in the influence of the magnetic field. They are also producing a force lateral to that field. The force is proportional to the current in the conductor. The remaining conductors that are carrying the same current but are not in the influence of the magnetic field do not produce any motor action. The use of Eq. (2-8$_{\text{cgs}}$), (2-8$_{\text{E}}$), or (2-8$_{\text{SI}}$) will be developed in later chapters, where the effect of armature radius and rotative speed will be shown. At this point, bear in mind that *the conductor that is generating a voltage is also subject to a mechanical force*. Both effects are present.

2-12 LEFT-HAND RULE OF MOTOR ACTION

Fleming's right-hand rule relates the direction of magnetic flux, the direction of motion, and the direction of the resulting induced voltage polarity. A similar mutually orthogonal relationship exists between the direction of the magnetic field flux, the direction of the imposed voltage and current, and the direction of the resulting mechanical force or motor action.

The left hand fits this force relationship if we again let the index finger point in the direction of the flux from north to south and let the middle finger point in

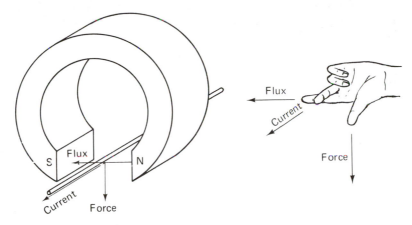

Figure 2-9. Left-hand rule.

the direction of the imposed voltage and its resulting conventional current flow. Under these conditions, the thumb will point in the direction of the force that is developed from the Biot–Savart relationship. See Fig. 2-9 and compare it with Fig. 2-2, which shows the Fleming rule relations.

In the usual situation both rules are valid. In a generator, the current resulting from the generated voltage flowing through the load circuit will produce a force that opposes the motion that caused the voltage generation. This opposing force will be proportional to the load current, and is the reason a generator requires more driving force as its electrical load is increased. In a motor, the motor torque or turning effort is proportional to the current that is flowing from the electrical power source. At the same time, the motor is generating a voltage according to Fleming's right-hand rule, which opposes the line voltage direction. A motor will then run *just fast enough* to develop just enough opposing voltage, or *back emf*, that it will limit its current to just enough to supply the required torque.

Example 2-6. Each armature winding conductor in a direct current generator is subject to the influence of the magnetic field for 10 in. of length (0.254 m). If, as a result of the generated voltage, a current of 42.5 A flows, how much force is felt by the conductor if the field is 102 000 lines/in.[2] (1.581 Wb/m²).

Without attempting to convert this force into power at this stage it is still useful to realize that we can directly do so when more of the relationships have been discussed.

Solution:

$$F = \frac{BIL}{1.13} \times 10^{-7} \text{ lb, from Eq. (2-8}_\text{E})$$

$$= \frac{102\ 000 \times 42.5 \times 10 \times 10^{-7}}{1.13} \text{ lb}$$

$$= 3.84 \text{ lb}$$

In SI, this is

$$F = \beta Il \text{ newton, from Eq. (2-8}_{\text{SI}})$$

$$= 1.581(42.5)0.254$$

$$= 17.1 \text{ newton}$$

As a check, there are 4.4476 newtons/lb. Therefore,

$$3.84 \ \cancel{lb} \times 4.4476 \ \frac{\text{newton}}{\cancel{lb}} = 17.1 \text{ newton}$$

2-13 OTHER RELATED LAWS

The great fundamental Ohm's law is of course basic to all kinds of electrical and electronic work, and the relation of $E = IR$ will be used again and again as this text develops. This equation holds true for all dc situations and for resistive ac conditions.

Ampere's circuital law is analogous to Ohm's law of magnetism or magnetic circuits, and this will be developed in simplified practical detail in later chapters after the fundamentals of motor and generator mechanical construction are developed.

James Watt's work on power will be developed in later discussions on motors and their power relations. It will be shown that the same laws hold for generators as for motors, but this belongs later on.

The armature circuit that is used to develop the voltage in a generator or torque in a motor always has some internal resistance. This resistance in turn causes a voltage drop under either a motor or generator situation. This means a generator actually generates more voltage than it delivers to a load. It is easier to see this relationship if we use E for a voltage that is generated and can be calculated, and V for a voltage that can be measured with a voltmeter. For a *generator*,

$$V_t = E_g - (I_a R_a)$$

Eq. (2-9)

where t = terminal, g = generated, and a = armature.

In a *motor*, the internally generated voltage cannot exceed the line voltage or no motoring current will flow and the unit is then generating. It is acting truly as a motor if it is drawing current from the line and producing torque. It will then show the following relation:

$$V_t = (I_a R_a) + E_g$$

Eq. (2-10)

It turns out that the same unit, without changing connections, may serve as either a motor or a generator depending upon circumstances.

Example 2-7. A dc machine is generating 125 V while delivering 8 A to a load. If its armature circuit total resistance is 1.35 Ω, what voltage must be generated internally in the armature?

Solution:

$$V_t = E_g - (I_a R_a), \text{ from Eq. (2-9)}$$

Thus

$$E_g = V_t + (I_a R_a)$$

$$= 125 \times (8 \times 1.35) = 135.8$$

$$= 135.8 \text{ V}$$

This generator must have been operating with a field flux and rpm such that it generated 135.8 V. If either its rpm or its field flux (or both) change, it will deliver a different current or voltage (or both) depending on the line conditions.

Example 2-8. The same machine in Example 2-7 is momentarily turning at only 90 percent of the rpm that it had in Example 2-7, and the line voltage is held at the same 125 V by external means such as batteries or other generators. The same line voltage will mean the same field flux.

    (a) What is the internal generated voltage now?
    (b) Is the unit acting as a motor or as a generator?
    (c) What armature current is flowing?

If the machine is turning 90 percent as fast, it is generating 90 percent as much voltage since the generated voltage is directly proportional to rpm.

Solution:
    (a) $E_g = 135.8 \times 0.9 = 122.2$ V.
    (b) Since the internally generated voltage is *less* that the line voltage, the unit draws current rather than supplying current. *The unit is acting as a motor.*
    (c) $125 - 122.2 = 2.8$ V drop in the armature. From Ohm's law, $I = E/R$. Therefore,

$$\frac{2.8 \text{ V}}{1.34 \ \Omega} = 2.07 \text{ A}$$

$$I_a = 2.07 \text{ A}$$

The current is flowing in the opposite direction to the conditions in Example 2-7. This can be checked using Eq. (2-10):

$$V_t = (I_a R_a) + E_g$$

Thus

$$I_a R_a = V_t - E_g \quad \text{and} \quad I_a = \frac{V_t - E_g}{R_a}$$

$$I_a = \frac{125 - 122.2}{1.35} = 2.07 \text{ A}$$

These two situations ignore the relatively smaller current that must also flow in the machine's field coils. In this situation, this field current might be around 0.6 or 0.7 A. Our example is true if the field were separately excited, that is, if the field were supplied independently from a separate line source. This will be explored in later chapters.

2-14 UNIT CONVERSIONS

The scientific and engineering world has long worked in a welter of engineering units that can perhaps be partially explained by historical facts. The very fundamental units of engineering measurement that we use in the United States and the English-speaking world came from the archaic units of measure dating from the Middle Ages.

On the other hand, the basic electrical scientific work was largely done on the European continent using the various metrically derived units that date from the Napoleonic era. Finally, it was in the United States that practical commercial electric machinery was first manufactured and then gradually engineered and scientifically systemized. The early work was done in pounds and inches mixed with volts, amperes, and ohms and flavored with lines or maxwells.

Finally, and at long last, order is appearing by international agreement, and linear measure is more and more rapidly becoming metrically based. The simple conversion of 2.54 cm = 1 in. comes from the practical pioneering work by Johansson on his gage blocks. Since at the time he was the only worker capable of measuring in millionths of an inch, he decided quite logically to cut off the equivalence to three significant figures and there it has remained.

A concerted international effort has resulted in Systeme Internationale or System International, or simply SI. It is only fair to state that this coordinated related system could not have been originally foreseen, since the scientific knowledge needed was not available. On the other hand, it is about 50 years overdue.

It is some comfort to note that the various scientifically advanced European nations themselves are having to accommodate considerable adjustments. This is because they variously worked in cgs, mks, and rationalized mks (RMKS) systems and some worked in centimeters, millimeters, or even decimeters.

SI is a clean sweep to work in meters and various multiples of 10^3—thus 10^{-6} meters, the micrometer; 10^{-3} meters, the millimeter; meters; and 10^3 meters, which is the kilometer. Similar unit work in simplifying exponents will cause some perfectly good metrically based terms to disappear. For example, the basic flux unit of the weber has a one-to-one relation with the newton of force, the meter of

TABLE 2-   UNIT CONVERSIONS USEFUL IN MACHINE CALCULATION

Units	cgs	English	English	RMKS or SI
Length	l'	L	L	l
	1 cm =	0.393 70 in. =	0.032 808 ft =	10^{-2} m
	2.54* cm =	1 in. =	0.083 333 ft =	0.0254 m
	30.48 cm =	12 in. =	1 ft =	0.3048 m
	10^2 cm =	39.370 in. =	3.2808 ft =	1 m
Area	a'	A	A	a
	1 cm^2 =	0.155 00 in.^{2} =	0.001 076 4 ft^2 =	10^{-4} m^2
	6.4516 cm^2 =	1 in.^{2} =	0.006 944 4 ft^2 =	6.4516×10^{-4} m^2
	929.03 cm^2 =	144 in.^{2} =	1 ft^2 =	0.092 903 m^2
	10^4 cm^2 =	1550.0 in.^{2} =	10.764 ft^2 =	1 m^2
Volume	u'	U	U	u
	1 cm^3 =	0.061 024 in.^{3} =	$0.353\ 15 \times 10^{-4}$ ft^3 =	10^{-6} m^3
	16.387 cm^3 =	1 in.^{3} =	5.7870×10^{-4} ft^3 =	16.387×10^{-6} m^3
	283 17 cm^3 =	1728 in.^{3} =	1 ft^3 =	0.028 317 m^3
	10^6 cm^3 =	610 24 in.^{3} =	35.315 ft^3 =	1 m^3
Force	f'	F	f	Obsolete metric f''
	1 dyne =	$0.224\ 81 \times 10^{-5}$ lb =	10^{-5} newton =	$0.101\ 97 \times 10^{-5}$ kgf
	4.4482×10^5 dyne =	1 lb =	4.4482 newton =	0.453 59 kgf
	10^{5*} dyne =	0.224 81 lb =	1 newton =	0.101 97 kgf
	9.8066×10^5 dyne =	2.2046 lb =	9.8066 newton =	1 kilogram force
Flux	ϕ'	Φ		ϕ
	1 line =	1 line	=	1 weber $\times 10^{-8}$
	10^{8*} line =	10^8 line	=	1 weber
Flux Density	β'	B		β
	$\dfrac{1 \text{ line}}{\text{cm}^2}$ =	$\dfrac{6.4516 \text{ line}}{\text{in.}^2}$	=	$\dfrac{10^{-4} \text{ weber}}{\text{m}^2}$
	$\dfrac{0.155\ 00 \text{ line}}{\text{cm}^2}$ =	$\dfrac{1 \text{ line}}{\text{in.}^2}$	=	$\dfrac{0.155\ 00 \times 10^{-4} \text{ weber}}{\text{m}^2}$
	$\dfrac{10^4 \text{ line}}{\text{cm}^2}$ =	$\dfrac{6.4516 \times 10^4 \text{ line}}{\text{in.}^2}$	=	$\dfrac{1 \text{ weber}}{\text{m}^2}$

*Definitions:
2.540 000 0 cm = 1 in.
10^5 dyne = 1 newton

10^8 lines = 1 weber = $\dfrac{1 \text{ newton meter}}{\text{ampere}}$

$\dfrac{1 \text{ weber}}{\text{meter}^2} = \dfrac{1 \text{ newton}}{\text{meter ampere}}$

NOTE: 1 line = 1 maxwell.
  1 line/cm^2 = 1 gauss.

distance, and the ampere. The line that is 10^{-8} Wb will probably ultimately disappear. If historical coincidence had resulted in a 10^3 or 10^6 or 10^9 relationship, it might still be used, but 10^8 does not fit. The mixed flux density unit of lines per square inch has always been awkward and will speedily disappear.

However, at this time and for a substantial part of the working lifetime of those who read this text, work with existing apparatus that was calculated and rated in various unit systems will be required. It is only prudent to be able to move from one system to another.

For convenience, relationships are shown in Table 2-1. Working from a typical line in the table of equivalents, suppose that it is desired to convert 1.273 ft to meters. Entering the third line under length, where 1 ft appears, find 1 ft = 0.3048 m. Since the original data that are to be converted are, in effect, *numerators*, the 1 ft of the equivalent must be a *denominator* in order that feet will *cancel*. 1.273 f̶t̶ × (0.3048 m/1 f̶t̶) = 0.3880 m.

Working in this fashion, and being careful *always to write down the units* and the *conversion fraction chosen* such that the *unwanted units cancel*, a reliable conversion will result. Inspection will show that the 1 ft denominator term is unnecessary if you are familiar with the particular conversion. In fact, the unit conversion table is written in such fashion that one can always find the given unit as a unity value. However, write it out in *any* new situation until you are sure!

QUESTIONS

2-1. Give an example of why electricity is an intermediate energy form.

2-2. Name a direct use of electricity.

2-3. Define energy conversion.

2-4. What is a reversible energy-conversion process?

2-5. What was Michael Faraday's basic discovery?

2-6. Name one archaic unit that was in regular use when Neumann found the quantity relations in Faraday's law of induction.

2-7. What is the reason for the differences between the two forms of Eq. (2-1)?

2-8. Why is Eq. (2-1) given in E_{av} and Eq. (2-2) given in e_{inst}?

2-9. What is the specific difference between Eqs. (2-3$_E$) and (2-2$_E$)?

2-10. How is the type of difference in Question 9 avoided in SI unit formulation?

2-11. How is the velocity function accounted for in Eq. (2-1)?

2-12. What change in velocity function is made from Eq. (2-1) to Eq. (2-2)?

2-13. What further change in velocity function is used in Eqs. (2-5), (2-6), and (2-7)?

2-14. The constant K or k in Eq. (2-7) in either unit system contains what specific design quantities?

2-15. What factors are contained in Z in Eq. (2-6)?

2-16. In a dc machine, what is the armature?

2-17. What is the function of a commutator?

2-18. What are the field poles in a dc machine?

2-19. Why is the outer main frame shell so heavily proportioned in a dc machine?

2-20. What is the function of the brushes?

2-21. Why is the armature magnetic structure laminated?

2-22. What are three functions of the armature shaft bearings?

2-23. State Eq. (2-8$_{SI}$) in sentence form accounting for all of the units involved.

2-24. When is motor action present in a generator?

2-25. When a unit is acting as a generator, in which direction does the motor action take place?

2-26. Why does the terminal voltage of a generator differ from the internally generated voltage when the unit is loaded?

2-27. Why can the numerator and denominator of a unit conversion relation be interchanged in position?

PROBLEMS

2-1. What is the average voltage generated in a one-turn loop of wire under the following conditions: the loop is originally linked with a magnetic flux having 3.75×10^6 lines and then removed from the flux in 0.12 sec?

2-2. How many volts are generated in each turn of a coil of wire that is moved into linkage with a magnetic field of 0.0535 Wb in 0.203 sec?

2-3. A single conductor is moved across a magnetic field of 43 200 lines per in.^{2}; 4 in. of the conductor is influenced by the field and it is moved at 60.5 in./ sec. What instantaneous voltage is generated?

2-4. If a conductor of 13.3 cm length is entirely immersed in a magnetic field of 9235 lines/cm^2 and moved across the field at a speed of 193 cm/sec, what voltage is generated over the length of the conductor?

2-5. A conductor that is 35.3 mm long is moving at a velocity of 2.33 m/sec across a magnetic field of 0.883 Wb/m^2. What voltage is developed?

2-6. A magnetic field of 8325 gauss (lines/cm^2) that is 1.12 ft wide is swept transversely by a conductor at a speed of 36.3 in./sec. Find the generated voltage using SI units.

2-7. A gaussmeter shows a magnetic field to be 9275 gauss. A conductor with 6.38 in. effective length is moved orthogonally across the field at 885 ft/min. Find the instantaneous voltage using Eq. (2-3$_E$). Note that 1 gauss = 1 line/ cm^2.

2-8. A magnetic velocity transducer is mounted on a test rig so that the turns of its pickup coil sweep through the magnetic field at an angle of 30.00°. The magnetic field strength is 10 250 lines/in.^{2}, and the effective conduction length per turn of the coil is 1.000 in. How much voltage is shown at the instant the coil velocity is 22.30 m/sec? Solve in both English and SI units.

2-9. A two-pole dc generator has individual armature coils with 12 turns of wire each. The field poles have a total effective flux of 403 000 lines/pole and the armature turns at 20.3 rev/sec. Find the average voltage per coil.

2-10. An armature has three turns/coil and rotates at 188.5 rad/sec. The two field coils have 0.0333 webers flux. What average voltage is developed in a coil?

2-11. A dc generator has the following specifications: total flux per pole is 778 000 lines, 6 poles, 72 coils of 4 turns each. The windings are simplex lap so that there are six parallel paths, and the machine rotates at 1800 rpm. What voltage is generated?

2-12. Using the data of Problem 2-11, convert the data to its SI equivalent and solve for the generated voltage using Eq. (2-6$_{SI}$).

2-13. It is desired to check the voltage output of the generator in Problems 2-11 and 2-12 over a range of speeds and field flux. Find the appropriate K and k for use in repetitive calculations using Eqs. (2-7$_E$) and (2-7$_{SI}$).

2-14. A meter movement has its moving coil conductor immersed in a magnetic field of 2210 lines/cm². The conductor length in the field is 1.67 cm and the current in the coil is 35 mA. Find the developed force per conductor.

2-15. An armature conductor carries a current of 12.5 A and has an effective length of 6.63 in., which is acted upon by a magnetic flux of 62 800 lines/in.². What is the lateral force on the conductor?

2-16. The same armature and field in Problem 2-15 produces what SI unit force when the current is 10.3 A and the flux is 0.608 Wb/m²?

2-17. A generator produces a generated voltage of 125.0 A. It delivers 10.6 A to the load through its armature circuit resistance of 1.22 Ω. What terminal voltage results if there is no other voltage loss than the armature circuit resistance?

2-18. If the same machine as in Problem 2-17 is being run as a motor with the same flux and at the same speed, what terminal voltage would be needed to supply the same current?

Answers. **2-1**, E_{av} = 0.313 V; **2-2**, E_{av} = 0.264 V; **2-3**, e_{inst} = 0.105 V; **2-4**, e_{inst} = 0.237 V; **2-5**, e_{inst} = 0.072 V; **2-6**, e_{inst} = 0.2620 V; **2-7**, e_{inst} = 0.676 V; **2-8(a)**, e_{inst} = 0.0450 V, **(b)** e_{inst} = 0.0450 V; **2-9**, $E_{av/coil}$ = 3.93 V/coil; **2-10**, $E_{av/coil}$ = 12.0 V/coil; **2-11**, E = 134 V; **2-12**, E = 134 V; **2-13(a)**, K = 9.6 × 10^{-8}, **(b)** k = 91.673; **2-14**, F = 12.9 dyne; **2-15**, F = 0.461 lb; **2-16**, F = 1.05 newton; **2-17**, V_t = 112.1 V; **2-18**, V_t = 137.9 V.

3

Actual Machine Construction, dc Dynamos

Chapter 2 introduced the required geometric shape and arrangement of the principal parts of a dc generator or motor. We shall now expand that description so that actual construction, which is always constrained by available materials and machine tools, may be understood.

3-1 MECHANICAL CONSTRUCTION

The simple illustration in Fig. 2-4 must be expanded to include all required practical details. Conventional construction of dynamos will be discussed first.

3-1-1 Main Frame. The *main frame* is required to serve as a return path for all the circulating magnetic flux that passes from the field poles to the armature. This flux-carrying requirement determines the needed cross section of magnetic material, usually carbon steel. As a result of the needed flux-carrying ability, there is usually far more metal than is needed for structural strength requirements. The most usual construction is a rolled ring structure with the end closure butt welded in automatic machinery which, in turn, leaves a visible weld. This ring structure must be lathe turned so that its inside surface is a true cylinder and so that its ends are square with the bore. These steps are required to maintain the geometric dimensions that are necessary so that the finished machine will assemble without unnecessary adjustment, and so that the field pole shoes will fit concentrically around the armature. Some form of mounting feet are usually welded onto this frame structure. See Fig. 3-1 for a representative main frame unit. The alternative mounting is to use the end bells for bolting surfaces. This structure is

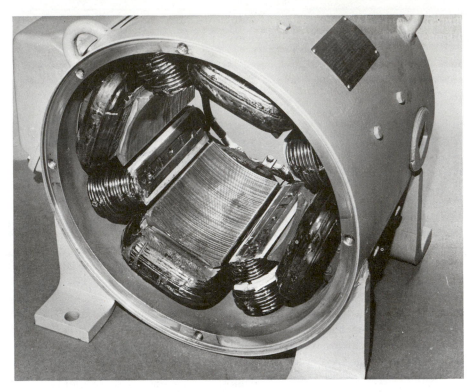

Figure 3-1. Representative dc main frame unit. (*Courtesy General Electric Company*)

seen at its simplest in medium sizes in an automobile starter motor where the butt weld clearly shows. The field poles are usually bolted to the frame. These bolts may appear as regularly spaced hexagon heads around the middle of the frame. Where space is critical, these fastenings may be countersunk. Again, this countersinking is usual on an automobile starter because, quite literally, external bolt heads would prevent mounting the unit as closely as is required.

There are other constructions seen on dc machine main frames depending upon make, size, and application. Some units are of cast iron or cast steel where the field poles and mounting feet are integral. This makes a very neat unit, but the machining required largely cancels any advantages. Cast-iron construction limits the flux density that may be achieved and is largely obsolete.

On very large units the main frame is split into an upper and lower half with a bolted flange joint on the horizontal centerline. This construction appears when the armature is too large and heavy to insert without a hoist. Figure 3-2 shows the split construction of a large machine. On the largest sizes the field poles and field coils also require crane handling, and therefore the separable structure serves two main purposes.

Small to medium units may have their main frame structure of punched laminations assembled in stacks. This structure makes an integral unit of the frame and the field poles. It can produce an excellent magnetic circuit and very sound

Figure 3-2. Split main frame construction for large dc machine. Note six field poles, six commutating field poles, and compensating windings. (*Courtesy General Electric Company*)

structure, but the punching die costs are high. Reliance motors are currently made in this fashion. See Fig. 3-3 for an example of punched main frame construction.

3-1-2 Field Poles. *Field poles* are usually, but not always, made of thin laminations of highly magnetic steel alloys. Laminated construction is necessary on the inner or *pole shoe* end of the field poles. This is because of the pulsations of field strength that result when the notched armature rotor magnetic structure passes the pole shoe. Variations in field strength result in internal eddy currents being generated in a magnetic structure. These eddy currents are losses; they may be largely prevented by having laminated magnetic structures. Laminated structures allow magnetic flux to pass along the length of the laminations, but do not allow electric eddy currents to pass across the structure from one lamination to another. The assembled stack of laminations is held together as a unit by appropriately placed rivets. The outer end of the laminated pole is curved to fit very closely into the inner surface of the main frame. A typical laminated field pole and pole shoe is shown in Fig. 3-4. Any break whatever in a magnetic structure causes significant reluctance, which is roughly analogous to resistance, so that more ampere turns are necessary to make a given magnetic flux flow in the structure as a whole. More ampere turns means more heat, which is a loss, so that the pole to main

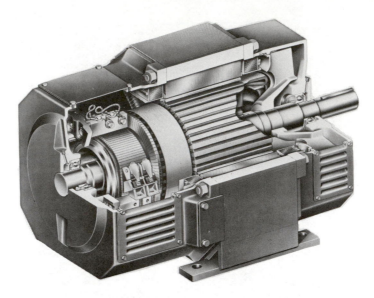

Figure 3-3. Direct-current machine with punched laminated main frame construction. (*Courtesy Reliance Electric Company*)

frame joint is usually quite tightly clamped by the field pole mounting bolts. The field coils needed will be discussed later.

3-1-3 Armature Structure. The *armature* structure serves a dual purpose in that it is the support for the winding conductors that pass through the magnetic field and is also a substantial portion of the magnetic flux circuit. Since any part of the armature magnetic structure sees cyclic reversals in magnetic flux direction, it is potentially subject to more severe eddy current losses than the field pole shoes and is, therefore, invariably laminated.

The usual construction, from the very smallest sizes up into the integral horsepower or kilowatt range, is a stack of discs of magnetic alloy steel. These discs are notched or perforated on the outer periphery to accommodate and support the armature windings. The required stack size is held as a unit by appropriate rivets, which are parallel to the shaft. The rivets themselves are a path for eddy currents and are a short-circuit path for generated voltages, so they must be insulated. Either that or their function must be taken by adhesively bonding the lamination stack by use of insulating varnish. However the stack is originally held, the windings themselves reinforce the unit as a whole. On the larger sizes the joint between the laminations and the shaft is keyed in order to transmit the torque forces. On smaller units, a press fit is sufficient. The very large units are complicated by the need to keep the lamination sizes within the range of reasonable punch and die tooling. In these cases the core of the armature is called a *spider*; it is a structural member that fills the radial space between the shaft and the built-up lamination stack, which now takes the form of a magnetic ring. Practical considerations limit the number of sizes of armature laminations to a few modular

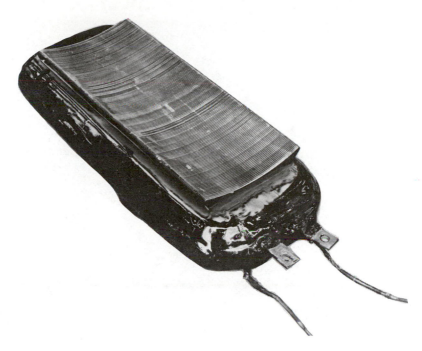

Figure 3-4. Typical laminated field pole and pole shoe with field coil. (*Courtesy General Electric Company*)

sizes, and also limit the range of the number of available winding slots in any one diameter.

3-1-4 Armature Winding Coils. The *armature winding coils* are placed in these slots in various arrangements that will be discussed later. However these coils are configured electrically, they must be so arranged mechanically that they may be mounted in the lamination slots. This mounting must be mechanically secure against centrifugal and torque forces, must be adequately insulated, and must allow the simplest possible assembly. In the larger sizes, these coils become very difficult to form, handle, and assemble in position.

3-1-5 Commutator. The windings must terminate in the rotary switching unit known as the *commutator*. This unit is almost always made of wedge-shaped segments of hard-drawn copper. The copper segments are insulated from each other and from their end clamps by strips of mica. The use of copper on a wearing part seems somewhat anachronistic but no better material has been found. The requirements of low resistance, excellent conductivity, and good wear resistance are conflicting. Figure 3-5 shows a typical armature rotor stack and commutator assembled to a shaft and wound with coils.

3-1-6 Brushes. The switching function is shared by the commutator and the *brushes*. These brushes are normally made of a carbon or carbon graphite or a

Figure 3-5. Direct-current armature rotor complete with coils, commutator, fan, and bearings. (*Courtesy General Electric Company*)

copper-filled carbon mixture. Here low contact resistance, some controlled internal resistance, and good wearing qualities are required. The actual contact surface is between the brush and a copper oxide surface on the commutator. If the rubbing friction and electrical sparking are extreme, the copper oxide is not renewed as fast as it wears and poor life results. This whole subject can be troublesome, but good life is achieved under proper design conditions. A commutator is shown assembled in Fig. 3-6 to show the segments. The commutator assembly is pressed or pressed and keyed to the shaft at a predetermined distance from the armature lamination stack. An armature is electrically completed by attaching the appropriate winding coil ends to the commutator segments. This joint is made with solder or even a high-temperature braze for heavy duty.

Finally, the whole assembly is strapped for strength to resist centrifugal stresses, dipped in insulating material and baked, balanced, and equipped with a fan for cooling. Figure 3-7 shows a completed armature in position in a cutaway motor.

3-1-7 Armature Shaft Bearings. *Bearings* are needed so that the armature assembly can rotate freely and yet be confined to its proper location.

3-1-8 End Bell Structure. The geometric location of the bearings in relation to the main frame is controlled by *end bells*, which are structural covers that enclose the machine.

Figure 3-6. Commutator assembly. (*Courtesy of General Electric Company*)

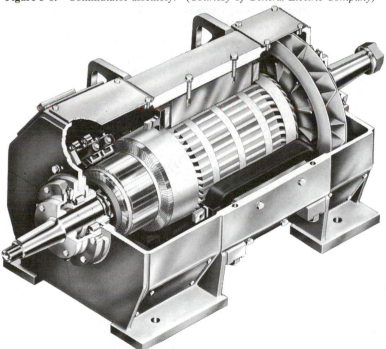

Figure 3-7. Complete armature assembly in position in a cutaway motor. (*Courtesy of General Electric Company*)

3-1-9   Brush Rigging.   In appropriate locations the brushes are supported by *brush rigging* from one end bell.   Various types of brush rigging are used depending upon size, number of brushes, ventilation, and service access requirements.   Brushes are supported in insulated tubes in smaller machines and rectangular sleeves or hinged mechanisms on larger units.   A large variety of different degrees of openness or protection are available in end-bell construction according to service uses.

3-1-10   Field Coils.   Our description was built inwardly from the main frame originally with magnetic consideration, but no means of creating the magnetic field within the magnetic structure has yet been provided.   This is performed by the use of *field coils*, which surround the field poles.   These coils are normally held in place between the pole shoe and the main frame.   Various types and combinations of coils are used.   Coil type is determined by whether there are many turns of relatively small wire designed to produce the required ampere turns with a small current from the line voltage, or a few turns of relatively larger wire designed to produce the needed ampere turns with a small voltage drop.   The first type of coil is known as a *shunt* coil; the second is a *series* coil, which is connected in series with the main armature line.   When both are present, the combination is known as a *compound* field.   The needed flux may be supplied by permanent magnets.

3-1-11   Commutating Fields.   On larger dc machines, from about 1 horsepower (hp) or 1 kilowatt (kW) up, *commutating fields* are provided between the main field poles.   These are smaller than the main field poles but are built similarly.   They are always connected in series.   The function of commutating field poles will be explained later when the electrical problems of commutation are discussed.

3-1-12   Compensating Windings.   On the larger very heavy duty units only, *compensating windings* may be present.   These are installed in the pole shoe faces of the main fields.   The compensating windings are needed in addition to the commutating fields to provide spark-free commutation under very high current conditions or transient conditions.   Again, this type of winding will be discussed later under commutation.   Figure 3-2 shows a large unit with its main frame separable at the horizontal parting line.   Both commutating fields and compensating windings are shown.

3-2   ARMATURE WINDINGS

The real working part of a motor or generator, whether alternating or direct current, is the *armature windings*.   These are where the voltage is generated and where the force that results in turning effort or motor action is developed.   The field windings serve to produce the magnetic field that is required; they carry only from 2 to 10 percent of the current in the machine if they are shunt windings.   Similarly, if the field is series wound, it will carry full armature current but will have only a few percent of the voltage that is present across the armature. Either way, the armature windings have a much larger wattage and are a more critical part of the design.

3-2-1 Winding Types. Only two basic configurations of windings are used, *lap* winding and *wave* winding. Some larger machines use a combination of these two basic types, known as *frog-leg* winding because of the appearance of the coil before it is installed. There are further subdivisions of each type, having relation to the number of conductors that are brought along in parallel such that a winding may be said to be *simplex* if a single conductor, *duplex* if doubled, *triplex* if tripled, and so on. An understanding of these winding configurations is necessary to determine the number of parallel paths that exist and to identify the currents that each conductor is required to carry. The number of parallel paths is the a term in Eq. (2-6). It is a direct divisor in the equation, so it is necessary to be able to identify or recognize whether a is 2, 4, 6, 8, or whatever. It will be a small even integer in any case.

The same purpose is behind all the various winding configurations: to take a conductor path from one polarity of brush, via a commutator segment, up through the magnetic field, around the back end of the armature structure, back the opposite way through the opposite field, and back finally to another commutator segment. This process is repeated around the commutator and the armature lamination stack until the winding is complete, symmetrical, and balanced.

Certain requirements must be carried out:

(1) Every coil must be of such a configuration that when seen as going away from the commutator it passes through the influence of one field polarity and when coming back it passes through the influence of the other field polarity. In this way the voltages generated are additive. The coils may be of one or many turns.

(2) The coils must be interconnected at the commutator such that all conductors under the influence of one magnetic pole are connected so that their incremental voltages are additive.

(3) The whole winding must be configured such that, regardless of the angular position of the armature, the commutator–brush connection relation to the field pole influence is maintained in the same sense.

Further requirements are less vital but certainly important:

(4) The coil shapes and end configuration must make maximum use of the copper so as to minimize resistance loss, cost, and weight.

(5) The manufacturing process should be of minimum difficulty as far as accessibility and possible damage in handling. It is difficult enough at best because of conflicting requirements.

(6) The windings should be as strong as possible and braced and/or supported

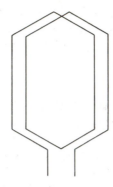

Figure 3-8. Lap armature winding coil.

where required. Copper is a superior electrical conductor, but it is a poor structural material for a highly stressed location, such as a rotating armature winding.

 (7) The finished armature assembly should be in good dynamic balance, and the balance must be maintained in service.

 All these requirements are met more or less in modern lap, wave, or frog-leg windings. In fact, an armature winding is a remarkable packaging job as far as density and efficiency are concerned.

3-2-2 Lap Windings. The differences among winding types arise from the way the coil ends are configured. A *lap* winding may have one or more turns of an approximately trapezoidal shape, with the two ends close to each other so that they can be connected to adjacent commutator segments. There are small variations in a duplex or higher-plex lap coil, but the coil ends are near each other (see Fig. 3-8).

3-2-3 Wave Windings. The *wave* winding has the same basic trapezoidal appearance in the coil body, but the coil ends are spread apart (see Fig. 3-9). The coil ends are connected to commutator segments that are one less or one more segment than the angular distance between two like polarity field poles. A wave

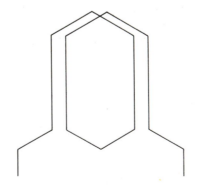

Figure 3-9. Wave armature winding coil.

winding must go at least twice around the armature before it closes back where it started.

3-2-4 Frog-Leg Windings and Equalizer Connections.
A *frog-leg* winding coil is a combination of both lap and wave coils.

The choice of winding types is by no means unanimous. Usually a wave winding is used because it is a bit easier to install and shows slightly better commutation, which is less brush sparking during operation. However, a limit of practical commutation seems to be reached at about 250 A/parallel path, or 500 A total. This current is reached in automotive starters, for example, and would be reached under heavy acceleration in an electric automotive vehicle motor even if a line voltage of over 100 V were used. Larger currents require lap winding, where more parallel paths can be used. The possible borderline situation to be found in a modern road vehicle motor will serve to illustrate one type of choice. If 500 A must be dealt with in a four-pole motor, for example, 250 A would flow in each path in a simplex wave winding. This is true because there are only two parallel paths in any simplex wave winding regardless of how many poles are used. A four-pole lap winding would have 500/4 = 125 A/path under the same conditions, because the lap winding inherently has the same number of parallel paths as the machine has field poles. When duplex or triplex windings are used, this current per path problem is not comparably helped in wave windings because perfect balance is not achieved between sections.

The heavy-current machine exaggerates a problem that exists in any machine. This problem is that the same voltage is not generated in each parallel path. This results in heavy circulating currents around the various parallel paths. These circulating currents move out and in the brush–commutator connections, causing heating, losses, and short commutator and brush life. One answer is the use of equalizer connections in a high-current lap winding. These connections connect winding parts that should be at exactly the same polarity.

The reason for having different voltages generated in apparently geometrically similar parallel paths is that the various field poles do not usually have the same amount of magnetic flux. The inequality is due to small air-gap differences caused by imperfect armature centering or unequal magnetic path reluctance resulting from poor field pole to main frame fit. Bearing wear compounds this problem.

Various winding types may be determined by inspection. In some cases this can be done by counting armature winding slots and commutator segments. The simplest situation is where there is the same even number of slots and segments. This must be a simplex lap-wound armature. If there were one more or one less commutator segment, it would be a simplex wave-wound armature.

3-2-5 Multielement and Multiplex Coils.
This easy situation is immediately complicated when there are many more commutator segments than armature slots. Many small armature slots make for a relatively fragile armature with limited winding space, so slot numbers are limited by mechanical reasons. However, the voltage that may exist between commutator segments or bars is limited to not over 10 to 15 V/bar. This means that a high-voltage dc machine must have many

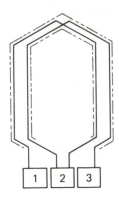

Figure 3-10.   Multielement single-turn lap coil.

segments per parallel path.   Since these two requirements are apparently conflicting, they are made compatible by the introduction of another factor.   This is the
multielement coil, which may contain two, three, four, or more elements.   This is
best understood by visualizing a situation where there might be 96 commutator
segments and 48 armature slots.   Obviously, there can only be 48 coil assemblies,
since each assembly has two sides and will therefore occupy two slots, and each
slot has portions of two different coils in it.   As a result, the winding is schematically
made as if there were 96 separate coils, but two adjacent coils must occupy the
same slot in every case.   Therefore, two complete coils are bound, insulated, and
installed as one.   Hence the term "multielement coil."   If these coils are connected
as shown in Fig. 3-10, the unit is used as a multielement coil.   If they are connected
as in Fig. 3-11, they are used as duplex connected coils.   The duplex situation will
result in half the voltage and twice as many parallel paths as the multielement
situation, and similarly if triplex or more.

When an armature is assembled, bound, or wrapped and impregnated with
insulation, it is almost impossible to tell one situation from the other.   On the

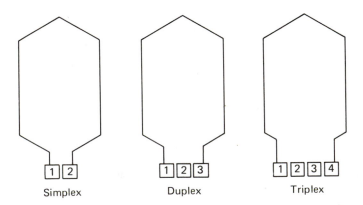

Figure 3-11.   Multiplex coil connections.

other hand, the two-to-one voltage difference and one-to-two current situation make all the difference in the world. This is one of the many situations that specifically point out that the subject of armature windings cannot be fully covered in a short text. On the other hand, the technician or engineer must know what he is dealing with to interpret tests correctly.

The usual situation with wave windings is that the commutator will have one or more or one or less segments than an integer multiple of the number of slots.

If one knows what he or she has on hand, all is simplified, and very simple rules may be followed where P = poles and a = parallel paths:

(1) Simplex lap windings have as many parallel paths as main field poles:

$$\boxed{a = P} \qquad \text{Eq. (3-1)}$$

(2) Simplex wave windings have two parallel paths regardless of the number of poles:

$$\boxed{a = 2} \qquad \text{Eq. (3-2)}$$

The amount or degree of multiplicity or plex determines the number of parallel paths in the following manner:

(3) A lap winding has poles times the degree of plex parallel paths:

$$\boxed{a = P \times \text{plex}} \qquad \text{Eq. (3-3)}$$

(4) A wave winding has two times the degree of multiplex parallel paths:

$$\boxed{a = 2 \times \text{plex}} \qquad \text{Eq. (3-4)}$$

(5) The frog-leg winding is a combination of both lap and wave winding, and they are made compatible by the relation between the number of parallel paths. The wave portion and the lap portion must have an equal number of parallel paths. In the usual situation, the lap portion is simplex. Therefore, a six-pole frog-leg winding would need a triplex wave section that has six parallel paths to match its simplex lap portion with six parallel paths. This would result in 12 parallel paths for the winding. Thus, in general, a frog-leg machine will have

$$\boxed{a = 2 \times P} \qquad \text{Eq. (3-5)}$$

In summary, this subject can be very complex, and added references must be used for any full understanding. However, a technician or engineer should recognize the various combinations that are possible so that he or she may, for example, be able to intelligently guide the selection of a rewinding pattern to match a machine to a unique problem.

3-3 COMMUTATOR AND RELATED ELEMENTS

The discussion of winding types has shown a common feature of all dc motors and generator windings: the windings are closed and continuous around the armature. The windings come out to a commutator bar and immediately return to the magnetic structure. All commutator bars are treated alike in any one machine. They will usually have two connections per bar, one in and one out at two different levels. This winding-to-commutator connection is the most critical electrical joint in the machine and must be made as perfectly as possible.

3-3-1 Brush Spacing. Brushes are held in contact with the commutator every 180 electrical degrees. These brushes are of alternate positive and negative polarity. There are as many brushes as there are poles in the main field; thus, an eight-pole machine will have eight brushes or brush sets equally spaced around the commutator. In some cases this rule will not be followed when a wave winding is used. This is because the wave construction is such that internal connection exists in the winding, where every positive point is connected to every other positive point, and every negative point is similarly connected to every other negative point. Thus, sometimes when space is at a premium not all brushes need be used. A lap winding needs all brushes.

Usually, because of the angular throw of the coil ends as they go from the armature lamination stack to the commutator, the electrical take-off points at the commutator are midway between the field poles, and the whole arrangement is compact.

3-3-2 Brush Neutral Position. If the voltages at the commutator are measured while the armature is rotating and while the main field poles are separately excited, a regular pattern can be seen. When measuring voltage progressively around a commutator starting from negative brush location, a gradually more positive voltage is seen. Then, as the region of the commutator that connects to windings that are now in the magnetic field and working is traversed, regular substantial increments of voltage are added. Finally, the rate of increase of voltage is reduced as the commutator segments that connect to windings that are leaving the far side of the working field region are connected. A peak positive voltage is finally seen, and further angular motion around the commutator in the same direction as before will disclose a gradual reduction of voltage. This pattern of increasing, then decreasing, voltage continues around the commutator for as many poles as exist in the machine. The highest positive voltage locations and lowest negative voltage locations are where the centers of the brushes should be placed. These are *brush neutral positions*.

When brushes are placed in these locations or neutral points, the maximum available voltage from the machine is connected and can be used. Trouble begins, however, when substantial currents are drawn from the brushes and therefore from the armature windings. The troubles stem from *armature reaction* and *armature reactance*. The names are similar, but the troubles are substantially different and must be treated in different ways.

3-4 ARMATURE REACTION

The form and shape of the magnetic field caused by the main fields becomes distorted when current is present in the armature windings. This is because the armature windings themselves are wound on a magnetic structure. The magnetic field thus produced in the armature is 90 electrical degrees from that produced by the fields. These two fields become vectorially combined in a distorted result. Figure 3-12 shows a typical undistorted field pole magnetic field that is present with little or no current in the armature. Then Fig. 3-13 shows a visualization of the armature-produced field, which is present to a degree approximately proportional to the armature current. Figure 3-14 shows the result of the combination of the two fields. The magnetic field on one side of the field pole is in effect swept aside and reduced. The mid-pole field is about the same as when no current is present in the armature. But the other side of the field pole now has a substantially

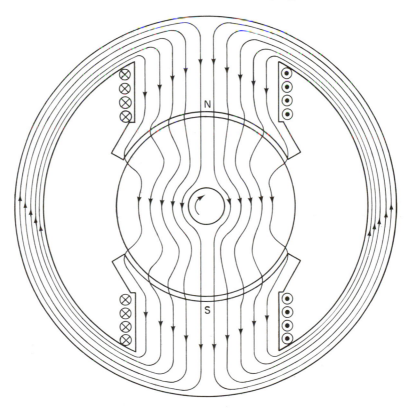

Figure 3-12. Magnetic flux distribution due to field poles only.

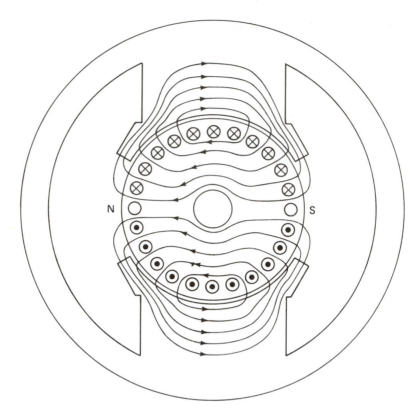

Figure 3-13. Magnetic flux distribution due to armature excitation only.

larger field than before. The voltages generated in the windings are, unfortunately, generated in direct proportion to the actual field that is present. Followed to its ultimate conclusion, this armature reaction results in the neutral voltage point being appreciably moved in relation to the brush position. As a result, the commutator and brush switching function is no longer spark free, and resulting brush and commutator life is drastically reduced.

3-4-1 Correcting Armature Reaction Effect. There are four main ways of combating the armature reaction problem:

(1) Rotate the brush hanger mechanism to find the correct but distorted neutral point. This is very effective at a fixed current load. On the other hand, the brushes must be moved each time the load is changed. This was the early answer, but it has been obsolete for over 50 years. It has the merit that it worked and is easily understood, but that is about all.

(2) Shape or otherwise modify the ends of the field pole shoes so that high flux cannot exist on the ends because of the high path reluctance. This is customarily done on almost all dc machinery built today. It reduces but does not eliminate the problem.

(3) Add *interpoles* or *commutating poles* to the field structure. These units

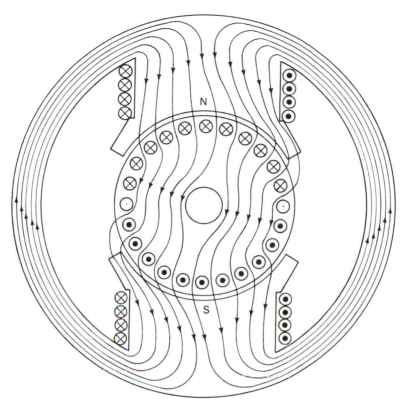

Figure 3-14. Combined magnetic flux distribution due to armature and field.

resemble small main field poles and are installed midway between the main field poles. They function by locally modifying the resultant of the main field pole and armature-caused distorted magnetic field. This distorted field has the local effect, around the brushes, of causing some voltage generation to take place in the armature coil that is undergoing switching by the commutator and brush. The result of the combined field caused by the main field, the armature reaction field, and the commutating field is that locally around the brushes there is no effective field. Therefore, there is no unwanted local voltage generation to spoil the required commutator and brush switching process. Since the effect of armature reaction is related to armature current, its counteraction by commutating fields is also needed in relation to armature current. This is readily accomplished by connecting the commutating field windings in series with the armature so that the variable requirement is automatically met in service. Commutating fields are so effective that they are invariably used in medium and large dc machines.

Once experienced, it is immediately obvious when commutating fields are incorrectly adjusted or wrongly connected. The commutator will spark viciously; there is a sound much like frying bacon and a sharp smell of ozone. Here immediate correction is necessary, because commutator and brush life may be reduced by more than 1000 to 1.

(4) Very large machines or machines with very severe duty cycles, such as in

some large machine tools or especially in rolling mill drives, require the last feature, which is *compensating windings*. These windings are laid directly into the pole shoe face of the main fields. They are parallel to the armature shaft and carry current in the opposite direction to the armature windings immediately adjacent to them. The result is that the main field flux symmetry is no longer distorted, since the armature reaction magnetic flux is equally and oppositely opposed by the compensating winding flux. The function of the commutating fields is still required, but to a reduced degree. Compensating windings are expensive and sometimes cumbersome, but their use results in reliable spark-free commutation under conditions that would be impossible without them. Their use will probably be necessary in automotive electric propulsion motors if the American desire for rapid acceleration combined with long life is to be met. The author predicts that manufacturing methods will be developed so that compensating windings will be used, and therefore compact motors will be easily adapted to severe use.

3-5 ARMATURE REACTANCE

Armature reactance, a similar sounding but different phenomenon, is the result of the inductive reactance of the particular armature winding that is undergoing commutation. The process of commutation of a rotating armature circuit involves the reversal of the current in each coil in turn as it passes the brushes. This can be seen when it is realized that the coil circuit on each side of a given brush is contributing current that flows toward the positive brushes and away from the negative brushes in a generator. The reverse is true in a motor, but the problem is related. If the last coil approaching the brush has current in it flowing toward the brush contact at the commutator segment, this current must abruptly reverse when the segment passes the brush and is moving away. This current reversal takes place during the time that both ends of the coil are short circuited by the brush. This effect is most easily understood with a simplex lap winding, but is related on all types of windings.

3-5-1 Brush Short-Circuiting Effect.
From the instant that the segment carrying the second end of the coil touches the brush, the polarities of each end must rapidly approach each other. This is because the two ends of the coil are shorted together by the low but real resistance of the body of the brush itself. The current rapidly drops to zero, and then as rapidly starts the other way. This happens as the originally leading segment passes the brush, and the reversed voltage and current on the other side take their effect. This process cannot take place instantaneously because the armature coil has significant inductance. It is easy enough to see that a coil wound as it is around a magnetic core must have real inductance. This inductance is related to the number of turns in the coil. It is also mutually related to other coils in the armature winding, and especially so to the coils that occupy the same slots in the armature structure. This serious problem is compounded by high rotative speed, because there is less time to affect the reversal. The inductive reactance of the coil is then related to the rotative speed or switching time and the number of turns in the coil.

Practical brushes overlap significantly more than one commutator segment, usually from two and one-half to three and one-half segments.

By a judicious balance of design, good commutation is achieved. This starts with coils with as little self-inductance and mutual inductance as reasonably possible. Then the brush overlap width and brush internal resistance are experimentally varied so that the coil can discharge its stored energy during the commutation cycle. Ideally, there will be a uniform change in coil current as it is swept by the brush. When the coil emerges from the influence of the brush, the current reversal should be fully established.

This current reversal process is also aided by adjusting the commutating field strength so as to cause the shorted coil to generate just sufficient voltage to aid the current reversal.

This whole process means that there is considerable energy dissipation both on the contact surface of the brush and within its bulk. A specific machine uses a specific brush with experimentally determined parameters.

3-5-2 Opposing Effects of Design Features.

Finally, it can be seen that armature reaction and armature reactance are separate phenomena. Either condition can cause brush sparking and perhaps conceal that the other condition is well handled. Three quarters of a century of development have produced very reliable long-lived motors and generators, but the commutation problem continues to limit motor or generator size to power relations. There is much development work currently on "commutatorless" units in which the problem is handled by various solid-state switching circuits that are phased in their operation by armature position. Reliable operation has been achieved in control mechanism sizes, but usually at a considerable increase in cost and some loss of efficiency.

There exists a constant conflict of opposing design features in any type of device or machine. This conflict is especially severe in dc motor or generator design. There has been slow but steady development resulting in more power in a given-sized machine or a smaller machine for a given power output. A smaller machine with smaller winding wire size will always show some increase in winding resistance. This increased resistance results in the development of more internal heat. The increased heat then results in increased winding temperatures. Higher winding temperatures require insulations that will still remain as insulations for a long useful life. However, almost always higher temperature means increased chemical activity in whatever it is that is heated. Insulation breakdown as a result of elevated temperature is one limiting factor in motor and generator development. Here much progress has been made, but much more is needed.

The coil end attachment to the commutator bar is a particularly vulnerable point in dc machine design. Aircraft and automotive starter motors are strictly limited as to the amount of time that they may be used continuously. Many aircraft starters are limited to around 30 sec of cranking before they must be allowed to cool for around 30 min. If this is not observed, the commutator connections will melt, resulting in expensive failure.

Another material limitation is in the magnetic character of the armature and field lamination steel that is used. A higher magnetic flux density will allow less armature coil turns for the same performance. This would result in lower winding resistance and at the same time less coil inductance. Both results are highly beneficial in reducing losses and improving commutation. There has again been steady progress in magnetic steel quality, but more progress is needed.

Very much longer life and better performance have been secured by improved motor or generator bearings. Thousands of hours of service can be realized from modern ball bearings without any requirement for lubrication or cleaning. This improvement did not come easily any more than the others did. Progress has seemed slow in this field of design, but by looking back in 10-year steps it can be seen that much has been accomplished. It seems certain that much future progress will be seen, and it will come none too soon. The increasingly severe energy shortages of the 1980s really require the expanded use of electric motors and generators, because they already have achieved such relatively high efficiencies. In this sense every development that increases efficiency in service reduces the energy needed. Less obvious but probably of equal importance, a reduction in size for equal power usually means that less energy is consumed in building the device because of less material used.

3-5-3 Desired Improvements. It would then appear that, if a significant increase in power output with a given size is to be realized with new dc motor and generator design, we need at least one of the following breakthroughs:

(1) A low-cost practical winding material with less resistance than copper. Of the pure metals, only silver has a lower resistivity, and only 5 or 6 percent lower at that. Alloying of metals always seems to *increase* their resistivity.

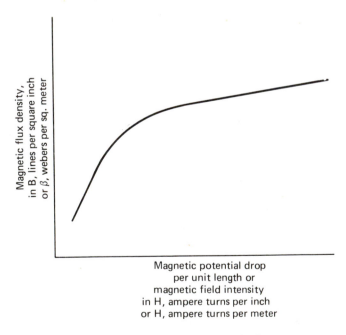

Figure 3-15. Form of B-H or magnetization curve.

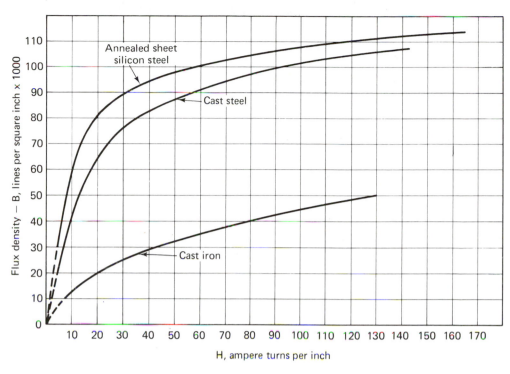

Figure 3-16. Magnetization curves, English units.

(2) A low-cost practical magnetic lamination material with a higher relative permeability than present magnetic lamination steels. The magnetization curve, or *BH* curve, or saturation curve, as it is variously known, shows the relation of flux per unit area versus magnetization force in ampere turns per unit length. In general, each type of magnetic material has a particular curve relation. Figure 3-15 shows a typical form of saturation curve for a modern magnetic circuit steel.

Note from Figs. 3-16 and 3-17 that a flux density of as much as 120 000 lines/in.² (1.86 Wb/m²) is obtained with great difficulty. 100 000 lines/in.² (1.550 Wb/m²) is around the knee of the curve. This is the present metallurgically limited flux range. If double as much flux could be reached before significant saturation began to be a limit, much smaller machines could be built. Either that or one half as many turns could be used per coil to generate the same voltage in the same physical size magnetic structure. Half as many turns would allow twice the cross section of copper in the same space. This would result in one quarter as much winding resistance, which would mean one quarter as much heat loss in the windings. Or the gain could be split in both a somewhat smaller size and lower losses. Since the heat loss in the armature windings turns out to be the largest single loss in the whole dc machine, one can immediately see the gain that would result in the use of better magnetic material.

Less winding turns also means less coil inductive reactance, and therefore eases the design and adjustment for spark-free commutation.

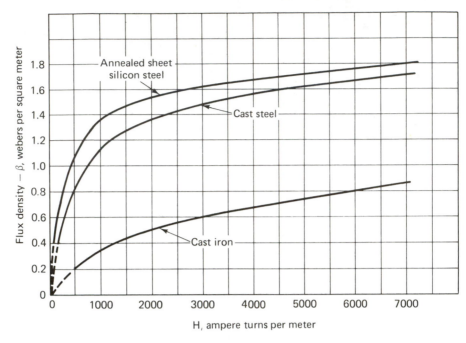

Figure 3-17. Magnetization curves, System International.

3-6 MAGNETIC CIRCUIT

We shall not attempt to present enough material to actually design a motor or generator magnetic circuit. That would be beyond the intended scope of this book. However, it is felt that if a simplified calculation is developed the student can readily see the importance of some of the various factors. Remember that the flux density that counts is that which exists through the air gap between the field pole faces and the armature coils that are directly adjacent. It is this flux that is cut by the moving armature coil windings. All other flux that exists around the machine is necessary to close the magnetic circuit. The design purpose is to achieve a reasonably high air-gap flux without any high losses due to forcing part of the circuit into saturation.

The calculations will show that the majority of the magnetic circuit reluctance is in the air gap. Therefore, about three quarters of the ampere turns of magnetizing force are needed to overcome the air-gap reluctance and force a reasonably large flux to circulate.

The magnetizing force needed to achieve a given field intensity in a known geometric shape of magnetic circuit must be developed in separate steps for the separate geometric elements in a magnetic circuit.

Since the desired end point is the field strength in the air gap, that value of flux will usually be the starting point for the design calculations.

Magnetic field intensity in air or in most nonferrous circuit materials is directly related to the magnetizing force. Therefore, the more magnetizing force, the more field flux as far as you care to go. This simple linear relation is not true in any ferrous magnetic material, even though a far higher field intensity can be achieved for a given magnetizing force.

Remembering from basic electricity course work that the following units are used:

The *magnetizing force* is proportional in ampere turns per unit length and is symbolized as H. Thus

$$H \text{ is in } \frac{\text{ampere turn}}{\text{inch}} \text{ in English units}$$

$$H \text{ is in } \frac{\text{ampere turn}}{\text{meter}} \text{ in SI units}$$

The magnetic field intensity is another name for the magnetizing force.

The *magnetic flux density* is a measure of flux in lines or webers per unit area and is labeled B (or β). Thus, flux density is measured as shown in Chapter 2 and is given as

$$B \text{ in } = \frac{\text{line}}{\text{in. (in.)}} \text{ in English units}$$

$$\beta \text{ in } \frac{\text{weber}}{\text{meter (meter)}} \text{ in SI units}$$

The relation between B and H for *nonferrous materials*, or *free space*, or, in a practical case, *air* was defined in basic cgs metric units originally; in SI it becomes

$$\frac{\beta}{H} = \mu_{0SI} = 4\pi \times 10^{-7} \frac{\text{weber}}{\text{meter ampere turn}}$$

As a result,

$$H = 0.795\ 77 \times 10^6 \beta \frac{\text{ampere turn}}{\text{meter}} \qquad \text{Eq. (3-6}_{SI})$$

This constant μ_{0SI} (read mu subzero) is called the *permeability of free space*; it is converted to the English units as follows:

$$\mu_{0E} = 4\pi \times 10^{-7} \frac{\text{weber}}{\text{meter amp turn}} \frac{10^8 \text{ line}}{\text{weber}} \frac{\text{meter}}{39.37 \text{ in.}}$$

which is then

$$\frac{40\pi}{39.37} = \mu_{0E} = 3.1919 \frac{\text{line}}{\text{inch ampere turn}}$$

Therefore, in English units

$$H = 0.313\ 30B\ \frac{\text{ampere turn}}{\text{inch}}$$ Eq. (3-6$_E$)

In either units, the appropriate quantity μ_{0SI} or μ_{0E} is a constant for air in the air gap. This is not true in the case of the magnetic portions of the circuit, where μ becomes μ_r or relative permeability compared to air, and is *not* a constant.

This variability of the μ_r factor is normally shown in *BH* curves of specific materials that are normally used in magnetic circuits. The μ_r may be from 100 to more than 1000 times that of air, but it is not a constant value. This high permeability means that many fewer ampere turns of magnetizing force are required to maintain a useful force in magnetic materials, even though the magnetic materials are the longer part of the circuit.

Typical *BH* curves are shown in Figs. 3-16 and 3-17.

3-6-1 Simplified Magnetic Circuit Analysis. For a study of a typical magnetic circuit relation in simplified form, first see the dimensional relations of a particular four-pole machine as shown in Fig. 3-18. Note that dimensions are given in both English and SI.

If we assume that the total flux ϕ per pole has been found from the use of Eq. (2-6) as 1 650 000 lines (0.0165 Wb), the flux density is $\phi/A = B$ (or β), where A is one effective pole shoe area. In this case

$$\frac{1\ 650\ 000}{5.5 \times 6} = 50\ 000\ \frac{\text{line}}{\text{in. (in.)}} = B$$

or

$$\frac{0.0165}{0.1397 \times 0.1524} = 0.775\ \frac{\text{Wb}}{\text{m (m)}} = \beta$$

Here we use meters instead of millimeters to get square meters. Note that the effective area per pole is the net effective pole shoe arc length times the axial length of the magnetic lamination stack of the armature and field. In this treatment we will avoid the issue of calculating the spread of the magnetic field from the poles. This phenomena, which is known as *fringing*, is treated in more advanced design texts, such as references (1) and (2).

The problem is much easier to follow if it is broken down into elements of the magnetic circuit. In this way each portion can be treated as its specific material and geometry requires.

Some thought while looking at the magnetic circuit diagram in Fig. 3-18 will show that each pole is similar. Also, as the centerlines show, the magnetic circuit divides each pole into two equal portions. An individual magnetic circuit then involves facing halves of two adjacent poles and the connecting portions of the exterior main frame and the armature lamination stack.

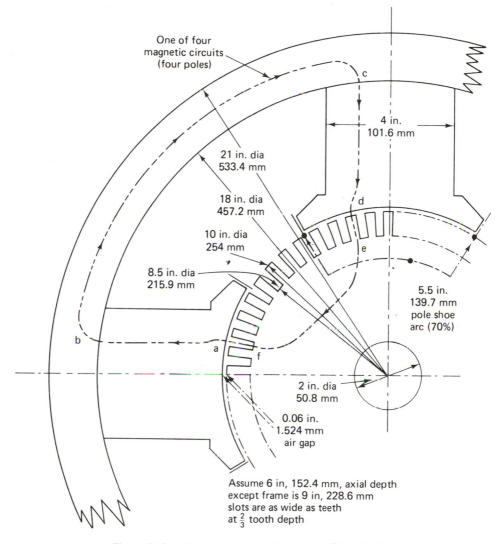

One of four
magnetic circuits
(four poles)

4 in.
101.6 mm

c

21 in. dia
533.4 mm

18 in. dia
457.2 mm

d

10 in. dia
254 mm

e

8.5 in. dia
215.9 mm

5.5 in.
139.7 mm
pole shoe
arc (70%)

b

a   f

2 in. dia
50.8 mm

0.06 in.
1.524 mm
air gap

Assume 6 in, 152.4 mm, axial depth
except frame is 9 in, 228.6 mm
slots are as wide as teeth
at $\frac{2}{3}$ tooth depth

Figure 3-18. Magnetic structure dimensions, Example 3-1.

By this type of division, each magnetic circuit path is acted upon by two different field coils. At the same time, each coil acts upon two different circuits, and each magnetic circuit can be considered to be acted upon by the full ampere turns of each coil. The ampere turns do not subdivide because they are not consumed. This may be verified by Kirchhoff's second law. Therefore, in computing the ampere turns required by each magnetic circuit, we can divide the required amount in half and each field coil can then be configured to supply one half the ampere turns needed.

From a study of the physical dimensions of Fig. 3-18, the various elements of the magnetic circuit can be determined. The procedure is normally simplified

TABLE 3-1 REQUIRED AMPERE TURNS FOR MAGNETIC CIRCUIT
Refer to Fig. 3-18 for dimensions

English Units

① Part	② Φ = path flux (lines)	③ Path area (in.²)	④ B = lines/in.²	⑤ Material used	⑥ H = AT/in. (Fig. 3-16)	⑦ Path length (in.)	⑧ Amp turns
Core *ab*	970 600	11.04	87 920	Silicon steel	27.1	3.94	107
Frame *bc*	970 600	13.5	71 900	Cast steel	26.0	13.3	346
Core *cd*	970 600	11.04	87 920	Silicon steel	27.1	3.94	107
Gap *de*	825 000	16.5	50 000	Air	$0.313\,30B$ = 15 660	0.072	1127
Arm. *ef*	825 000	12.9	63 950	Silicon steel	11.9	7.85	93.0
Gap *fa*	825 000	16.5	50 000	Air	$0.313\,30B$ = 15 660	0.072	1127
					Total ampere turns for two half-poles		2907
					Total ampere turns *per pole*		1454

SI Units

① Part	② ϕ = path flux (Wb)	③ Path area (m²)	④ webers/m²	⑤ H = AT/m used	⑥ H = AT/m (Fig. 3-17)	⑦ Path length (m)	⑧ Amp turns
Core *ab*	0.009 706	0.007 12	1.363	Silicon steel	1030	0.100	103
Frame *bc*	0.009 706	0.008 71	1.114	Cast steel	980	0.338	331
Core *cd*	0.009 706	0.007 12	1.363	Silicon Steel	1030	0.100	103
Gap *de*	0.008 25	0.0106	0.775	Air	$0.795\,77 \times 10^6\,\beta$ = 0.6167×10^6	0.001 83 or 1830×10^{-6}	1128
Arm. *ef*	0.008 25	0.008 32	0.991	Silicon Steel	443	0.199	88
Gap *fa*	0.008 25	0.0106	0.775	Air	$0.795\,77 \times 10^6\,\beta$ = 0.6167×10^6	0.001 83 or 1830×10^{-6}	1128
					Total ampere turns for two half-poles		2881
					Total ampere turns *per pole*		1440

by using each part in the magnetic circuit that has a specific cross section, length, and material as an element.

The required number of ampere turns to magnetize each element in the circuit are then determined. All the individual element ampere turn requirements are then totaled to find the ampere turn requirement of the complete circuit.

Depending on the degree of detail and precision of the calculations, certain allowance factors may be added so that the design level of flux may be met and perhaps exceeded.

With simplifying assumptions instead of the true path lengths and cross sections, the field ampere turn requirements for the machine shown in Fig. 3-18 are summarized in Table 3-1. Since all the magnetic flux in the outside frame and field poles does not actually flow through the air gap as desired, owing to leakage directly between field pole tips, about *15 percent more* flux must be figured in the external part of the circuit. The air gap itself is assumed to be *20 percent larger* than its true minimum dimension to allow for the shape of the inner face of the pole shoe and the reduced magnetic material area in the armature teeth. The magnetic path cross section in the armature lamination stack is examined at different parts, and the least area found is taken as the ruling dimension.

Working with a tabulation such as Table 3-1, the individual path lengths and cross section areas are tabulated as they are calculated. Note that the usual English and SI units are both shown. The use of the table is as follows:

(1) The first column is merely element identity.

(2) Column 2 is the required total flux Φ (or ϕ).

(3) Column 3 is the area of each element of the path.

(4) Column 4 is the flux density from $B = \Phi/A$ *or* $\beta = \phi/a$.

(5) Column 5 is the specific material for that part.

(6) Column 6 is the required magnetizing force H in ampere turns for each linear inch (or linear meter) of path length of the specific material at the particular flux density. These quantities are extracted from appropriate BH curves of the material in question. In the case of the air gaps, H is from Eq. (3-6$_E$) or (3-6$_{SI}$).

(7) Column 7 is the individually determined path lengths.

(8) Column 8 is simply column 6 times column 7.

(9) In each of the unit systems involved, the individual ampere turn requirements are then totaled.

The final figure for the ampere turns *per pole* leads to the design requirements for a specific field coil. Obviously, the basic flux requirement that was used as 1 650 000 lines (or 0.0165 Wb) in this case has to be determined for the maximum needed.

Some thought on this process will allow the *rough* calculation of any dc motor or generator magnetic path ampere turn requirement. The division of the summary by 2 is not invariably true since low-priced automobile starter motors, for example, only wind every other pole. Thus, in *that specific case* there is only one coil per magnetic circuit.

3-6-2 Magnetic Circuit Design Observations.
This process was intended to develop a feel for the magnetic requirements; certain things should therefore be observed:

(1) The two air gaps per path require the majority of the ampere turns, about 78 percent of the total in this case.

(2) If the air-gap flux is 50 000 lines/in.^{2} (0.775 Wb/m^2), the flux density in the armature teeth must be much higher. If there is 50 percent tooth and 50 percent slot and almost all the flux is in the root of the tooth, the flux density would be twice as high. A flux of 100 000 lines/in.^{2} (or 1.55 Wb/m^2) is around the knee of the *BH* curve. The teeth are then seen to be near saturation and if pushed to a much higher flux density would begin to account for substantially more ampere turns.

(3) The main frame flux density is held below the knee of the curve, because that is the long part of the path and would require much more ampere turns if its flux density were increased. This is the reason for the ''armor plate'' proportions of the main frame. Speculate for a moment on the weight that could be saved if the magnetic path material could be substantially reduced. Note in Fig. 3-18 that the material is 1.5 in. thick (38.1 mm) and half again wider axially than the rest of the magnetic elements. A glance at the *BH* curves shows the design penalties necessary for cast iron.

(4) We have used simplifying assumptions in Sect. 3-6-1: 15 percent flux side leakage loss, and an 8 percent loss of net lamination stack thickness. Also, we have for simplicity ignored the locally higher ampere turns that are required to make the flux flow through the reduced area of the armature teeth. Finally, we simply stated that the *effective* pole arc length was 5.5 in. (or 139.7 mm) without actually stating the mechanical dimensions of the pole shoe. The mechanical-to-effective-pole-arc dimension relationship requires relatively elaborate field plotting techniques that are beyond the scope of this text. One recognized assumption sometimes used is that the total effective pole shoe arc is the mechanical arc length plus twice the air gap. When working with one half the pole area and air-gap area, the difference between the one-half pole shoe arc and one-half effective arc would be one air-gap length.

The assumptions above are reasonably conservative and will result in a good first approximation. Thus, this procedure would be good enough to determine what part of a circuit would saturate first if, for example, a motor or generator were to be rewound for a special purpose.

In any case, more field coil turns would be provided so that enough field would be achieved. In the example provided, at least 1500 ampere turns would be built into the machine. The actual number of turns would depend upon whether it was a series or a shunt coil. It is necessary to jockey the coil wire size and the number of turns until the desired current will flow through the chosen number of turns to produce the ampere turns. Coils for any reasonable voltage or current can be designed if the ampere turn requirement is known.

The previous discussion has shown at least some of the inherent problems in a dc machine magnetic circuit. However, in Sect. 3-5-1 the magnetic field pole coverage was simply assumed as 70 percent. This variable of pole coverage requires qualitative understanding.

There are some constraints that need to be explored first. If the field poles were made so large in physical scope that they covered all, or nearly all, of the periphery of the armature, they would approach too close to each other. If the end of a north magnetic pole approached the end of a south magnetic pole to within a small fraction of an inch (or a few millimeters), substantial magnetic flux would flow directly from pole to pole without entering the armature. One boundary condition is where the poles are separated far enough so that most of the magnetic flux that is created by the ampere turns of the field coils will flow from one pole down into the armature, through the armature, and back out to the opposite magnetic pole. *Only the flux that enters the armature structure and is cut by the armature windings is effective.* All other flux is wasted.

Another constraint on closeness of field pole shoe tips is that there must be enough room for a commutating pole between adjacent field poles if a commutating pole is required.

The other boundary condition that determines field pole width is that, if the pole coverage of the surface of the armature is too small, the required flux density becomes impractically high.

Equation (2-6) shows that the generated voltage is directly dependent on the total flux per pole Φ (or ϕ) in either lines or webers. It is obvious that, if a certain peak value of B (or β) in flux per unit area cannot be reasonably exceeded, the area must be as large as reasonably practical.

As a result of these two constraints, the pole coverage of the surface of the armature is usually around 70 percent of the total surface. This is not an absolute value, but most dc machinery is built at or around this range of pole coverage.

The *effective pole coverage* is a bit more of an arc around the armature surface than the actual physical dimensions of the pole itself. Actual determination of the effective magnetic field spread is beyond the scope of this book.

A reliable design approximation for the spread of this field fringing is that the effective pole arc coverage will extend beyond each end of the pole shoe itself by the amount of the minimum air-gap dimension. A further simplification is to ignore it entirely on a first approximation, since the error is only a few percent. A worked example will show the relationships between pole coverage and field flux density.

Example 3-1. A dc machine runs at 1250 rpm (130.9 rad/sec). Its armature diameter is 14 in. (355.6 mm), and its field pole faces total 10.5 in. long (266.7 mm) in an axial direction after allowing for cooling slots. The radial gap between the field poles and the rotor is 0.125 in. (3.175 mm). The field pole effective coverage is 75 percent, including fringing. The armature has 96 coils of three turns per coil and is wound duplex lap. The machine has six poles. It generates 250 V no load.

(a) What is the flux per pole?
(b) What is the effective pole arc length, neglecting pole tip flux fringing?
(c) What is the average air gap or pole shoe surface flux density?

Solution:

(a) Equation (2-6) may be used directly since all needed quantities are supplied. Only rearrangement is needed.

$$E_g = \frac{\Phi ZSP}{60a} \times 10^{-8} \frac{\text{volt}}{\text{gen}}, \text{ from Eq. (2-6}_E); \qquad \Phi = \frac{E_g 60a \times 10^8}{ZSP} \text{ line}$$

$$= \frac{\phi Z\omega P}{2\pi a} \frac{\text{volt}}{\text{gen}}, \text{ from Eq. (2-6}_{SI}); \qquad \phi = \frac{E_g 2\pi a}{Z\omega P} \text{ weber}$$

$$= 250 \text{ V}$$

$$S = 1250 \text{ rpm (or } \omega = 130.9 \text{ rad/sec)}$$

$$P = 6$$

$$a = P \times \text{plex} = 6 \times 2 = 12 \text{ parallel paths} \quad [\text{use Eq. (3-3)}]$$

Use the appropriate equation depending on winding type (lap) and plex (duplex or 2). Equations (3-1) through (3-5) are related types.

$$Z = 96 \times 3 \times 2 = 576$$

Remember that there are always *two* conductors per turn (out and back).

$$\Phi = \frac{(250)60(12)10^8}{576(1250)6}$$

$$= 4\ 167\ 000 \text{ lines}$$

or

$$\phi = \frac{250(12)2\pi}{576(130.9)6}$$

$$= 0.041\ 67 \text{ Wb}$$

(b) The pole arc length involves a visualization of the shape of the parts and recognition of what the nomenclature means when considered geometrically.

The *net* field pole flux coverage is 75 percent of one-sixth (six poles) of the circumference of a circle 14 in. in diameter (355.6 mm). Remember that the air gap is *all around* the armature, and thus the diameter of the *inside* of the pole shoes is the armature diameter plus two air gaps. This is true *regardless* of the number of poles. The flux coverage arcs per pole are then

$$\frac{[14 + (0.125 \times 2)]\pi(0.75)}{6} = 5.596 \text{ in. arc length}$$

or

$$\frac{[0.3556 + (0.000\ 175 \times 2)]\pi(0.75)}{6} = 0.1421 \text{ m arc length}$$

The arc calculated above is a circumferential distance at the inner surface of a field pole shoe.

(c) The flux density portion of the problem requires that the pole *area* be known. This involves a length times a width. Since we have an *arc* length, we need only multiply by the other dimension in the axial direction to get area.

$$5.596 \times 10.5 = 58.76 \text{ in.}^2/\text{pole}$$

or

$$0.1421 \times 0.2667 = 0.037\ 90 \text{ m}^2/\text{pole}$$

Again the dimensions were in millimeters, but in SI, meters are used to be coherent. This is a portion of a cylindrical surface but is still an area.

The flux density is then easily found by dividing the total flux per pole from part (a) by the area per pole, keeping the units consistent.

$$\frac{4\ 167\ 000 \text{ line}}{58.76 \text{ in.}^2}$$

$$B = 70\ 920\ \frac{\text{line}}{\text{in.}^2}$$

or

$$\frac{0.041\ 67 \text{ Wb}}{0.037\ 90 \text{ m}^2}$$

$$\beta = 1.099\ \frac{\text{Wb}}{\text{m}^2}$$

As a check,

$$1\ \frac{\text{line}}{\text{in.}^2} = 0.000\ 015\ 5\ \frac{\text{Wb}}{\text{m}^2}$$

$$70\ 920 \times 0.000\ 015\ 5 = 1.099 \text{ Wb/m}^2$$

3-7 CORELESS MOTORS

Recently, a new class of motor construction has emerged in the physically smaller sizes. The *coreless* motor has no magnetic core to the rotating armature. There are two fundamentally different forms of construction, depending upon the physical form chosen. The advantages to be gained are (1) reduced inertia of the armature rotor, which allows extremely rapid acceleration, and thus a short time constant of response in a control mechanism; and (2) little or no cogging, i.e., the tendency the motor has to move in short angular jerks as the torque just overcomes the field to armature magnetic attraction forces. Cogging is caused by the normally non-uniform or toothed structure of the armature magnetic core. It may be reduced by skewing or introducing a small helical angle in the armature teeth, and may be eliminated altogether if there is no armature magnetic core. The two major types are considered separately.

3-7-1 Disc or Printed Circuit Rotors.

If the total armature is made in a flat disc form, with only enough thickness to give some structural strength and to enable attaching armature circuit conductors to both faces, then a natural form of construction is to use a printed circuit. In this case the structural base is the nonconducting and also nonmagnetic laminate material. The separation between armature conductors is etched into the original copper-coated faces of the basic laminate board material. In smaller sizes even the commutator segments may be made from the printed circuit. Variations include separate commutators that are attached to the rotor circuit and even separate punched and/or formed windings that may be adhesively attached to the basic board in the intermediate sizes. The end result is that the armature is thin in an axial dimension and thus does not need any magnetic core of silicon steel to keep the air gap between field poles down to a reasonable dimension. The field poles are arranged parallel to the armature shaft instead of radially. This results in flux lines parallel to the shaft through the armature conductor area in the region where the conductors are essentially radial. See Fig. 3-19 for a disc motor construction.

Flux will pass from the commutator side of the armature winding disc to the back side in one pole pair and from the back side to the commutator side in an adjacent pole pair. This means that an individual armature winding coil must go radially outward through the field flux on one face of the disc and then be connected around the outside of the disc through a portion of the circumference to connect up with the other face of the disc. The other working part of this coil then passes radially inward under the influence of the next adjacent field pole pair. A lap or a wave winding may be used. The outer circumferential part of the winding here corresponds to the back end of a normal winding coil, as used in a usual laminated core armature. The effective length of the winding conductor is the radial dimension of the field poles.

As motor size increases, the usual developmental problems of any inherently

Figure 3-19. Disc rotor motor. (*Courtesy of PMI Motors Division, Kollmorgen Corp.*)

new construction become apparent. Armature winding heat, caused by normal I^2R loss, has a tendency to warp the disc and to loosen the bond holding the windings. This must be overcome by careful selection of materials and equally careful proportioning of size and field gap spacing. These motors may be built in a range of sizes and with any normal field pole construction of shunt, series, or compound winding. However, present construction usually uses permanent magnet fields.

One limitation is that it is difficult to get more than one winding turn per coil. As a result, the motors are limited to low voltage in the sizes usually built. On the other hand, most servocontrols use both relatively small motors and low supply voltages. The inherent advantages of low inertia and low threshold starting voltages make this type of motor suitable for control system service.

3-7-2 Can Motors. This name comes from the physical appearance of these small motors. Plain cylindrical motors had existed before, as in automobile starters, but very small motors had usually been of open frame construction with an end lump for their field magnet. Thus, a small round case motor was named a can motor. The shape is based on the construction in which the exterior case is a simple magnetic steel return path for the field circuit. The field is usually a two−pole slug of Alnico or ceramic magnetic material and is placed concentrically *inside* the can. Samarium Cobalt supermagnetic material will be used on an increasing basis.

The cup-shaped coreless armature revolves in the concentric airspace between the exterior can and the interior field magnet. Here, even more than in the disc shaped rotors of the previous section, the rotor windings are self-supporting. The cup-shaped winding assembly has a configuration similar to that of a conventional motor winding that uses a magnetic lamination core, except that it is cantilevered out from the commutator. In order to make this construction practical, the active winding conductors are skewed in relation to the shaft, one way on the exterior of the cup, and the other way on the interior. The whole winding is moulded into a cup form with an epoxy or other adhesive matrix. The shaft is frequently single ended, going only as far as the commutator assembly. This is perfectly adequate in the miniature sizes that are produced. End bells are required, one at the commutator end to support the bearing and the simplified brush rigging, and one at the other end to keep the field magnet and the exterior can concentrically located.

These motors also have very low inertia and no tendency toward cogging at low speeds and powers. Their present uses are in instrument servocontrols and, by the thousands, in scale models. The performance is exceptional for their size, since the cogging tendency of conventional small motors with as few as three armature slots was and is severe. This factor could have been minimized by introducing more armature slots and commutator segments. However, in small sizes the armature coil winding wires are hairlike in size, and to make more coils of still smaller size would be impractical.

On the other hand, can motors seem at present to have an upper size limit owing to the difficulty of adequately supporting the armature windings against torque and centrifugal stresses. Remember, as a wire diameter increases, its cross section, and thus its weight per unit length, increases as the square of the diameter. At the same time the surface area increases only linearly with diameter. As a result, an adhesive matrix has a linearly increasing area for its strength bond at the same time that the loads imposed on that bond are increasing by a square law.

The need for a support cup for the windings that is in itself a nonconductor complicates the manufacturing process, which is already difficult.

At present, this coreless can motor construction has a specific, if limited, niche in the overall field of motor sizes. The very superior performance will cause its niche to expand, since it is limited only by competitive cost factors and practical construction methods. The present range of sizes is from a little over an inch in diameter (25–30 mm) on down.

3-8 BRUSHLESS MOTORS

All motors and generators considered so far in this chapter have had their armature built as a rotating element, or rotor. This has been convenient because all machines so far considered needed a rotor angle switching mechanism to select the polarity of the armature coil connections. Motor and generator developers have long dreamed of eliminating the commutator and brushes which, up to the last few years, has been the only practical means of performing this required switching.

Only recently have there appeared on the market various practical means of switching coil polarities without the use of brushes and commutators. Some, but not all, of these commutatorless or, more simply, brushless, motors will be considered here.

Most of these new types take advantage of the fact that no rotary mechanical switch is needed in the motor by placing the armature windings in a stationary location in the stator. The field is generally a permanent magnet, which is mounted on the shaft and serves as the *rotor*. This inside-out construction is then much like the various basic ac armatures, and for the same reasons. (See Chapter 11, Sect. 11-1 for an exposition of the reasons for inverting a motor construction.)

3-8-1 Armature Circuit Switching Mechanisms. A variety of means are now, or have been recently, used to reverse the direction of current flow in a coil of a brushless dc motor. Most of the following methods need amplification for practical operation.

(1) Magnetic reed switches have been used successfully in low-cost brushless units. The switches are actuated by part of the rotor field magnet or by a separate small timing magnet on the rotor shaft. This mechanism *does not* eliminate the radio frequency interference that results from mechanically moving commutator-to-brush contacts.

(2) Light and photodiode switches have been used in which the light is covered and uncovered appropriately by a rotor appendage or mask. The resulting photodiode output signal is amplified by suitable solid state circuits to operate the switching elements.

(3) Hall Effect devices are used to sense rotor position. These elements use a

current–bias signal and are affected by changing magnetism. They produce small voltage signals that can be used to trigger transistors or other solid state devices, such as silicon-controlled rectifiers (SCRs). The change in flux linkage from the rotating field magnet is sensed by the Hall Effect device. Various numbers of Hall Effect units are placed between appropriate pairs of stator coils to sense the rotor position.

Note that all of these devices are used in smaller numbers than the commutator segments that they replace. As a result, the armature coils exist in reduced numbers, such as four or six. The armature coil configuration then resembles a two-phase or three-phase ac stator. In fact, the whole motor then operates much like an ac synchronous motor, where the frequency of the ac signal is controlled by the shaft position and speed rather than the line supply. This ac analogy is carried even further in that some dc brushless motors are really ac synchronous motors that are driven by a dc supply that is electronically chopped into a fixed frequency alternating current. This type of motor is used when constant output speeds are desired and is *not* the true controllable, variable-speed, brushless dc motor that is considered here.

At present, in the late 1980s, the Hall Effect switching device seems to be more and more the choice of brushless motor designers. The motors exhibit very nearly the same external characteristics as a comparably sized, otherwise conventional permanent magnet field dc motor. They counterbalance their higher cost by being highly reliable and by being absolutely free of radio frequency interference and explosion hazard.

QUESTIONS

3-1. On a dc machine, why is heavy main frame construction required?

3-2. Is an odd number of field poles possible? Why?

3-3. Name two main functions for the cylindrical laminated steel structure of the main portion of a dc armature.

3-4. Where are armature winding coils placed in a dc machine?

3-5. To what are the ends of the armature winding coils connected?

3-6. What is the usual material for the commutator bars or segments?

3-7. What is the function of the commutator?

3-8. What determines the quantity and spacing of the brushes?

3-9. Name three major functions of the armature shaft bearings.

3-10. What is the purpose of the end bells?

3-11. Where are the main field coils located?

3-12. How are the field coils mounted or attached?

3-13. What is the purpose of commutating fields?

3-14. Where are commutating fields mounted?

3-15. What is the purpose of compensating windings?

3-16. Where are compensating windings placed?

3-17. What is armature reaction and what is its most serious effect?

3-18. How do commutating field windings cope with the effect of armature reaction?

3-19. What is armature reactance?

3-20. Why do the brushes normally cover more than two commutator segments?

3-21. Describe the flow of magnetic flux in the closed magnetic circuit in a dc machine.

PROBLEMS

3-1. An armature is wound with a simplex lap winding for eight poles. How many parallel paths does it have?

3-2. If the armature for the same eight-pole machine were wound with a simplex wave winding, how many parallel paths would exist in the winding?

3-3. A six-pole machine has duplex lap armature windings. How many parallel paths exist in the armature?

3-4. A four-pole dc machine is wound with a triplex wave armature coil. How many parallel paths exist?

3-5. A machine with eight poles makes use of frog-leg windings. How many parallel paths must exist with windings of the least complexity?

3-6. A large dc machine has a 0.375 in. air gap. How many ampere turns are required to overcome the air-gap reluctance to maintain a magnetic flux of 43 200 lines/in.^{2}?

3-7. A similar large dc machine has a 10 mm air gap. How many ampere turns are required to maintain a 0.7513 Wb/m^2 air-gap flux?

3-8. The dc machine magnetic circuit shown in Fig. 3-18 is subdivided into various geometric parts, each of which has an individual cross section and a specific material. Determine the *cross-sectional area* in English units for the pole core, *ab*. Remember to allow for the stacking factor of the laminations and the fact that only half the total pole core area is assumed to be in path *ab*.

3-9. Determine the *length* of path *ab* in Fig. 3-18 using English units.

3-10. Determine the *cross-sectional area*, path *ab*, of Fig. 3-18, in SI units.

3-11. Find the length of path *ab* in Fig. 3-18 in *meters*.

Note that problems of this type and those following can be profitably designed around actual measurements of whatever dc machines are used in the school laboratory. The actual flux required can be calculated from Eq. (2-6) if turns and coil information is available. Alternatively, gaussmeter readings of a separately excited but not rotating machine can be taken. The student can progressively assemble real data.

3-12. Again from Fig. 3-18, find the effective length of the magnetic path *bc*. Simplify the calculations, but allow for the fact that *bc* is not a full one quarter of the circumference. Defend your path-length simplifying assumptions. Exact agreement is not expected, but 10 or 15 percent correspondence is.

3-13. Find path length *bc* in meters. The same simplification as in Problem 3-12 is reasonable.

3-14. From Fig. 3-18, find the cross-sectional area of path *bc* in English units.

3-15. Repeat Problem 3-14 to find area *bc* in square meters.

3-16. Determine the *simplified* magnetic path length through the armature of Fig. 3-18 in English units. Note that the magnetic flux must travel radially through the tooth structure and at some depth in the core between *e* and *f*. Many degrees of approximation are possible, but use the simplest that seems plausible. Identify as path *ef* even though it includes the teeth.

3-17. Determine path length *ef* of Fig. 3-18 by the same definition as Problem 3-16, but in meters.

3-18. Determine the magnetic cross section of the armature rotor in Fig. 3-18 in English units. Note that at least two different cross sections should be investigated, and the smallest area chosen for further magnetic structure calculations. This is a laminated unit.

3-19. Repeat Problem 3-18 in SI units.

3-20. Find the cross-sectional area of the air gap *de* in Fig. 3-18 in English units. Be consistent and use half of one pole air-gap area if half a pole core area was used in Problems 3-8 and 3-10.

3-21. Find the air-gap cross-sectional area *de* in Fig. 3-18 in SI units.

3-22. Determine the probable air-gap length *de* in Fig. 3-18 in inches.

3-23. Determine the probable air-gap length *de* in Fig. 3-18 in meters.

3-24. If air-gap flux density in Fig. 3-18 is 50 000 lines/in.^{2}, find ampere turn requirements of the gap *de* by using the gap length in Problem 3-22.

3-25. Repeat Problem 3-24 using $\beta = 0.775$ Wb/m^2, and find air-gap ampere turn requirements using the gap length found in Problem 3-23.

3-26. Assuming the 1 650 000 lines *total* flux per pole used in the Fig. 3-18 design in Sect. 3-6-1, and the *BH* curve of silicon steel in Fig. 3-16, calculate the ampere turns requirement for the pole core path *ab*. Use the cross-sectional area from Problem 3-8 and length from Problem 3-9. Assume that 15 percent of flux generated is lost.

3-27. Assuming $\phi = 0.0165$ Wb total flux per pole in the design of Fig. 3-18, and using the *BH* curve for silicon steel of Fig. 3-17, calculate the ampere turns requirement for the pole core path *ab*. Use the cross-sectional area from Problem 3-10 and the length from Problem 3-11. Again assume 15 percent flux loss.

3-28. Determine the ampere turns requirement for the main frame segment *bc* of Fig. 3-18. Use the cast steel *BH* relation from Fig. 3-16. Use the cross-sectional area from Problem 3-14 and length from Problem 3-12. Use the flux requirement of Problem 3-26 plus 15 percent leakage allowance.

3-29. Determine the ampere turns requirement for the main frame segment *bc* of

Fig. 3-18 using SI units and data from Problems 3-13, 3-15, and 3-27, and the *BH* curve for cast steel in Fig. 3-17.

3-30. Determine the ampere turn requirement for the armature core *ef* of Fig. 3-18. Use the cross-sectional area from Problem 3-18 and path length from Problem 3-16. The flux requirement is from Problem 3-26. The armature core is sheet silicon steel.

3-31. Determine the ampere turn requirement for the armature core *ef* of Fig. 3-18 using SI units in the calculations. Use the cross-sectional area from Problem 3-19 and length from Problem 3-17. Flux requirement is from Problem 3-27. Silicon sheet steel is used.

Answers. **3-1**, 8 paths; **3-2**, 2 paths; **3-3**, 12 paths; **3-4**, 6 paths; **3-5**, 16 paths; **3-6**, 5075 AT; **3-7**, 5979 AT; **3-8**, 11.04 in.^{2}; **3-9**, 3.94 in.; **3-10**, 0.007 12 m^2; **3-11**, 0.100 m; **3-12**, 13.3 in.; **3-13**, 0.338 m; **3-14**, 13.5 in.^{2}; **3-15**, 0.008 70 m^2; **3-16**, 7.85 in.; **3-17**, 0.199 m; **3-18**, 12.9 in.^{2}; **3-19**, 0.008 32 m^2; **3-20**, 16.5 in.^{2}; **3-21**, 0.0106 m^2; **3-22**, 0.072 in.; **3-23**, 0.001 82 m; **3-24**, 1128 AT; **3-25**, 1128 AT; **3-26**, 107 AT; **3-27**, 103 AT; **3-28**, 346 AT; **3-29**, 331 AT; **3-30**, 93.0 AT; **3-31**, 88.0 AT.

4

Direct-Current Generator Characteristics

The dc generator has been of great importance since the early commercial development of electric lighting under Edison and early urban and interurban trolleys following the developments of Sprague. Both of these uses flowered in the last 20 years of the 1800s. Direct current is now long obsolete for major commercial lighting, and the urban trolley has about disappeared. Urban electric transportation faces some major forms of resurrection as inner-city traffic problems become increasingly intolerable and as pressures for less wasteful use of energy become more insistent.

The development of electric starting and lighting for automobiles required the building of around 100 million or so small dc generators over about a 50-year period. The development of this form of generator stopped around 1965 with the widespread use of small ac generator rectifier units, which had a number of advantages. The fashion of major uses of large quantities of dc power has thus risen and fallen on various product lines, but other compensating uses have and will continue to appear.

Starting in the 1920s and increasing with a rush during and after World War II, the diesel electric locomotive has effectively entirely replaced the steam locomotive on the railroads in the United States and most of the more developed parts of the world. In the diesel electric, the diesel engine is the prime mover and is directly connected to a large dc generator on a one-to-one basis. The locomotives are really electric locomotives with their own portable dc supply.

Urban subways and the few remaining elevated railways are all dc-driven, usually from local district power substations where large ac-to-dc rotary converters supply the direct current to the rails. Industry uses major amounts of direct current in plating processes and especially in the reduction of bauxite to alumina in the primary production of aluminum.

New installations that require direct current in small to moderate amounts will probably use rectified alternating current when it is conveniently available. On the other hand, where the prime mover is a rotating engine, dc generators will continue to be used since they are efficient devices.

The modern trend to urban electric automotive vehicles will require vast amounts of dc power for battery charging. In fixed installations this direct current will probably be supplied by transformer rectifier units. The on-board battery charging to be used during braking of the vehicle will require that each vehicle be able to use its main motor or motors as a generator in order to convert kinetic energy to usable electric energy.

At the time this is written, the forms of motors and generators to be used are very experimental and hotly debated. Ultimately, there will be whole families of closely related designs that will be built in the millions. Rarely has a future product need been so universally foreseen and yet not at all agreed upon. As late as 1986 there was no commercially available motor generator that was optimized for the electric automobile.

The area of use of dc generators will undoubtedly continue to change, and thus the requirement for understanding of the various units and their limitations will continue to grow.

4-1 SEPARATELY EXCITED GENERATOR

There are three basic types of dc generators, with subtypes, and they have distinctly different characteristics. To see how each type and subtype gains its characteristics, it is logical to see how the generator is affected by variations of field strength. Then the variations of field strength that result from various field-to-armature connections can be better understood. It is in the various ways of connecting the field circuit that the generator types develop their individual characteristics.

If the main fields are supplied from an outside source, their ampere turns of excitation can be independently controlled. We shall see in a practical generator that the output voltage is controlled by many factors, as shown in Eq. (2-6). The parameters of number of conductors Z, number of field poles P, and number of parallel paths in the armature windings a are designed and built-in features of the particular unit. Once built, then one way to control the generated voltage of a given generator is to vary its rotative speed S in revolutions per minute (or ω in radians per second). The other way is to vary its field flux per pole Φ in lines per square inch (or ϕ in webers per square meter).

The rotative speed is controlled by the characteristics of the prime mover that is coupled to the generator and any gearing or belt drive that may be interposed between the prime mover and the generator.

The field flux is determined by the overall magnetic path characteristics, which are a designed and built-in feature much as the armature windings and number of poles. The field coils are also designed and built with a particular number of winding turns of a particular wire. The ampere turns that are present in the coil are determined by the physical number of turns and the current that flows as a result of the applied field voltage. The applied field voltage can of course be of almost any value within reason when supplied separately.

Chapter 3 focused on the relationship between the H value of ampere turns per inch of length of a magnetic path (or ampere turns per meter of length) and the resulting B magnetic flux per square inch (or β flux per square meter). The end result is that a particular field coil which is mounted around a particular magnetic circuit having a built-in field pole area, by virtue of its size, will then have a Φ (or ϕ) value that is related to its H value according to the overall BH curve of the magnetic circuit. This of course depends on the specific materials used.

4-1-1 Generator Magnetization Curve. Thus, generated volts E_g are directly related to Φ (or ϕ), since it is then the only variable when the rotative speed is held constant. Then the output voltage will be related to the field excitation current in a curve that appears much like a BH curve with different abcissa and ordinate scales. In fact, there is a whole family of generated voltage curves, all of related shape, with one for each particular rpm (or radian per second).

This *internal characteristic* of a generator is known as its magnetization curve. A specific sample curve as measured under school laboratory conditions is shown in Fig. 4-1. Here only one specific rotative speed of 1800 rpm (or 188.5 rad/sec) is shown. Twice the speed would result in twice the voltage for any particular field strength. Similarly, 0.9 the speed would result in 0.9 the voltage, and so on.

The polarity of the output voltage depends upon the direction of rotation and the direction of the current in the field. Either of these may be controlled, so the output polarity may be controlled. The actual direction of field flux is of course dependent on the absolute direction of the winding rotation for any one connection. The actual direction of winding of the coils is another of the design features of a generator. It is not often possible to determine which way the coils are wound by inspecting the parts, owing to external insulation and protection. However, if it does make a difference to know polarity prior to an actual test, it is easy enough to check coil by coil with a small excitation voltage and a small magnetic compass. If the marked or colored end of the compass needle is attracted to a particular pole shoe with a known winding excitation polarity, a north pole is present; if the other end of the compass needle is attracted, a south pole. It is usually easier to rotate the machine and test output polarity. If the result is not the wanted polarity, reverse the field winding connections. This polarity problem is easy on a separately excited generator, but is much more involved on a self-excited shunt generator, as will be seen later.

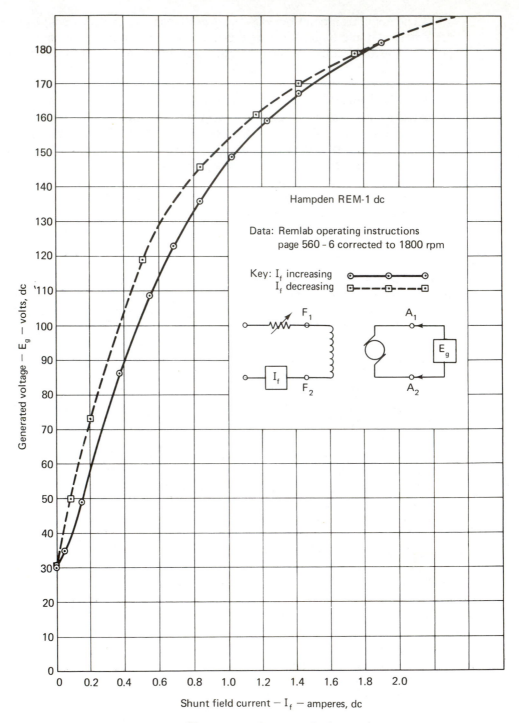

Figure 4-1. Direct current dynamo no load saturation test.

4-2 BASIC GENERATOR TYPES

With the knowledge that the generated voltage depends upon the field magnetic flux, we can now endeavor to understand the ways in which this field flux can be supplied from the generator itself.   The basic ways are as follows:

(1) The field connections can be tapped directly from the armature.   In this way the voltage that drives the current through the field coil is the full armature circuit voltage.   The current can be made less by the use of a series resistance in the field circuit if desired.   This connection is called a *self-excited shunt* generator, or simply a *shunt* generator.   Ordinarily, if a generator is separately excited, it is labeled as such; but if self-excited, the term may be only implied.   This is usually the same field used for separate excitation.   The word "shunt" here is synonymous with parallel, but historically the term "parallel generator" has not been used.

(2) The field connections can be connected in series with the armature circuit. In this way the field ampere turns will be primarily controlled by the resistance of the connected load.   There will be no field excitation if no current flows to a load. This is known as a *series* generator.   Its output characteristics will be seen to be very different from the shunt generator.   A series-connected field coil is composed of relatively few turns of large wire, whereas a shunt-connected field coil is composed of many turns of finer wire.   The same field ampere turns can be produced, for example, with 120 turns $\times$ 10 A = 1200 ampere turns in a series coil or 1200 turns $\times$ 1 A = 1200 ampere turns in a shunt coil.   The coils are very different, but the ampere turns are the same.

(3) The third type combines a shunt and a series field and is called a *compound generator*.   As might be expected, this type combines the features of the shunt and series generators.   There is a further subdivision according to whether the shunt field is connected across the armature alone, called *short-shunt compound,* or connected across both the armature and series field, called *long-shunt compound*. These different varieties of compound generators have slightly different characteristics but normally perform the same tasks.

4-3 SHUNT GENERATOR BUILDUP AND POLARITY

Self-excitation implies a process taking place within the generator.   This process starts when the machine is rotated in its normal direction at or near its normal speed.   Initially, and briefly, a voltage E_g will exist that is the result of moving the armature conductors through the small residual magnetic field that remains in the magnetic circuit after the last previous operation.   This voltage will exist steadily if the field circuit is inadvertently or deliberately left open circuited.   Figure 4-1 shows this voltage as the curve intercept on the vertical or ordinate axis, about 30.5 V.   This may be larger or smaller depending on the particular machine.

With a closed field circuit, this small voltage will now exist across the field and a small current will flow.   If this current produces enough ampere turns, and in the correct polarity, the field magnetism will be increased, and a higher voltage will be generated.   This process is known as *buildup*.   The increasing voltage

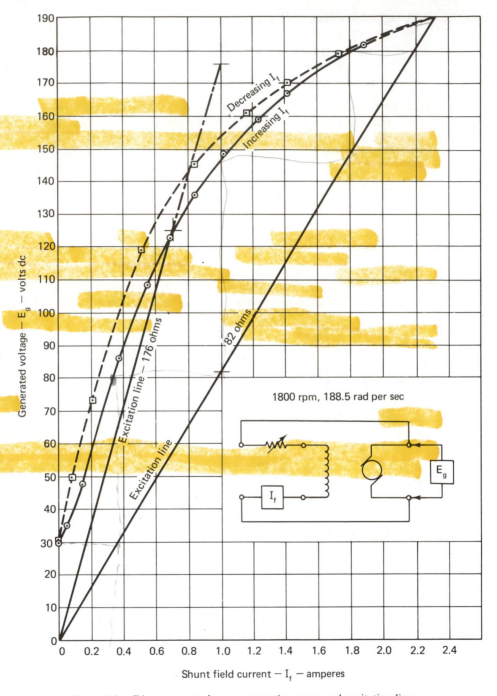

1800 rpm, 188.5 rad per sec

Figure 4-2. Direct current dynamo saturation curve and excitation lines.

produces increasing field flux until, by the curvature of the magnetization line, further voltage produces less and less additional flux.

4-3-1 Field Excitation Line. Finally, an equilibrium is reached when the magnetization curve crosses the field *excitation line* (see Fig. 4-2, which is the same as Fig. 4-1 except that the excitation line has been added).

Since the abcissa of Fig. 4-2 is field current in amperes and the ordinate is armature circuit voltage E_g in volts, the slope of the excitation line is $E_g/I_f = R_f$, or the resistance of the field circuit in ohms (Ω). This R_f is the sum of the field winding resistance and any field resistance or rheostat that may also be connected in the field circuit. Again in Fig. 4-2, with the machine set to $E_g = 125$ V at 1800 rpm (188.5 rad/sec), the field current is around 0.709 A. This requires 125/0.709 = 176 Ω in the field circuit. Since this particular machine has a field circuit resistance of 82 Ω, there was then 176 − 82 = 94 Ω of field rheostat resistance set in for the test shown. In this case, then, the excitation line is simply the 176 Ω slope.

With a given field circuit total resistance, here 176 Ω, a higher voltage than at the intersection of the magnetization or saturation curve and the excitation line cannot be generated. Using Fig. 4-2, visualize a higher voltage, such as 130 V. From the curve, this can be seen to require enough ampere turns that a field current of about 0.768 A is needed. From Ohm's law, this would require $R = E/I = 130/0.768 = 169 \Omega$ in the field circuit. With a setting of 176 Ω, the required 0.768 A cannot flow with only 130 V. In fact, no higher than 125 V E_g can exist unless the field circuit resistance is reduced.

If all added field circuit resistance is eliminated and only the 82 Ω shunt field is present, this machine will build up to the intersection of an 82 Ω excitation line and its saturation curve. At this rotational speed this machine would build up to 190 V. Otherwise identical machines would have approximately but not exactly the same characteristics, since no two machines of the same make and model are exactly identical in the reluctance of their magnetic circuits. They therefore require different ampere turns for the same flux.

The buildup process takes enough time to be visually seen by the action of a voltmeter after the field circuit is closed. Sometimes quite a few seconds elapse during buildup. This is more prominent when the field circuit resistance setting is high and when the field inductance is high.

4-3-2 Generator Build Down. Buildup is not an invariant or automatic process, however. If the machine is incorrectly connected, or if it is rotated in reverse, buildup will not take place. The opposite condition, or *build down*, occurs when the voltage that is reached is nearly zero instead of the 3 to 15 percent of final voltage that is the result of the residual magnetism in the field.

Build down takes place when the voltage that is generated by the residual magnetic field action on the armature windings is of the wrong polarity for the field circuit as it is connected. The current that flows in the field as a result of the residual voltage will produce ampere turns of magnetizing force regardless of the current direction. However, if the residually produced ampere turns tend to produce a magnetic field that is in the opposite polarity to the residual field, the two field forces cancel. The resulting lesser field produces less rather than more voltage, and the machine builds down. There is no harm done; the machine will simply not build up to voltage until the field connections are reversed.

4-3-3 Residual Magnetic Field and Field Flashing. There may still be no buildup, even with the correct polarity of field connection, if the residual field is weak or nonexistent. A new machine that has never been tested, or a machine that has been disassembled or perhaps affected by mechanical shock, will have insufficient field for buildup. The remedy for no buildup is to *polarize the field* or *flash* the field in the correct polarity with a separate excitation supply of the correct voltage.

Even this step must be performed thoughtfully, because a strong residual field of the wrong polarity will still require field connection reversal for satisfactory operation. The flashing process is well named because of the vigorous inductive arc that is produced when the excitation circuit is disconnected. Care must be observed, because in large sizes the arc may be dangerous.

4-3-4 Generator Polarity. A generator may satisfactorily build up to a useful voltage but with the wrong or opposite polarity from the desired polarity. The easiest correction is to shut down and reverse the external connections, leaving the internal connections as they were. This simple remedy may not be feasible, however, because of the inconvenience of lead lengths or because one end of the circuit may be grounded to the frame internally, as it was in the usual automotive generator.

The first operation of a new generator connection may be quite exasperating until connections are exchanged or the field is flashed or both.

This whole relation is much easier to follow and understand when it is done step by step in the laboratory. The author strongly recommends a step-by-step laboratory sequence of buildup, build down, change of rotation, reconnection, polarity change, and field flashing until the whole relationship is seen and satisfactorily understood. A self-excited shunt generator is really a simple and reliable device once set up. The familiar internal combustion engine is far more cantankerous unless it is intelligently timed and all its interrelated requirements for air, fuel, rotation, spark, etc., are met.

4-4 SHUNT GENERATOR EXTERNAL CHARACTERISTICS

The generated voltage relationship shown in Fig. 4-2 is correct for no-load conditions only. When an external load is connected and current is drawn from the generator, the characteristic is modified.

There are two reasons for the voltage reduction that takes place when a shunt generator is loaded. The major reason is that there is a finite resistance to the armature circuit. Any current then causes a voltage drop within the armature circuit analogous to a simple line drop in wiring of any kind. At the very least, the armature circuit consists of the armature windings, the commutator, and the

brushes. There may also be present commutating fields and perhaps compensating windings. All these vital parts are connected in series, and their combined resistances, even though as low as possible, are additive and significant. Second, any armature reaction that is uncompensated causes a loss of total field flux and a resulting loss of generated voltage. The loss of terminal voltage due to armature circuit resistance drop and armature-reaction-caused flux reduction both show as less voltage applied to the shunt field. This lesser field circuit voltage causes fewer ampere turns and thus lower flux, and finally lower voltage unless adjusted away by a change in field rheostat setting.

4-4-1 Internal Voltage Loss in a Shunt Generator.
There are three parallel circuits in a shunt generator when working as shown in Fig. 4-3. These are the armature circuit in all its parts, the field circuit, and the external load. The common junction of all the branches or the paralleling point is at the terminals of the generator. The voltage that is measured at this point is usually identified as V_t or terminal voltage. V_t and E_g are not the same value. In a generator $V_t = E_g - (I_a R_a)$ as was shown in Eq. (2-9), where

E_g = internally generated voltage that results from the factors in Eq. (2-6)

I_a = current that flows in the armature circuit

R_a = total series resistance of all the elements present in the armature circuit:

   (1) Armature resistance
   (2) Brush resistance
   (3) Commutating field resistance
   (4) Compensating field resistance

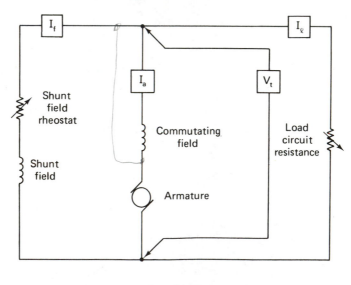

$$I_a = I_\ell + I_f$$

Figure 4-3. Parallel circuits in a shunt generator.

From Fig. 4-3 and Kirchhoff's current law, it can be readily seen that the armature circuit current is the sum of the load current and the field circuit current. The armature circuit is the source branch of the total circuit, and its current direction is from − to + while the field and the external load have current directions from + to −:

$$I_a = I_f + I_l$$
Eq. (4-1)

Also, since these three circuits are in parallel, from Kirchhoff's voltage law we can see that the voltages are alike:

$$V_a = V_f = V_t$$
Eq. (4-2)

Example 4-1. A 100 kW, 240 V shunt generator has a field circuit resistance, including a rheostat, of 55 Ω and a total armature circuit resistance of 0.067 Ω. Calculate:

  (a) The full load line current flowing to the load.
  (b) The field circuit current.
  (c) The armature circuit current.
  (d) The full-load generated voltage that is required to deliver rated voltage to the load.   Ignore armature reaction.

Solution:

(a) $I_l = \dfrac{\text{kW} \times 1000}{240} = \dfrac{100 \times 1000}{240} = 416.7 \text{ A}$

(b) $I_f = \dfrac{E_l}{R_f} = \dfrac{240}{55} = 4.36 \text{ A}$

(c) $I_a = I_f + I_l$ [from Eq. (4-1)] $= 4.36 + 416.7 = 421.1 \text{ A}$

(d) $V_t = E_g - (I_a R_a)$,   [Eq. (2-9) transposed as $E_g = V_t + (I_a R_a)$]
$E_g = 240 + (421.1)0.067 = 268.2 \text{ V}$

To meet these conditions, the generator must be set to a bit more than 268.2 V at open circuit.   This is true because when delivering 240 V to a load the field sees the same 240 V.   At this point the field flux is adjusted to produce a generated voltage of 268.2 V.   If the unit were set no load, the field circuit would see the full 268.2 V.   What is wanted is that the same ampere turns will exist in the field at 240 V as was required to produce 268.2 V open circuit.   This means that, if 55 Ω were required in the field circuit under full-load conditions, the open-circuit field current would be (268.2/240) (4.36) = 4.87 A.   A higher voltage than 268.2 would be generated.   Not as much as (4.87/4.36) 268.2 = 299.5 V, however, since that would presume that a linear relation existed between ampere turns and field flux.   The generator is probably deliberately operated well up into the knee of its *BH* curve or saturation curve.   Equation (2-9) is only a *first approximation*.

4-4-2 Internal and Related External Characteristics. With a separately excited field, the simple relation of Eq. (4-1) is nearly true. Even in the case of a *self-excited* shunt generator, it is a fair first approximation. However, as seen above, the loss of field flux that results from the lower V_t causes a still lower V_t to exist when self-excited. This is best seen from Fig. 4-4, which is derived from a saturation curve of the type of Figs. 4-1 and 4-2. The internal characteristic will be shown to govern largely the external characteristic of a shunt generator.

The form of Fig. 4-4 relates a magnetization curve–excitation line relationship as in Fig. 4-2, but mirror imaged so that the I_f or field current increases to the left. The right side uses the same ordinate scale of voltage, but the right side of the plot is considered as V_t rather than E_g.

The right-side abscissa scale is drawn as a convenient scale for the load current range of the machine in question, with I_a being the sum of the load and field currents according to Eq. (4-1). I_a is a simplified label for $I_{a\ ckt}$ and is the same.

The construction is performed by projecting a vertical line up from the intersection of the excitation line and the magnetization curve. On this vertical line, armature circuit voltage drops are plotted to the same scale as the V_t ordinate scale's armature voltage. The armature circuit voltage drops are calculated at convenient increments of *load* current, usually at 0.25, 0.5, 0.75, 1.0, and 1.25 times rated load. Here the armature current used should be according to Eq. (4-1), and the armature circuit voltage drops will then fit:

$$V_{a\ drop} = I_a \times R_a$$

Eq. (4-3)

Be careful not to confuse Eqs. (2-9) and (4-3), since Eq. (4-3) is only the right term of Eq. (2-9) or the internal voltage drop portion.

When points are marked on the vertical voltage drop line for each load chosen, projections are made to the magnetization curve that are parallel to the excitation line. The intersections of these projections are then dropped vertically to the excitation line. Next, a horizontal projection from these last intersection points is made across to the right side of the plot. Each specified armature load current trace is then intersected with a vertical projection from the external load current that would correspond with the particular armature current on the load current abcissa scale. This requires reverting to the original load current data that were used in Eq. (4-1) to get armature current. The intersections of these horizontal and vertical lines are then points on a continuous external characteristic line. The value of this plot form is that it compensates for the nonlinearity of the magnetization curve (saturation curve).

These types of plots can be carried out for increasingly realistic conditions, but that again is beyond the scope of this text. Siskind (reference 3) explores these procedures very well, as do others, and they may be consulted if a particular problem requires it.

Figure 4-4 also shows the simplified external characteristic due to Eq. (2-9) alone, which, as has been stated, is a good first approximation. The difference can be readily seen.

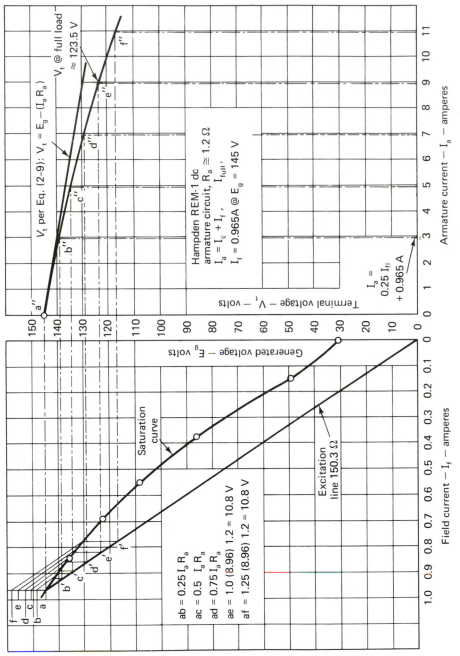

Figure 4-4. External versus internal characteristic curves.

The following labels and annotations appear within the figure:

V_t per Eq. (2-9): $V_t = E_g - (I_a R_a)$

V_t @ full load ≈ 123.5 V

Hampden REM-1 dc
armature circuit, $R_a \cong 1.2\ \Omega$
$I_a = I_\ell + I_f$, I_{full},
$I_f = 0.965A$ @ $E_g = 145$ V

$I_a = 0.25\ I_{fl} + 0.965$ A

Terminal voltage — V_t — volts

Generated voltage — E_g — volts

Armature current — I_a — amperes

Field current — I_f — amperes

Saturation curve

Excitation line 150.3 Ω

$ab = 0.25\ I_a R_a$
$ac = 0.5\ \ I_a R_a$
$ad = 0.75\ I_a R_a$
$ae = 1.0\ (8.96)\ 1.2 = 10.8$ V
$af = 1.25\ (8.96)\ 1.2 = 10.8$ V

When actual tests are made and carried out to very substantial overloads, the external characteristic curve is seen to droop steeply and finally reach a maximum current. This is due to very significant demagnetizing effects due to armature reaction and saturation at very high current. This effect will be discussed when other types of dc generators are developed further in this text.

4-4-3 Generator Regulation. Regulation is a term having to do with voltage change under load. The external characteristic curve is really a plot of the regulation of a machine. The American Standards Association defines regulation such that:

$$\text{Percent voltage regulation} = \frac{E_g - V_t}{V_t} \times 100 \qquad \text{Eq. (4-4)}$$

where:

E_g = voltage generated at no load (same as before)

V_t = terminal voltage at *full load* this time

Example 4-2. A self-excited shunt generator is rated at 240 V full load. At no load the voltage is 252 V. What is the percent regulation?

Solution:

$$\frac{252 - 240}{240} \times 100 = 5.00\% \text{ regulation}$$

Note that positive regulation is defined as a higher voltage no load than at full load. We shall find later that certain compound or series generators can deliver a higher voltage at full load than at no load. This opposite effect is defined as negative regulation. The agreed standardization and use of terms in this fashion saves much difficulty and even litigation in business use.

If the regulation problem had been stated another way, this same procedure would be modified.

Remember that a percentage is *not* the same amount of change going either way as shown above. The percent regulation (such as 8 percent) is a percentage of the rated *full-load* voltage. Another way assumes that the full-load voltage is 92 percent of the open-circuit voltage, or 8 percent less than the open-circuit voltage. The second way is *wrong* and should be avoided.

The operation of a generator at a point around the knee of its saturation curve results in better regulation (smaller percentage of regulation). This is because a reduction in V_t, caused by high load current will result in less loss of flux in the field than if operating on a point below the knee of the curve. Until other steps are added by the use of a series field, the best regulation that can be obtained is when the loss of terminal voltage is that due to the armature circuit resistance alone. Obviously, then, low armature circuit resistance is a desirable feature, especially when low resistance benefits efficiency also.

It should be kept in mind, however, that, when operating within a reasonable load range, zero regulation (or constant voltage) can be held with appropriate field adjustment.

4-5 SERIES GENERATOR

When a generator has its main field coils connected in series with the armature circuit and the load, it is then known as a *series generator*. This type of generator is not often used by itself now, but series fields are very much used in combination with shunt at the present. To understand the operation of a generator that contains a series field as one of its elements, it is necessary to first understand the operation of a series field acting alone.

4-5-1 Series Generator Buildup and Polarity. During initial operation, with no current yet flowing, a residual voltage will be generated exactly as in the case of a shunt generator. The field magnetic path structure may be similar or even identical to the structure of a shunt generator, and so a comparable residual magnetism is held after operation.

The residual voltage will cause a current to flow through the whole series circuit if the circuit is closed. There will then be a buildup of voltage to an equilibrium point exactly analogous to the buildup of a shunt operator. In this case, with a series field alone, buildup will not take place with an open load circuit, and the machine will idle unloaded at its residual voltage level indefinitely.

The voltage that will be reached can be seen from a plot such as Fig. 4-5, which is similar to the Fig. 4-2 except that the abcissa scale which was field circuit current alone is now the *load current*, with all current going through the series field.

4-5-2 Series Generator Characteristics and Adjustment. In a practical sense, all the current may not flow in the series field because the field usually contains some percentage more turns than are needed for the intended task. The field current is then usually adjusted with a series field *diverter* resistor or rheostat. Figure 4-6 shows the complete circuit of a series field generator and load. Note that there is a parallel circuit between the series field and the series field diverter. This parallel region of the circuit carries the whole armature current and load current. Usually a substantial portion of the load current but not all of it passes through the series field. Wattage in the diverter is all loss.

The actual external characteristic of a diverter-adjusted series generator will then be a very similiar curve, but with a lower slope as shown dotted on Fig. 4-5.

The basic solid curve on Fig. 4-5 is really the saturation curve of the generator dropped by the amount of the armature $I_a R_a$ voltage drop. Usually there will also be some armature-reaction-caused voltage drop present.

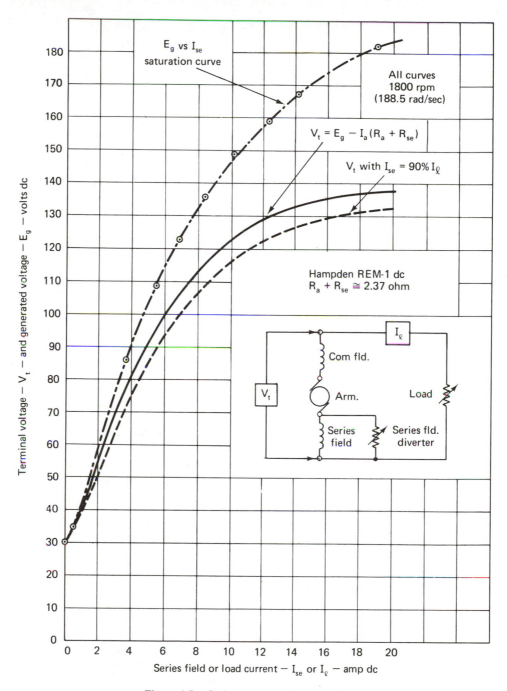

Figure 4-5. Series generator characteristics.

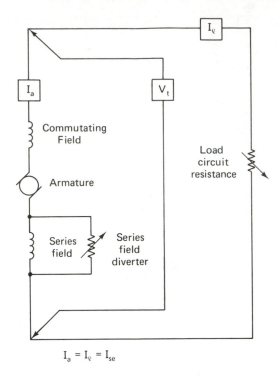

$$I_a = I_\ell = I_{se}$$

Figure 4-6. Series generator circuit.

4-5-3 Series Generator Polarity and Build Down. A series generator is subject to the same possible difficulties in buildup as a shunt generator. Again, if the series field is connected in reverse, the machine will build down for the same reasons. Still similar to a shunt generator, if the field magnetic polarity is reversed from that desired, the achieved voltage may be in the wrong polarity. The same corrections are used by flashing the field to get the desired polarity if a simple change of external connections is not feasible. It is usually much easier to flash a machine's field if the machine has a shunt winding that can be used. The same magnetic circuit is affected by either a series or a shunt coil, but the series coil needs *high* current.

4-6 SERIES GENERATOR EXTERNAL CHARACTERISTICS

Figure 4-5 shows a typical external characteristic curve for a series generator. Here the slope of the curve may be raised or lowered by adjusting the diverter up to a point. The high slope limit is determined by the turns in the series field. As a result, if a higher voltage is desired at a particular current load, a different field coil must be used. This range of effect may be readily seen with some school laboratory machines.

Example 4-3. A typical widely used school laboratory machine is the Hampden REM-1, which has both a 1200-turn shunt field and a 120-turn series field. Figure 4-1, which is the saturation curve of this type of machine, shows that 125 V can be reached with 0.709 shunt field amperes when running no load. As a first approximation, these machines have about 1.33 Ω resistance in the armature, commutating field, and brushes at 8 A load. The series field has about 1.04 Ω cold. Thus, there is around 1.33 + 1.04 = 2.37 Ω in the machine's interior series circuit. Determine if this machine can deliver 125 V at 8 A or its 1 kW rating when running series only. See Fig. 4-5 for series field conditions.

Solution: This will produce about 2.37 × 8 = 18.96 V or around 19 V drop at the rated 8 A load. Thus, to deliver 125 V to the load the machine must generate 125 + 19 -- 144 V internally (E_g = 144 V); 144 V requires about 0.922 A on the shunt field as determined from Fig. 4-1. Since the series field has one tenth as many turns as the shunt field, the series field will need 10 times the current, or 9.22 A, to supply enough ampere turns to produce 144. It can then be seen that this machine *cannot* deliver its rated 1 kW load 125 V (125 V × 8 A = 1000 W) when operating as a series machine. It could easily produce 125 V at a higher current such as 10.5 A or so, or exactly 1 kW to a load at a lower voltage of about 116.7.

This limitation is, of course, not a shortcoming of the machine but rather a designed-in limitation. A usual machine will have proportionately less series field than this when designed for compound operation, as will be seen in the next sections.

4-6-1 Series Generator Internal Voltage Loss. It is well to recognize that a series generator has two major voltage drops between the generated voltage E_g and the terminal voltage V_t. Therefore, a modified version of Eqs. (2-9) and (4-3) may be used when

$$\boxed{V_t = E_g - I_a(R_a + R_{se})}$$

Eq. (4-5)

when

  R_a includes the armature and brushes along with commutating fields and perhaps compensating windings. R_{se} includes the resistance of the series field, which may be in parallel with a series sales diverter.

Note that the armature circuit includes whatever parts need to be considered at the moment. Sometimes R_a includes everything, even a series field, sometimes not; but be aware and consider the situation.

  The series generator always has negative voltage regulation when operated in its rated range. The output voltage then is less at low load than it is at high load when within range. At high overloads the approach of saturation in the field magnetic structure reduces the increased voltage for each increase in load current.

Finally, the effect of increasing armature reaction with higher current actually reduces the output voltage, and positive regulation begins to appear. This effect is beyond the steady-state range of the usual machine. In the past this characteristic has been used by operating so far beyond the knee of the saturation curve that the machine is nearly constant current. This feature was useful as a welding generator. However, the low residual voltage makes it difficult to strike the arc to start the weld. Some degree of compounding is usually used now.

The ultimate reduction of voltage at very high loads also, to some degree, protects the machine from the full violence of a solid short circuit. On very large machines with very low armature circuit resistances, a short circuit can be both spectacular and destructive unless properly protected by breakers.

4-7 COMPOUND GENERATOR OPERATION

A compound generator is a dc machine that contains both shunt and series fields. As might be expected, compound generator characteristics combine the features of the shunt and series components. The usual compound machine does not have as many ampere turns of series field as it has shunt field. However, there will be enough to produce some degree of rising voltage with increasing load. Figure 4-7 shows a compound generator connection. Compare it with Fig. 4-3 and see the added series field. If the shunt field connects between the armature and the series field, it is connected *short shunt*. If the shunt field connects on the line side of the series field, it is connected *long shunt*. There is not much advantage one way or the other.

4-7-1 Cumulative Compound Operation.
When a compound generator has its series field flux aiding its shunt field flux, the machine is said to be *cumulative compound*. The cumulative effect, if held very small by a series field diverter rheostat current bypass, will produce a lesser slope to the characteristic curve. This will then produce lower percentage, but still positive, voltage regulation. This condition is *undercompounding*.

A frequent use of a compound generator is when the series field adjustment just compensates for the voltage loss due to internal $I_a R_a$ drop and armature reaction drop. This condition is called *flat compound,* because the characteristic curve is approximately flat. There will be some higher voltage at mid-load range due to the bending of the internal characteristic curve. This bend is usually a very few percent. A flat compound generator is then adjusted to a zero percent regulation when considering full rated load and no load points. In actuality, there may be 1 percent or so difference between no load and full load due to increments of adjustment available at the diverter, or to integer number of turns in the field windings if no diverter is used. Flat is then approximately flat rather than absolutely flat.

When the series field effect is larger than the machine internal voltage loss, the output voltage increases with load. This form of cumulative compounding is called *overcompound*. A frequent form of adjustment is to overcompound just

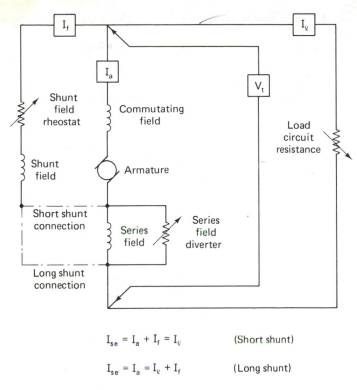

$$I_{se} = I_a + I_f = I_\ell \qquad \text{(Short shunt)}$$

$$I_{se} = I_a = I_\ell + I_f \qquad \text{(Long shunt)}$$

Figure 4-7. Compound generator circuits.

sufficiently to compensate for a significant transmission-line voltage drop. In this fashion the voltage at the distant *load* can be held nearly constant from no load to full load. If there were a 15 V line drop at full load, an overcompound machine might be set at 140 V at full load and 125 V no load. The load would then see 125 V under *either* condition.

4-7-2 Differential Compound Operation. When the series field is connected in reverse so that its field magnetic flux opposes the shunt field magnetic flux, the generator is then *differential compound*. This connection is useful when a significantly lower voltage is wanted under high load conditions. Another use is when an *approximately* constant current is wanted; still another is when a particular maximum current should not be exceeded. Although not specifically used as such now, a differential compound generator can serve as a dc welding generator where current is wanted rather than voltage. Its advantage over a series generator, which can serve for a constant current source, is that at no load there is sufficient voltage to strike an arc and start the weld. Similarly, an arc light requires nearly constant current even though the resistance of the arc varies widely.

Do not fall into the mental trap that a device is thought to be obsolete just because its principal historic use is obsolete. The device has fundamental characteristics that may be just what is needed to make a technical and commercial success of some totally new endeavor. This type of serendipity is constantly occurring in engineering.

In this vein, Henry Ford's planetary transmission as used in the ancient Model T Ford was widely considered as hopelessly obsolete well before it went out of production in 1927. By the late 1930s when General Motors first produced an automatic transmission that became the pioneer hydramatic of the 1940s and 1950s, the band clutched planetary gear set had found a new home. Since then and on into the 1980s, every successful automatic transmission has used one or more planetary stages. The production of the second reincarnation of the automotive planetary transmission has far exceeded the first in quantity produced, dollar value, and universality of application.

Like both the shunt and series dc generators, the compound generator is subject to buildup and polarity problems for the same reasons. The easiest way to establish the needed buildup and desired polarity in a compound generator is to start under no load conditions. At no load, only the shunt field is effective. Shunt field reversal and/or field flashing may have to be resorted to to get proper operation. Since the field magnetic structure is the same, any field flashing for the shunt field then affects both fields. When successful, no-load buildup is achieved, and with the desired polarity; then loaded operation is tried. If under load application the voltage rises, the series field is then connected cumulative. If the voltage drops a small amount, perhaps increasing the series diverter resistance to increase the series field effect may be necessary because the previous adjustment may have been undercompound. If the voltage drops significantly, the connection is differential compound. Great care must be used, because usually the commutating fields, and compensating windings, if present, are connected at the factory and should not be disturbed. If these windings are inadvertently reversed, the commutation will be disastrously affected.

Within the *same direction* of rotation, *any* dc generator may have its *polarity* changed by flashing the field without any internal connection changes.

If the *direction* of rotation is changed, the machine will not build up unless the field connections are changed, and *then* it will have the opposite polarity. The original polarity with opposite rotation will require field flashing and also reconnection.

In summary, either rotation direction and, in each direction, either polarity may be achieved. If a compound machine has the type of compounding desired and, because of opposite rotation, requires field connection change, both series and shunt field must be changed.

In final relief, however, if a machine is operating correctly and neither reverse rotation nor polarity change is desired, operation may be confidently expected. The machines are not "moody" or capricious. They merely have requirements that must be obeyed.

4-8 COMPOUND GENERATOR CHARACTERISTICS

From the previous discussions it can be seen that a compound generator can be wound and adjusted for a very broad range of external characteristics. The most

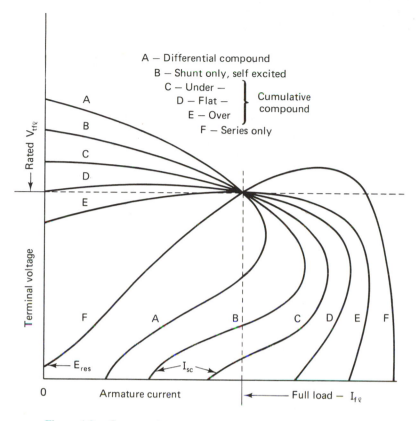

Figure 4-8. Compound generator external characteristics adjustments.

negative regulation available is that of a pure series generator; the most positive regulation is available with a strongly differentially compound arrangement. Literally any arrangement in between can be achieved. Some extreme types may not be useful or desirable, but many of the more normal adjustments are widely used. A compound generator can be used to compensate for undesirable but otherwise unavoidable circuit conditions due to long connecting lines, as one example.

Figures 4-8 and 4-9 show related families of generator external characteristics drawn from two different viewpoints. Figure 4-8 compares the range of adjustments possible within a single compound machine. All different curves intersect at the rated voltage and rated load current. This situation results in widely different no-load voltages and overload voltages. Figure 4-9 compares the range of adjustments possible when no-load voltages are all alike. Here the different voltages show under loaded conditions with the differences becoming more accented as the load increases. The plain series field generator cannot produce a high adjusted voltage at no load and, if the same unit, has the same curve on both types of plots.

The differentially compounded generator, when adjusted so that the series field effect is very heavy, will not normally reach the same load current as it would if not connected in that fashion. Its purpose is not to hold a particular voltage at a particular current, but to hold a relatively constant current over a range of voltage.

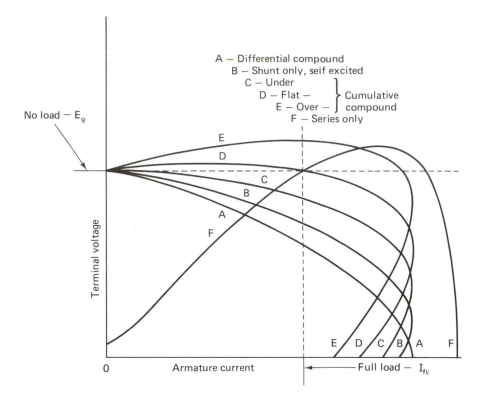

Figure 4-9. Compound generator external characteristics adjusted for common no load voltage.

4-9 COMPOUND GENERATOR ADJUSTMENTS

A compound generator is usually built and supplied with more series field capability than needed in normal operating conditions. It is then trimmed or adjusted to have just the degree of compounding that the specific use requires. Among other things, the degree of compounding that a particular series field will supply will depend upon the basic rpm (or radians per second) adjustment. After rotative speed or speed range is set, the resulting no-load voltage adjustment will determine the location of the operation point in relation to the knee of the saturation curve. The point is that the ratio of series field ampere turns to change of output voltage is not fixed.

Trimming is normally done by bypassing or diverting some of the load current around the series field. If the series field sees only 0.6 of the load current, its effect is the same as if it had 0.6 times the number of turns and had all the current pass through it. The determination of the correct diverter resistance is usually done by experiment. This diverter resistance is only a variable rheostat in the smaller sizes of machines, since it may carry a very high current and have a very low resistance. After repeated trials, the adjustment may have been bracketed with too high and too low diverter values. At this point a linear interpolation may be made, and a fixed value chosen and built into the machine connection. Since the diverter may be only a few thousandths of an ohm, the actual adjustment is a painstaking procedure, as Example 4-4 will show.

Example 4-4. An 1100 A short-shunt overcompound generator is equipped with a series field diverter of $R_{div} = 0.0088 \ \Omega$. The basic series field has an operating temperature resistance of $R_{se} = 0.0027 \ \Omega$. The series field has six turns per pole.
    (a) Calculate series field ampere turns when operated with the diverter connected.
    (b) Calculate ampere turns with diverter unconnected.
    (c) Calculate ampere turns with the series field short circuited.

Solution:

(a) $I_{se} = 1100 \times \dfrac{0.0088}{0.0088 + 0.0027} = 842$ A (parallel current rule);

    $6 \times 842 = 5052$ ampere turns (diverted).

(b) $6 \times 1100 = 6600$ ampere turns (series only).

Short-circuited conditions are not what they seem to be, since it is extremely difficult to make a good joint much under 0.001 Ω. If this is true, apparently short-circuited conditions are really *about* the same as with a 0.002 Ω diverter! If so,

(c) $I_{se} = 1100 \times \dfrac{0.002}{0.002 + 0.0027} = 468$ A and then $468 \times 6 = 2808$

    ampere turns.

and to *open circuit* the series field. Actually, the apparently short-circuited series field is the operation with the lowest resistance diverter that can be installed. This still allows a significant current in the series field. A variation in the length of an apparent short circuit may be just what is needed. Even in school laboratory machines of only 2 or 3 kW output, the variations of series diverter that are produced by varying lengths of heavy-gage copper plug in leads will produce a visible effect.

    Another way of looking at this problem of adjusting diverters is to consider that the original given $R_{div} = 0.0088 \ \Omega$ might install as around 0.01 Ω with clamped end connections. The measurement of these very low resistances usually requires a Kelvin bridge and very careful technique. A final thought is that preparation of an adjustment to resistances in increments of around 0.001 Ω is quite uncertain. In practice, adding a fourth significant figure after the decimal point is still more uncertain.

    An adjustment problem is more complicated if the generator is short shunt. Remember that

$$\boxed{I_{se} = I_a \quad \text{(long shunt)}}$$
                                                       Eq. (4-6)

and

$$\boxed{I_{se} = I_l \quad \text{(short shunt)}} \qquad \text{Eq. (4-7)}$$

where

$$I_l = I_a - I_{sh}$$

which is, $I_a = I_f + I_l$ transposed as shown in Eq. (4-1).

Direct-current generator characteristics are explored in a number of texts, but for the level next above this, where the math is still algebra, references (4) and (5) give excellent detail.

QUESTIONS

4-1. What is meant by a separately excited generator?

4-2. Why does the magnetization curve or saturation curve bend and tend to flatten?

4-3. What is meant by a self-excited generator?

4-4. What does the term generator buildup mean?

4-5. Name three basic dc generator types.

4-6. What determines the output voltage polarity of a dc generator?

4-7. What is shown by a field excitation line on a generator magnetization curve?,

4-8. What is the difference between a generator internal characteristic curve and its external characteristic curve?

4-9. Name two major causes of internal voltage loss in a dc generator under load conditions.

4-10. What does the term generator regulation mean?

4-11. What output voltage is available in a series generator under no-load conditions?

4-12. Does a series generator output voltage continue to increase even with a very serious current overload?

4-13. What does the term generator build down mean and is it harmful to the generator?

4-14. What two basic internal connections that have a major effect on the external characteristic curve are available in a compound generator?

4-15. What two different connections are recognized for the shunt field in a compound generator?

4-16. Is residual magnetic field polarity an important consideration in a compound generator?

4-17. How is a compound generator adjusted to a particular characteristic curve?

PROBLEMS

4-1. A shunt field generator has a field current of 1.13 A and a full-load current 16 A. What is the armature current?

4-2. The same generator as in Problem 4-1 has a load voltage of 125 V. What is (a) the armature circuit voltage? (b) the field circuit voltage?

4-3. If the same generator as in Problem 4-1 has an armature circuit resistance of 0.693 Ω, what is the armature circuit voltage drop at full load using armature current from Problem 4-1?

4-4. Under no-load conditions, what would be the terminal voltage of the generator in Problems 4-1 through 4-3 if no voltage drop other than armature circuit resistance is considered?

4-5. With the no-load voltage of Problem 4-4, what is the voltage regulation of the generator in Problem 4-1?

4-6. Using the generator saturation curve or excitation curve shown in Fig. 4-2, which applies to a particular model of 1 kW rated dc generator, what no-load voltage could be expected with a 150 Ω total field circuit adjustment?

4-7. Using the same excitation curve in Fig. 4-2, what field circuit resistance would result from a 160 V no-load adjustment?

4-8. The same generator in Problems 4-1 through 4-4 is connected with a series field only, which has a resistance of 0.322 Ω. When the normal full-load current is delivered at the rated 125 V, (a) what total armature circuit voltage drop is developed? (b) what generated voltage must be produced in the armature?

4-9. Again the same machine as in Problems 4-1 through 4-4 is connected as a compound generator with a series field of lower resistance than in Problem 4-8 (or the same series field bypassed so that R_{se} is now 0.105 Ω). If the rated current is delivered to a load, what is the series field current if (a) the machine is connected long shunt? (b) the machine is connected short shunt?

Answers. **4-1**, 17.1 A; **4-2(a)**, 125 V, **(b)** 125 V; **4-3**, 11.9 V; **4-4**, 137 V; **4-5**, 9.52%; **4-6**, 145 V; **4-7**, 128.5 Ω; **4-8(a)**, 16.2 V, **(b)** 141 V; **4-9(a)**, 17.1 A, **(b)** 16 A.

5

Paralleling of Direct-Current Generators

There are many reasons for paralleling generators, especially when it is recognized that this usage of the word parallel means *duplicate* or *multiple*.

(1) Power sources such as generators are frequently primary *safety* items and are therefore duplicated or paralleled for reliability.

(2) Many major types of machinery, again such as generators, run most *efficiently* when *loaded* to their design rating. Electric power costs less per kilowatt hour when the generator producing it is efficiently loaded. Therefore, when the load is reduced, one or more generators can be shut down and the remaining units kept efficiently loaded.

(3) Breakdown or routine maintenance frequently requires that the device being worked on be *isolated* from its work and *shut down*. Therefore, if power sources are paralleled, the routine or emergency operations can be performed without disturbing the load conditions. This affects both safety and economy.

(4) In the modern world of expanding population, goods, and services, the use of electricity is constantly increasing. When *added capacity* is required, the new equipment can be simply paralleled with the old. This frequently means more flexible operation since it increases the choices available for reasons 2 and 3.

(5) In many situations not confined to generators, the equipment available to do a particular task may not be available in a sufficiently large *capacity* or *size* in a single unit. Here paralleling must be a design feature just to meet original load requirement. An absolute limit to the size and output capacity of a dc generator does not seem apparent, but in any endeavor a new largest size is always more expensive and usually has unforeseen "bugs," which may be ruinously costly.

5-1 MODERN USES OF PARALLELING

Power sources are rarely duplicated in home or automobile service, but usually are in aircraft, marine, rail, and industrial use.

In a marine situation, for example in a Navy or Coast Guard ship, there is much use of dc power. Here the dc generators are frequently diesel driven and the whole diesel-electric generator set is duplicated. In larger ships the generators may well be driven by small steam turbines. In these situations the duplication is for reliability, even to the extreme case of battle damage. Usually the load can be carried by one unit, but overload situations or transitions from one generator to the other call for paralleling if no loss of service is to be experienced.

In aircraft service there is usually one generator (or alternator) for each engine. In addition, there will be an auxiliary power supply (APS) on larger aircraft for engine starting, ground standby, and flight emergency.

The growth of multiengine aircraft with paralleled dc main generators started in a modern sense in the 1930s. The enormous expansion in size and number of aircraft through World War II was electrically powered with paralleled dc generators. This trend lasted through the postwar reciprocating engine transports. Starting with the U.S. Air Force B-36 bomber, prime power became paralleled ac alternators, even with all the complexity that this necessitated. Most small and medium aircraft of twin-engine design, of which there are thousands, are still built with dc prime power with paralleled generators.

The trend in the very latest designs of military aircraft in medium multiengine sizes is now back to 28.5 V dc prime power. The reason is that on a military aircraft the major use of electricity is for various electronic devices. Communications equipment, electronic countermeasures, various radars, and such specialized gear as antisubmarine warfare equipment are all now based upon solid-state electronics. Thus, the equipment uses low-voltage dc power and relatively less high-voltage power.

In modern transport aircraft there also is much elaborate communication and navigation equipment, which can use the normal aircraft 28.5 V dc power. At the same time, many kilowatts of power are used for seat lighting and galley food warming for which 115 V and 208 V ac is perfectly applicable. As a result, prime ac power will certainly continue into the foreseeable future, but the use of prime dc power and its requirements for paralleled generators will expand. It certainly behooves a modern technician and engineer to understand and be able to work with parallel generator situations.

5-2 PARALLEL DIRECT-CURRENT GENERATOR REQUIREMENTS

The principal types of situations where paralleling of dc generators is required are parallel shunt generators of the same or varying sizes and parallel compound generators of the same or varying sizes.

In all different situations there are certain requirements that must be met for successful electrical paralleling. A parallel circuit is defined as one in which the *same* voltage exists across each unit at the paralleling *point*. This is absolutely required by Kirchhoff's voltage law.

If the *generated* voltages of the individual generators are not all the same, and they are paralleled, three different conditions may be met:

(1) When a generator is producing an internally generated voltage E_g that is appreciably *above* the voltage at the paralleling point, *generator action* is taking place and the unit is delivering current to the load.

(2) If the generator is developing the same voltage as that existing at the paralleling point, no effective generating action is taking place and no current is flowing to the load. The generator is said to be *floating* on the line. It is neither contributing or drawing current and is still being rotated by its own prime mover.

(3) If the generator is set so as to develop less internal E_g than the voltage at the paralleling terminals, it will draw current *from* the paralleling point and will be operating as a *motor*. The development of motor action will be explored in the next chapter. In this case, though, the unit is drawing power from, rather than contributing power to, the paralleling junction point.

These three situations are in entire agreement with Kirchhoff's current law as any parallel circuit must be. These situations will be explored with an example after other conditions of paralleling are described.

(1) The *polarities* of the generators *must be the same* or the connections must be interchanged until they are. A typical situation, as studied in an electrical circuit or network course, is where two or more voltage sources are in series with each other and with their polarities both aiding a circulating current. A condition of this type would be intolerable with parallel generators, because the voltages are high in relation to the internal armature circuit resistances. With aiding polarities, the circulating current will meet Ohm's law by which the voltage is the sum of both machines' voltages and the current-limiting resistance is only the armature circuit resistances. The circulating current would then equal a direct short circuit, and there would be no useful energy for the load from either generator.

(2) The *voltages* should be nearly if not actually *identical* so that each machine will contribute. Again, worked examples will show the problems.

(3) The change of voltage with change of load should be of the *same character*. A positive regulation machine cannot usefully combine with a negative regulation machine. Circulating currents would dominate the situation. An exact match of characteristics is desirable but not always achieved.

(4) The prime movers that drive the generators should have similar and stable rotational speed characteristics. The prime movers should either all be such that they have constant or *flat* rotational characteristics or should all *droop* in speed with increasing load. A *rising* speed characteristic with increasing load is *unstable* and will cause the affected machine to take more than its share, or even all, of the load.

These conditions will be explored in more detail as they apply.

5-3 PARALLEL SHUNT GENERATORS: IDENTICAL UNITS

The simplest paralleling situation is between two or more identical shunt generators with identical prime mover characteristics and identical field settings. This situation is comparable to paralleling identical batteries in comparable stages of their life cycle. Each battery or generator takes its fair share of the load over the whole normal load range. The only problem occurs if incorrect polarities are inadvertently used.

If one or more generators is carrying a substantial load and a third is started up in preparation to cutting it into the parallel combination, there may be substantial transient currents when it is switched into the combined circuit. The problem occurs because the loaded generator or generators have a substantial $I_a \times R_a$ drop in their armature circuits, while the unloaded one does not. They are then operating on different parts of their characteristic curve. The switching transient current takes place during the process of equalization when the unloaded machine "pulls down" to the new combined voltage. If the units are large, and the switching current is then heavy, the field current of the unloaded generator should be adjusted down until the incoming unit voltage matches that of the paralleling point. The unit is then switched in and floats on the line with little or no circulating current. Next, the field adjustment is returned to its previously known and marked position, whereupon the new unit will assume its share of the load.

If the units are not identically adjusted, they will not share the load equally. Perhaps the easiest way to see this is to work Example 5-1.

Example 5-1. Three identical 5 kW shunt dynamos are driven by a single prime mover. Each machine has a field resistance of 60 Ω and an armature circuit resistance of 0.18 Ω. The three machines are set to no-load voltages of (A) 124 V, (B) 120 V, and (C) 115 V. They are then connected to a bus that is maintained at 120 V by other machines. Determine for *each* of the three machines:

(a) The line current sent into or drawn from the bus.
(b) The armature current.
(c) The power drawn from or delivered to the bus.
(d) The power generated.

Solution: Using Eq. (2-9) transposed, we may approximate:

(a) $$I_a = \frac{E_g - V_t}{R_a} = \frac{124 - 120}{0.18} = 22.2 \text{ A} \quad \text{machine A}$$

$$= \frac{120 - 120}{0.18} = 0 \text{ A} \quad \text{machine B}$$

$$= \frac{115 - 120}{0.18} = -27.8 \text{ A} \quad \text{machine C}$$

Machine A is supplying 22.2 A and acting as a *generator*. Machine B is *floating* on the line. Machine C is drawing 27.8 A *from* the bus and is thereby acting as a *motor* rather than being driven by the prime mover.

(b) Since each machine is connected to the line, each field is "seeing" 120 V and thus drawing current that the armature is presumed to supply; thus

$$I_f = \frac{V_t}{R_f} = \frac{120}{60} = 2 \text{ A}$$

Then   armature A = 22.2 + 2 = 24.2 A   machine A

B = 0 + 2 = 2.0 A   machine B

C = −27.8 + 2 = −25.8 A   machine C

The machines are very much unbalanced and yet *each* has a current within its safe range, since from basic power law,

$$\frac{P}{E} = I = \frac{5 \times 1000}{120} = 41.7 \text{ A} \text{ rated}$$

(c) $P = IE = 22.2 \times 120 = 2664 \text{ W}, \text{ A}$

$0 \times 120 = 0 \text{ W}, \text{ B}$

$-27.8 \times 120 = -3336 \text{ W}, \text{ C}$

Machine C is drawing more power from the bus than A is delivering, while B is floating. The bus is not aided by the three improperly adjusted machines.

(d) Each machine delivers power to its field. Therefore, $P = IE = 2 \times 120 = 240$ W to each field.

$$2664 + 240 = 2904 \text{ W generated}, \text{ A}$$

$$0 = 240 = 240 \text{ W generated}, \text{ B}$$

$$-3336 + 240 = -3096 \text{ W motor power}, \text{ C}$$

Machine C will deliver 3096 W motor power back to its prime mover if we decide to neglect all losses but the power to the field. It will be seen later that this is only an approximation. The individual machine losses are a subject in themselves.

If each machine had been set to at least the voltage of machine A, there would be a reasonable purpose to the parallel combination. Full rated load would require that each machine deliver 41.7 A, which with the same procedure as (a) would require

$$E = 120 + (I_aR_a) = 120 + [(41.7 + 2)0.18] = 127.9 \text{ V}$$

This is 7.9 V greater than the bus voltage. This example is only a first approximation since the machine performance is not linear, as can be seen by the curve in the typical external characteristic curves.

5-4 PARALLEL SHUNT GENERATORS: DIFFERENT UNITS

In the more general case of paralleling between units of different output capacity, and perhaps different brand and with different character prime movers, generalized requirements become apparent. If the units in question are all plain shunt generators, they must all have a drooping external characteristic curve. This, of course, means that they all will have some degree of positive voltage regulation. The ideal situation is where each no-load voltage is the same, and each machine has the same output voltage at its own particular full-load current. In this case, the external characteristic curves could be redrawn with the horizontal or abscissa scale as percentage of output current. In this fashion the various units in question would have the same external characteristic curves except as they might curve to a different degree at the midpoints. See Fig. 5-1 for two characteristic curves that are drawn first versus current output and then versus percentage of current output. The differences are apparent at the midrange. The effective share of the load that each generator will assume can be determined by striking a horizontal line in the appropriate part of the range as shown in Fig. 5-1b. In this case, one generator is producing 58 percent of its rated load while the other is producing only 42 percent. The reason for this imbalance is that the voltage at the paralleling point must be identical, as Kirchhoff's law requires.

The different shapes for the external characteristic curves, even when plotted on a percentage basis, are due mostly to the degree of saturation that exists in the particular machine's magnetic field. One machine may be operating at the midpoint of the knee of its saturation curve, and the other above the knee. The generators may be more effectively balanced by changing the angular velocity of one or the other and then readjusting the field so that the output voltages are as desired. This will straighten or bend the characteristic curve as the case may be, but it will also change the spread from no-load to full-load voltages a small amount. It is better to favor the end-point agreements if the setup is to operate over the whole load range. If operation is to remain in the upper load ranges, an unbalance of voltages at zero load can be tolerated. This would result in an unbalanced no-load situation; one unit would generate and the other would motorize with a substantial circulating current. These are some of the reasons why original installations of parallel shunt generators are preferably designed with identical machines. In a practical sense, an added machine may simply be chosen from what is readily available at the time. Reference (3) (in Chapter 4) gives an excellent treatment of the imbalance problem if more depth is desired.

5-5 PARALLEL CUMULATIVE COMPOUND GENERATORS: IDENTICAL UNITS

A very large share of dc power generation is performed with parallel cumulative compound units. This is subject, of course, to divisions between flat compound groups and overcompound groups. While the uses are different, the operational

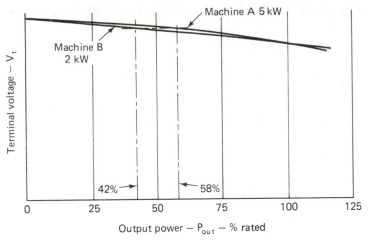

(a) Dissimilar characteristic curves

(a)

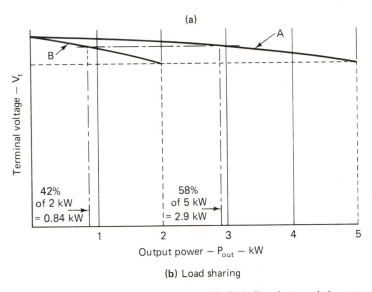

(b) Load sharing

Figure 5-1. Parallel shunt generators with dissimilar characteristic curves.

problems are similar. The major problem in common with all devices is where their output level increases with load. In the case of cumulative compound generators, it is easy to understand the problem of instability if we first postulate operation without a special provision to ensure stability. If two or more generators are assumed to be operating at a midrange load and are exactly equally sharing the load, all is briefly within reason. If any form of transient upsets the perfect balance for a short while, trouble begins. Even if we assume that the generators are driven by the same prime mover at the same speed, a transient can exist if the brush contact on one machine is momentarily disturbed for any reason.

When one machine output changes in relation to the other for whatever reason, one machine delivers a greater current that the other. The machine with the greater current then has a greater current in its series field, which will increase its output voltage because of an increase in overall field flux. The higher voltage causes a greater current to flow, which further increases the field flux, and on and on. At the same time, the other machine or machines see a lower current through their series field and thus have less total field flux. Their voltages, and therefore their share of the load, reduce. This takes place even if there are more machines going down than going up. The high-load generator takes more load and the low-load generator(s) takes less. The process continues until the high-load machine saturates its field structure, whereupon it cannot further significantly increase its voltage. The other machine or machines may actually see a reversal of current and be driven as a motor or motors.

A typical situation such as this will drive a nominally 125 V generator up to 180, 200, or more volts. The other units then see a reversal of current as mentioned above. If the reversal is great enough, the reverse ampere turns of the series field will exceed the ampere turns of the shunt field, which is still working in the original direction. This overcoming of the field flux actually flashes the field of the low generator or generators and causes a reversal of their output polarity. At this point the machines are now series aiding instead of in parallel, and are drawing short-circuit currents.

Somewhere in this process there must be relief. Either breakers will go out, fuses will blow, or somewhere there will be a burn-out failure. The situation after a late breaker outage is that the remaining machine or machines will be operating at the *opposite* polarity. Obviously, this state of affairs *cannot* be allowed to happen or it would completely nullify any hope of using parallel cumulatively compound generators.

5-5-1 Equalizer Bus Connection. The means of successful parallel cumulative compound operation is really quite simple. Since the series fields, which are necessary to cumulative operation, are the seat of the trouble, the cure must be applied there. What is needed is to assure that the series fields all carry the same current. This is easily accomplished by the provision of a third bus wire, which connects the armature ends of the various fields.

With both ends of the series fields now connected to bus conductors, the fields are in simple parallel. If they are in parallel, the voltage across the series fields, and thus their currents, must be the same. The schematic diagram of parallel cumulatively compound generators complete with the necessary equalizer bus is shown in Fig. 5-2.

The voltage across the series fields may vary but they must vary all together and in the same sense. With this third conductor properly applied, paralleling of cumulatively compounded generators becomes reliable and routine. This equalizer bus must be properly breaker protected. It must be so arranged that a routine overload will reliably open the line at the other end of the armature *before* the equalizer bus is open circuited. Frequently, the equalizer and the other end of the series field are protected with a double-pole breaker so that, if the equalizer goes, it will take the whole machine off the line. The equalizer breaker should

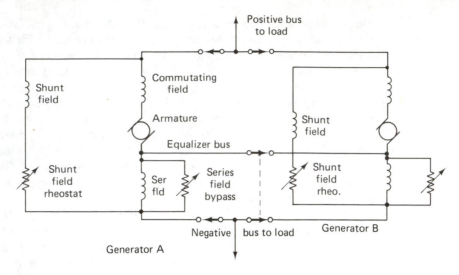

Figure 5-2. Parallel cumulatively compound generators: identical units.

be very much higher in rating than the main-load breaker on the other end of the armature circuit. If the equalizer is inadvertently open circuited on one machine, the whole arrangement is then unstable and will lead to the succession of unstable events described before. This whole process is easily demonstrated in a school laboratory electrical machine setup. One demonstration makes "true believers" from a group of students. Finally, it is reassuring to realize that a large installation will usually have reverse current relays that will prevent the polarity-reversing current surge from taking place.

It is entirely possible to adjust parallel generators incorrectly so that they are not equally sharing the load, as in Example 5-1, and yet not have unstable consequences. The equalizer bus will carry enough circulating current that all series fields are operating at the same current. The degree of balance is then set by the shunt field adjustments or the prime mover conditions. Operation will be stable and reliable.

This section has been devoted to the operation of identical machines, which then would not require separate series field diverters. If diverters were present, they would all need to be alike, since a different diverter adjustment will affect all series fields. The next section is devoted to this problem.

Ordinarily, cumulatively compound generators are put in parallel by first closing the series breakers. This causes sharing of the series field currents, even though the machine in question may not yet have its third breaker in and thus not be loaded. Conservative operation is to then reduce the output voltage of the incoming machine by use of its shunt field rheostat until it matches or nearly matches the bus voltage. After paralleling, the adjustment of the field rheostat is then returned to a predetermined position. If this is the first operation, ammeters are used to assure full balance of loads.

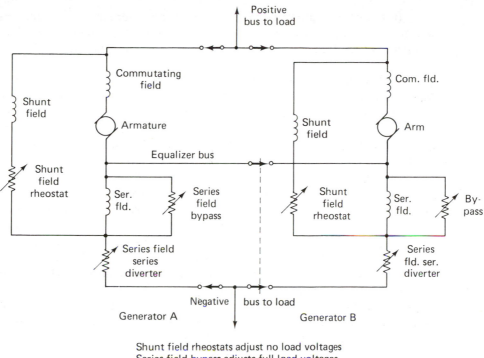

Positive
bus to load

Shunt field

Commutating field

Armature

Shunt field rheostat

Ser. fld.

Series field bypass

Series field series diverter

Equalizer bus

Com. fld.

Arm

Shunt field

Shunt field rheostat

Ser. fld.

By-pass

Series fld. ser. diverter

Negative | bus to load

Generator A

Generator B

Shunt field rheostats adjust no load voltages
Series field bypass adjusts full load voltages
Series field series diverter compansates for
different series field voltage drops

Figure 5-3. Parallel cumulatively compound generators: different units.

5-6 PARALLEL CUMULATIVE COMPOUND GENERATORS: DIFFERENT RATINGS

In this final and more universal situation, another problem arises. Machines that have similar characteristic curves so as to be reasonably paralleled are not necessarily compatible. If they are of different size or different make, they may well have different series field voltage drops. Since it is necessary that bus connections parallel both ends of the series fields for successful operation, it is required that they *all* have similar voltage drops. This important conflict is easily resolved with the use of series field *series diverters*. The addition of a series diverter, as shown in Fig. 5-3, allows all the individual series fields to have the same bus-to-bus voltage drop and yet carry their individually required field currents. If there is a large discrepancy in the field voltage drops, there will be a significant power loss in the series diverter itself. It is undesirable to have any more voltage drop across a diverter than is necessary for proper load sharing. Their use on identical machines as a vernier trim adjustment is then not required and probably wasteful. The usual school laboratory is not equipped to demonstrate the *requirement* for series diverters, but can obviously demonstrate their *function*. Where different machines are available, this makes a very interesting and wholly realistic experiment.

In industry or aircraft or on shipboard, an emergency or standby generator is not usually the same sized unit as the regularly paralleled machines. This problem is then frequently met and is entirely practical in nature.

5-6-1 Transposed Equalizer Connections. The catalog of problems is not finally complete until the transposed equalizer connection is understood. This condition should be avoided at all costs, since it constitutes another form of violent short circuit. A machine connection is never designed this way, but it may easily happen by inadvertence. Figure 5-4 shows the condition in question. One can readily understand how this can happen in an original setup or especially on school laboratory machines, where field reversals and perhaps field flashing are required in order to complete the setup. The voltages that exist at each end of the equalizer bus in Fig. 5-4 are drastically different, and yet individual machine operation *before* paralleling may be perfectly normal. Before any parallel setup is placed in parallel operation, careful step-by-step checks must be made. Many a transistor circuit has been ruined in a silent and unspectacular failure because sufficient checks were not made before closing the switch. Here, or on any large machine of any kind, a mistake can be costly and even hazardous. One reliable check is to carefully follow the leads to both ends of each series field. They must all be on the same end of the armature, and there must be nothing else in between these two bus lines but the series fields and any needed series diverters.

Some larger machines have two series fields, one on each end of the armature. They may only be paralleled with machines with similar fields, and then two equalizer bus circuits are required. Consult more detailed references before embarking on complicated parallel combinations.

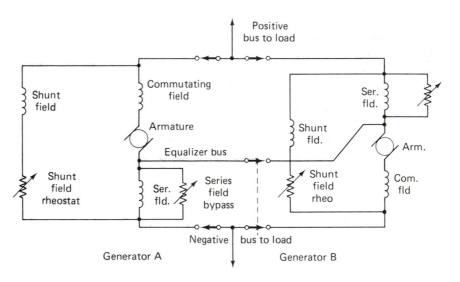

Note that equalizer bus connections
are transposed to opposite brushes

Figure 5-4. Transposed equalizer bus: cumulative compound generators.

QUESTIONS

5-1. Why do many ships use multiple dc generators in parallel?

5-2. Why do many aircraft use one generator per engine?

5-3. Name the electrical requirements for successful paralleling of dc generators.

5-4. Why are identical shunt generators comparatively easy to parallel?

5-5. What happens if an incoming generator is set substantially above the bus voltage to which it will be paralleled?

5-6. What happens if an incoming generator is set to deliver exactly the same voltage as the bus to which it will be paralleled?

5-7. What happens if a generator is set to a no-load voltage that is below the bus voltage to which it is paralleled?

5-8. What happens if an incoming generator positive terminal is joined to the bus negative, and the incoming negative terminal is joined to the bus positive terminal?

5-9. What is the effect if the generators to be paralleled have different percentage voltage regulation?

5-10. Can separate but identical generators that are driven from the same shaft be successfully paralleled if adjusted to different voltages?

5-11. What circuit feature is required for successful paralleling of compound generators?

5-12. What happens if the circuit feature in Question 5-11 is omitted or becomes disconnected?

5-13. Are there special connection requirements for an equalizer bus?

5-14. What additional feature is necessary for successful paralleling of different make or rating compound generators?

5-15. What switching sequence may be required when entering compound generator paralleling?

5-16. How may some generator polarities become reversed in parallel compound generator operation?

5-17. Can a compound generator with two series fields at opposite ends of the armature be successfully paralleled with units having one series field?

5-18. Can generators that are being driven in opposite directions be successfully paralleled?

5-19. Is it necessary that each generator take the same percentage of its rated load as other generators in a parallel group?

PROBLEMS

5-1. A 7.5 kW dc generator is to be paralleled with a 250 V bus. If its voltage is set to 265 V no load, and it has an armature circuit resistance of 0.523 Ω, (a) what current will it deliver? (b) is this current within the machine's rating?

5-2. What current would the same generator as in Problem 5-1 deliver to the same 250 V bus if (a) it has been set at 259 V? (b) it has been set at 245 V?

5-3. A 250 kW dc generator has been carrying its full rated load when attached to a 600 V bus. The machine has an armature circuit resistance of 0.083 Ω. If the bus breaker is suddenly opened, what no-load voltage will be found to be produced by the generator if change of field current is neglected?

Answers.   **5-1(a)**, 28.7 A, **(b)** rating 30.0 A;   **5-2(a)**, 17.2 A, **(b)** −9.56 A; **5-3**, 635 V.

6

The Direct-Current Motor

Since some segments of the population rarely see a dc motor, or for that matter recognize one if they see it, it is widely felt that dc motors are unimportant. This error is compounded by the fact that in the last few decades many machine tool drives have changed from variable-speed dc motors to later types of ac motors that have extensive variable-speed capability. In reality, the importance of dc motors has been steadily increasing in recent years.

All automotive motors are direct current, from the starting motor to all accessory motors, including heaters and power seats. The simplest automobile has at least a starter and a combined heater and defroster motor. Many have power seats, power tops, air-conditioner blowers, windshield wipers, power rear windows on station wagons, and power side windows on all types. The list is very extensive. There are probably as many as 300 or 400 million dc motors in service in the United States alone.

The dc traction motor is continuously being developed in railroad service. As discussed under generators, almost all the trains in the major industrial countries of the world are either diesel electric or straight electric. Only a relative handful of these are ac powered. Most actually drive the wheels with geared series dc traction motors. A modern diesel electric locomotive will have one motor per axle so that each section of a diesel locomotive will have at least four traction motors. Some have six and, occasionally, eight. There are frequently four or more sections for each locomotive, with all the multiplication of quantity that is implied.

Subways and various other types of rapid transit trains are dc motor powered with usually two to four traction motors per train car. Even the few remaining urban trolley cars are dc motor powered.

There probably will be some, perhaps extensive, development of linear ac induction motors in the new resurgence of urban high-speed transit systems. This is only viewed as a specialized fraction of the total new construction, however.

It seems inevitable that there will be a significant new production of electric and composite powered electric-driven private automotive vehicles in the next few decades. The original push for this, starting in the late 1960s and early 1970s, was to find a nonpolluting vehicle type for urban use. In the mid-1970s, the widespread and long overdue awareness of the shortage of fuel energy has added emphasis to electric vehicle development because the dc motor is an efficient user of energy.

The energy that is used in all these motors comes from diverse sources. In present automobiles the energy comes from the liquid fuel that powers the reciprocating or rotary gasoline engine. The engine, even if it is a rotary Wankel, turns an alternator or a generator to supply electric power. When the vehicle is stopped and the main engine is shut off, the electric power comes from the storage battery.

In locomotives the dc power usually comes from the dc generator that is driven by the diesel engine or engines, so liquid fuel is used. Urban electric vehicles are usually supplied dc power from a third rail near ground level or from an overhead wire system. This dc power comes from converters or rectifiers in substations which, in turn, get their ac electricity from central station electric power plants, which may be fossil fueled by natural gas, by heavy relatively unrefined oil, or by coal. Some are nuclear powered.

In the final analysis, almost all railroad and rapid transit wheels are ultimately turned by dc motors. It seems very clear then that a modern technician and engineer should understand these flexible and efficient forms of motors.

6-1   MEANING OF POWER

Many people find it is difficult to relate electrical power in watts to rotating mechanical power in horsepower. The usual reason is that one or both types of power are not clearly understood. The root difficulty seems to be in appreciating the distinct scientific differences between force, work, and power.

6-1-1   Force.   *Force is a tendency to cause motion*, such as a pressure on a surface, the tension on a cable, or the pull of a magnet. It need not actually cause motion in order to be a force. Force is measured in dynes in the cgs system, in pounds in the English system, and in newtons in SI. They interrlate as shown in Table 2-1 and as partially shown in Table 6-1. See Table 6-4 for more detail.

In line with downgrading the use of cgs and emphasizing SI, we will only need the units shown in Table 6-1. It is good practice, however, to remember the location of a conversion to dynes and other cgs units, because many outstanding references were printed in cgs units and they should not be considered useless. As long as we can multiply or divide by 10^5, we can readily move from dynes to newtons or back.

6-1-2   Work.   *Work is a force taken through a distance*. Motion must take place in a mechanical consideration of work, such as a rotating electric motor. If you

TABLE 6-1 FORCE UNITS

English		SI
1 pound (lb)	=	4.4482 newton (N)
0.224 81 lb	=	1 newton (N)

TABLE 6-2 WORK UNITS

English	SI
1 ft-lb	= 1.3558 newton-meter (N·m)
0.737 55 ft-lb	= 1 newton-meter

are standing and holding a heavy weight, you are doing no work in a scientific sense even though your muscles are fully tensed.   You were doing work when you picked up the load, say 100 lb raised 3 ft high. $100 \times 3 = 300$ foot-pounds (ft-lb) of work was done raising the load.

The cgs unit of work is the erg or dyne-centimeter.   The foot-pound is used in English units and the newton-meter (N·m) or joule (J) is used in SI.   The kilogram meter is obsolete, but some equipment is calibrated this way.   The work units then relate as shown in Table 6-2.   See Table 6-4 for more detail.

The units shown should become familiar, but the remaining units in Table 6-4 will be necessary in working with older references.   Many smaller sizes of rotating machines used in control system work have their torque or turning effort rated in dyne-centimeters so it is not just a historical curiosity. 10^{-6} newton-meter or 10^{-6} joule will become an auxiliary SI unit in order to preserve the 10^{-3}, 10^{-6}, etc., relationships.   Some confusion will of course appear, so be warned.   Incidentally, some references use foot-pounds and some use pound-feet.   They are of course the same magnitude.   The inch-pound or pound-inch is also used and is simply a 12-to-1 relation to foot-pounds because of 12 in. in 1 ft.   Small units calculations may also use the ounce as force.

6-1-3 Power.   The concept of *power* then simply becomes the *rate of doing work*.   Because of historical precedent, the English units of power are not simply the foot-pound per second or per minute.   The original unit of power was developed from James Watt's determination of horsepower.   His identification of the problem and determination of the magnitude of a 1 horsepower was and is such a valuable concept that the basic electrical unit of power was named the Watt, even though James Watt did no electrical work himself.   Watt defined 1 horsepower as 550 ft-lb/sec or 33 000 ft-lb/min.

Since power is work in a given time, the units then become dyne-centimeters per second or ergs per second in cgs, foot-pounds per second or foot-pounds per minute or horsepower with proper multipliers in English units, and newton-meter per second or joule per second or simply the watt in SI.   Since the SI units are coherent, $1 \text{ N·m/sec} = 1 \text{ W} = 1 \text{ V} \times 1 \text{ A}$.   The basic unit of power in SI is then

TABLE 6-3 POWER UNITS

	English		SI
$1\dfrac{\text{ft-lb}}{\text{sec}}$	$= 1.8182 \times 10^{-3}$ hp	$= 1.3558\dfrac{\text{N·m}}{\text{sec}}$	$= 1.3558$ W
$550\dfrac{\text{ft-lb}}{\text{sec}}$	$= 1$ hp	$= 745.70\dfrac{\text{N·m}}{\text{sec}}$	$= 745.70$ W
$0.737\ 55\dfrac{\text{ft-lb}}{\text{sec}}$	$= 1.3410 \times 10^{-3}$ hp	$= 1\dfrac{\text{N·m}}{\text{sec}}$	$= 1$ W
$737.55\dfrac{\text{ft-lb}}{\text{sec}}$	$= 1.3410$ hp	$= 10^{3}\dfrac{\text{N·m}}{\text{sec}}$	$= 1$ kW

the watt, whether used mechanically, electrically, chemically, thermodynamically, or however. This is *coherency*, and it is enormously convenient when one can learn to trust the simple relations involved.

The power units are shown in Table 6-3.

English unit mechanical power is then measured in horsepower, and the comparable SI power unit of mechanical power is then the kilowatt. They are comparable sized units and no confusion should arise after practice.

6-2 RELATION OF TORQUE AND POWER

In the measurement of mechanical power, in either horsepower or kilowatts, certain further concepts need to be clarified. Torque or rotational energy, or, more simply, turning effort, is a force acting at a right angle at a radial distance from a center of rotation. One pound acting at a crank radius of 1 ft is 1 lb-ft of torque. Similarly, 1 N acting at 1 m crank radius is 1 N·m of torque. Of course, fractional or multiple forces and radii may be used as well.

Either of these torque forces when rotating must move through 2π times its radius in one complete circle of motion.

Then 1 hp of rotating power can be derived as follows:

F = force in pounds

D = radial distance of force action in feet

S = time rate of rotary motion in $\dfrac{\text{revolutions}}{\text{minute}}$

2π = the multiplier for radians per revolution, but radian is nondimensional. Thus

$$\frac{2\pi \text{ radians}}{\text{revolution}}$$

By definition: 1 horsepower $= \dfrac{33\ 000 \text{ ft-lbs}}{\text{min}}$

Since this may be treated as any other pair of equivalent units, either quantity may be the numerator or the denominator.

We may state:

$$\frac{1 \text{ hp min}}{33\ 000 \text{ ft-lbs}}$$

$$\frac{F \cancel{\text{lbs}} \ D \cancel{\text{ft}} \ S \cancel{\text{rev}} \ 2\pi \text{ rad hp } \cancel{\text{min}}}{\cancel{\text{min}} \ \cancel{\text{rev}} \ 33\ 000 \ \cancel{\text{ft-lbs}}} = \frac{2\pi \ FDS}{33\ 000} = \text{hp}$$

The use of radians then allows rotary motion in time to be equivalent to linear motion in time. The term radian is nondimensional and so disappears.

This simplifies to $(FDS/5252.1)$ = horsepower, or if torque T in foot-pounds is entered directly instead of force F times distance D, it becomes

$$\boxed{\frac{TS}{5252.1} = \text{horsepower}} \qquad \text{Eq. } (6\text{-}1_E)$$

Some texts use N for rpm; some use rpm directly. Other labels are also used. Some leave the $2\pi/33\ 000$ without further cancellation, but all of the forms are equivalent and should be recognized as such.

A similar derivation for SI units follows. Rotating mechanical power in watts or kilowatts is

f = force in newtons

d = radial distance of force action in meters

ω = time rate of rotary motion in radians per second but radian is non-dimensional: $\dfrac{\text{radians}}{\text{second}}$

Since the angular rate is stated in radians per unit time, the 2π conversion is not required in SI. Then when combined and canceled, we have

$$f \text{ newtons } d \text{ meters } \omega \frac{\text{radians}}{\text{second}} = fd\omega = \text{watts}$$

since 1 N·m/sec is 1 W, and the term radian drops out. In larger sizes this becomes

$$\langle fd\omega \times 10^{-3} = \text{kilowatts} \rangle \qquad \text{Eq. } (6\text{-}1_{SI})$$

or if $f \times d = t$ in newton-meters,

$$\langle t\omega \times 10^{-3} = \text{kilowatts} \rangle \qquad \text{Eq. } (6\text{-}1_{SI})$$

When System International is finally adopted, we may expect that motor power will be rated in watts or kilowatts much as generators have always been. When using nonstandard force or torque units, convert by using Table 6-4; then find power from Eq. (6-1).

These rotary power formulas hold for *any rotating device* whether electrical, mechanical, hydraulic, wind driven, or whatever.

Starting in the late 1970s, certain European automotive manufacturers were scheduled to rate gasoline automobile engine power in kilowatts. The term horsepower is finally approaching an honorable retirement after more than two centuries of worldwide use. It is over 50 years since the widespread replacement of the horse as a draft animal, and the modern student hardly recognizes that the term horsepower actually meant the time–work capability of a living animal.

Again, in the derivations above the strict mathematical use of dimensional analysis has not been conventionally observed. Here words were introduced that are not truly dimensions, such as radian and revolution. Since they help explain the combinations of actual units and since they cancel in the end, this unorthodox approach was used and will be used. The author has little patience with methods of derivation that do not explain where each number comes from.

6-2-1 Development of Torque. Since an electric motor, or any rotary-power-producing device, is a producer of continuously rotating torque, and since we can equate torque and angular velocity to get power, we need to know how torque is developed and how it is measured.

From Chapter 2 we find that Eqs. (2-8$_E$) and (2-8$_{SI}$) have covered the development of force on a conductor that is immersed in a magnetic field:

$$F = \frac{BIL \times 10^{-7}}{1.13} \text{ lb, from Eq. (2-8}_E),$$

and

$$f = \beta Il \text{ newton, from Eq. (2-8}_{SI}).$$

Remember that L was in inches and l was in meters in these equations. These equations can be related to the torque in a motor situation by multiplying as follows: (1) by the number of conductors Z that are in the armature; (2) by the decimal equivalent of the percent of effective pole arc coverage to find the numbers of the conductors that are in the magnetic field and thus are functioning; (3) then multiplying by the effective radius of the conductors D in *feet* (or d in meters) to convert the force to a torque; and (4) by converting the current I in amperes to I_a, the total armature current, and dividing by a the number of parallel paths in the winding, which yields the actual current per conductor.

$$T = \frac{BI_aLZ \text{ (\% cov.) } D \times 10^{-7}}{1.13a} \text{ ft-lb} \qquad \text{Eq. (6-2}_E)$$

$$t = \frac{\beta I_a lZ \text{ (\% cov.)}d}{a} \text{ N·m} \qquad \text{Eq. (6-2}_{SI})$$

THE DIRECT-CURRENT MOTOR   CHAP. 6

These torques can then be substituted into Eq. (6-1$_E$) or Eq. (6-1$_{SI}$), as the case may be, to get gross power developed. See also Eq. (6-8$_E$) and Eq. (6-8$_{SI}$).

Obviously, Eqs. (6-1) and (6-2) could be further combined to get horsepower in one grand equation. This is risky because one tends to lose track of the whole picture. Equation (6-2) in its various unit systems is already a combination of Eqs. (2-6) and (2-8).

Note also that no mention is made of how many poles are present, since the percentage of the total armature circumference is the required factor. All magnetically covered conductors are working to produce torque, and all poles normally have the same flux density. The torques that are developed here are gross torques rather than net delivered torque. The study of the losses between gross and net will be developed later.

Example 6-1. A dc motor takes a total line current of 135 A while rotating at 2550 rpm (267.0 rad/sec). The field pole flux density is 52 300 lines/in.^{2} (0.8106 Wb/m^2) and there is 72 percent field pole arc coverage. There are four parallel paths and 96 total conductors, which are effectively in the magnetic structure for 6 in. (152.4 min). The effective radius of action is assumed to be the outside of the armature, which has a diameter of 7.375 in. (187.3 mm). Determine (a) the gross developed torque, and (b) the power that would result if all this torque were available on the rotating shaft.

Solution:

(a) Using Eq. (6-2$_E$),

$$T = \frac{BI_aLZ \ (\% \ \text{cov.}) \ D \times 10^{-7}}{1.13a} \ \text{ft-lb}$$

The effective radius D is

$$\frac{7.375}{2(12)} = 0.3073 \ \text{ft}$$

Note that Eq. (2-8$_E$) delivered a force in pounds when the lengths and areas were in inches, but we want torque in *foot*-pounds, so D must be in *feet*.

$$T = \frac{52 \ 300(135)6(96)0.72(0.3073)10^{-7}}{1.13(4)}$$

$$= \frac{5.23(1.35)6(9.6)0.72(0.3073)}{1.13(4)}$$

$$= 19.91 \ \text{ft-lb}$$

In SI the length and radius must be in *meters*, even though mechanical dimensions are traditionally given in millimeters for design drawings. Thus,

$$l = 152.4 \times 10^{-3} = 0.1524 \ \text{m}$$

$$d = \frac{187.3}{2} \times 10^{-3} = 0.0936 \ \text{m}$$

In SI,

$$t = \frac{\beta I_a l Z \; (\% \; \text{cov.}) \; d}{a} \qquad \text{Eq. (6-2}_{\text{SI}})$$

$$= \frac{0.8106(135)0.1524(96)0.72(0.0936)}{4}$$

$$= 26.97 \; \text{N·m}$$

By now it should be obvious that, in order to develop skill in working with SI units, examples have been done in both English units and SI with the same absolute data values.

This again gives a golden opportunity to check; since 0.737 56 ft-lb = 1 N·m, then 26.97 × 0.737 56 = 19.89 ft-lb; check.

(b) Using

$$\frac{TS}{5252.1} = \text{horsepower} \qquad \text{Eq. (6-1}_{\text{E}})$$

$$\text{hp} = \frac{19.91(2550)}{5252.1} = 9.667 \; \text{hp}$$

and

$$t\omega \times 10^{-3} = \text{kW} \qquad \text{Eq. (6-1}_{\text{SI}})$$

$$\text{kW} = 26.97 \times 267.0 \times 10^{-3} = 7.201 \; \text{kW}$$

Since 1 hp = 746.7 W or 0.7457 kW, then 9.667 × 0.7457 = 7.208 kW; check.

6-3 MEASUREMENT OF TORQUE

With an understanding that torque is simply turning effort measured in force times radius, we then require an understanding of how the torque quantity is measured so that realistic problems may be studied.

6-3-1 Prony Brake. At first thought it seems a bit obscure as to how a moving force such as a torque can be accurately measured. The initial method, and still a practical device, is the *prony brake*, named for its inventor Baron de Gaspard Clair Francois Marie Riche Prony (1755–1839). In its usual form, a prony brake is a screw-actuated clamp brake band, which is wrapped around a rotating water-cooled brake drum. The tendency for the brake mechanism to travel around with the drum is due to the friction-transmitted torque force from the moving mechanism. The friction between the drum and the brake shoe is related to the band clamping pressure. This force, which is attempting to rotate the brake with the moving drum, is resisted with a brake arm, which is connected to a fixed anchorage through a calibrated spring scale. Obviously, any force units can be used at the brake scale and any convenient arm length and length unit can be used (see Fig. 6-1 for a typical prony brake installation).

$Power = \frac{Work}{time} = Rate\ of\ work\ doing$

$work = (Force)(distance)$

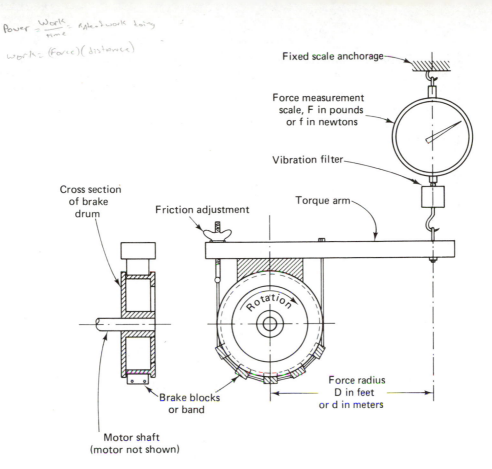

Torque T = F x D in pounds feet
or t = f x d in newton meters

Figure 6-1. Typical prony brake for motor torque measurement.

    With the device under measurement rotating, the brake clamp is adjusted until the desired rpm (or radians per second) is held. Under this condition, all the torque developed by the prime mover is then absorbed by the brake and shown by the scale reading times the arm length. In large sizes the brake requires constant cooling, since all the prime-mover output power is converted to friction-caused heat at the brake surface. The prony brake may vibrate or oscillate badly unless carefully prepared, but it is simple and useful. Reference (5) has an excellent reference section on the prony brake, but in English units only.

Example 6-2. A dc motor is tested with a prony brake loading device and the following data are taken: brake load arm radius from center, 2.0 ft (0.6096 m). Zero scale reading before testing (tare reading) was 3.13 lb (13.92 N). Operating brake scale reading during test is 20.3 lb (90.3 N). The rotating speed is 855 rpm (89.53 rad/sec).

    (a) What torque is developed?
    (b) What power does this represent?

Solution: The net scale force is the operating force less the tare force. Sometimes the units are statically balanced or the scale adjusted to zero so that there is no tare force, but this is not usual. Tare scale reading cannot be ignored if it is present.

(a) $20.3 - 3.13 = 17.17$ lb net scale force or $90.3 - 13.92 = 76.38$ N net scale force.

$$T = F \times D = 17.17 \times 2 = 34.34 \text{ ft-lb torque}$$

$$t = f \times d = 76.38 \times 0.6096 = 45.56 \text{ N·m torque}$$

(b) Since the torque is now available, the short form of Eq. (6-1) can be used:

$$\frac{TS}{5252} = \text{hp, from Eq. (6-1}_\text{E})$$

$$= \frac{34.34 \times 855}{5252}$$

$$= 5.59 \text{ hp}$$

$$t\omega \times 10^{-3} = \text{kW, Eq. (6-1}_\text{SI})$$

$$= 46.56 \times 89.53 \times 10^{-3}$$

$$= 4.17 \text{ kW}$$

It is always worthwhile to check; here 1 hp = 0.746 kW, so $5.59 \times 0.746 = 4.17$ kW; check.

6-3-2 Two-Scale Prony Brake. In small sizes a *two-scale prony brake* may be used, as shown in Fig. 6-2. In this device the effective radius arm is the *radius* of the grooved pulley, or in very small sizes the radius of the motor shaft. The friction is varied by moving both spring scales and thus tightening or loosening the nylon cord "brake band." The net torque reading is the *difference* between the two scale readings. Since no cooling is provided, this system is used in the smaller sizes of servomotors, such as may be used in instrumentation system operation. The usual scale calibrations are in milligrams, grams, or ounces, and due care must be taken to keep units straight.

6-3-3 Dynamometer. The most versatile torque measurement device is the *dynamometer.* Unfortunately, it is also the most expensive. A dynamometer may be used to either *absorb* or *produce* torque, so it is more flexible in use than a prony brake or other allied devices, which belong to the class of *absorption* dynamometers.

A dynamometer is usually a dc shunt field machine of appropriate size and rpm capability. The key to the operation of a dynamometer is the fact that, in any loaded rotating electrical machine, motor action and generator action take place simultaneously. When absorbing power, a dynamometer serves as a gen-

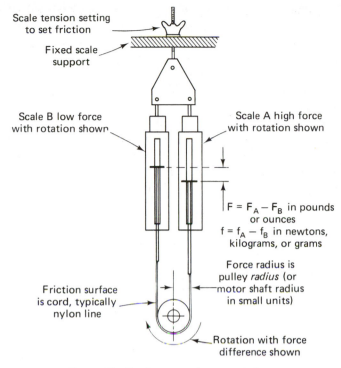

Scale tension setting
to set friction

Fixed scale
support

Scale B low force
with rotation shown

Scale A high force
with rotation shown

$F = F_A - F_B$ in pounds
or ounces
$f = f_A - f_B$ in newtons,
kilograms, or grams

Force radius is
pulley *radius* (or
motor shaft radius
in small units)

Friction surface
is cord, typically
nylon line

Rotation with force
difference shown

Torque $T = F \times D$ converted to pounds feet
or $t = f \times d$ converted to newton meters

Figure 6-2. Typical two-scale prony brake for motor torque measurement.

erator, and the generated voltage is connected to a resistive *load bank*. The ohmic adjustment of the load bank, together with the voltage that the dynamometer is adjusted to produce, determines the current that flows. Since the current that is flowing in the dynamometer armature produces motor action according to Eqs. (6-2) and (6-1), the restraining torque that resists the rotation exists in the armature windings. Any torque force that exists in the armature is there by action with the field magnetic flux. Therefore, when a dynamometer (or for that matter any generator) is absorbing a torque, its field tends to be pulled around equally by the motor action that exists.

The stator and surrounding frame structure of a dynamometer are mounted on low-friction ball bearings, which are concentric or coaxial with its armature rotor bearings. The whole field and frame or stator assembly are accurately balanced by appropriate weights to remove any gravity torque effects.

During operation, the dynamometer stator tends to rotate in its support bearings and is restricted from doing so by a torque arm and an appropriate scale, much as in the prony brake. The length from the bearing centerline of the stator assembly to the scale attachment point is the torque moment arm. It is usually marked on a conspicuous label attached somewhere on the unit.

It is not necessary to measure the electric quantities in the dynamometer in order to measure torque and power. Only the scale arm length, the scale force, and the rotative speed need to be recorded. The speed is usually conveniently read by a directly attached tachometer. Any brush, magnetic, and bearing drags also are shown on the spring scale, so the dynamometer is an accurate device. Its great versatility lies in its ability to also function as a motor and to drive *any* rotating device within its capability and, at the same time, measure the power produced at the shaft coupling.

The bulk of the energy produced by a dynamometer is normally dissipated in a resistive load bank, which is remotely located and may be appropriately cooled. In very large sizes the generated power is used to turn a dc motor, which in turn drives an ac generator. This alternating current is then synchronized with the commercial power lines and used to pump power back into the lines. Substantial

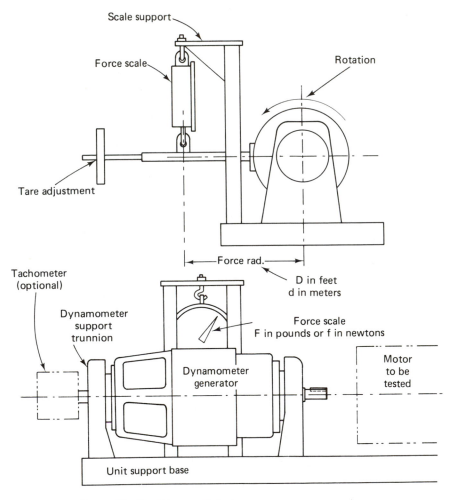

Torque T = F x D in pounds feet
or t = f x d in newton meters

Figure 6-3. Typical electric dynamometer for motor or generator torque.

portions of the electrical power used in automotive engine plants have been derived in this fashion rather than being wasted in a load bank.

Modern gas turbines are tested by dynamometers, but special precautions must be observed. Here the prime-mover rotative speed usually is reduced through gearing to a speed that is compatible with the dynamometer. Special precautions must be taken to guard against torsional vibrations at the speeds that are used. Even with the largest sizes, the use of the equations developed in this chapter hold.

The dynamometer-developed torque scale force and torque arm length are used exactly as in the case of a prony brake (see Fig. 6-3 for a typical dynamometer).

Example 6-3. It is desired to test a 2500 hp (1865 kW) mill motor on a dynamometer at both its low and high rated speeds of 225 and 450 rpm (23.562 and 47.124 rad/sec). If a dynamometer rig is available that has a 5 ft radius arm (1.524 m), what force scale range must be available in pounds or newtons?

Solution: From $TS/5252$ = horsepower [Eq. (6-1$_E$)],

$$T = \frac{\text{hp}(5252)}{S} = \frac{2500(5252)}{225} = 58\ 350 \text{ ft-lb}$$

$$= \frac{2500(5252)}{450} = 29\ 180 \text{ ft-lb}$$

$$T = F \times D$$

so

$$F = \frac{T}{D} = \frac{58\ 350}{5} = 11\ 670 \text{ lb, at 225 rpm}$$

$$= \frac{29\ 180}{5} = 5836 \text{ lb, at 450 rpm}$$

From the size of these forces we can see that, in very large sizes, a dynamometer becomes unwieldy. In fact, in the upper ranges performance testing is usually done by calculation by logical means that are developed in the next chapter. Motors exist with 20 or 30 times this power.

In SI, we use $fd\omega \times 10^{-3}$ = kW [Eq. (6-1$_{SI}$)]. Here

$$f = \frac{\text{kW} \times 10^3}{d\omega} = \frac{1865(10^3)}{1.524(2356)} = 51\ 940 \text{ N, at 23.56 rad/sec}$$

$$= \frac{1865(10^3)}{1.524(47.12)} = 25\ 970 \text{ N, at 47.12 rad/sec}$$

Since pounds $\times$ 4.448 = newtons,

$$11\ 670 \times 4.448 = 51\ 910 \text{ N}$$

$$5836 \times 4.448 = 25.960 \text{ N}$$

Note that in SI this problem was done in one step by using the formula involving force and distance.

6-4 BACK ELECTROMOTIVE FORCE IN A MOTOR

When a motor armature is rotating as a result of the torque that is produced by motor action, *it is also acting as a generator at the same time.* Since there must be a magnetic field from the field poles in order that motor action can take place, that same field then generates voltage in any passing armature conductor. By Lenz's law and the relation between Fleming's right-hand rule of generator action and the left-hand rule of motor action, it can be seen that the generated voltage *opposes* the current produced by the applied voltage that causes the motor action. Review the appropriate sections in Chapter 2 to clarify these relations. This generated voltage that opposes the applied line voltage is known as *counter emf* or as *back emf.* The two terms mean the same thing and are sometimes used interchangeably.

Since the back emf entirely follows the rules and formulas of generators, as discussed in previous chapters, its magnitude is a linear function of rotating speed if flux is held constant. As a motor comes up to speed, its back emf increases until it is a substantial part of the imposed line voltage. This back emf is the very necessary and beneficial effect that regulates the armature current drawn from the lines.

The torque that a motor produces is a linear result of the armature current that flows if the field flux is constant. The I_a current is initially limited only by the armature circuit resistance. We have seen that the armature circuit resistance R_a is as small as it can reasonably be made. As a result, the initial current that is drawn by a typical dc motor would be unacceptably high if it were not limited by either a reduced starting voltage or added starting resistance. This problem will be discussed further on.

The initially high armature current in a motor will cause sufficiently high motor torque, due to motor action, that the motor armature will accelerate. As the motor angular velocity increases, its back emf also increases. Soon a point is reached where the difference between the back emf and the line voltage becomes small and the armature current is reduced. The final balance is reached, usually after only a few seconds, where the difference between the two voltages just allows enough current to pass to meet the torque requirement and the angular velocity (rpm or radians per second) stabilizes at a steady state.

As a result, the voltage across the armature circuit V_a less the back emf or counter voltage E_c just equals the IR drop across the armature, from Ohm's law:

$$V_a - E_c = I_a R_a \qquad \text{Eq. (6-3)}$$

Since E_c opposes the current flow in the armature, it has the *same* polarity as the line voltage V. If the motor were to be turned fast enough that its developed E_c were equal to the line voltage V_l, there would be no current flow. The motor would then have no torque and would be *floating on the line.* Note that V_l is *probably the same* as V_a unless there is a voltage drop in a starting resistance. If the motor were rotated still faster, it would generate a higher voltage than the line voltage and is then acting as a generator in parallel with the line. It obeys all applicable *generator* equations.

Another useful relation is found by multiplying Eq. (6-3) by the equilibrium current that flows in the armature:

$$V_a I_a - E_c I_a = I_a^2 R$$

This is easily recombined algebraically to

$$\boxed{E_c I_a = V_a I_a - I_a^2 R_a}$$

Eq. (6-4)

6-4-1 Armature Power Developed.

Equation (6-4) shows that the *armature power* developed in the motor, which we may call P_d, is equal to $E_c I_a$. This can be seen because $V_a I_a$ is the total impressed armature power (voltage times current), and $I_a^2 R_a$ is the armature circuit copper loss, a simple $I^2 R$ relation. Note that in the equations above every a subscript means armature circuit, so that a *ckt* could have been correctly used. Visualize that the effective R_a resistance includes everything in the armature circuit: armature, brushes, commutating fields, series field, and compensating winding if present, and also any starter or controller resistance.

It appears then that the lower the R_a, the less the power loss, and this is true although there are also other power losses, which will be discussed in the next chapter.

Motor rotational speed is self-adjusting until just enough current flows to meet the torque requirement. If the load torque increases, the motor will slow enough so that the resulting drop in E_c will allow I_a to increase sufficiently to carry the load. Conversely, if the load decreases, the motor will speed up enough so that the resulting higher E_c will limit the I_a to a lower value which just meets the needs. When a motor is entirely unloaded, the speed will increase enough to allow just sufficient current to pass to meet all the various internal torque losses. On the other hand, the load may become so high that not enough current can pass to meet the torque requirement, and the motor will stall. A stall current, incidentally, is equal to a full line voltage starting current and will be very high. This will be explored under the motor starting requirements.

It is sometimes difficult to realize that a voltage times a current, which equals electrical power in watts, will appear, if it is not otherwise accounted for, as mechanical power in watts in a motor.

As a useful first approximation then, the $E_c I_a$ term in Eq. (6-4) is *mechanical power* developed:

$$\boxed{E_c I_a = P_d}$$

Eq. (6-5)

The $I_a^2 R_a$ term is a power *loss* that appears directly as heat in the various elements in the armature circuit that are involved. A worked example will show this.

Example 6-4. A dc motor while carrying its normal load draws 22.5 A at 125 V from the line. It has an armature circuit resistance of 0.45 Ω. Determine the following in both English and SI units where applicable: (a) the back emf being developed, and (b) the gross developed mechanical power.

Solution:

(a) Using Eq. (6-3), $V_a - E_c = I_a R_a$. Transposed algebraically, $E_c = V_a - I_a R_a$.

$$E_c = 125 - (22.5)0.45 = 114.9 \text{ V back emf}$$

This holds for either system of units since basic volts, amperes, and ohms are the same in each.

(b) Using Eq. (6-5), $E_a I_a = P_d$.

$$114.9 \times 22.5 = 2585 \text{ W} \quad \text{or} \quad P_d = 2.585 \text{ kW}$$

Since there are 746 W/hp, this is

$$\frac{2585}{746} = P_d = 3.465 \text{ hp}$$

The motor is then *developing* 3.465 hp or 2.585 kW *gross* mechanical power. The net useful power is a bit less, as will be seen later. We will be able to show also that motor power as calculated from more detailed equations, such as Eqs. (6-1) and (6-2), will agree with the power as determined by volts and amperes.

6-4-2 Equilibrium Motor Speed. It should be realized then that the rotating speed of a dc motor is the equilibrium result of back emf due to the speed allowing just enough current to pass to meet the gross torque requirements. The gross power developed is the equilibrium result of the gross armature circuit wattage from the lines minus the armature circuit $I^2 R$ copper loss. Any necessary shunt field wattage is also a loss, as will be seen, as are also some magnetic and mechanical losses.

The back emf or counter emf developed is then a major factor in determining motor performance. The equations that determine this back emf are the basic generator relations culminating in Eq. (2-6) in its various forms. However, the Φ or φ flux terms that are used are dependent on the manner in which the various shunt and/or series field coils receive their excitation. The various types of motors are then fundamentally dependent on how their field winding circuits are arranged.

In summary, we may use Eq. (6-3) and transpose it to $E_c = V_a - I_a R_a$. Then for a given motor, where its mechanical arrangement is fixed, we may use a modification of Eq. (2-6$_E$) as

$$E_c = \frac{\Phi Z S P}{60a} \times 10^{-8} \text{ V}$$

It can be seen that, once constructed, Z, P, and a are fixed quantities. Similarly,

10^{-8} and 60 are unit arrangement constants in the English system and where used are constants. Thus we may say that

$$\boxed{E_c = K\Phi S} \qquad \text{[this is close to Eq. (2-7}_E\text{), where } E_c = E_g\text{]} \qquad \text{Eq. (6-6}_E\text{)}$$

and

$$\langle E_c = k\phi\omega \rangle \qquad \text{[this is close to Eq. (2-7}_{SI}\text{), where } E_c = E_g\text{]} \qquad \text{Eq. (6-6}_{SI}\text{)}$$

where K and k are *fixed constants*. Equating Eq. (6-6) and the transposed Eq. (6-3)

$$K\Phi S = V_a - I_a R_a$$

and

$$k\phi\omega = V_a - I_a R_a$$

These can then be transposed as

$$\boxed{\text{rpm} = S = \frac{V_a - I_a R_a}{K\Phi}} \qquad \text{Eq. (6-7}_E\text{)}$$

and

$$\langle \text{radians per second} = \omega = \frac{V_a - I_a R_a}{k\phi} \rangle \qquad \text{Eq. (6-7}_{SI}\text{)}$$

These equations both show that the speed of a motor is *directly* proportional to the counter emf ($V_a - I_a R_a$) and is *inversely* proportional to whatever the fixed specific K (or k) is and to the field flux Φ (or ϕ). The load on a motor changes I_a and Φ or ϕ in different ways depending upon the manner of connecting the field circuits. The three types of dc motors are shunt, series, and compound. Just as with dc generators, their *speed* versus *load* characteristics are very different.

6-5 SHUNT MOTOR

A shunt motor is connected exactly as the shunt generator is connected. In the usual sense the line voltage is constant or nearly so. This means that the field flux Φ (or ϕ) will be a constant value. The field may or may not use a field rheostat to change field current and thus the flux value. Where the field flux change served to vary the generated *voltage* in a shunt generator, it now affects the rotating *speed* in a shunt motor. This is true because the change of flux changes the back emf volts to rotating speed relation. *Weakening* the field requires a higher rotative speed to produce the same required back emf and thus increases the effective speed. Conversely, *strengthening* the field reduces the rotative speed required to produce the required back emf, which consequently reduces the effective speed of the motor.

The process has a limited effective range, since with a very weak field the motor will tend to be unstable with a high load. There is also a top limit to the field due to saturation. Any one motor may have a 2-to-1 on up to a 4-to-1 effective range of speed by field control. Wider speed ranges require compensating windings and elaborate controls.

6-5-1 Shunt Motor Speed Characteristics. When once adjusted for a particular speed and when holding the same field adjustment, the shunt motor then is a *relatively constant speed* motor over its full normal load range. This follows from Eq. (6-7), where the field flux Φ (or ϕ) is held nearly constant. The only variable then is I_a, the armature circuit current. If the current increases owing to an increasing load, the $I_a R_a$ term linearly increases. The effect is that S (or ω) will drop a small amount over the normal load range.

The speed characteristics of the shunt motor directly follow from Eq. (6-7) as shown in Sect. 6-4-2.

6-5-2 Shunt Motor Torque Characteristics. The torque characteristics of the shunt motor follow from Eq. (6-2), but these may be conveniently modified as shown:

$$\boxed{T \;=\; C\Phi I_a}$$

Eq. (6-8$_E$)

and

$$\langle\, t \;=\; c\phi I_a \,\rangle$$

Eq. (6-8$_{SI}$)

Since the physical pole dimensions are known in a motor, the constants C (and c) are rearranged so that Φ (and ϕ) can be used instead of B (and β). A bit of study will show that all other factors in Eq. (6-2) are constants. The torque of a shunt motor is then *directly* proportional to armature current.

Figure 6-4 shows a typical set of shunt motor characteristic curves. The rpm (or radians per second) is seen to be quite flat. The torque is very nearly a straight line increase with horsepower (or kilowatts). The current rises concavely upward in a gradual curve from a minimum no-load level. The shape of the efficiency curve will be discussed in Chapter 7. The characteristic curves of a motor are usually presented in this form in the manufacturer's specification literature. A larger motor will, in general, be more efficient and a smaller less efficient than the typical 5 hp (3.73 kW) motor shown here.

6-5-3 Speed Regulation. Electric motor speed regulation is expressed in percentages in much the way that generator voltage regulation is handled. Again, as in generators, the rated load situation is viewed as the standard condition, so that regulation may be stated as follows:

$$\boxed{\frac{(S_{nl} - S_{fl})}{S_{fl}} \, 100 \;=\; \text{percent regulation}}$$

Eq. (6-9$_E$)

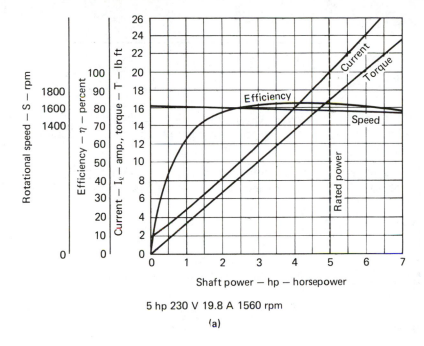

5 hp 230 V 19.8 A 1560 rpm

(a)

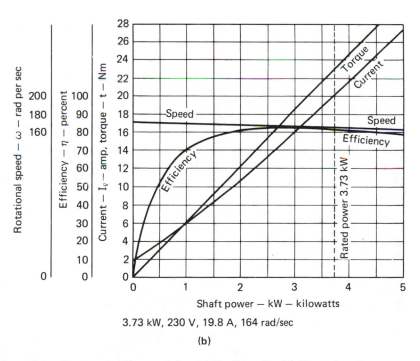

3.73 kW, 230 V, 19.8 A, 164 rad/sec

(b)

Figure 6-4. Shunt motor characteristics: (a) English units; (b) SI metric units.

or in SI

$$\left\langle \frac{(\omega_{nl} - \omega_{fl})}{\omega_{fl}} 100 = \text{percent regulation} \right\rangle \qquad \text{Eq. (6-9}_{\text{SI}})$$

Here again *positive* regulation means a higher no load speed than full load speed. The use of these regulation equations is shown in Example 6-5.

Example 6-5. A dc shunt motor is rated at 1800 rpm (188.5 rad/sec) under full load. Its no-load speed is 1910 rpm (200.0 rad/sec). Find its speed regulation.

Solution: Using Eq. (6-9$_E$),

$$\frac{1910 - 1800}{1800} \times 100 = 6.11\%$$

and using Eq. (6-9$_{\text{SI}}$),

$$\frac{200.0 - 188.5}{188.5} \times 100 = 6.10\%$$

6-5-4 Standard Power Ratings. Standard sizes and ratings of dc motors are shown in the chart of Table 10-1. Be aware that these chart values are not absolute and that internationally standardized sizes for motors have not yet been agreed on in SI. Agreement may not be reached until the late 1980s or early 1990s. Table 10-1 presents values and sizes that may possibly become standard.

Standard shunt motors are widely used in the machine tool industry for their characteristic ability to be adjusted to a range of different speeds and then to reasonably closely hold that speed over a wide range of load.

Example 6-6. A shunt motor at its rated conditions develops 1.340 hp (1.000 kW) at 1800 rpm (188.5 rad/sec). Its line voltage V_l = 125 V, and its line current I_l = 10.67 A. The machine has a shunt field resistance of 110 Ω and a total armature circuit resistance R_a = 1.233 Ω. If the motor torque load increases by 20 percent, determine (a) its probably new rpm (or radians per second), (b) the power developed under these new conditions, and (c) the line current that will be drawn.

Solution:

(a) Determine the rpm (or radians per second) by use of Eq. (6-7). In this case it can be done conveniently two ways:

$$S_2 = S_1 \frac{V_a - I_{a2}R_a}{V_a - I_{a1}R_a}$$

This form of proportion is set up because $K\Phi$ is the same in each case if we

142   THE DIRECT-CURRENT MOTOR   CHAP. 6

neglect any change due to armature reaction. Note that $I_{fld} = 125$ V/110 $\Omega =$ 1.136 A, and thus $I_{a1} = 10.67 - 1.14 = 9.53$ A. I_{a2} must be 20 percent larger since the torque is *directly* related to I_a from Eq. (6-8), and since in a shunt motor $C\Phi$ (or $c\phi$) is unchanged unless the line voltage changes. $I_{a2} = 9.53 \times 1.2 = 11.44$ A.

$$S_2 = \frac{1800[125 - (11.44)1.233]}{125 - (9.53)1.233} = \frac{1800(110.9)}{113.2} = 1763 \text{ rpm}$$

The other approach is to find the existing $k\phi$ term (or $K\Phi$ in the above) by using Eq. (6-7$_{SI}$):

$$\omega = \frac{V_a - I_a R_a}{k\phi}$$

so we invert to find $k\phi$:

$$k\phi = \frac{V_a - I_a R_a}{\omega} = \frac{125 - (9.53)1.233}{188.5} = 0.6008$$

We can ignore the rather complicated units in $k\phi$ since we find that they cancel out and we will not change the value.

$$\omega_2 = \frac{V_a - I_{a2}R_a}{k\phi} = \frac{125 - (11.44)1.233}{0.6008} = 184.6 \text{ rad/sec}$$

Since to perform the problem in the different ways may seem roundabout, a check is especially desirable: rpm $\times$ 0.104 72 = rad/sec. Thus, 1763 $\times$ 0.104 72 = 184.6 rad/sec; check. When working by yourself, perform a problem in the way that you see as the most straightforward, and then try alternative ways as a check.

(b) $TS/5252$ = horsepower; Eq. (6-1$_E$) must be transposed to find the original torque, which was not specified: $T = \text{hp}(5252)/S$. Another form is $T = 7.0438P/S$, where P is in watts. 1.340(5252)/1800 = 3.910 ft-lb (originally), and this must be increased by 20 percent (or 1.200). Then

$$\frac{TS}{5252} = \frac{3.910(1.200)1763}{5252} = 1.575 \text{ hp}$$

Note that the loaded rotational speed is now used. $t\omega \times 10^{-3}$ = kilowatts; Eq. (6-1$_{SI}$) must also be transposed. $t = (\text{kW} \times 10^3)/\omega = 1000/188.5 = 5.305$ N·m originally. Thus

$$t\omega \times 10^{-3} = 5.305(1.200)184.6 \times 10^{-3} = 1.175 \text{ kW}$$

Since hp $\times$ 0.746 = kW, 1.575 $\times$ 0.746 = 1.175 kW; check.

(c) The line current is simply the I_{a2} new armature current *plus* the field current:

$$11.44 + 1.14 = 12.58 \text{ A}$$

This holds for both unit systems.

A shunt motor can be operated separately excited, which has the benefit of somewhat wider speed range and a bit better speed regulation. This type of motor will also be discussed under special vehicle motors when the series motor is presented.

6-5-5 Permanent Magnet Field Motors. Another subvariety of the shunt motor is the permanent magnet field motor. With a fixed line voltage these act much like a shunt motor, but have slightly smaller overall size and lower cost. Their speed control is through varying the *line* voltage. This type of motor is very widely used in scale models and toys.

Permanent-magnet field motors are also applicable to automotive accessory drives when cost and size are paramount considerations. These motors are a more recent development due to high-flux, low-cost Alnico magnets and, most recently, to ceramic magnet development.

All dc motors have related starting and reversing requirements. These problems will be discussed at the end of this chapter.

6-6 SERIES MOTOR

A series motor is internally connected with the main field coils in series with the armature circuit. This means that all armature circuit current passes through the fields. The field coils are then designed and built with relatively few turns of large cross section. The few field coil turns supply the needed ampere turns with the heavy armature circuit current. The large cross section results in low resistance and therefore low wattage field losses, even with high current.

In a series motor the field flux is entirely dependent upon the armature current. When operating below the knee of the saturation curve, the field magnetic flux is then almost directly proportional to the armature current.

6-6-1 Series Motor Speed Characteristics. In a series motor situation, the speed versus back emf relations that were shown in Eq. (6-7) for a shunt motor need to be modified:

$$\text{rpm} = S = \frac{V_a - I_a(R_a + R_{se})}{KK'I_a}$$

Eq. (6-10$_\text{E}$)

and

$$\text{radians per second} = \omega = \frac{V_a - I_a(R_a + R_{se})}{kk'I_a}$$

Eq. (6-10$_\text{SI}$)

In the series case the armature circuit has the added resistance of the series fields, which increases the voltage drop due to resistance and reduces the back emf by that amount. The K (or k) terms are modified by another K' (or k') term and I_a.

The field flux is now proportional to the fixed number of turns in the field and the particular saturation curve, which we represent as K' (or k'), and the field excitation current, which is now I_a. K' (or k') is then *not constant over the full range of operation*.

The speed of the motor is still *directly* proportional to the counter emf although the counter emf numerator term is slightly different. The speed is still *inversely* proportional to the denominator term, but now the denominator has an I_a factor. Armature current I_a is a direct multiplier to the whole denominator, and thus the I_a effect is far larger in the denominator. Since the flux Φ (or ϕ) is now almost directly related to I_a, being the $K'I_a$ (or $k'I_a$) terms, an increase in I_a must inversely *reduce S* (or ω). The speed curve closely resembles a hyperbola.

6-6-2 Series Motor Torque Characteristics. The torque of a series motor is also drastically different than that of a shunt motor.

Equations (6-2) show the detailed torque relations of a shunt motor, and they also hold for a series motor except that the B (or β) term, like the Φ (or ϕ) terms above, is also dependent on I_a. As in the speed equations, most of the terms are fixed by design and construction, and the equations can be simplified as shown. Equations (6-8) are a basis:

$$T = C\Phi I_a \quad \text{and} \quad t = c\phi I_a$$

These may be further modified to

$$\boxed{T = CC'I_a^2} \qquad \text{Eq. (6-11}_E\text{)}$$

same as
ιs ιs'

and

$$\langle t = cc'I_a^2 \rangle \qquad \text{Eq. (6-11}_{SI}\text{)}$$

C' (or c') now contains the field terms that affected Eq. (6-9) and the armature circuit current I_a.

The torque then *increases* approximately as the square of the armature current up to the onset of saturation, where the field flux no longer increases as rapidly. The torque then gradually becomes *linear* with I_a.

Figure 6-5 shows a representative set of series motor characteristic curves. Notice that the torque *increases* nearly *parabolically* and then blends into a straight linear increase with horsepower (or kilowatts). The speed *decreases* approximately hyperbolically with increased power. The current curve is also concave upward but does not increase as steeply as the torque. Efficiency will be discussed in Chapter 7. The characteristics cannot usually be safely determined for very low power because the speed may well be destructively high. As a result, a series motor of any size beyond the smallest is never disconnected from its load. Belts or change gears that can become inadvertently disconnected cannot be safely used. Direct couplings or fixed gear ratios must be used.

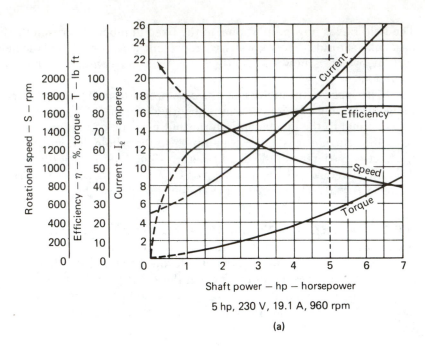

5 hp, 230 V, 19.1 A, 960 rpm

(a)

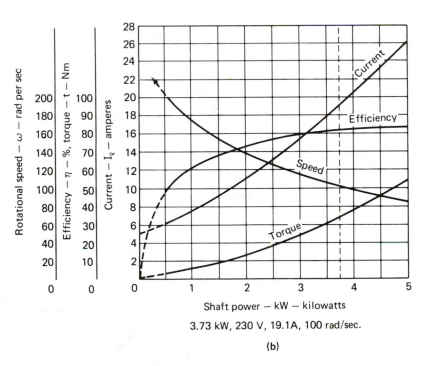

3.73 kW, 230 V, 19.1A, 100 rad/sec.

(b)

Figure 6-5. Series motor characteristics: (a) English units; (b) SI metric units.

THE DIRECT-CURRENT MOTOR   CHAP. 6

Example 6-7. A 120 V, 1750 rpm (183.3 rad/sec) series motor takes 37.8 A at its rated load of 4.692 hp (3.500 kW). It has an armature circuit resistance including all but the series field of 0.277 Ω and a series field resistance of 0.163 Ω. If the load is reduced until the rated angular velocity reaches its safe maximum of 3500 rpm (366.5 rad/sec), at that speed the line current is 10.2 A.

(a) Calculate the percent of change in field *flux*.

(b) Calculate the power at this speed.

(c) If the load is increased so as to pull down the speed to 1200 rpm (125.7 rad/sec) and the field flux cannot increase, owing to saturation beyond normal field current, what current is then drawn?

(d) What power is developed under the conditions of part c?

Solution:

(a) Find the KK_1' and KK_2' under the two conditions from Eqs. (6-10$_E$) and (6-10$_{SI}$).

$$S = \frac{V_a - I_a(R_a + R_{se})}{KK'I_a}$$

from Eq. (6-10$_E$) tranposed to

$$KK_1' = \frac{V_a - I_{a1}(R_a + R_{se})}{SI_{a1}}$$

when $R_a + R_{se} = 0.277 + 0.163 = 0.440$ Ω:

$$\frac{120 - (37.8)(0.440)}{(1750)37.8} = KK_1' = 0.001\ 563$$

$$\frac{120 - (10.2)(0.440)}{(3500)10.2} = KK_2' = 0.003\ 236$$

The KK' factor is really the rate of making flux per ampere and is *not* a constant due to the nonlinearity of the saturation curve. In reality, K is a constant, but K' is not except in a region of a small change where it would be a function of the slope of the saturation curve at that point. However, since $K'I_a$ represented actual flux in the derivation of Eq. (6-10), if we multiply KK' by the I_a at that point we will have a quantity *proportional* to flux. Thus,

$$KK_1'I_{a1} = 0.001\ 563 \times 37.8 = 0.059\ 08$$

and $$KK_2'I_{a2} = 0.003\ 236 \times 10.2 = 0.033\ 01$$

Then in the same fashion as finding the percent of change in speed for regulation we may equate as shown:

$$\frac{0.059\ 08 - 0.033\ 01}{0.059\ 08} \times 100 = 44.1\%\ \text{reduction in flux}$$

The same problem in the same way in SI is, from Eq. (6-10$_{SI}$),

$$\omega = \frac{V_a - I_a(R_a + R_{se})}{kk'I_a}$$

$$kk'_1 = \frac{V_a - I_{a1}(R_a + R_{se})}{\omega I_{a1}}$$

$$\frac{120 - (37.8)(0.440)}{(183.3)37.7} = kk'_1 = 0.014\ 92$$

$$\frac{120 - (10.2)(0.440)}{(366.5)10.2} = kk'_2 = 0.030\ 90$$

Then

$$0.014\ 92 \times 37.8 = 0.5640$$

and

$$0.030\ 90 \times 10.2 = 0.3152$$

Again, these are *proportional* to flux, so that

$$\frac{0.5640 - 0.3152}{0.5640} \times 100 = 44.1\% \text{ reduction in flux}$$

Here, as in other series motors, a large change in current, which we may take as a $[(37.8 - 10.2)/37.8]\ 100 = 73$ percent drop in current, has resulted in a 44.1 percent drop in flux. Equations (6-10) are not exact but only *illustrate* what happens in a series motor. They would be more nearly exact if we used $K\Phi$ or $k\phi$ in the denominator, because the angular velocity must represent the change in flux in the series field rather than the change in current in the field.

(b) The power can be found from the torque and the angular velocity. Since the angular velocity is given, we need to determine the probable torque under the new conditions. The problem is to determine which torque equation to use. Equations (6-11) will be accurate only over a small region owing to saturation effects. Returning to Eq. (6-8), we may see the following:

$$T = C\Phi I_a \hspace{3cm} \text{Eq. (6-8}_E)$$

and

$$T = c\phi I_a \hspace{3cm} \text{Eq. (6-8}_{SI})$$

The torques are proportional to built-in mechanical constants C (and c) and to the armature current I_a. We know that Φ and ϕ *reduce* about 44.1 percent from case 1 and case 2 and the current change is specified. Working backward from Eq. (6-1), we may find the *original* torques. From horsepower $= TS/5252$ transposed to $T = \text{hp}(5252)/S$, we get

$$\frac{4.692(5252)}{1750} = T_1 = 14.08 \text{ ft-lb torque}$$

And from kilowatts $= t\omega \times 10^{-3}$ transposed to $t = (\text{kW}/\omega) \times 10^3$, we find

$$\frac{3.5 \times 1000}{183.3} = t_1 = 19.09 \text{ N·m torque}$$

Then substituting into a transposed version of Eq. (6-8$_\text{E}$), we find

$$C\Phi_1 = \frac{T_1}{I_a} = \frac{14.08}{37.8} = 0.3725$$

and

$$c\phi_1 = \frac{t_1}{I_a} = \frac{19.09}{37.8} = 0.5050$$

In these cases C (and c) is truly constant. Therefore, if we reduce the $C\Phi$ (and $c\phi$) term by 44.1 percent, we will have a correct value of $C\Phi$ (and $c\phi$) for further steps.

$T_2 = C\Phi_1(1.000 - 0.441)I_{a2}$. This is because a percentage reduction then *leaves* 1 minus the decimal equivalent of the percentage.

$$0.3725(1.000 - 0.441)10.2 = T_2 = 2.124 \text{ ft-lb}$$

and

$$0.5050(1.000 - 0.441)10.2 = t_2 = 2.879 \text{ N·m}$$

With the new torques and using the new rotating speeds, we can find power:

$$\frac{TS}{5252} = \frac{2.124(3500)}{5252} = 1.415 \text{ hp}$$

$$t\omega \times 10^{-3} = 2.879(366.5)0.001 = 1.055 \text{ kW}$$

1.415 hp $\times$ 0.746 kW/hp $= 1.055$ kW; check.

(c) In the increased load situation, the data state that no further increase in flux is achieved. If so, then the $K'\Phi$ or $k'\phi$ terms that have been found can be used in transposed versions of Eq. (6-10), where the *whole denominator* is then constant. For convenience then, Eq. (6-10) becomes

$$S = \frac{V_a - I_a(R_a + R_{se})}{K''}$$

where we let K'' stand for $K''I_a$, which was shown to be proportional to flux in part (a).

$$K'' = \frac{V_a - I_{a1}(R_a + R_{se})}{S}$$

We have already determined the values to be

$$K'' = 0.059 \ 08$$

and

$$k'' = 0.5640$$

Since S and ω are given, we may transpose in a different fashion:

$$V_a - I_a(R_a + R_{se}) = K''S$$

so that then

$$I_a(R_a + R_{se}) = V_a - (K''S)$$

and $\qquad I_a = \dfrac{V_a - (K''S)}{R_a + R_{se}} = \dfrac{120 - [0.059\ 08(1200)]}{0.440} = 111.6\ \text{A}$

and similarly for SI units; as a check we see that $k''\omega = 0.5640(125.7) = 70.89$, which is the same as $K''S = 0.5908(1200) = 70.89$. Thus the results will be the same, barring algebraic error, when using SI since

$$I_a = \dfrac{V_a - k''\omega}{R_a + R_{se}}$$

(d) The current will enable the determination of torque and then power. Remember that flux has *not* increased in this part of the problem. Torque is then found using Eq. (6-8) as a basis. Since $C\Phi$ and $c\phi$ are constants, the torque is simply increased by the ratio of the I_a current. Reaching back to part b, we find the original full load torques to be

$$T_1 = 14.08\ \text{ft-lb} \quad \text{or} \quad t_1 = 19.09\ \text{N·m}$$

then $\qquad T_3 = T_1 \times \dfrac{I_{a3}}{I_{a1}} = 14.08 \times \dfrac{111.6}{37.8} = 41.57\ \text{ft-lb}$

and $\qquad t_3 = t_1 \times \dfrac{I_{a3}}{I_{a1}} = 19.09 \times \dfrac{111.6}{37.8} = 56.36\ \text{N·m}$

The power is then straightforward from

$$\dfrac{TS}{5252} = \dfrac{41.57(1200)}{5252} = 9.50\ \text{hp}$$

$$t\omega \times 10^{-3} = 56.36(125.7)0.001 = 7.08\ \text{kW}$$

$9.50\ \text{hp} \times 0.746\ \text{kW/hp} = 7.08\ \text{kW}$; check.

This problem is obviously a long, detailed procedure. It was handled in this fashion to show that formulas must be treated with caution and used only where they apply or else modified with reason to suit the situation. It is entirely reasonable that a technician or engineer will be faced with apparently simple problems that actually require considerable thought. The nonlinearity of the magnetization curve causes all kinds of difficulty, of which this is only one example. Just such a problem as this might well be involved in a motor determination for a hoist or other material-handling device. A detailed study of currents, torques, and powers might be required in order to size the motor or determine the gear ratio or even the current protective device ratings.

6-6-3 Series Versus Shunt Motor Usage. The series motor is almost invariably used in railroad, rapid transit, or traditional types of service trucks such as fork lifts, because the torque curve shape closely fits the requirements.

A typical railroad or any of the other tasks mentioned requires very high torque for starting and acceleration in the low ranges. Railroad equipment does not require rapid acceleration in the cruising range of speeds.

The highway passing problem with its high acceleration in the mid-speed range is entirely foreign to railroad practice. Another consideration is that, during the early period of electric railway motors and their applications, no change speed gear boxes (transmission), such as are required in automotive work, were possible with the state of the art of gearing. Although most authorities state that the series motor is the ideal motor for vehicle service, that is really a traditional view only. A study of highway vehicle torque versus speed requirements shows that most modern gasoline-powered automobiles that have been considered as having satisfactory performance in the passing range are so geared that their maximum torque is not reached until at or above the usual highway speeds. Most experimental electric automotive vehicles that have so far been built in the modern search for low pollution and low energy consumption have been very unsatisfactory in highway passing. The same vehicle may have comparatively outstanding acceleration below 30 mph (48.2 km/h) due to its high series motor torque at low speed.

The shunt motor, or even the separately excited shunt motor if used, with a normal manual transmission or a specially valved automatic transmission looks as if it would be a more satisfactory combination. This is just another example of where knowledge of device characteristics may contribute to a change of previously accepted values. More on this later.

6-7 COMPOUND MOTOR

When a dc motor has both a shunt and a series field, it is known as a *compound* motor. If the motor is connected so that the series field *aids* the shunt field, it is known as a *cumulative compound* motor. When the series field is connected to *oppose* the shunt field, it is then a *differential compound* motor. The terms are identical to those used in identifying dc generators. In fact, the same machine may be used as a motor or generator.

In the same fashion as with generators, the compound motor acts with a combination of the characteristics of shunt and series motors. The larger the effect of the series field, the more the characteristics resemble a series motor. Although almost any range of performance characteristics is potentially available, only certain regions of the potential range are normally used. We may eventually see devices, as yet unknown, that need the special features of compound motors.

The cumulative compound motor develops a high torque to match an increase in torque load, as does a series motor. However, the cumulative compound motor has a definite and controllable no-load speed so that there is not a "runaway" problem. This makes the type particularly adaptable to uses requiring sudden applications of heavy load. Some uses are for rolling mill drives and heavy shears or large punches. A particular advantage under sudden but short-duration heavy loads is that, when the motor drops in rotative speed as it is loaded, it gives up a portion of its stored kinetic energy to drive the load. If the speed were to be held more closely, the transient would have to be met by high peak currents from the supply line.

Cranes, hoists, and elevators use cumulative compound motors since they can smoothly start a heavy load and yet not overspeed when unloaded. The series field is frequently cut out of the circuit automatically when the hoist is up to speed. The steady-state situation is then handled as if by a shunt motor. A further advantage is the ability to use the motor as an adjustable brake by using it as a generator with a descending load, since it has the shunt field available.

The differential compound motor is not so widely used, but it has some special characteristics that lend it to some special services. When the series field opposition effect is adjusted so that the loss of flux effect just cancels the loss of speed of a shunt motor when loaded, a substantially constant speed results. There are two major performance problems, however, in a differential compound that severely limit its use: (1) when set for a flat or even a rising speed with load characteristic the motor will tend to "run away" or severely increase its speed with a high load; and (2) this same effect of the series field tending to "take over" will cause a differential compound motor to start in the opposite direction from that desired unless the starting current is carefully held down. Usually, the series field is cut out by special switching during starting. Much as the differential compound generator is a special-purpose machine, the differential compound motor is even more so.

6-7-1 Compound Motor Speed Characteristics.
The speed characteristics of compound motors can be developed from Eq. (6-7). They take the following form when *cumulative*:

$$\boxed{\text{rpm} = S = \frac{V_a - I_a(R_a + R_{se})}{K\Phi_{sh} + KK'I_a}} \qquad \text{Eq. (6-12}_\text{E})$$

and

$$\left\langle \omega = \frac{V_a - I_a(R_a + R_{se})}{k\phi_{sh} + kk'I_a} \right\rangle \qquad \text{Eq. (6-12}_\text{SI})$$

Here the numerator is the same as the series motor situation, and the denominator is the sum of the shunt and series cases.

For the *differential* compound motor,

$$\boxed{S = \frac{V_a - I_a(R_a + R_{se})}{K\Phi_{sh} - KK'I_a}} \qquad \text{Eq. (6-13}_\text{SI})$$

and

$$\left\langle \omega = \frac{V_a - I_a(R_a + R_{se})}{k\phi_{sh} - kk'I_a} \right\rangle \qquad \text{Eq. (6-13}_\text{SI})$$

These equations differ from Eqs. (6-12) only by the minus sign in the denominator. It can then be seen that if the I_a becomes large enough the second half of the denominator will overcome the first half. In this case there is no effective field and a full short-circuit current will flow. As the $KK'I_a$ term (or $kk'I_a$) increases, the total effective flux becomes less and less and the motor accelerates dangerously.

6-7-2 Compound Motor Torque Characteristics. The torque of compound motors is also a combination of the shunt and series torque characteristics. For cumulative compound motors,

$$T = C\Phi_{sh}I_a + CC'I_a^2$$ Eq. (6-14$_E$)

and $$t = c\phi_{sh}I_a + cc'I_a^2$$ Eq. (6-14$_{SI}$)

and for differential compound motors,

$$T = C\Phi_{sh}I_a - CC'I_a^2$$ Eq. (6-15$_E$)

and

$$t = c\phi_{sh}I_a - cc'I_a^2$$ Eq. (6-15$_{SI}$)

These various compound motor characteristics can be seen in Fig. 6-6. In practical compound motors the characteristics bracket a performance region on both sides of those of a shunt motor with the same armature circuit. A shunt motor may in fact be a compound motor with the series field windings simply not used in a particular application.

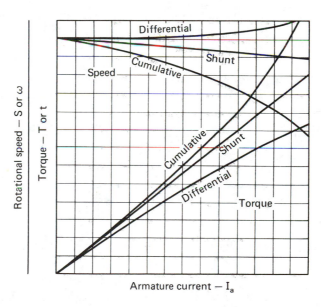

Figure 6-6. Speed and torque relations for shunt and compound fields with the same armature.

Compound motor torque change can be expressed as a ratio:

$$T_2 = (T_1) \frac{I_a \times \Phi_2}{I_a \times \Phi_1}$$

Eq. (6-16$_E$)

$$t_2 = (t_1) \frac{I_a \times \phi_2}{I_a \times \phi_1}$$

Eq. (6-16$_{SI}$)

There is a distinction between a compound motor and a *stabilized shunt* motor. The stabilized shunt unit has a series field, but with so few turns that its only purpose is to stabilize the shunt characteristics rather than add compound characteristics. It is schematically a compound, but practically a shunt motor. The small series field compensates for armature reaction.

6-8 COMPARISON OF DIRECT-CURRENT MOTOR CHARACTERISTICS

After the initial discussions of each type of dc motor, a comparison of all three is indicated. Figure 6-7 compares series, shunt, and compound performance when the three types are rated at the same horsepower (or kilowatts) and rpm (or radians per second). In this type of plot the characteristics all cross at the design load point.

Figure 6-6 shows the comparison among shunt, series, and compound operation for the *same motor* with its armature being acted upon by either or both fields. Note that here the outputs vary at the same armature current.

Figures 6-6 and 6-7 are drawn with the motors operating on their full rated voltage. Different voltages will move the curves, but the relative shapes will remain similar.

From Figure 6-7 we may deduce the following: (1) between no load and full load the curves diverge, and if the load torque were around mid-load, the shunt motor would draw the least current and the series the most; (2) at overloads the opposite is true. The series motor can develop an overload torque of around double the rated torque without drawing excessive current.

Figure 6-6 shows the other part of the story in that the speed characteristics for the same armature are very much different. Therefore, the ability to take increasing torque loads that is the special contribution of a series field is accompanied by a much greater speed change.

6-9 DIRECT-CURRENT MOTOR STARTING PROBLEMS

There have been a number of references in past pages to the desirability of keeping the armature circuit resistance of a dc motor as low as it reasonably can be built. Under normal operation conditions, this low resistance is entirely beneficial. However, during starting and acceleration, excessive current would flow if the full line

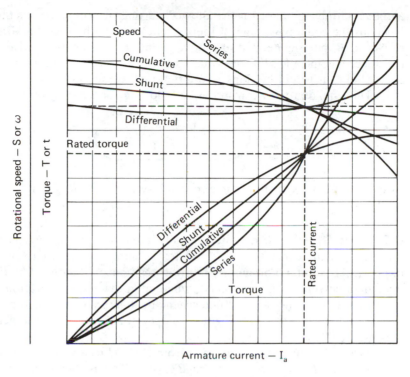

Figure 6-7. Speed and torque relations for shunt, series, and compound fields. Motors of same rated speed, torque, and current.

voltage were placed across the armature circuit. This is true because of Ohm's law, where, in a complete circuit, current equals voltage divided by resistance: $I = E/R$. In normal operation the current is adjusted, as stated earlier, by the presence of the counter or back emf. Thus, the normal current is limited by the fact that the effective voltage across the low armature resistance is a self-adjusting result of $I = (V_l - E_c)/R_a$. Since the E_c factor is not available at the start, the R_a must be *increased* by added resistance or the V_l must be *reduced* in some fashion.

In the usual industrial situation the line voltage is fixed, or nearly so. Such fluctuations as may take place are due to the source characteristics and the total line voltage drops between the source and the motor. It is usually desirable to limit the maximum starting current to 125 or 150 percent of the normal running current. This value is not a rigorous requirement, and occasionally a value of 200 percent or more may be used under special conditions. The desired conditions can be equated by the following modification of Ohm's law, which is a general case:

$$I_r \times M = \frac{V_l - E_c}{R_a + R_s} \qquad \text{Eq. (6-17)}$$

Equation (6-17) shows that the full-load *rated* motor current, I_r, times an agreed multiplying factor for the allowed starting current M equals the line voltage V_l minus the counter voltage (if there is one) divided by the *total* armature circuit resistance R_a plus the needed starting resistance R_s. Here the total armature circuit resistance will include *all* series windings, such as a series field, unless it is specially shorted out, as would be done in a differential compound motor.

This equation then transposes as follows to find the R_s term:

$$R_s = \left(\frac{V_l - E_c}{I_r \times M}\right) - R_a \qquad \text{Eq. (6-18)}$$

Equation (6-18) fits problems in either unit system since R, E, and I are universally used.

6-9-1 Dc Starting Switch. Usually, the motor is allowed to accelerate until the high starting current drops down to the normally rated current by the effect of a partial counter voltage. Then the resistance is reduced until the desired maximum current is again reached, whereupon further acceleration takes place.

Figure 6-8 shows a simplified schematic diagram of the current-limiting parts of a typical starting switch. Figure 6-8 does not show any of the actuating mechanisms.

When the current has dropped from $I_r \times M$ to a value at or near I_r, which may be called I_l, some countervoltage exists. This is shown in general as

$$E_c = V_l - I_l(R_a + R_s) \qquad \text{Eq. (6-19)}$$

Here all the terms have been defined before.

The intermediate speeds that are reached when the current decays to the I_l arbitrary lower limit are approximated as

$$S = S_{\text{rated}} \times \frac{E_c}{E_l - (I_l \times R_a)} \qquad \text{Eq. (6-20}_\text{E})$$

A related equation when working in SI is

$$\omega = \omega_{\text{rated}} \times \frac{E_c}{E_l - (I_l \times R_a)} \qquad \text{Eq. (6-20}_\text{SI})$$

These formulas are seen to be the rated angular velocity multiplied by the ratio of the counter or back emf, divided by a term that is the normal operating countervoltage. This approximation is invalid for currents that are much different from the rated I_l in a motor with a series field. With these equations an example will show how starting resistances are determined. Refer to Fig. 6-8 for detail nomenclature at intermediate points.

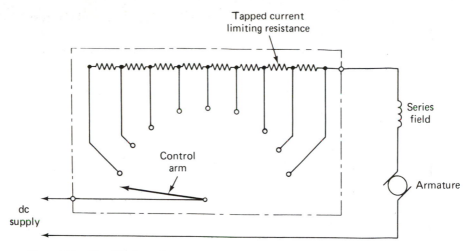

Figure 6-8. Simplified starting switch.

Example 6-8. A dc motor is rated at line conditions of 230 V and 27.5 A at normal full load. It turns at 1750 rpm (183.3 rad/sec) under rated conditions. It has a total armature circuit resistance of 0.803 Ω, and it is desired to hold its maximum starting current to 150 percent of normal full-load current. Determine the following:

(a) The current that would flow if there were no added starting resistance.

(b) A total starting resistance to meet specified conditions.

(c) The rotational speed that may be expected when the motor has accelerated sufficiently to reduce the line current to the normal rated value.

(d) Intermediate values of starting resistances that will allow 150 percent of rated current to flow for further acceleration and the speeds reached when the current decays to 100 percent.

Solution:

(a) There is no added resistance and no back emf, so Ohm's law can be used directly.

$$I = \frac{E}{R} = \frac{230}{0.803} = 286.4 \text{ A}$$

This current is obviously excessive, and high currents such as this are the reason why starting resistances are used. It is $(286.4/27.5) \times 100 = 1041$ percent of the specified current, or $1041/150 = 6.94$ times the desired starting current, which is in itself an overload.

(b) Use Eq. (6-18):

$$R_s = \left(\frac{V_l - E_c}{I_r \times M} \right) - R_a$$

$$\frac{230 - 0}{27.5 \times 1.5} - 0.803 = 4.77 \text{ Ω} = R_s \text{ total}$$

Note that in a real sense high accuracy is not needed and the resistance would be satisfactory if between 4.5 and 5.0 Ω.

(c) First determine countervoltage at *rated current* with $R_{s\ \text{tot}}$ in the armature circuit. Use Eq. (6-19):

$$E_c = V_l - I_l(R_a + R_{s\ \text{tot}})$$

$$230 - 27.5(0.803 + 4.77) = 76.7 \text{ V} = E_{c1}$$

Then using the back or counter emf and Eq. (6-20)

$$S = S_{\text{rated}} \times \frac{E_c}{E_l - (I_l \times R_a)}$$

$$1750 \times \frac{76.7}{230 - (27.5 \times 0.803)} = 645.5 \text{ rpm} = S_1$$

or in SI

$$\omega = \omega_{\text{rated}} \times \frac{E_c}{E_l - (I_l \times R_a)}$$

$$183.3 \times \frac{76.7}{230 - (27.5 \times 0.803)} = 67.61 \text{ rad/sec} = \omega_1$$

Rpm $\times$ 0.104 72 = rad/sec = 645.5 $\times$ 0.104 72 = 67.6 rad/sec; check.

(d) Since a back emf and a partial speed now exist, the starting resistance is now reduced so that 150 percent of rated current again flows and the process is repeated. Use Eq. (6-18)

$$\frac{230 - 76.7}{27.5 \times 1.5} - 0.803 = 2.91 \text{ } \Omega = R_{S2}$$

The starting resistance is then reduced from 4.77 Ω to 2.91 Ω at the second point on the starter.

The second rpm is now determined using Eq. (6-19) again

$$230 - 27.5(0.803 + 2.91) = 127.9 \text{ V} = E_{c2}$$

and then Eq. (6-20)

$$1750 \times \frac{127.9}{230 - (27.5 \times 0.803)} = 1076 \text{ rpm} = S_2$$

The motor increases from 645.5 up to 1076 rpm

$$183.3 \times \frac{127.9}{230 - (27.5 \times 0.803)} = 112.7 \text{ rad/sec} = \omega_2$$

Repeating the process as many times as necessary, we find for steps 3, 4 and 5

$$R_{s3} = 1.67 \ \Omega, \qquad S_3 = 1363 \text{ rpm}, \qquad \omega_3 = 142.8 \text{ rad/sec}$$

$$R_{s4} = 0.845 \ \Omega, \qquad S_4 = 1554 \text{ rpm}, \qquad \omega_4 = 162.8 \text{ rad/sec}$$

$$R_{s5} = 0.295 \ \Omega, \qquad S_5 = 1682 \text{ rpm}, \qquad \omega_5 = 176.1 \text{ rad/sec}$$

Checking the necessity for another resistance step, we find that Eq. (6-18) shows that an R_{s6} resistance would be $-0.0708 \ \Omega$. The minus value indicates that no additional resistive steps are needed to avoid a current of over 150 percent of the rated value once a velocity of 1682 rpm (or 176.1 rad/sec) is reached. Ohm's law checks this as follows, using the calculated value of E_{c5} = 199.8 V.

$$I = \frac{E}{R} = \frac{230 - 199.8}{0.803} = 37.6 \text{ A}$$

which is *less* than $27.5 \times 1.5 = 41.25$ A, so the sixth stage of the starting sequence puts the motor directly across the line with safety.

This six-step starting sequence, which involves five different values of R_S, from $R_{S \ tot}$ to R_{S5}, was necessary because of some arbitrary constraints, which are as follows:

(1) A maximum starting current of 150 percent or 1.5 times the normal full-load operating current was specified. This could have been different.

(2) It was assumed that the motor was driving a nearly constant torque load so that the starting current was not allowed to drop below the normal full-load range. Implications of this will be expanded.

(3) The line voltage was assumed to be constant. This might not be true at 150 percent rated current.

(4) An armature circuit resistance of 0.803 Ω was specified, which is a specific peculiarity of the chosen motor. This value will change in approximate inverse relation to motor size or rated power.

(5) All current and rpm (or radians per second) relations are assumed to be linear. This is nearly true for a shunt motor, but less so for compound or series.

If the motor was connected to a load whose torque characteristics varied with rotational speed, the process would be considerably different. If the torque were to increase linearly with speed or even to increase as the square of speed, much further acceleration could be allowed for each step. There might be effective acceleration in the lower ranges even though the line current dropped as far as 50 percent of normal. Then a second step might allow a drop from 150 down to 75 percent, and so on. By judicious choice perhaps only one-half as many steps would be needed.

Obviously, if higher than 150 percent of rated current could be used, more acceleration per step could be achieved even with the 100 percent rated current lower limit.

There are many choices to be made if an optimized starter is to be designed for a particular situation. This introductory situation does not attempt to show how the step changes are performed, nor does it show the various protective devices that are normally supplied. The whole subject of starters and motor controllers will be covered in more detail in Chapters 8 and 9. At this point, rest assured that except in the very small sizes a dc motor always needs some form of starting device and frequently needs a speed control. Reference (3) (Chapter 4) has good detail on this subject.

6-10 DIRECT-CURRENT MOTOR REVERSING

The direction of rotation of any dc motor depends upon the magnetic polarity of its main fields and the direction of the conventional current that flows in the armature windings that are immersed in the fields. The direction that results is then in accordance with the left-hand rule of motor action as described in Sect. 2-12 and Fig. 2-9. The left-hand rule relationship is met under all the field poles in any motor. The methods of winding the armature, whether lap, wave, or frog-leg coils, all accomplish the same thing in a motor: the current is directed oppositely under north and south poles, so each region pulls in the same direction. A bit of manipulation with the left hand shows that when the flux (as shown by the index finger) flows inward to the armature, as under a north pole, the current (as shown by the middle finger) must go parallel to the assumed shaft direction. If an adjacent and therefore south pole is simulated, the flux points out from the shaft. Then if the hand is further rotated such that the current is also opposite to its original relationship, the thumb (which represents motion) is pointed in the *original* direction.

All this means that rotational direction is determined by four factors:

(1) The direction of the field coil winding, which is built in.
(2) The connection polarity of the whole field, which may be changed or switched.
(3) The direction of the armature coil winding, which is also built in.
(4) The connection polarity of the brush group, which is the fixed access point to the armature winding and which may be changed or switched.

Factors 2 and 4 are accessible modifications and may be manipulated. Conversely, factors 1 and 3 are fixed by the construction of the unit and cannot realistically be changed.

If the overall terminal connection polarities of a dc motor are changed, the current direction of *both* the armature and the field are changed. Since this, in effect, does the same thing as moving from one pole to another locally, the rotating torque direction is not changed. The exception to this is a permanent magnetic field motor where only the armature has coil windings. In this case reversal can be produced only by reversing the line connections.

It can be seen then that reversal is accomplished by changing the polarity of *either* the armature or the field, but *not* by changing *both*.

6-10-1 Direct-Current Motor Reversing Circuit Connections.

In a shunt motor the field normally has much less current than the armature, and so it would seem that the field is the logical place to switch. This is not usually the best practice since the field is a highly inductive circuit. Thus, if any switching is performed before the field current is fully decayed, the switch points will arc viciously or even dangerously in large sizes. Furthermore, since any switching point is usually the most unreliable part of an electric circuit, it is unwise to switch where the basic action desired contributes to further unreliability. Finally, upon failure of a field reverse switch contact, the next start will take place with little or no field, which causes high-speed runaway. For these reasons the armature circuit is usually the reversed part.

In a series motor it does not make much difference since the field coil is much less inductive with its fewer winding turns. The field also has the same current as the armature. A compound motor must have *both* fields changed if field re-

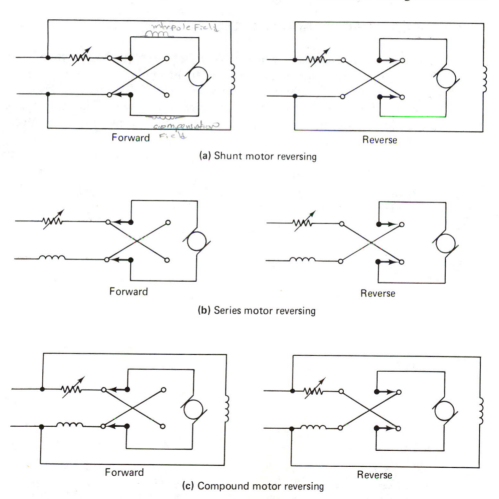

Figure 6-9. Direct-current motor reversing.

versing is used, so armature switching is less complex. If only one field were reversed, a compound motor would be changed from cumulative to differential, which changes the whole character of the motor, as has been seen.

Figure 6-9a shows shunt motor reversing connections schematically, Fig. 6-9b shows the series motor reversing problem, and Fig. 6-9c shows the compound motor schematic for reversing.

Note that the commutating fields and the compensating windings, if present, *must be left in their original relationship* with the armature, so the reversal must take the *armature circuit as a whole*. If the commutating field or the compensating winding were inadvertently switched in relation to the armature, the commutation quality would be destroyed. Taking either of these windings out of their proper relationship is worse than if they were not there at all.

Both the starting and reversing problems will be covered in practical hardware detail in Chapter 9.

6-11 FORCE, TORQUE, AND SPEED UNIT CONVERSIONS

There are many instances when working with real hardware where either rating information or test equipment will give data in archaic or nonstandard units. Small motors, for example, may have torque information provided in gram-centimeters or dyne-centimeters. Existing small force scales may be calibrated in grams or larger units in kilograms. Some motor ratings are provided in ounce-inches or pound-inches.

In any such case use Table 6-4 to convert the undesired ratings to either foot-

TABLE 6-4 FORCE UNIT CONVERSIONS USEFUL WITH MOTORS
(For length, area, and flux density units see Table 2-1)

Unit:	cgs	English		SI	Obsolete metric
Quantity:	f'	F	F'		f
1 dyne $= 1.0197 \times 10^{-3}$ g		$= 3.5969 \times 10^{-5}$ oz $= 0.224\,81 \times 10^{-5}$ lb		$= 10^{-5}$ N $= 0.101\,97 \times 10^{-5}$ kgf	
980.66 dyne $= 1$ g		$= 35.274 \times 10^{-3}$ oz $= 2.2046 \times 10^{-3}$ lb		$= 9.8066 \times 10^{-3}$ N $= 10^{-3}$ kgf	
27.802×10^{3} dyne $= 28.350$ g		$= 1$ oz $= 0.0625$ lb		$= 0.278\,02$ N $= 28.350 \times 10^{-3}$ kgf	
4.4482×10^{5} dyne $= 453.59$ g		$= 16$ oz $= 1$ lb		$= 4.4482$ N $= 0.453\,59$ kgf	
10^{5} dyne $= 101.97$ g		$= 3.5969$ oz $= 0.224\,81$ lb		$= 1$ N $= 0.101\,97$ kgf	
9.8066×10^{5} dyne $= 10^{3}$ g		$= 35.274$ oz $= 2.2046$ lb		$= 9.8066$ N $= 1$ kgf	

NOTE: Force includes relationship to obsolete gram and kilogram force, which are still used in instruments.

TABLE 6-4 (contd)   WORK OR TORQUE UNIT CONVERSIONS USEFUL WITH MOTORS

Unit:	cgs	English	SI	Obsolete metric
Quantity:	f' d' or t'	FD, F'D', or T, T'		fd or t

1 dyne cm
$= 1.0197 \times 10^{-3}$ g cm
$= 1.4161 \times 10^{-5}$ oz in.
$= 7.3755 \times 10^{-8}$ lb ft
$= 10^{-7}$ N·m
$= 1.0197 \times 10^{-8}$ kgf m

980.66 dyne cm $= 1$ g cm
$= 0.013\,887$ oz in.
$= 7.2329 \times 10^{-5}$ lb ft
$= 9.8066 \times 10^{-5}$ N·m
$= 10^{-5}$ kgf m

7.0615×10^{4} dyne cm
$= 72.008$ g cm
$= 1$ oz in.
$= 5.2083 \times 10^{-3}$ lb ft
$= 7.0615 \times 10^{-3}$ N·m
$= 0.720\,08 \times 10^{-3}$ kgf m

1.3558×10^{7} dyne cm
$= 1.3826 \times 10^{4}$ g cm
$= 192$ oz in. $= 1$ lb ft
$= 1.3558$ N·m
$= 0.138\,26$ kgf m

10^{7} dyne cm
$= 1.0197 \times 10^{4}$ g cm
$= 141.61$ oz in.
$= 0.737\,55$ lb ft
$= 1$ N·m $= 0.101\,97$ kgf m

9.8066×10^{7} dyne cm
$= 10^{5}$ g cm
$= 1.3887 \times 10^{3}$ oz in.
$= 7.2329$ lb ft
$= 9.8066$ N·m $= 1$ kgf m

ANGULAR VELOCITY OR SPEED UNIT CONVERSIONS

Unit:	cgs	English	SI
Quantity:	s or rps	S or rpm	ω or rad/sec
1 rev/sec		= 60 rev/min	= 2π or 6.2832 rad/sec
0.016 667 rev/sec		= 1 rev/min	= 0.104 72 rad/sec
0.159 15 rev/sec		= 9.5493 rev/min	= 1 rad/sec

pounds (pounds-feet) or to newton-meters, depending upon the unit system desired. The rotating speed figures will be needed in either revolutions per minute or radians per second. Conversions from one to the other or from revolutions per second to either rpm or radians per second are convenient using Table 6-4.

The existing unit can be found in a unity situation with the desired conversion as a factor in the same line of the table. One gram-centimeter of torque is then 9.8066×10^{-5} N·m or 7.2329×10^{-5} lb-ft, whichever is desired. The equations in this chapter may then be used.

QUESTIONS

6-1. Discuss the difference between the force, work, and power.

6-2. What is meant by saying that SI units are coherent?

6-3. In what way did James Watt establish a unit system for power measurement?

6-4. What is torque?

6-5. What is the source of the torque force in a dc motor?

6-6. Why is percent of coverage involved in Eq. (6-2)?

6-7. What is the purpose of the term a in Eq. (6-2)?

6-8. How does a prony brake measure torque?

6-9. What is the mechanism of torque measurement with a dynamometer?

6-10. What is the needed data from a prony brake or a dynamometer to measure rotating shaft power?

6-11. What is back emf or counter voltage?

6-12. How is back emf beneficial?

6-13. What is the effective voltage across a dc motor armature?

6-14. What is the meaning of armature power?

6-15. Is electrical power related to mechanical power?

6-16. What is meant by equilibrium motor speed?

6-17. What is meant by the term shunt motor?

6-18. What is the dominant speed characteristic of a shunt motor?

6-19. What is meant by speed regulation?

6-20. What is meant by the term series motor?

6-21. What relationship exists between speed and load torque in a series motor?

6-22. What is meant by a compound motor?

6-23. What conditions require the use of a compound motor?

6-24. What is one major problem in dc motor starting?

6-25. How is a dc motor reversed?

PROBLEMS

6-1. A dc shunt motor is rotating at 2550 rpm and developing 42.2 lb-ft of torque. What horsepower is developed?

6-2. A dc shunt motor is rotating at 267.0 rad/sec and developing 57.2 N·m of torque. How many kilowatts of mechanical power are developed?

6-3. A dc motor has a flux density of 23 200 lines/in.² through its field poles. The total armature current is 8.00 A, and there are two parallel paths in the simplex lap armature winding. The effective winding length that is immersed in the field flux is 3.83 in. Seventy-two percent of the armature periphery is covered by the field pole shoes. The armature is 5 in. in diameter, and there are 864 conductors in the armature winding. What torque is developed?

6-4. A similar dc motor to that of Problem 6-3 has a winding length of 97.3 mm and a rotor diameter of 127 mm. If the current is the same 8.00 A, what torque is developed in SI units if the flux is 0.3596 Wb/m²?

6-5. A 125 V dc motor for an electric automobile has an armature circuit resistance of 0.042 Ω. It is operating at a steady speed and drawing 135 A armature current. What is its back emf?

6-6. The same motor as in Problem 6-5 is operated under the same conditions. What is its gross developed armature power?

6-7. If the same vehicle motor in Problems 6-5 and 6-6 operates at 2550 rpm with 50 000 lines/in.^{2} field flux density and under the volt and current conditions of Problem 6-5, what would be its speed if the field flux were reduced to 43 250 lines/in.^{2}?

6-8. Using the same motor as in Problems 6-5 and 6-6, if it operates at 267 rad/sec with 0.775 Wb/m^2 flux density, what would be its speed if the field flux were reduced to 0.670 Wb/m^2?

6-9. The dc motor of Problem 6-3 develops 4.08 lb-ft of torque, with field flux Φ = 23 200 lines/in.^{2} and 8.00 A armature current. What will be its torque if its field flux is increased to 28 400 lines/in.^{2}?

6-10. The same motor as in Problem 6-3 develops 5.53 N·m of torque, with field flux Φ = 0.3596 Wb/m^2 and 8.00 A armature current. What will be its torque if its field flux is increased to 0.4402 Wb/m^2?

6-11. A dc shunt motor on a machine tool runs at 1800 rpm when fully loaded and 1925 rpm at no load. What is its speed regulation?

6-12. What is the speed regulation of a motor that runs at 188.5 rad/sec loaded and 201.6 rad/sec unloaded?

6-13. A dc series motor operates at its full rated load at 1500 rpm. The line conditions are 125 V and 10 A. The motor has an armature circuit resistance, including the commutating field, of 1.25 Ω and a series field resistance of 0.425 Ω. Assuming that the field flux changes linearly with field current, what will be the motor speed if the load is such that the line current drops to 6.28 A?

6-14. The same series motor as in Problem 6-13 carries its full rated load at 157.1 rad/sec. What would be its speed if the line current drops to the same 6.28 A?

6-15. The same series motor as in Problem 6-13 develops 7.08 ft-lb torque at 1500 rpm. If its current drops from 10 A at full load to 6.28 A, what torque is then developed?

6-16. The same motor as in Problem 6-13 develops a torque of 9.60 N·m at 157.1 rad/sec and 10 A line current. If its current drops to 6.28 A, what torque is developed?

6-17. It is desired to start the series motor of Problem 6-13 with a starting switch resistance that will limit the starting current to 175 percent of rated current. What starting circuit resistance must be used?

6-18. If the motor in Problem 6-13 is equipped with the starting circuit resistance of Problem 6-17 and is then allowed to accelerate to the point where the line current drops back to the rated current, what back emf exists?

6-19. With the motor in Problem 6-13 at the equilibrium conditions of Problem 6-18, what speed exists?

6-20. If the motor of Problems 6-13 through 6-19 is to be rated in SI units, its normal speed is 157.1 rad/sec. What will be its equilibrium speed if it has the starting resistance from Problem 6-17 and is at the equilibrium conditions of Problem 6-18?

Answers. **6-1**, 20.5 hp; **6-2**, 15.3 kW; **6-3**, 4.08 ft-lb; **6-4**, 5.53 N·m; **6-5**, 119 V; **6-6**, 16 100 W or 16.1 kW; **6-7**, 2948 rpm; **6-8**, 308.8 rad/sec; **6-9**, 4.99 lb-ft; **6-10**, 6.77 N·m; **6-11**, 6.94%; **6-12**, 6.95%; **6-13**, 2526 rpm; **6-14**, 264.6 rad/sec; **6-15**, 2.79 ft-lb; **6-16**, 3.79 N·m; **6-17**, 5.47 Ω; **6-18**, 53.6 V; **6-19**, 743 rpm; **6-20**, 77.8 rad/sec.

7

Efficiency of Direct-Current Machinery

The efficiency of any device or process is simply a measure of the *ratio of its useful output to its gross input*. The output and input must be measured in the same units, and agreed procedures must be used so that the results are comparable to other tests. The study of efficiency in devices and processes has received a great deal of scientific attention but not much recognition from the general public.

Since the operation of any device requires an energy input in order to produce its desired output, and since the amount of energy lost in a process is inversely related to efficiency, the newly recognized worldwide shortage of energy makes this subject extremely important. The scientific and engineering world was and is well aware of an increasingly serious international problem of energy availability. However, U.S. citizens as a whole suddenly became aware of the nightmare proportions of the problem when the Arabic nations of the Middle East in late 1973 abruptly shut off the supplies of oil that had normally flowed to the rest of the industrialized world. The published figures at the time did not agree on whether the United States received 5 or up to 15 percent of its total oil supply from the Middle East, but with that fraction missing we no longer had enough to satisfy our increasingly wasteful use.

Chapter 6 introduced the concepts of force, work, and power. Therefore, we now recognize that forces must move through distances within desired times to perform work at desired rates. However, when *work* is performed, *energy* must be consumed since work and energy are synonymous. For example, 1 N·m = 1 J of energy or 0.737 56 ft-lb. Force times distance is work *or* energy. Power brings in a time relation. The rate of performing work determines the rate of the use of energy or would if there were a 1-to-1 ratio.

Unfortunately, there is more to it than that simple relation. The rate of doing work depends upon the rate of consumption of energy *times* the *efficiency* of the *process*. If a process is 50 percent efficient, it then takes *twice* as much energy to perform the process as if it were 100 percent efficient. The difference is *lost*, most in unrecoverable heat and a part, at least, in difficult-to-recover heat.

The necessity for balanced thinking and careful attention to efficiency can be seen in our nationwide desire for *convenience*. In addition, our highly commendable desire for *clean air* has run head on into the problems of energy and efficiency.

Convenience in such an everyday item as a refrigerator has led to increased energy use. A refrigerator with a self-defrosting freezer compartment, which is such a time-saver for the housewife, uses added energy. This is because the cooling coil in the freezing compartment must operate at *below* 0°F (− 17.7° Celsius) if it is to maintain the freezer at 0°F or thereabouts. This is sufficiently low that even when the refrigeration device is not operating the cooling coil surfaces are well below the freezing point of 32°F (0°C). Therefore, any frost that has come from the moisture in the air remains frozen. The self-defrosting units have timer-operated heating coils, which typically *heat* the cooling coil area to a few degrees above the freezing point in order to melt off and drain away accumulated frost. The unit must then operate in its refrigeration cycle long enough to dissipate the added defrosting heat and thus keep the food at the desired cold point. The end result is that the overall consumption of power and thus of energy of a self-defrosting freezer unit seems to be at least 30 percent greater than a nonautomatic defrost unit of comparable size and temperature. Automatic defrost refrigerator-freezer units now dominate the market; thus, higher energy consumption is in use in millions of units.

The clean air standards of the Environmental Protection Agency that have been imposed upon the automobile industry have unfortunately had a major effect on energy consumption. It has been widely stated by automotive industry authorities that the added fuel cost of the less efficient but beneficially less polluting automobiles has required about 7 percent additional fuel each year compounded since 1970. This trend was not allowed to continue, and automobiles have been reduced in size, weight, and aerodynamic drag. All of these factors have reduced energy requirements. Engines have been reduced in size or displacement and, for any given size, have become more efficient.

Since the electric-generating industry has to meet similar clean air requirements, it must of necessity compete for the clean-burning liquid fuels until new processes have been developed and put into commercial practice that will allow clean burning of the plentiful high-sulfur-content coal supplies.

The result of all this enormous political and economic pressure will be to competitively force, if not to legally force, the development and marketing of more efficient appliances and devices. It seems probable that, all else being nearly equal, the device which is more efficient and thus uses less precious energy will have a strong competitive advantage. Thus, the newly graduated technician or engineer will be faced with all the problems of efficient designs that have been only a minor consideration in the past. The consumer who hardly cared what wattage his air conditioner used in the past will now largely make his choice upon power consumption, if for no better reason than the size of his electric bill.

A motor or a generator has a physical size that is nearly directly related to the amount of heat that it has to dissipate in operation. This heat dissipation, of course, depends upon the losses within the machine. Since the size of a machine, of any sort, is a major factor in its weight and cost, the smallest machine that will perform a given task usually has an advantage in cost. The smaller machine may be more efficiently loaded and, if so, will use less energy. This needs investigation for each task.

7-1 BASIC EFFICIENCY RELATIONSHIPS

The efficiency of any device, such as a motor or generator, is then simply its output power divided by its input power when they are in the same units. The output is the input *minus* the various accumulated losses. Similarly, the input is the output *plus* the same losses. For dc motors and generators, these losses turn out to be regular and predictable. Some losses can be nearly eliminated by careful design, but all can be reduced. Efficiency formulates as follows:

$$\text{efficiency} = \frac{\text{output}}{\text{input}} \times 100 = \eta\%$$

Eq. (7-1)

(η is the lowercase Greek letter eta).

$$\text{output} = \text{input} - \Sigma \text{ losses}$$

Eq. (7-2)

$$\text{input} = \text{output} + \Sigma \text{ losses}$$

Eq. (7-3)

Σ is the Greek uppercase sigma, which is used to mean "the summation of." As a result, either of the relations shown below holds where it may be easier to measure the output or input, but not both:

$$\text{efficiency} = \frac{\text{input} - \Sigma \text{ losses}}{\text{input}} \times 100 = \eta\%$$

Eq. (7-4)

or

$$\text{efficiency} = \frac{\text{output}}{\text{output} + \Sigma \text{ losses}} \times 100 = \eta\%$$

Eq. (7-5)

Example 7-1. A 20 hp (14.92 kW) motor operating on 125 V takes 144 A when operated at rated conditions. Determine (a) the losses involved, and (b) the efficiency.

Solution:

(a) Input power is $125 \times 144 = 18\ 000$ W or $= 18.00$ kW. From Eq. (7-3), input $=$ output $+ \Sigma$ losses, so that input $-$ output $= \Sigma$ losses:

$$18.00 - 14.92 = 3.08 \text{ kW losses}$$

(b) Since various values are available, efficiency may be determined in a number of ways:

Equation (7-1) is easiest and gives

$$\eta = \frac{\text{output}}{\text{input}} \times 100 = \frac{14.92}{18.00} \times 100 = 82.9\%$$

As a check we can use Eq. (7-4):

$$\eta = \frac{\text{input} - \Sigma \text{ losses}}{\text{input}} \times 100 = \frac{18.00 - 3.08}{18.00} \times 100 = 82.9\%$$

Equation (7-5) gives the same result:

$$\eta = \frac{\text{output}}{\text{output} + \Sigma \text{ losses}} \times 100 = \frac{14.92}{14.92 + 3.08} \times 100 = 82.9\%$$

Note that the two simple checks would have caught an error in part (a) also.

7-1-1 Convenient Data Sources. Usually, with a dc *generator* the *output* can be measured with a voltmeter and an ammeter. The input is a mechanical power, and either the device driving it must be *calibrated* such that its power relationship is known or a dynamometer must be used.

Similarly, with a dc *motor* the *input* can be measured with a voltmeter and an ammeter. The output is then a mechanical power, and a prony brake or dynamometer is used.

If the summary of all the losses can be accurately determined, reasonable power determinations can be made with Eqs. (7-4) and (7-5). All work in this field is performed in watts or kilowatts, and therefore we are really working in SI. A motor output, for example, may be determined in horsepower but will be converted to watts (or kilowatts) for efficiency calculations. Remember, "the court of last resort" for mechanical power is a dynamometer, but a dynamometer may not be available. After a discussion of the identity and measurements of the various losses, a method will be developed in this text whereby the determination of motor output power or generator input power may be reliably determined without recourse to a dynamometer. These methods will have advantage in many school laboratories where the usual multistation setups are a multipurpose ac machine connected to a multipurpose dc machine. With these methods, direct power may be calculated even though a dynamometer is not available. In the usual industrial situation dynamometers are not available unless the laboratories are unusually well equipped or unless a major part of the product line is a power-generation device of some sort.

7-2 TYPES OF LOSSES IN DIRECT-CURRENT MACHINES

For simplicity, generator loss will be discussed first; then similarities of motor and generator losses will be shown. With a generator the difference between mechanical power input and electrical power output is composed of a family of losses. The mechanical input power is normally stated and measured in horsepower in the English system of units. Even prior to the use of System International units, the input power was and is converted to watts or kilowatts, and all further measurements followed in watts or kilowatts. The relationship of Eq. (6-1$_E$), which is $TS/5252.1$ = horsepower and the equivalent of horsepower $\times$ 0.745 70 = kilowatts, are used here. The ASME Guide SI-1 gives the full equivalent as: horsepower $\times$ 745.6999 = watts, so 745.7 is good to seven significant figures with a single digit roundoff. By far the most widely used equivalent is hp $\times$ 0.746 = kW, and it can be seen that very little error is involved. Probably other loss data are not good beyond three significant figures, so use judgment.

7-2-1 Rotational Losses.
Following Fig. 7-1 step by step, we find that the mechanical input power is first reduced by the *rotational losses* or the *stray power* losses. That is, any power required to turn the armature is a direct subtraction from the input power. Therefore, the rotational power is not available for the development of electrical power. Rotational losses are a summary of bearing friction, brush mechanical friction, cooling fan power, windage losses of the armature as a whole, and magnetic circuit drag loss. Since the armature rotates within the magnetic field, the magnetic flux path in the armature laminations is required to change continually. Any one portion of the armature must charge and discharge cyclically. The *hysteresis losses* that result show as consumption of mechanical rotation power.

In addition to hysteresis losses, there is another magnetic-related loss called *eddy current losses*. This loss occurs because *any* conductor that moves in a magnetic field has a voltage generated within it. Since the laminations themselves are conductive, they have internal voltages that circulate in whatever path is available. These losses would be very substantial without a laminated armature structure and would cause much heating in the core. Even with laminations they are not entirely negligible, but they are small.

All these losses can be isolated and measured, but ordinarily they are taken as a whole, where

$$\boxed{P_{\text{rot}} = I_a V_a} \qquad\qquad \text{Eq. (7-6)}$$

The voltage and current are taken when running at no load. The procedures will be described later.

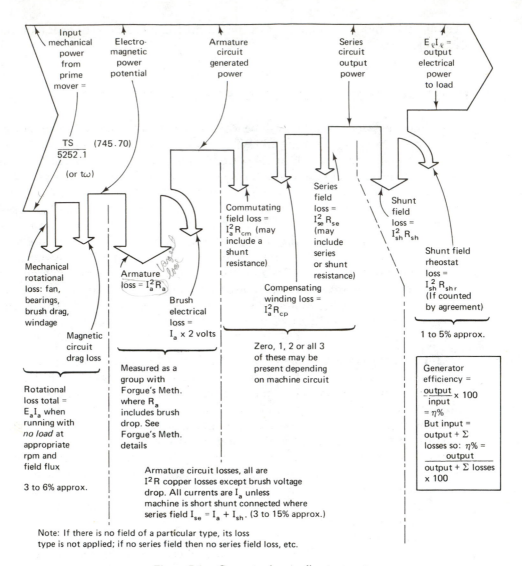

Figure 7-1. Generator losses, direct current.

The figure contains the following labels and text:

Input mechanical power from prime mover =

$\dfrac{TS}{5252.1}$ (745.70)

(or $\tau\omega$)

Electromagnetic power potential

Armature circuit generated power

Series circuit output power

E_vI_v = output electrical power to load

Mechanical rotational loss: fan, bearings, brush drag, windage

Magnetic circuit drag loss

Rotational loss total = E_aI_a when running with *no load* at appropriate rpm and field flux

3 to 6% approx.

Armature loss = $I_a^2R_a$

Brush electrical loss = I_a x 2 volts

Measured as a group with Forgue's Meth. where R_a includes brush drop. See Forgue's Meth. details

Commutating field loss = $I_a^2R_{cm}$ (may include a shunt resistance)

Compensating winding loss = $I_a^2R_{cp}$

Armature circuit losses, all are I^2R copper losses except brush voltage drop. All currents are I_a unless machine is short shunt connected where series field $I_{se} = I_a + I_{sh}$. (3 to 15% approx.)

Series field loss = $I_{se}^2R_{se}$ (may include series or shunt resistance)

Zero, 1, 2 or all 3 of these may be present depending on machine circuit

Shunt field loss = $I_{sh}^2R_{sh}$

Shunt field rheostat loss = $I_{sh}^2R_{shr}$ (If counted by agreement)

1 to 5% approx.

Generator efficiency = $\dfrac{output}{input}$ x 100 = $\eta\%$ But input = output + Σ losses so: $\eta\%$ = $\dfrac{output}{output + \Sigma\ losses}$ x 100

Note: If there is no field of a particular type, its loss type is not applied; if no series field then no series field loss, etc.

7-2-2 Winding Resistance Losses.

The potential electromagnetic power that is available after the rotational losses are subtracted is subject to reduction by various electrical power losses owing to the resistances of the various circuits within the machine.

The electrical power losses are mostly due to the various resistances that are present in the various parts of the windings. There is some exception to this, since the armature windings are undergoing reversals of polarity and current and thus have some inductive effect, as has been discussed before. This inductive effect is difficult to measure and calculate, but it is present. The armature resistance R_a, as separate from the total armature circuit resistance $R_{a\,ckt}$, is measured by a number of procedures. The alternatives and their various reasons for use will be discussed later. All following equations use R_a as if it were $R_{a\ ckt}$.

The commutating field, the compensating windings, and the series field, if any are present, are also the locations of resistive losses. Any of these resistances can be easily measured by voltmeter–ammeter methods or by a Kelvin bridge. The various winding resistances, R_a, R_{cm}, R_{co}, and R_{se}, may be summarized as R_a if convenient. Since all these windings are subject to the same current, the losses are then

$$P_a = I_a^2 R_a$$

Eq. (7-7)

Armature circuit
$P_{??} = R_a + R_a + R_{sert} R_{interpole}$

These losses vary as the load current squared and are thus known as variable losses. The $I_a^2 R_a$ is the largest single loss in the whole machine. $I_a^2 R_{a\ ckt}$ is then larger yet, since there are more resistive components present. Be sure that you account for all elements of R_a.

After the rotational losses and all the armature circuit losses are subtracted, the series circuit output power remains.

7-2-3 Shunt Field Loss. The final generator loss is the winding loss in the shunt field. This is normally considered as a fixed loss, since for any particular excitation adjustment of the shunt field rheostat, it is fixed if the line voltage is fixed. When the shunt field circuit resistance is known, this is simply

$$P_{sh} = I_{sh}^2 R_{sh}$$

Eq. (7-8)

If the field circuit voltage is known, the loss is

$$P_{sh} = V_{sh} I_{sh}$$

Eq. (7-9)

In a shunt generator or a long-shunt compound generator

$$V_{sh} = V_l$$

Eq. (7-10)

In a short-shunt situation, the V_{sh} voltage is subject to a series field voltage drop first, where

$$V_{sh} = V_l - I_l R_{se}$$

Eq. (7-11)

All these are subject to prior agreement as to whether the power loss in a shunt field rheostat is to be counted or not. If not, V_{sh} will vary with rheostat adjustment. Under this circumstance the relation in Eq. (7-8) is easier to use. The shunt field rheostat losses are *not* included in manufacturers' published efficiency data, by industry agreement. However, in a special situation where an installation efficiency is needed this rheostat loss must be counted to be realistic. As in so many situations, use the formulation that fits the situation and, most important, use your head.

Example 7-2. A 75 kW dc generator is operated at 230 V. Tests have shown that the rotational loss is 1810 W and the shunt field circuit draws 5.35 A. The armature circuit has a resistance of 0.035 Ω and the brush drop is 2.2 V. Calculate (a) the rated current delivered, (b) the total losses, (c) the input power required, and (d) the efficiency at the rated load.

Solution:
(a) From basic power law, $P/E = I$. Thus,

$$\frac{75\ 000}{230} = 326.1\ \text{A}$$

(b) From power law, again $P = IE$:

$$5.35 \times 230 = P_{sh} = 1230\ \text{W}$$

The rotational loss is given as 1810 W. The variable losses are found from the armature current, which is $326.1 + 5.35 = 331.5$ A. Therefore, the copper loss in the armature circuit is $I_a^2 R_a = (331.5)^2 0.035 = P_a = 3846$ W. The brush voltage drop is given as 2.2 V, which implies that the 0.035 Ω armature circuit resistance did *not* include the brush resistance. If the resistance had included the brushes, the given data would normally state the fact. The brush power is then $P_b = I_a E_b = 331.5 \times 2.2 = P_b = 729.3$ W. The total losses then are

$$1230 + 1810 + 3846 + 729 = P_{\Sigma loss} = 7615\ \text{W}.$$

(c) Input power from Eq. (7-3) = output + Σ losses = 75 000 + 7615 = 82 615 W. A value of this size would normally be stated in kilowatts, as 82.62 kW. In horsepower, this is $82.62 \times 1.341 = 110.8$ hp.

(d) The efficiency can now be found in various ways if input, output, and total losses are known. From Eq. (7-1),

$$\frac{75\ 000}{82\ 615} \times 100 = \eta = 90.8\%$$

Note that this is a generator and that the input *must be* larger than the output power.

Direct-current *motor* losses are seen in Fig. 7-2. They are the same types of losses and may be measured in the same way, but they are subtracted in the opposite sequence. *In a generator the input mechanical power must be larger than the output electrical power* to compensate for the losses. The motor situation is opposite in that the *input electrical power must be larger than the output mechanical power* to compensate for losses. These two statements are equivalent word relations for Eqs. (7-2) and (7-3).

The reasons for the opposite sequence for a motor's losses are made clearer perhaps by the following *two* statements: in a self-excited generator the shunt field

EFFICIENCY OF DIRECT-CURRENT MACHINERY   CHAP. 7

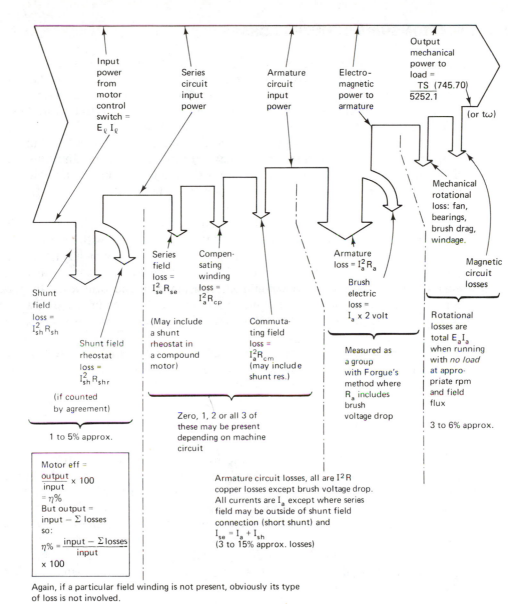

Figure 7-2. Motor losses, direct current.

cannot be excited by electrical power until that power is generated by the armature. Conversely, in a motor the input power is not all available for the armature. The field power is taken as it is available.

By similar reasoning, in a generator the various armature circuit series windings must extract their power from the armature circuit generated power before any power gets "outside" to a load. By the same reasoning, in a motor any armature circuit series winding power requirements must be satisfied before the remaining electrical power gets "inside" and is converted to mechanical power.

7-3 RELATIONSHIPS BETWEEN LOSSES

Motor and generator losses are of two types, fixed losses and variable losses. The fixed losses are not really fixed since they vary with adjustments, but they do not differ to a major degree with load current. These are the rotational losses and the shunt field loss. Obviously, the rotational loss is much different at 2000 rpm (209.4 rad/sec) than at 1000 rpm (104.7 rad/sec). However, at a single field setting a shunt machine may vary only a few percent in rotational speed. Similarly, a shunt field rheostat adjustment range may vary the field current and thus the loss by a factor of 3 or 4. If the field rheostat is set and not modified, the field current would only vary as the line voltage changed. These categories are then *fixed* losses.

Variable losses are the armature circuit resistance losses, since their power loss varies as the *square* of the current, and the current varies with the load. These are all I^2R losses, with the exception of the armature itself, which is not quite, but nearly, an I^2R loss. The problem is that the R_a resistance changes a bit as the current changes owing to the nonlinear resistance of the brushes. In general, however, the series field, the commutating field, and the compensating windings, if they are present, are variable I^2R losses. The armature itself is also considered to be an I^2R loss.

The machine reaches its maximum efficiency when the variable losses equal the fixed losses. The reasons for this statement are not obvious, and it requires a simple calculus proof to show that it is true.

$$\boxed{P_{\text{rot}} + V_l I_{sh} \cong I_a^2 R_a \quad \text{at } \eta \text{ max}} \qquad \text{Eq. (7-12)}$$

Equation (7-12) is a general case for any motor or generator. However, some of the terms may not be present in a specific case. For example, in a series motor the only fixed loss would be the mechanical parts of the rotational loss. Even the magnetic part of the rotational loss would then be a variable. In addition, if a particular type of series winding is not present in a machine, it is not a part of the variable losses.

A useful relation arises from Eqs. (7-4), (7-5), and (7-12). This is based upon the fact that, if the fixed and variable losses are equal at maximum efficiency, the following is also true:

$$\boxed{\eta_{\max} \cong \frac{\text{output}}{\text{output} + 2(\text{variable losses})} \cong \frac{\text{output}}{\text{output} + 2(\text{fixed losses})}} \qquad \text{Eq. (7-13)}$$

and

$$\boxed{\eta_{\max} \cong \frac{\text{input} - 2(\text{variable losses})}{\text{input}} \cong \frac{\text{input} - 2(\text{fixed losses})}{\text{input}}} \qquad \text{Eq. (7-14)}$$

where η is in *decimal* form.

EFFICIENCY OF DIRECT-CURRENT MACHINERY   CHAP. 7

Since the output or input power can be either mechanical or electrical, $V_l I_l$ = P_{out} for a generator and $V_l I_l = P_{in}$ for a motor. Here P_{out} or P_{in} are in watts or kilowatts, whether mechanical or electrical. These relations lead to the following useful equations:

$$\eta_{max} \cong \frac{P_{out}}{P_{out} + 2I_a^2 R_a}$$ Eq. (7-15)

and

$$\eta_{max} \cong \frac{P_{in} - 2I_a^2 R_a}{P_{in}}$$ Eq. (7-16)

where η is in *decimal* form in either case.

A few progressively more detailed examples will show how these various equations are put to practical use.

Example 7-3. A 60 hp (44.74 kW) rated, 230 V shunt motor has an armature circuit resistance, including brush resistance, of 0.052 Ω. The field resistance is 48.7 Ω. The motor is stated to have a maximum efficiency of 90.9 percent. Calculate (a) the line current carried at the maximum efficiency, and (b) the stray power loss (rotational loss).

Solution:

(a) If we assume $I_a \cong I_l$ by neglecting the smaller field current, Eq. (7-15) becomes

$$\eta_{max} \cong \frac{P_{out}}{P_{out} + 2I_l^2 R_a}$$

The *output* power is specified by the motor power rating. If we modify this by using the appropriate $I \times V$ instead of P, we will have a common I_l term that will cancel out, thus aiding the solution:

$$\eta_{max} \cong \frac{I_l V_l}{I_l V_l + 2I_l^2 R_a} = \frac{\cancel{I_l}(V_l)}{\cancel{I_l}(V_l + 2I_l R_a)}$$

$$\cong 0.909 \cong \frac{230}{230 + 2(I_l)0.052} = \frac{230}{230 + 0.104 I_l}$$

$$230 + 0.104 I_l = \frac{230}{0.909} = 253$$

$$0.104 I_l = 253 - 230 = 23$$

$$I_l = 221 \text{ A}$$

Since horsepower and wattage are both power, and since volts times amperes equals watts, volts times amperes can be substituted for horsepower and the dimensions of the equation are still true.

(b) The total fixed losses will match the total variable losses at the peak efficiency by Eq. (7-12); thus the relation $I_a^2 R_a = P_{rot} + P_{sh}$. The power in the shunt field is from power law: $P_{sh} = E_l^2/R_{sh} = (230)^2/48.7 = 1086$ W, and $I_a = I_l - I_{sh} = 221 - (230/48.7) = 216.3$ A. Then

$$P_{rot} = I_l^2 R_a - P_{sh} = (216.3)^2 0.052 - 1086 = 1347 \text{ W}$$

Example 7-4. It is desired to determine the conventional efficiency of a machine that is to be operated as a 2 kW shunt generator. The machine is operated at 125 V dc. Its armature circuit resistance at normal operating temperature is 0.453 Ω, including the commutating field. A brush voltage drop of 1.79 V was determined experimentally. The shunt field current is held at 1.125 A. The rotational loss at this field excitation when tested in three different fashions gave 94.2, 90.0, and 92.8 W when operating at the rated 1800 rpm (188.5 rad/sec). Determine the following:

(1) Efficiency at 0, $\frac{1}{4}$, $\frac{1}{2}$, $\frac{3}{4}$, full, and $1\frac{1}{4}$ loads. Tabulate all steps and assume the rotational loss and shunt field losses to be constant over the operating range.

(2) Load for maximum efficiency and the percent of efficiency at maximum.

Note that in a plain shunt generator the output voltage will change owing to the normal voltage regulation; therefore, the shunt field loss and the rotational losses would show some change. This effect is to be ignored here and will be explored in a later example.

Solution:

(a) The three values of rotational loss have no specified preference, so a simple average is used:

$$\frac{94.2 + 90.0 + 92.8}{3} = P_{rot} = 92.3 \text{ W}$$

The field power loss is

$$P = IE = 1.125 \times 125 = P_{sh} = 140.6 \text{ W}$$

The *fixed losses* are then

$$92.3 + 140.6 = P_f = 232.9 \text{ W}$$

The rated load current is

$$\frac{P}{E} = I = \frac{2000}{125} = 16.0 \text{ A}$$

Therefore, the various output currents to the load will be simple fractions of this, or 0, 4.0, 8.0, 12.0, 16.0, and 20.0 A. *Armature* currents will always be sufficiently more to supply the shunt field, or 1.12, 5.12, 9.12, 13.12, 17.12, and 21.12 A.

TABLE 7-1   SOLUTION TO 2-kW SHUNT GENERATOR EFFICIENCY

Load (%)	0	25	50	75	100	125
Load current (A)	0	4.00	8.00	12.00	16.00	20.00
Field current (A)	1.12	1.12	1.12	1.12	1.12	1.12
Armature current (A)	1.12	5.12	9.12	13.12	17.12	21.12
Arm. ckt. I_a^2	1.25	26.21	83.17	172.1	293.1	446.0
Rotational loss, P_{rot} (W)	92.3	92.3	92.3	92.3	92.3	92.3
Shunt field loss, P_{sh} (W)	140.6	140.6	140.6	140.6	140.6	140.6
Total fixed losses (W)	232.9	232.9	232.9	232.9	232.9	232.9
Armature loss, $I_a^2 R_a$ (W)	0.6	11.9	37.7	78.0	132.8	202.0
Brush loss, $E_b \times I_a$ (W)	2.0	9.2	16.3	23.5	30.6	37.8
Total losses (W)	235.5	254.0	286.9	334.4	396.3	472.7
Output to load (W)	0	500	1000	1500	2000	2500
Input = output + Σ losses = (W)	235.5	754.0	1286.9	1834.4	2396.3	2972.7
η = eff% = output/input	0	66.3	77.8	81.8	83.5	84.1

From this point a tabular solution is preferable for ease of following the steps and to avoid errors (see Table 7-1). Notice that the tabulated efficiency is still increasing at the overload of 125 percent. Part b will explore where the peak efficiency will probably be found.

Figure 7-3 is a plot of the results of this problem. It should be followed to see how the efficiency changes. Note also the position of part b.

(b) The maximum efficiency can be found from Eq. (7-12) in its second portion because the fixed losses are known; at η_{max},

$$P_{rot} + V_l I_{sh} \cong I_a^2 R_a$$

where $232.9 \cong I_a^2 \times 0.453$. Transposing, we find

$$I_a^2 \cong \frac{232.9}{0.453} = 514.1 \quad \text{or} \quad I_a \cong \sqrt{514.1}$$

Remember, *this is an approximation*: $I_a \cong 22.7$ A. This is $I_l = I_a - I_{sh} = 22.7 - 1.1 = 21.6$ A, which is

$$P = IE = 21.6 \times 125 = 2700 \text{ W} = P_{max} = 2.70 \text{ kW}$$

The efficiency at this load can be found in a number of related ways, but

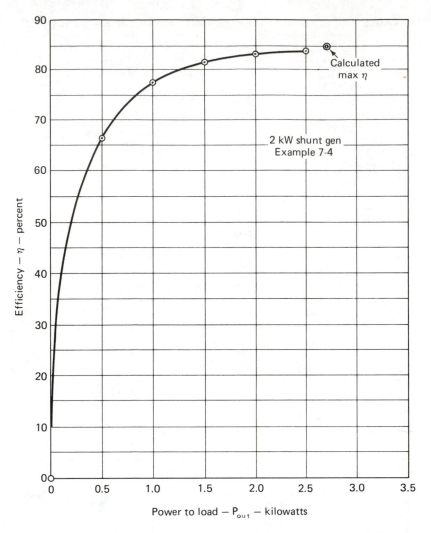

Figure 7-3. Efficiency versus output, 2 kW shunt generator, Example 7-4.

if we develop the total losses as twice the fixed losses, we can use the second part of Eq. (7-13):

$$\eta_{max} = \frac{\text{output}}{\text{output} + 2(\text{fixed losses})}$$

$$= \frac{2700}{2700 + 2(232.9)} \times 100 = 85.2\%$$

This problem was developed from actual student data on a school machine. Although the process is reasonable, the results should be inspected for reasonableness with the following considerations: the machine involved was a General Electric unit with a number 216 frame size. This size machine can normally *absorb*

5 hp (3.73 kW) input as a generator or *produce* 3 hp (2.24 kW) as a motor when operating at this range of rpm (radians/second) and voltage. The maximum efficiency, which is calculated to be at 2.70 kW output when acting as a generator, then looks reasonable.

7-4 ISOLATION AND DETERMINATION OF LOSSES

This chapter has, so far, defined and used the various losses without attempting to show how their required numerical inputs were obtained. This section will concentrate on how the needed values are determined without involvement in how they are used.

7-4-1 Self-Powered Rotational Loss Tests.
The total rotational loss, which is composed of a number of mechanical and magnetic losses, can be determined in a number of ways. The actual method used is dictated by the equipment available and the required accuracy. Usually, the various incremental losses are not needed unless some redesign is contemplated or unless some operating range that is very different from normal is required.

Accurate overall rotational loss determination requires operating the machine at its desired rotating speed or range of speeds. This is because the mechanical losses vary nearly directly with the speed, and the magnetic losses (hysteresis and eddy current) vary as the square of the speed.

Rotational loss determination also requires that the actual operating magnetic flux be present during measurement since the loss varies with flux. The hysteresis losses vary from the 1.6 power to the square of the flux density, depending upon the magnetic properties of the lamination steel that is used. The eddy current losses are also a function of the flux squared.

Direct measurement of rotational power loss is surprisingly simple and straightforward. A dc generator or motor is operated as a *motor* while running free or disconnected from a load. Under these conditions, all the electrical power that reaches the *armature* is consumed in turning the armature. Then if the operation is adjusted so that the field flux and speed are as desired, the rotational loss is, from Eq. (7-6), $P_{rot} = I_a V_a$.

This voltage V_a results from the armature motion through the field flux, and it must be adjusted to a different value from the normal line voltage. The voltage V_a in a generator must conform to Eq. (2-9), which is $V_t = E_g - I_a R_a$. Here V_a must be adjusted to the E_g in the above formula. The reason for this is that during a rotational test very little current flows in relation to normal full-load operation. Therefore, in a rotational test the $I_a R_a$ voltage drop is insignificant. Then during a rotational test the line voltage applied, which is the V_a above, must be higher than normal line voltage if the unit is a generator or lower than normal line voltage if the unit is a motor. In each case the difference is the $I_a R_a$. So the voltage used during a rotational loss test must be for a *generator*

$$V_a = V_l + I_a R_a$$

Eq. (7-17)

and for a *motor*

$$\boxed{V_a = V_l - I_aR_a}$$ Eq. (7-18)

In each case, V_l is the normal operating line voltage for the unit in question. Current I_a is the normal full-load armature current if a full-load efficiency point is desired.   If not, intermediate values are chosen that are appropriate to the load in question.   This process means that the unit is operating at nearly the proper field flux.   The last statement is seen to be reasonable when it is realized that the back emf, when running light or free of load, *very closely* matches the supplied line voltage.   Furthermore, if the speed is adjusted by changing the field current, the flux is nearly the same as that which must have existed in normal operation to have the relation in Eq. (2-9) for a generator [or Eq. (2-10) for a motor].

Thus, the process of determining rotational loss requires the following steps:

(1) Determine the needed V_a voltage and adjust the power supply as required.

(2) Start the unit as a motor using an appropriate starting resistance or starting switch until up to speed.

(3) Adjust the speed to the desired rpm (or radians/second) by varying the field excitation.   A separate field supply may be needed.

(4) Measure the resulting I_a current under the conditions.

(5) The product of these V_a and I_a values is the rotational loss power from Eq. (7-6).

(6) Measure and record over as wide a range of V_a and speed as is needed for the efficiency calculation in question.

Example 7-5.   A 3.5 kW compound generator is normally operated at a line voltage of 125 V at full load.   The *total* armature circuit resistance is 0.403 Ω. The shunt field current is 1.72 A.   Under test an armature current of 2.65 A is required.   What is the rotational loss power?

Solution: The normal load current I_l is found from power law $P/E = I$ to be

$$\frac{3500}{125} = 28.00 \text{ A}$$

The shunt field current must also be supplied by the armature, so $I_a = 28.0 + 1.72 = 29.72$ A.   From Eq. (7-17), we find the required armature voltage:

$$V_a = V_l + I_aR_a$$

Then $125 + (29.72 \times 0.403) = V_a = 137.0$ V, from Eq. (7-6):

$$P_{rot} = I_aV_a = 2.65 \times 137.0 = 363.0 \text{ W}$$

A single data point results from a given V_a and speed. From step 6 above a whole family of data curves may be developed. Figure 7-4 shows a plot of a broad range of rotational loss tests that were performed on a school machine. Note that there are regions of operation which are "forbidden" in that the machine will not operate at a particular speed and line voltage. These curves may be conveniently simplified to straight lines if only a fixed speed and a normal *small range* of voltage are studied. This extreme range was developed as a part of a test procedure being developed for an automotive vehicle motor project.

7-4-2 Separately Powered Rotational Loss.

A serious problem arises if the unit in question is designed for high voltage and no suitable dc supply exists that has sufficient voltage to run a rotational loss test. In this case, the test may be done mechanically by use of a dynamometer as a driving source. Since a rotational loss is not often more than 10 percent of the normal power rating of a machine, a relatively small drive can be used. A power source of 5 hp (3.73 kW), for example, can be used to rotate a motor or generator whose normal load is in the 50 hp (37.3 kW) range. A school-size dynamometer that can *absorb* 5 hp or *deliver* 3 hp (2.24 kW) can drive a rotational loss test on a unit in the 30 to 50 hp range (22.4 to 37.3 kW). This presumes that the larger unit can be mechanically connected to the dynamometer. This same school dynamometer may be operated by a 125 V dc source, and the larger machine may need 230 V or more. When driven mechanically, the large machine can supply its own field by self-excitation and be adjusted to the terminal voltage V_t that matches its required V_a for the test. A dynamometer then can measure the mechanical power being supplied to turn the unit under test. If the V_t (or V_a) and speed are correct, the field flux is correct, and the power being supplied mechanically is the desired rotational loss power after the self-excited field power is subtracted.

Similarly, the external driving machine may be simply a *calibrated motor* with known losses, and able to produce known power that covers the range of the desired rotational loss power. More on the subject of calibrated motors and generators will be presented later in this chapter.

In the event that a *portion* of a rotational loss is desired, external drives must be used, since the unit cannot be self-powered at the same time that it is missing some of its vital parts. Incremental loss determination would not be pursued unless some special requirements were to be met or unless design troubleshooting were under way. The procedures are straightforward enough if a suitable dynamometer or a suitably calibrated motor is available.

7-4-3 Mechanical Brush Drag.

Mechanical brush drag can be determined by driving the unit in question while it is complete, and then again after removing its brushes. The *difference* in driving power required is the brush drag power that was used. This power should *roughly* conform to the following relationships since

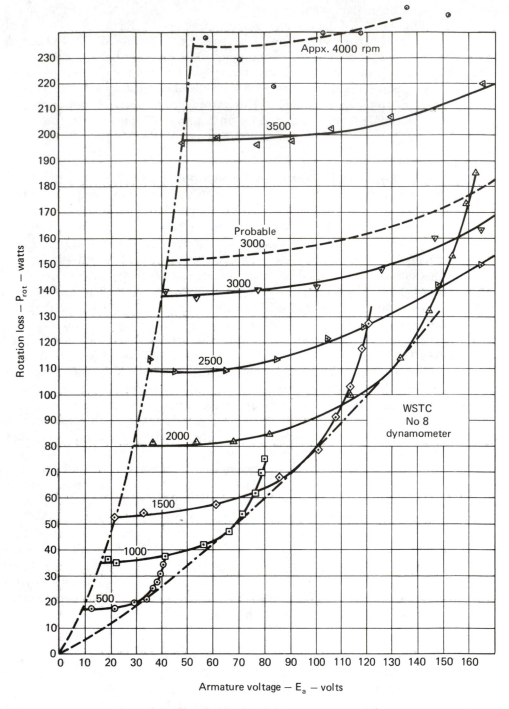

Figure 7-4. Rotational power loss versus armature voltage.

brushes work best at reasonable unit pressures and with fully broken in commutator and brush surfaces:

	Watts/in.^{2} of surface/1000 ft/min peripheral speed	Watts/cm^2 of brush surface/ m/sec peripheral speed
Carbon and graphite brushes	8.0	0.24
Metal–graphite brushes	5.0	0.15

Again, view the watts power here as a mechanical loss. Note that the English equivalents are given as two significant figure numbers. This form of power loss is only an average of many tests. Therefore, an SI equivalent should also be only two significant figures.

Brush friction is not only directly dependent upon the brush contact area and the rubbing velocity but also on the brush contact pressure. Brush pressure variations have a complicated effect. An increase in pressure will generally reduce sparking but will increase mechanically caused wear. It is literally like operating with the brakes on; the brakes (the brushes in this case of course) will heat and drag and wear in relation to the pressure applied to them. A usual pressure is *very approximately* 3 lb/in.^{2} (20 600 N/m^2) at 3000 ft/min peripheral velocity (15.2 m/sec).

A worked example is given to show how brush drag is evaluated on a large dc machine.

Example 7-6. A 300 hp (223.7 kW), six-pole dc motor operates at 750 rpm (78.54 rad/sec). Its commutator is 22.5 in. in diameter (571.5 mm), and the motor carries six individual brushes at each of six brush locations (36 brushes in all). Each brush is 0.75 in. (19.05 mm) in a circumferential direction and 1.625 in. (41.27 mm) measured parallel to the armature shaft. It is assumed that the brush pressure is such that when fully broken in the brush drag will be in the normal range for carbon and graphite brushes. Ignore the difference between the length of the curved contact arc between the brush and commutator and the 0.75 in. (19.05 mm) brush dimension. Calculate the brush friction loss in watts and horsepower.

Solution: Commutator rubbing velocity is found by determining circumference in feet (or radius in meters) and then multiplying the number of circumferences (or radians) per unit time:

$$\frac{\pi \times 22.5 \times 750}{12} = 4418 \text{ ft/min}$$

or

$$\frac{0.5715 \times 78.54}{2} = 22.44 \text{ m/sec}$$

Note that if the *diameter* is 571.5 mm then the *radius* is 0.5715/2 m, and the outer surface travels one radius of distance in one angular radian of motion. This is one of the simplifications of radian measure.

Again notice that dimensions are given in *millimeters*, but that SI calculations are determined in meters. In some cases areas are used in square centimeters although the centimeter is not used in linear dimensions when working in SI.

Brush area is then the total number of brushes times the area per brush:

$$6 \times 6 \times 0.75 \times 1.625 = 43.87 \text{ in.}^2$$

or

$$6 \times 6 \times 1.905 \times 4.127 = 283.0 \text{ cm}^2$$

Note: Here centimeters were used to get square centimeters.

Then the velocity times the area times the appropriate loss factor gives the loss in watts.

$$\frac{4418}{1000} \times 43.87 \times 8 = 1559 \text{ W brush friction loss}$$

$$22.44 \times 283 \times 0.24 = 1521 \text{ W brush friction loss}$$

Remember that the factors 8 and 0.24 are *approximations*, as shown before, and that these figures should be *rounded off* to 1500 or 1600 W (1.5 or 1.6 kW) in *either* situation.

This type of brush friction loss can be directly measured if a dynamometer is available, as has been mentioned. Similarly, *any* individual loss that can be present as a part of the whole and then can be eliminated can be evaluated.

7-4-4 Hysteresis and Eddy Current Losses. The magnetic circuit hysteresis and eddy current total drag can also be evaluated in this fashion. Using a dynamometer or a calibrated motor as a test drive source, the magnetic losses are determined as follows:

(1) Measure and record input mechanical power required to drive a motor or generator while unloaded, but with *separately excited* normal field excitation current and speed.

(2) Connect a low-current supply to the field in the opposite polarity so that the field coil can be excited in the reverse direction. The field excitation is gradually increased in the reverse direction until the electromagnetically created counterfield just counteracts the residual magnetic field. This is checked by observing the generated voltage as it is reduced to as near *zero* as can be measured. At this point there is no effective flux, so the input power can be measured again.

(3) The magnetic loss total is the *difference* between steps 1 and 2. This does not separate the hysteresis and eddy current losses but rather measures their total effect.

Hysteresis losses can be separately calculated, but are not easily separately measured in a rotating machine since both are present at the same time. The technique involves changing flux independently from other variables and then the performance of simultaneous equations. Then flux is held and frequency (speed) is changed, and further simultaneous equations followed. The technique is beyond the scope of this book, but it is good to know that something can be done. A form of this procedure is well shown in reference (5), which should be studied if the problem arises in the course of experimental work.

The actual numerical determination of eddy current losses is also considered to be beyond the scope of this text. It is well to understand the factors that are involved and a study of reference (5) or (3) will enable a specific test to be quantified if care is used in unit selection.

The mechanical loss total of the shaft bearing friction, cooling fan power, and armature surface windage are easily determined with a dynamometer or calibrated motor. Here if the field is bucked out until no voltage is generated and then the brushes are lifted while maintaining the field, all the power now required is the total of the mechanical losses above. If the cooling fan were then removed and the input power again determined, the *difference* would be the cooling fan power.

Further separate power loss identity by actual test would require special apparatus. To find the armature surface windage would involve the use of a special shaft that rotated the bearings in place. Here the remaining bearing friction would not be quite accurate since part of bearing friction is due to mechanical load. It would not be possible to make a small shaft that would similarly load the bearings unless the load were principally end thrust in the first place.

In summary, the overall rotational loss is the easiest to obtain experimentally and is usually the most useful measurement.

7-4-5 Winding Resistances. The electric power losses are mostly conventional I^2R losses. The determination of an individual electric loss means that the particular resistance must be known to the required accuracy at the normal operating temperature. If the current through that same element is also known, the power loss calculation is quite straightforward.

All the stationary winding resistances are relatively easy to find by the voltmeter–ammeter method. Usually, a simple ohmmeter reading is insufficiently accurate owing to the low resistance involved. A shunt field resistance is the only one that is normally high enough to use an ohmmeter. A typical school laboratory machine may have a shunt field resistance of the order of 80 Ω. The series fields, commutating fields, and compensating windings are all usually less than 1 Ω. If three-significant-figure accuracy is wanted, these resistances must be found by a Kelvin bridge or by the voltmeter–ammeter technique.

Usually, in the performance of an efficiency test the machine will have operated long enough that the windings will be nearly at normal operating temperature. In this case the various windings can be checked by connecting them individually to the power supply and measuring their voltage drop at various currents within their range.

7-4-6 Resistance from Wire Characteristics. Sometimes the best approach is to calculate the probable resistance from the known wire sizes and length of wire in a winding. Either use this method or its reverse process to determine the length and, therefore, the probable number of turns from a known resistance. Either way, basic wire resistance information is needed.

Any good basic electricity text will show that the resistance of a conductor is directly proportional to its length and inversely proportional to its cross-sectional area. The remaining term involved is related to the resistivity of the particular material. Therefore, resistance is conventionally expressed as

$$R = \rho \frac{L}{CM}$$

Eq. (7-19$_E$)

where

R = resistance in ohms

ρ = resistivity characteristic of the material at 20°C (ρ is the lower case Greek letter rho)

L = length of the conductor in feet

CM = cross section area in circular mils; a circular mil is an area relation that is conventionally found by squaring the diameter in *thousandths of an inch*; this is *not* a true area in square inches since that would be $\pi d^2/4$, where d is diameter in inches.

The relation is then

$$1 \text{ circular mil} = 0.785\,40 \times 10^{-6} \text{ in.}^2$$

The quantity ρ is 10.371 for annealed drawn copper wire in the English system of units which is

$$\rho = 10.371 \frac{CM\ \Omega}{ft}$$

or

$$\rho = 0.017\,241 \frac{\Omega(mm)^2}{m}$$

in an SI unit relation. In SI, as will be shown later,

$$R = \rho \frac{1}{a}$$

Eq. (7-19$_{SI}$)

Other useful materials are as shown in Table 7-2.

There are, of course, a great many other useful materials with resistivity available in the English ρ quantity or in $(\Omega \times 10^{-6} \text{ cm}^2)/\text{cm}$, where the basic cubic

TABLE 7-2 RESISTANCE CHARACTERISTICS OF CONDUCTORS

$$\frac{CM\ \Omega}{ft} = \frac{601.53\ \Omega\ (mm)^2}{m} \qquad T_i = \frac{1}{\alpha} - t \quad \alpha_t = \frac{1}{T_i + t}$$

Material	ρ at 20°C $\dfrac{CM\ \Omega}{ft}$	ρ at 20°C $\dfrac{\Omega\ (mm)^2}{m}$	Inferred absolute temp. T_i °C	Temp. coef. at 20°C = α_t $= \dfrac{\Omega}{\Omega°C}$
Advance (55 Cu, 45 Ni)	294	0.489	−50 000	0.000 02±
Alumel (94 Ni, 2 Al, 3 Mn, 1 Si)	200 at 0°C	0.333 at 0°C	−833 at 0°C	0.001 2 at 0°C
Aluminum	17.01	0.028 28	−236	0.003 9
Brass (66 Cu, 34 Zn)	23.5	0.039	−480	0.002
Carbon		1000–4000		
Chromel (90 Ni, 10 Cr)	421– 662 at 0°C	0.70– 1.10 at 0°C		0.000 0.000 at 0°C
Constantan (55 Cu, 45 Ni)	266	0.442	−4980	0.000 2
Copper (annealed)	10.371	0.017 214	−234.5	0.003 93
Germanium	266	0.45		
German silver (18% Ni)	200	0.33	−2480	0.0004
Gold	14.7	0.024 4	−274	0.003 4
Iron (pure)	58.4	0.097 1	−172, −141	0.005 2– 0.006 2
Iron (cast)	448–588	0.744–0.798		
Kovar A (29 Ni, 17 Co, 0.3 Mn, balance Fe)	1 708	2.84		
Lead	132	0.219	−230	0.004
Magnesium	28.0	0.044 6	−230	0.004 0
Manganin (84 Cu, 12 Mn, 4 Ni)	290	0.482		±0.000 015
Mercury (liquid)	576	0.958	−1 103	0.000 89
Monel (67 Ni, 30 Cu, 1.4 Fe, 1 Mn)	253	0.42	−480	0.002 0
Nichrome (65 Ni, 12 Cr, 23 Fe)	675	1.122	−2 250	0.000 17
Nickel	41.5	0.069	−193	0.004 7
Nickel silver (64 Cu, 18 Zn, 18 Ni)	168	0.28	−365	0.000 26
Phosphor bronze (4 Sn, 0.5 P, bal. Cu)	56.5	0.093 9	−314	0.003
Platinum	63.2	0.105	−314	0.003
Silicon	511 000	850		
Silver	9.79	0.016 28	−243	0.003 8
Steel (0.4 to 0.5 carbon)	78–132	0.13–0.22	−314	0.003
Steel, manganese (13 Mn, 1 C, 86 Fe)	421	0.70		
Steel, stainless (0.1 C, 18 Cr, 8 Ni, bal. Fe)	541	0.90		
Tin	68.6	0.114	−218	0.004 2
Titanium	330	0.549	−176	0.005 10
Tungsten	32.0–33.1	0.0548–0.0551	−202, −180	0.004 5– 0.005
Zinc	36	0.06	−250	0.003 7

centimeter sample size has an area of a square centimeter. These unit relations compare as shown:

$$\frac{\Omega \times 10^{-6} \, cm^2}{cm} = 6.0153 \, \frac{CM \, \Omega}{ft}$$

There is little published material available at this time in any System International agreed units and for agreed metrically based wire sizes. However, the actual resistivities of materials will not change even though the wire sizes change.

Wire resistivity would logically be listed in: $\Omega \, (mm)^2/m$, in any SI listing, and this relation compares with the older centimeter based ρ as shown:

$$1.7241 \times 10^{-6} \, \frac{\Omega(cm)^2}{cm} = 0.017 \, 241 \, \frac{\Omega(mm)^2}{m}$$

or a 10^{-4}-to-1 ratio.

Copper wire has long been available in the United States in the various American Wire Gage sizes. A study of the information in Table 7-3 will show that the *area* of a wire doubles for each decrease of three gage numbers, or halves for each increase of three gage numbers. For example, AWG No. 20 is 1 021.5 CM or 0.000 802 29 sq. in.² area; AWG No. 17 is 2 048.2 CM or 0.001 608 6 sq. in.² area. A jump of 10 AWG numbers changes area about 10 times. Of course, any *increase* in area is accompanied by a corresponding *decrease* in resistance.

Since these AWG wire sizes are not an even decimal dimension in inches, their use *may be* adopted as a tentative SI standard. Hence Table 7-3 shows corresponding dimensions for these AWG sizes in millimeters diameter, square millimeters cross-sectional area, ohms per kilometer of length, and weight in kilograms per kilometer or grams per meter.

In the event that wire sizes are ultimately standardized in decimal millimeter diameters that do not correspond to the AWG sizes, as seems probable, a tentative SI unit wire table is suggested in Table 7-4.

TABLE 7-3 STANDARD ANNEALED SOLID COPPER WIRE AMERICAN WIRE GAGE (AWG) SIZES

	Diameter		Cross section			Resistance			Weight
Gage No.	mils: 0.001 in.	mm	Circular mils: CM	Square inch: in.² × 10⁻³	Square mm: mm²	Ω/1000 ft at 20°C	Ω/km at 20°C	lb/1000 ft	kg/km or g/m
0000	460.0	11.68	211 600	166.19	107.22	0.049 01	0.160 8	640.6	953.1
000	409.6	10.40	167 810	131.80	85.032	0.061 80	0.202 8	508.0	755.8
00	364.8	9.266	133 080	104.52	67.432	0.077 92	0.255 6	402.9	599.5
0	324.9	8.252	105 530	82.883	53.472	0.098 27	0.322 4	319.5	475.4
1	289.3	7.348	83 694	65.733	42.408	0.123 9	0.406 5	253.4	377.0
2	257.6	6.543	66 373	52.129	33.632	0.156 2	0.512 5	200.9	298.9
3	229.4	5.827	52 634	41.339	26.670	0.197 0	0.646 3	159.3	237.0
4	204.3	5.189	41 742	32.784	21.151	0.248 4	0.814 9	126.4	188.1
5	181.9	4.620	33 102	25.998	16.773	0.313 3	1.028	100.2	149.1
6	162.0	4.115	26 250	20.617	13.301	0.395 0	1.296	79.47	118.2
7	144.3	3.665	20 816	16.349	10.548	0.498 2	1.634	63.02	93.77

TABLE 7-3 (contd)

| Gage No. | Diameter | | | Cross section | | Resistance | | Weight | |
	mils: 0.001 in.	mm	Circular mils: CM	Square inch: in.² × 10⁻³	Square mm: mm²	Ω/1000 ft at 20°C	Ω/km at 20°C	lb/1000 ft	kg/km or g/m
8	128.5	3.264	16 509	12.966	8.365 1	0.628 1	2.061	49.98	74.37
9	114.4	2.906	13 094	10.284	6.634 8	0.792 0	2.598	39.64	58.98
10	101.9	2.588	10 381	8.153 2	5.260 1	0.998 9	3.277	31.43	46.76
11	90.74	2.304	8 234.0	6.466 9	4.172 2	1.259	4.131	24.93	37.09
12	80.81	2.053	6 529.9	5.128 5	3.308 7	1.588	5.210	19.77	29.42
13	71.96	1.828	5 178.4	4.067 1	2.623 9	2.003	6.571	15.68	23.33
14	64.08	1.628	4 106.8	3.225 4	2.080 9	2.525	8.284	12.43	18.49
15	57.07	1.450	3 256.7	2.557 8	1.650 2	3.184	10.45	9.859	14.67
16	50.82	1.291	2 582.9	2.028 6	1.308 8	4.015	13.47	7.819	11.63
17	45.26	1.150	2 048.2	1.608 6	1.037 8	5.063	16.61	6.200	9.225
18	40.30	1.024	1 624.3	1.275 7	0.823 03	6.384	20.94	4.917	7.316
19	35.89	0.912	1 288.1	1.011 6	0.652 64	8.051	26.41	3.899	5.801
20	31.96	0.812	1 021.5	0.802 29	0.517 61	10.15	33.30	3.092	4.601
21	28.46	0.722 9	810.10	0.636 25	0.410 48	12.80	41.99	2.452	3.648
22	25.35	0.643 9	642.40	0.504 54	0.325 51	16.14	52.95	1.945	2.894
23	22.57	0.573 3	509.45	0.400 12	0.258 14	20.36	66.80	1.542	2.294
24	20.10	0.510 5	404.01	0.317 31	0.204 72	25.67	84.22	1.223	1.820
25	17.90	0.454 7	320.40	0.251 64	0.162 35	32.37	106.2	0.969 9	1.443
26	15.94	0.404 9	254.10	0.199 57	0.128 75	40.81	133.9	0.769 2	1.144
27	14.20	0.360 7	201.50	0.158 26	0.102 10	51.46	168.8	0.610 0	0.907 6
28	12.64	0.321 1	159.79	0.125 50	0.080 97	64.90	212.9	0.483 7	0.719 7
29	11.26	0.286 0	126.72	0.099 53	0.064 21	81.83	268.5	0.383 6	0.570 8
30	10.03	0.254 8	100.50	0.078 93	0.050 92	103.2	338.6	0.304 2	0.452 6
31	8.928	0.226 8	79.70	0.062 60	0.040 39	130.1	426.8	0.241 3	0.359 0
32	7.950	0.201 9	63.21	0.049 65	0.032 03	164.1	538.4	0.191 4	0.284 8
33	7.080	0.179 8	50.13	0.039 37	0.025 40	206.9	678.8	0.151 7	0.225 7
34	6.305	0.160 1	39.75	0.031 22	0.020 14	260.9	856.0	0.120 3	0.179 0
35	5.615	0.142 6	31.52	0.024 76	0.015 97	329.0	1079.	0.095 43	0.142 0
36	5.000	0.127 0	25.00	0.019 64	0.012 66	414.8	1361.	0.075 70	0.112 6
37	4.453	0.113 1	19.83	0.015 57	0.010 05	522.9	1716.	0.060 01	0.089 29
38	3.965	0.100 7	15.72	0.012 35	0.007 97	659.7	2164.	0.047 60	0.070 82
39	3.531	0.089 69	12.47	0.009 794	0.006 32	831.6	2728.	0.037 75	0.056 17
40	3.145	0.079 88	9.888	0.007 767	0.005 01	1049.	3442.	0.029 94	0.044 52

Factors

mil = defining dimension

mm = (0.0254 mm/mil)(mil)

CM = mil × mil = reference data

in.² = (π × 10⁻⁶ in.²/4 CM)(CM)

(mm)² = (645.16 (mm)²/in.²)(in.²)

(ohm/1000 ft) = (10.37 CMΩ/ft)(1000 ft/CM)

(ohm/km) = (ohm/1000 ft)(3.2808 × 10³ ft/km)

(lb/1000 ft) = [3854.4 lb/in.²(1000 ft)](in.²)

(kg/km) = (lb/1000 ft)(3.2808 × 10³ ft/km)(kg/2.205 lb)

TABLE 7-4 PROPOSED STANDARD ANNEALED SOLID COPPER WIRE PREFERRED DIAMETERS IN METRIC DIMENSION (MILLIMETERS) POSSIBLE ISO WIRE SIZE STANDARD

Preferred size mm	2nd preferred size, mm	AWG	Area square mm	Res Ω/km at 20°C	Weight kg/km
12		0000	113.10	0.1524	1005
	11	000	95.033	0.1814	844.7
10		00	78.540	0.2195	698.1
	9.0	0	63.617	0.2710	565.5
8.0			50.266	0.3430	446.8
	7.0	1	38.485	0.4480	342.1
	6.5	2	33.183	0.5195	295.0
	6.3		31.172	0.5531	277.1
6.0			28.274	0.6097	251.3
	5.5	3	23.758	0.7256	211.2
5.0		4	19.635	0.8780	174.5
	4.5	5	15.904	1.084	141.4
4.0		6	12.566	1.372	111.7
	3.5	7	9.621 1	1.792	85.52
3.0		8	7.068 6	2.439	62.83
	2.8	9	6.157 5	2.800	54.73
2.5		10	4.908 7	3.512	43.63
	2.2	11	3.801 3	4.535	33.79
2.0		12	3.141 6	5.488	27.93
	1.8	13	2.544 7	6.775	22.62
1.6		14	2.010 6	8.575	17.87
	1.4	15	1.539 4	11.20	13.68
1.2		16	1.131 0	15.24	10.05

Factors:

Reference dia in mm $= d$

$17.24\Omega(\text{mm})^2/\text{area} = \Omega/\text{km}$

$\pi d^2/4 = (\text{mm})^2 = \text{area}$

$[(8.889\ \text{kg}/(\text{mm})^2\text{km})] \times \text{area} = \text{kg/km}$

TABLE 7-4 (contd)

Preferred size mm	2nd preferred size, mm	AWG	Area square mm	Res Ω/km at 20°C	Weight kg/km
	1.1	17	0.950 33	18.14	8.447
1.0		18	0.785 40	21.95	6.981
	0.90	20	0.636 17	27.10	5.655
0.80		21	0.502 66	34.30	4.468
	0.70		0.384 85	44.80	3.421
	0.65		0.331 83	51.95	2.950
		—22—			
	0.63		0.311 72	55.31	2.771
0.60			0.282 74	60.97	2.513
	0.55	23	0.237 58	72.56	2.112
		24			
0.50		25	0.196 35	87.80	1.745
	0.45	26	0.159 04	108.4	1.414
0.40		27	0.125 66	137.2	1.117
	0.35	28	0.096 211	179.2	0.855 2
0.30		29	0.070 686	245.9	0.628 3
	0.28	30	0.061 575	280.0	0.547 3
0.25		31	0.049 087	351.2	0.436 3
	0.22	32	0.038 013	453.5	0.337 9
0.20			0.031 416	548.8	0.279 3
	0.18	33	0.025 447	677.5	0.226 2
0.16		34	0.020 106	857.5	0.178 0
	0.14	35	0.015 394	1120	0.136 8
0.12		36	0.011 310	1524	0.100 5
	0.11	37	0.009 503	1814	0.084 5
0.10		38	0.007 854	2195	0.069 8
	0.090		0.006 361	2710	0.056 5
0.080		39	0.005 026	3430	0.044 7
	0.070	40	0.003 848	4480	0.034 2

Factors:

Reference dia in
  mm = d

$17.24 \Omega(\text{mm})^2/\text{area} = \Omega/\text{km}$

$\pi d^2/4 = (\text{mm})^2 = \text{area}$

$[(8.889 \text{ kg}/(\text{mm})^2\text{km})] \times \text{area} = \text{kg/km}$

The wire sizes chosen are based on some reasonable assumptions as may be seen. Their basis is as follows:

(1) International Standards Organization (ISO) recommendation R 388 preferred round metal stock size dimensions.

(2) Metric stock sizes actually in use in presently metric countries.

(3) Rounded metric equivalents of frequently used U.S. inch sizes.

(4) Resulting American National Standard preferred metric sizes for round metal products based on items 1 to 3 above.

Both the previous American standards and this new tentative SI system are based on geometric series or modified geometric series.

The ISO basis is the Renard System of progression, which is based on roots of 10. A Renard R-20 system is then based on a size progression of the $\sqrt[20]{10}$, which is $\cong 1.122$; Renard R-10 is $\sqrt[10]{10}$, which is $\cong 1.259$. To illustrate the use of these factors, we see that if we multiply 1×1.122 and then $1.122 \times 1.122 \times 1.122, \ldots$, and so on, we build up geometric increases in sizes as seen in Table 7-5.

The American National Standard preferred stock sizes match these incremental steps with the exception of 6.3 mm, which is not used. 0.65 and 0.70 are substituted, as are 0.065 and 0.070, and 6.5 and 7.0 mm. This area is *not* yet standardized by wide agreement. General Motors for example has elected to go with the 6.3 increment in screw thread sizes simply because it appears logical and no decision had yet been made.

With this 6.3 exception, the preferred sizes used in Table 7-4 are Renard R-10 steps of size progression for the first preference and R-20 for the second preference.

Note again in Table 7-4 the wire cross section roughly doubles for each three-size increment increase. Similarly, it roughly halves for each three-size increment decrease. This is only approximately true since the basic wire size dimensions are based on two significant figures. If the size increments had been based on the

TABLE 7-5 RENARD R-20 SIZE INCREMENTS

Step factor	Calculated value	Standard roundoff	Step factor	Calculated value	Standard roundoff
1×1	$= 1.000$	$\cong 1.0$			
1×1.122	$= 1.122$	$\cong 1.1$	1×1.112^{11}	$= 3.547$	$\cong 3.5$
1×1.122^2	$= 1.259$	$\cong 1.2$	1×1.112^{12}	$= 3.980$	$\cong 4.0$
1×1.122^3	$= 1.412$	$\cong 1.4$	1×1.112^{13}	$= 4.466$	$\cong 4.5$
1×1.122^4	$= 1.585$	$\cong 1.6$	1×1.112^{14}	$= 5.011$	$\cong 5.0$
1×1.222^5	$= 1.778$	$\cong 1.8$	1×1.112^{15}	$= 5.622$	$\cong 5.5$
1×1.122^6	$= 1.995$	$\cong 2.0$	1×1.112^{16}	$= 6.308$	$\cong 6.3$
1×1.122^7	$= 2.238$	$\cong 2.2$	1×1.112^{17}	$= 7.077$	$\cong 7.0$
1×1.122^8	$= 2.512$	$\cong 2.5$	1×1.112^{18}	$= 7.940$	$\cong 8.0$
1×1.122^9	$= 2.818$	$\cong 2.8$	1×1.112^{19}	$= 8.910$	$\cong 9.0$
1×1.122^{10}	$= 3.162$	$\cong 3.0$	1×1.112^{20}	$= 9.997$	$\cong 10.0$

three- or even four-significant-figure results of a Renard progression, the effect of doubling or halving for three-size increments would be more pronounced. As it is, it is still useful. There is also a nearly 10-to-1 ratio of area or resistance for each change of 10 sizes. In both cases then rules do not hold through the missing 6.3 or 0.63 sizes owing to the break in progression. Finally, any equivalent use of circular millimeters to correspond with circular mils (CM) in inches is avoided.

With these wire table relations it is possible to calculate the resistance of any wire size and length combination at a nominal room temperature of 20°C.

7-4-7 Temperature Effect on Resistance.

When the resistance of a coil is calculated or measured at 20°C, the actual operating resistance must be determined at a measured or assumed operating temperature.

Most materials that are used as conductors show a regular increase of resistance with increase of temperature. This is especially true with copper, which is by far the most often used material. Metallic conductors of most materials show an experimentally determined change of resistance, which corresponds to the form shown in Fig. 7-5.

The experimental curve of Fig. 7-5 has a long straight line portion, which covers the normal range of temperatures used in electrical machines. This straight line projects down to the left of the figure and intersects the zero resistance abcissa base line at a point called the inferred absolute zero temperature. This point varies from one material to another, as shown in Table 7-2.

The slope of the inferred straight line is really a function of how much resistive material is under consideration and of the basic resistivity of the material. Using the known proportionate properties of similar triangles, and recognizing that with

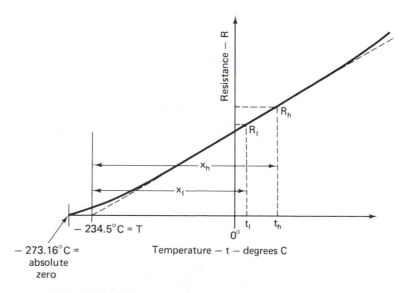

Figure 7-5. Resistance versus temperature diagram for copper.

copper the points R_h, t_h, $-234.5°C$ and R_l, t_l, $-234.5°C$ form similar triangles, we can set up the following relation:

$$\frac{x_l}{R_l} = \frac{x_h}{R_h}$$

Since x_l is the sum of 234.5°C and the temperature at t_l (read as temperature sub low), and x_h is the sum of 234.5°C and the temperature at t_h (read as t sub high), we obtain the following proportion:

$$\frac{234.5 + t_l}{R_l} = \frac{234.5 + t_h}{R_h} \qquad \text{Eq. (7-20)}$$

Since the length of the base line of the triangle is needed, we can ignore the minus sign in $-234.5°C$. Notice also that for copper the inferred absolute temperature is the easily remembered two–three–four–five sequence.

For any resistive material, this equation can be written as

$$\frac{T_i + t_l}{R_l} = \frac{T_i + t_h}{R_h} \qquad \text{Eq. (7-21)}$$

Here T_i, or temperature inferred, refers to the absolute inferred temperature in Table 7-2. Other materials can be found in various references.

Either Eq. (7-20) or Eq. (7-21) can be algebraically transposed so that any one quantity can be determined. Convenient forms are

$$R_h = \frac{R_l(234.5 + t_h)}{234.5 + t_l} \qquad \text{Eq. (7-22)}$$

and more generalized as

$$R_h = \frac{R_l(T_i + t_h)}{T_i + t_l} \qquad \text{Eq. (7-23)}$$

Obviously, these equations can be transformed so that the unknown is conveniently the lower temperature. In this fashion they appear as

$$R_l = \frac{R_h(234.5 + t_l)}{234.5 + t_h} \qquad \text{Eq. (7-24)}$$

Again, the generalized version is

$$R_l = \frac{R_h(T_i + t_l)}{T_i + t_h} \qquad \text{Eq. (7-25)}$$

Some worked examples will show their uses under both conventional and unusual conditions. These equations are quite universal in their utility.

Example 7-7. A coil of copper wire has a resistance of 100 Ω at 20°C. What will be its resistance at 80°C?

Solution: Using Eq. (7-22), substitute and solve:

$$R_h = \frac{R_l(234.5 + t_h)}{234.5 + t_l} = \frac{100(234.5 + 80)}{234.5 + 20}$$

$$= \frac{100(314.5)}{254.5}$$

$$= 123.6 \ \Omega$$

Example 7-8. Find the resistance of the same coil of wire at -20°C.

Solution: Equation (7-24) is convenient here since -20°C is now low.

$$R_l = \frac{R_h(234.5 + t_l)}{234.5 + t_h} = \frac{100[234.5 + (-20)]}{234.5 + 20}$$

$$= \frac{100(214.5)}{254.5}$$

$$= 84.28 \ \Omega$$

Example 7-9. A compound motor series field coil shunt diverter is to be wound with a short length of No. 12 AWG ($\cong$2.0 mm) Nichrome wire so that its resistance at 20°C will be 2.375 Ω.
(a) How much length of wire is required?
(b) What will be the resistance of this coil at 300°C?

Solution:
(a) From the AWG wire size, Table 7-3, we find that No. 12 AWG has 6529.9 CM cross section and from Table 7-4 that 2.0 mm diameter wire, which is the nearest metric size to AWG 12, has a cross-sectional area of 3.1416 (mm)^2.

Since Nichrome wire has different resistance properties than copper, we must compute its resistance per unit length from Eq. (7-19$_E$) or (7-19$_{SI}$). Table 7-2 shows that Nichrome has a ρ of 675 (CM Ω/ft) or 1.122 [Ω (mm)^2/m]. If we divide these values by the appropriate areas, we then have

$$\frac{675 \ \cancel{CM} \ \Omega}{6529.9 \ \cancel{CM} \ \text{ft}} = 0.1034 \ \frac{\Omega}{\text{ft}}$$

and in SI

$$\frac{1.122 \ \Omega \ (\cancel{mm})^2}{3.1416 \ (\cancel{mm})^2 \ \text{m}} = 0.3571 \ \frac{\Omega}{\text{m}}$$

Remember that here there *will not* be a direct check between English and SI units since the two wire sizes are *not* exact equivalents; 2.0-mm-diameter wire has an area of 3.1416 mm^2, whereas No. 12 AWG has an area of $0.7854(2.052)^2$ = 3.307 mm^2.

Since we need 2.375 Ω of wire, this will require that the quantities of ohms per foot and ohms per meter be inverted to produce feet per ohm and meters per ohm. Therefore, we have

$$\text{length} = \frac{1 \text{ ft}}{0.1034 \text{ }\Omega} \times 2.375 \text{ }\Omega = 22.97 \text{ ft}$$

and in SI

$$\text{length} = \frac{1 \text{ m}}{0.3571 \text{ }\Omega} \times 2.375 \text{ }\Omega = 6.651 \text{ m}$$

With some thought, this can and should be checked since neither size is a familiar quantity. The length in feet divided by 3.2808 ft/m times the ratio of wire areas between No. 12 AWG and 2.0 mm diameter should agree closely with the length in meters.

$$\frac{22.97 \text{ ft}}{3.2808 \text{ ft}} \frac{\text{m}}{} \frac{3.1416}{3.307} = 6.651 \text{ m; check}$$

(b) Nichrome would be a logical choice for this type of service because it is stable at high temperatures, so a small-sized element can be used. Furthermore, its resistance change at higher temperatures is reasonably small. Table 7-2 discloses that the Nichrome inferred absolute temperature is $-2250°C$. Using Eq. (7-23), we have

$$R_h = \frac{R_l(2250 + 300)}{2250 + 20} = \frac{2.375(2550)}{2270} = 2.668 \text{ }\Omega$$

Resistances at other temperatures may be readily computed where the inferred absolute temperature is not known if the resistance temperature coefficient is available. This coefficient is

$$\alpha_{20} = 0.003\ 93 \frac{\Delta\Omega}{\Omega°C}$$

for copper. This may be read as coefficient alpha at 20°C equals 0.003 93 Ω change in resistance per ohm per degree change of temperature in Celsius. As the statement infers, this coefficient is only 0.003 93 when the beginning temperature is a room temperature of 20°Celsius (which is 68°F). For other starting temperatures, a different α must be available. Some older references list a few relations for α_{25}. For a general case of any temperature, to any other temperature within reasonable working range, Eqs. (7-22) through (7-25) are most useful.

In the special case of starting at 20°C and with a known α_{20} for the material, the following relation is most convenient:

$$\boxed{\Delta R = \alpha_{20} R_{20} \Delta°C}$$ Eq. (7-26)

where α_{20} is from Table 7-2 for the material in question, R_{20} is the resistance of the sample at 20°C, and $\Delta°C$ is the change in temperature from the base 20°C condition in degrees Celsius.

7-5 ARMATURE CIRCUIT RESISTANCE

The largest single loss in a dc motor or generator when running at rated load is the $I_a^2 R_a$ copper loss of the armature windings and associated brush mechanism. At the same time, the resistance R_a is difficult to determine to the needed accuracy. In fact, some texts suggest ohmmeter or Ohm's law methods on a nonrotating machine and then suggest that the result be "multiplied by 1.2 to 1.6, depending on experience." Obviously, this is not satisfactory for inexperienced personnel, regardless of how well grounded they may be in theory.

The determination of the winding resistance is relatively easy, but the variable resistance of the brushes and the brush sliding contact are normally not measured at all. Instead, a conventional voltage drop value of 1 V positive and 1 V negative is allowed. This total 2 V brush drop holds for any number of poles or parallel paths, and would only be significantly different on a highly loaded short-duty machine that has copper composition brushes, such as an automobile starter motor.

Four methods of determining armature circuit resistance are discussed, starting with the simplest and, unfortunately, the least accurate and finishing with Forgue's method. This method has been developed at Waterbury State Technical College. It is both the most time consuming and complicated, but gives results that allow loss calculations which are accurate to a fraction of a percent error without juggling.

7-5-1 Ohmmeter or Ohm's Law Method. The resistance from brush to brush as determined by an ohmmeter is quick and easy and can be done without any disassembly, but it is not much better than one significant figure accuracy. While taking this value, a bit of careful rotation of the armature shaft will show a very large variation in resistance, especially with a two-pole, two-brush machine. The *lowest* obtainable resistance value is the best because it is obtained with the best brush contact conditions. This value includes the brush volume resistance of the carbon compound and a static contact resistance, which is less than the running resistance.

Very low resistance readings using an ohmmeter are nearly useless, especially when the resistance desired is of the order of 0.1 ohm or less. Therefore, it is desirable to measure as high a resistance as possible by opening up or uncombining parallel circuits if possible. This can sometimes be done by lifting the brushes and reading the resistance between commutator segments that are 180° apart. On larger units there may be equalizer connections within the armature windings which nullify any attempt to isolate the readings to any two points. Also, a commutator which has an uneven number of segments is a wave-wound unit with two parallel paths only and lifting brushes is not helpful. Any reading taken must be multiplied by two if four-pole, three if six-pole, and so on. Use caution because a false reading is possible. If the configuration is not known do not try to simplify.

The readings obtained on the commutator and corrected for number of poles and for winding configuration are for the armature copper only. When determining $I_a^2 R_a$ wattage with this form of R_a, one must also add the 2 V brush drop times the I_a.

7-5-2 Voltmeter–Ammeter Methods.

Either through the brushes or directly on the commutator, the readings are refined by passing a current through the brush-to-brush contact and reading the voltage drop. $R_a = E_a/I_a$ from Ohm's law. The brushes then show some voltage drop, but not as much as if they were moving at a regular rotational speed. If the voltage readings are taken on the commutator directly, the unknowns of brush contact resistance are eliminated, and the resistance of the armature copper only is found.

Since this method is better for finding the low resistance values of a typical armature, better results are obtained than by use of an ohmmeter.

Readings taken from a positive brush to a negative brush, or from the commutator segments that are centered under the brushes, show resistance of the whole armature because of brush-to-brush shunts and/or equalizer connections in multipole armatures.

Obviously, if the resistance of the armature only is determined, the brush drop of an assumed 2 V must be resorted to. Forgue's method shows that this is a reasonable *approximation* only since brush voltage drop varies.

If any readings are taken when the armature is slowly rotated by hand, they must be taken in both directions of rotation and the results averaged. This procedure is used so that the effect of the small generated voltage due to residual field magnetism may be eliminated.

Any of these methods must take into account whether or not commutating fields and/or compensating windings are present. One must simply sort out the connections in the armature circuit. Manufacturer's installation schematics are usually so generalized that they are not much use, and a visual search must be made by tracing and probing to see which connections are brought out and are available. If they are not accessible, current may be fed in at the brush holders for the armature determination and between one brush connection and an outside terminal to uncover whatever field resistances, if any, are present.

7-5-3 Resistance Calculation.

A tedious but reliable way to determine armature circuit resistance is by calculation from the cross section, length, and connection of the windings. This means that the number of coils, turns per coil, average coil turn length, and coil-to-coil connections must be determined by inspection and measurement. A little care can determine the number of turns; a measured piece of string wound as a coil is wound to get coil length and a direct micrometer measurement to obtain a winding cross section are needed. Then the wire size and resistance per foot (or meter) can be determined. Again the number of parallel paths resulting from winding configuration and brush location must also be determined. This procedure is the only reasonable one to follow *before* an armature design is fixed.

Finally, any of the above methods must be corrected for the resistance that will result at normal operating temperature. A reasonable way, if winding temperatures are not known, is to take the direct measurements of resistance by whatever method is chosen after at least a *half-hour* run at *rated load*.

The real difficulty of any of these methods is that the brush voltage drop is only approximated, whereas Forgue's method gives a direct result including whatever brush drop actually exists.

7-5-4 Forgue's Method.

Forgue's method requires that the machine to be measured be driven by a separate prime mover at whatever rpm the data are intended to represent. Thus, the method is only applicable where means of rotating the machine are at hand. In any usual school laboratory, connected pairs of machines are normally available, but they may not be available in an industrial situation. It is worth considering as to whether the benefit of the better result is worth the extra time to connect a suitable prime mover. Furthermore, if the armature circuit resistance is required over a range of rpm, the connected prime mover must be adjustable over the desired range.

The principle involved in Forgue's method is that the residual magnetism remaining in the tested machine must be carefully and entirely bucked out so that the machine may be rotated at its rated rpm and generate *no measurable voltage*. With the normal generator action thus prevented, the armature is simply a rotating resistance with complex rubbing contact through the brushes. The resulting resistance is then determined by the voltmeter–ammeter method. It is not so simple as the concept seems because the amperage required to buck out the field residual magnetism is not constant, because each time a different current is forced through this now passive rotating resistance, it generates a strong magnetic field, which is a cross field in relation to the main field poles. The field residual magnetism distribution is then affected by the armature cross magnetism. This has the effect of redistributing the remaining residual field and requiring a different field bucking current.

Procedure.

(1) Connect as shown in Fig. 7-6. The switching method is unimportant but the low-range voltmeter *must not* see the drop due to test current.

(2) Rotate the machine at the desired rpm or range of rpm.

(3) Buck out the field carefully until no detectable voltage is generated. This will require opposite to the usual field current polarity unless the magnetic field is driven too far negatively, where the original polarity may again be needed. Small field current changes usually suffice.

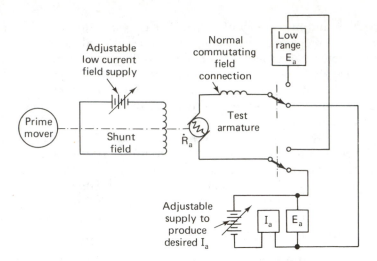

Figure 7-6. Forgue's method, test circuit.

(4) Apply an outside power supply to the *armature* and adjust for the first desired armature current. Record E_a and I_a.

If commutating fields are present they must be in the circuit and carrying the I_a current. Their presence or absence in the circuit makes around a 2-to-1 difference in the armature resistance, although they are of low resistance themselves.

(5) Disconnect armature power supply and recheck to see if the armature generates any voltage. If it does, readjust the field bucking current and retest because the data are invalid.

(6) Good data points have zero generated voltage (or nearly so), both before and after applying current.

(7) Develop data at five or more currents.

(8) Plot in the form shown in Fig. 7-7. Good data are a straight line after lowest currents and project to actual brush voltage drop, as shown.

This method of determining armature circuit resistance by dynamic procedures was first attempted by John Forgue of Waterbury State Technical College (WSTC), class of 1970, while he was engaged in some special laboratory tests. The need to buck out the field residual magnetism was suggested by Domenic Buonocore of WSTC, while Forgue and his lab partners, Ford and Schupack, were attempting some static and dynamic measurement of the armature circuit resistance of a small 1 hp dynamometer. Subsequent tests on all available machine sizes at WSTC have shown the method to give a substantially more accurate armature circuit resistance than more conventional methods.

While in direct search of further proof of Forgue's method, Donald Rosengrant (WSTC 1974) was able to show conclusively the effect of the commutating field on this test. A family of plots was developed under similar conditions. The commutating field was first omitted from the circuit; the commutating field was then placed in the circuit in the correct polarity; finally, it was connected backward.

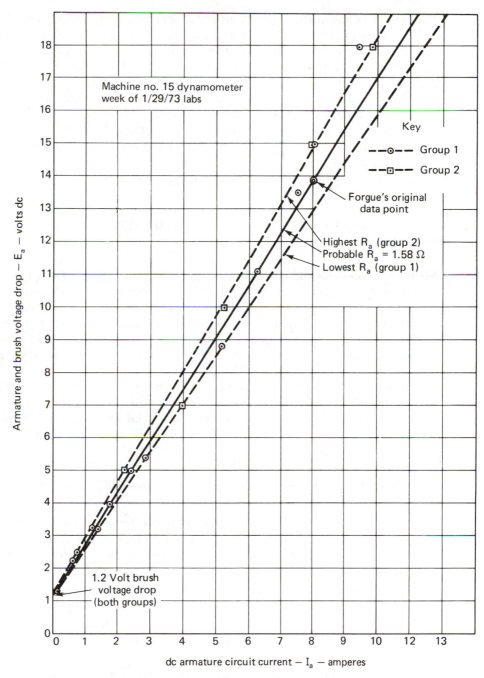

Figure 7-7. Armature circuit resistance tests. Forgue's method, WSTC power lab.

The lowest voltage drop or resistance was obtained using Forgue's method with the commutating field correctly connected and carrying the current. A distinctly higher resistance was obtained for the armature with the commutating field disconnected, and still higher and with scattering data when reversed. All three sets of E_a data were recorded across the brush terminals, regardless of the commutating field configuration. Only the low resistance line representing the correct commutating field circuit connection should be used. The others are false readings.

Since this chapter contains material not obtainable elsewhere, it has been covered in some detail. However, the whole subject of efficient use of energy is steadily increasing in importance. This requirement for critical analysis of losses will probably have a dominant effect on design decisions for the next generation.

7-6 CALIBRATING A DIRECT-CURRENT MOTOR OR GENERATOR

Many times an industrial research situation arises where the measurement of input power or output power in a rotating machine is needed. The deluxe equipment situation is through the use of an electrodynamometer of the appropriate size. A dynamometer is really a specially mounted dc generator in the most universal situation. When a special situation arises where, because of space, time, or budget restriction, a dynamometer cannot be used, a calibrated motor or calibrated generator can frequently be successfully used.

Calibrated here implies "known" or "measured," and in association with a motor or generator means known input versus output power relationships. If it can be shown that, when a motor draws a particular input current and voltage at a particular speed it has a specific output power, then that motor is calibrated under those conditions. Calibration can cover only a few specific operating points or the whole range of operation of the unit.

Many situations can be visualized where this ability to calibrate can be an asset:

(1) Machine tool drive power requirements can be determined over as wide a range as is needed. Here, if the normal motor is direct current, the procedures described below can be applied to the standard drive motor. Alternating-current motor calibrations will be described in later chapters.

(2) Pump or blower drive power requirements can be explored with a calibrated motor. This can be especially important when building service power requirements need to be efficiently applied.

(3) Internal combustion engine tests of efficiency or pollution output can be explored with a calibrated generator.

(4) The calibrated machine is always less costly and takes less space than the dynamometer. The operational flexibility is not much less than the dynamometer.

(5) The calibrated machine is especially useful for repetitive tests; even unskilled operators can record a few meter readings with minimum setup time. A test specification can easily be prepared to cover certain voltage and current require-

ments. This is as easy to work with as specified torque scale readings on a dynamometer.

The calibration of a specific dc machine requires a number of steps, which are covered in sufficient detail in the companion laboratory text that students can actually calibrate a machine. Here the process will be summarized so that it can be understood. Furthermore, an actual calibration will be shown that was the by-product of a student investigation.

Calibration of a dc motor or generator requires the following steps:

(1) Determine the range of speed and power to be investigated. Verify that the available machine is capable of this desired range.

(2) Measure rotational losses over the desired rpm (or radians per second) speed range. For each speed increment chosen, vary the armature circuit voltage so as to cover the expected range of $I_a R_a$ voltage drop. A motor will need rotational loss tests from the normal supply line voltage down to as far as the maximum I_a current will drop the required back emf voltage. Similarly, a generator will require rotational loss tests at higher than the expected output line voltage. This is because a generator is required to generate a high internal voltage so that the internal $I_a R_a$ voltage drop is made good. These effects have been discussed before. The rotational loss test method chosen is, of course, dependent on available equipment and power supply voltages. If a family of curves can be developed at available voltages, the needed higher internal voltage points can be carefully extrapolated. Figure 7-8 shows an actual rotational loss curve for the machine in the worked example. This was nominally a 125 V machine, but a range of from 90 to 130 V was tested. Projection of up to 150 V is shown and has proved to be acceptable.

(3) Measure armature circuit resistance over the full expected range of armature current needed. Cover the range of expected rotative speed. A family of curves may be the result if a wide range of speed is needed. Use the best method available. If another rotative prime mover of the right speed range is available, Forgue's method is the first choice because it produces clearly superior results.

(4) Measure all field windings by appropriate voltmeter–ammeter methods, including commutating fields, series fields, series field shunt bypass resistances, compensating windings, if any, and the shunt field.

(5) Correct all resistances to the expected winding temperatures, or measure them at a stabilized operating temperature, whichever can be reasonably achieved.

(6) Calculate and summarize as shown in Example 7-10.

(7) Prepare curves in the form that facilitates the intended use.

Example 7-10 will help clarify these procedures.

Example 7-10. A 1.5 kW, 125 V shunt generator is to be calibrated to cover an operation range of 1500 rpm (157.1 rad/sec) with load currents up to 10 A.

Solution:
(a) The rotational losses were determined by connecting and operating the unit as a motor. A range of line voltages was used. The speed was adjusted

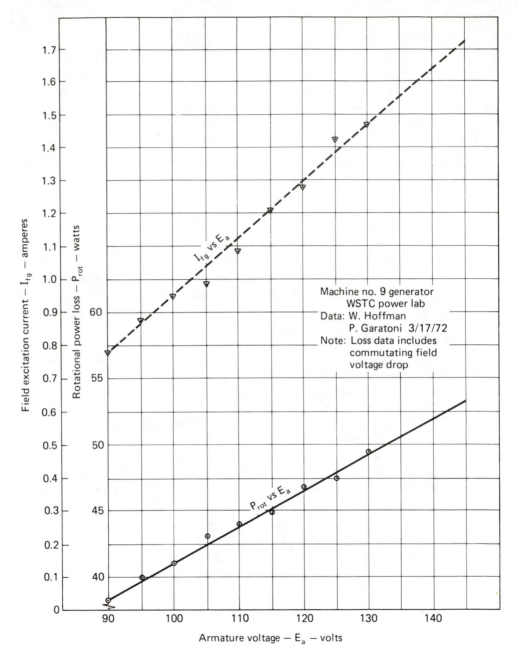

Machine no. 9 generator
WSTC power lab
Data: W. Hoffman
P. Garatoni 3/17/72
Note: Loss data includes
commutating field
voltage drop

Figure 7-8. Rotational loss.

EFFICIENCY OF DIRECT-CURRENT MACHINERY   CHAP. 7

to the desired rate by varying the field strength. The plotted results are shown in Fig. 7-8. These rotational loss data are considered to be a straight line relation, although it would have shown the curved form of Fig. 7-4 had tests been made at higher line voltages (E_a here).

(b) The armature circuit resistance was tested by Forgue's method. The generator was rotated by use of a separate motor, and tests here were made at 1500 rpm (157.1 rad/sec) only due to lack of time. A single data point was taken at 1900 rpm (199.0 rad/sec), which plotted on the same line as the lower base speed. Figure 7-9 is the plotted result of these data. The curve labeled "static" was taken without rotation. Forgue's method or dynamic data were taken over the same range of current but with the unit rotating and the field bucked out. Table 7-6 is smoothed data taken from Fig. 7-9 at 1 A increments.

(c) The commutating field resistance was taken by moving the voltmeter prods and recording the commutating field voltage drop at two different currents. This can be seen in the dotted line of commutating field resistance on Fig. 7-9.

There was no series field or compensating windings present and therefore no data were taken.

Shunt field current data were taken throughout the rotational test and recorded as I_{fg}, or current, field, generator. Since the calibration is intended to cover self-excited operation of the generator, the shunt field and shunt field rheostat losses will be included. In this case the field loss will be taken as $P = IE$ in watts. The E will be considered as line voltage, and I, the I_{fg} that is appropriate at the time. This will be developed later.

(d) The whole basis of calibration of a machine is, in the case of a generator, that the input power must equal the output power plus the total losses. From Eq. (7-3), input = output + Σ losses. Here the task is to tabulate what the losses must be at various output power levels and then calculate what the required input power must be.

From such a calculation and tabulation a curve or family of curves may be produced. In these the abcissa is the output watts at a particular voltage and speed, and the ordinate is total input power in either watts or horsepower as desired.

The procedure to produce these curves with the loss data available in this case is as follows:

Column (1) of Table 7-7, which is the calibrated generator output current in 1 A increments, was used as the first step in constructing the table. Here any reasonable step could be used, of course. The 10 A maximum was arbitrary and was determined by the proposed use of the machine.

Column (2) is the armature circuit voltage drop from Fig. 7-9.

Column (3) is an assumed average armature circuit voltage drop due to the midrange shunt field current; 1.55 A was chosen from a study of the field circuit curve, which is shown on Fig. 7-8. It is not unrealistic to assume this current, but it would obviously have been more accurate to have an actual field circuit current reading with the generator under actual load. Finding that some useful data are not available is not just the lack of student foresight. It happens

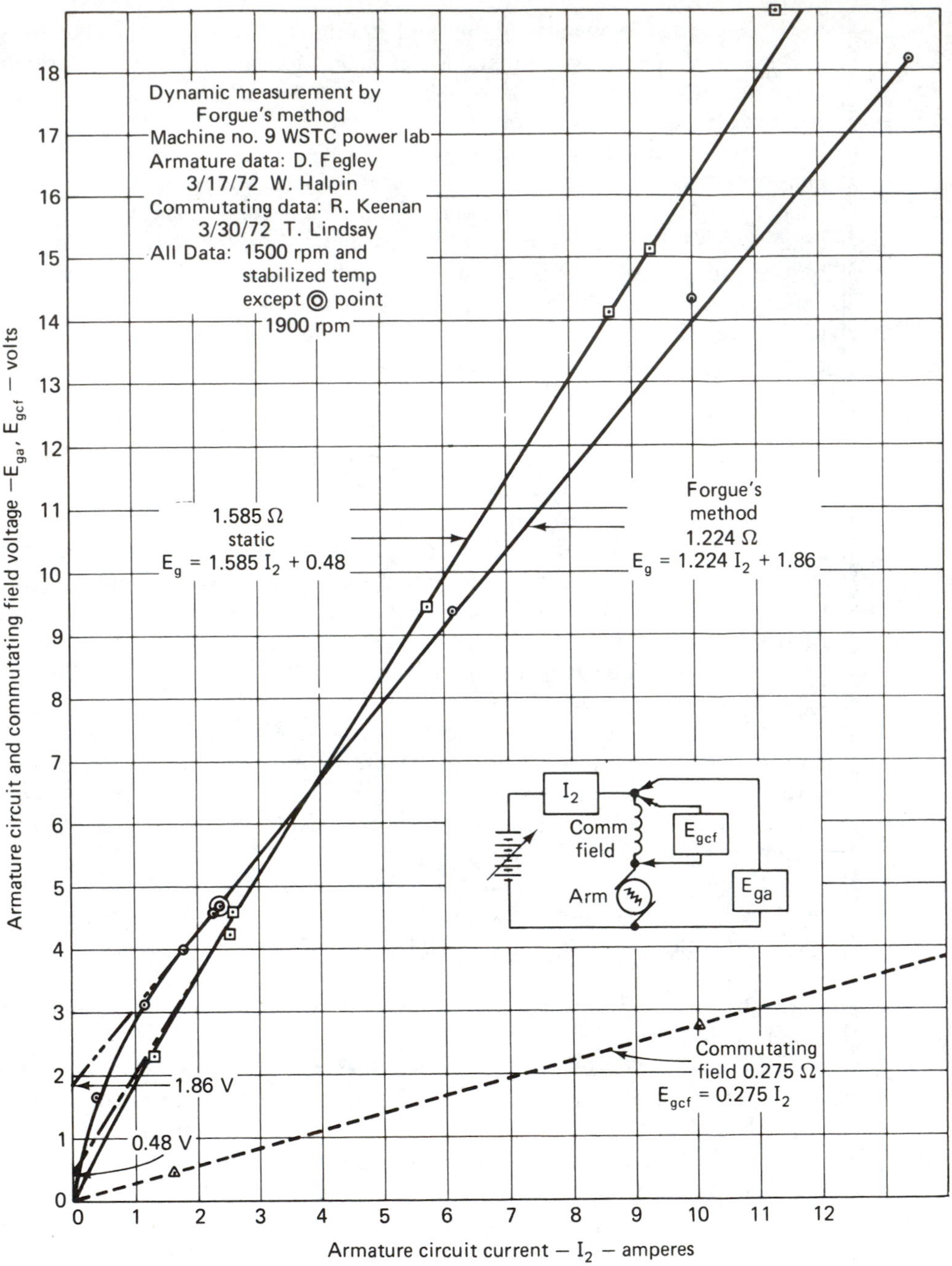

Dynamic measurement by
Forgue's method
Machine no. 9 WSTC power lab
Armature data: D. Fegley
3/17/72 W. Halpin
Commutating data: R. Keenan
3/30/72 T. Lindsay
All Data: 1500 rpm and
stabilized temp
except ◎ point
1900 rpm
1.585 Ω
static
$E_g = 1.585 I_2 + 0.48$
Forgue's
method
1.224 Ω
$E_g = 1.224 I_2 + 1.86$
I_2
Comm
field
E_{gcf}
Arm
E_{ga}
1.86 V
0.48 V
Commutating
field 0.275 Ω
$E_{gcf} = 0.275 I_2$
Armature circuit and commutating field voltage — E_{ga}, E_{gcf} — volts
Armature circuit current — I_2 — amperes

Figure 7-9. Generator armature circuit resistance tests.

TABLE 7-6 ARMATURE CIRCUIT RESISTANCE, STATIC AND DYNAMIC: DYNAMIC DATA BY FORGUE'S METHOD

Static Tests (smoothed)		Forgue's Method (smoothed)	
I_a (A)	E_{ga} (V)	I_a (A)	E_{ga} (V)
0.0	0.0	0.0	0.0
1.0	1.73	1.0	2.89
2.0	3.49	2.0	4.32
3.0	5.22	3.0	5.53
4.0	6.80	4.0	6.77
5.0	8.40	5.0	7.99
6.0	9.99	6.0	9.21
7.0	11.57	7.0	10.43
8.0	13.16	8.0	11.66
9.0	14.75	9.0	12.87
10.0	16.35	10.0	14.10

Static copper resistance = 1.585 Ω including commutating field.	Dynamic copper resistance = 1.224 Ω including commutating field.
Brush voltage drop = 0.48 V	Brush voltage drop = 1.86 V.

Commutating field = 0.275 Ω

NOTE: Data extracted from Fig. 7-9.

frequently in industrial situations. Here we have a satisfactory estimate of some data that has a secondary effect. More on this later.

Column (4) is the sum of the 125 V output, plus the voltage drop due to the armature circuit resistance at the particular load current, plus the voltage drop due to the shunt field current.

Column (5) is the rotational loss that would exist at the particular field flux that is required to develop the voltage in column (4). This rotational loss was extracted from Fig. 7-8. The curve was entered at the abcissa scale reading for the voltage of column (4). Figure 7-8 experimental data stopped at 130 V due to power supply limits at the time. The straight line extrapolation to 141 V is not out of line with the fact that the data taken show no evidence of curve. If higher load currents had been required, a greater spread of loss data conditions would be required.

Column (6) is the actual shunt field amperes required to develop the voltage needed in column (4). This line is shown dotted on Fig. 7-8. Note that this line will show a distinct curve if the generator is approaching saturation. Here a straight line extrapolation to 1.65 A appears safe.

Column (7) is the shunt field circuit loss in watts. Here it is the line voltage times the field current in column (6). This means that the shunt field rheostat loss is included. Again, this is a reasonable situation for a self-excited generator. Some generator specification efficiencies omit the losses in a rheostat by agreement. This item must be considered in light of the intended use.

TABLE 7-7 CALIBRATION FOR 1.5 kW GENERATOR: 1500 rpm (157.1 rad/sec) and 125 V output

(1) I_{out} (A)	(2) Arm. ckt volt. drop from I_{out}	(3) Avg. $E_{a\ ckt}$ drop from $I_f \cong 1.55$	(4) Total arm. volts $E_g = 125 + (2) + (3)$	(5) Rot. loss (W)	(6) Shunt field amps. at E_g total	(7) Shunt field loss (W)	(8) $I_{out} + I_f = I_a$ (A)	(9) Arm. ckt volt. drop at I_a	(10) Arm. ckt power loss (W)	(11) Σ losses (5) + (7) + (10) = (W)	(12) Output power $I_{out} \times 125$ V = (W)	(13) Input power = output power + Σ loss (W)	(14) Input power hp = W/746
0	0.0	1.89	126.9	47.7	1.418	177.3	1.42	3.53	5.0	230.0	0	230.0	0.308
1	2.89	1.89	129.8	49.2	1.466	183.3	2.47	4.88	12.1	244.6	125	369.6	0.495
2	4.32	1.89	131.2	49.6	1.491	186.4	3.49	6.13	21.4	257.4	250	507.4	0.680
3	5.53	1.89	132.4	50.0	1.512	189.0	4.51	7.38	33.3	272.3	375	647.3	0.868
4	6.77	1.89	133.7	50.3	1.533	191.6	5.53	8.63	47.7	289.6	500	789.6	1.058
5	7.99	1.89	134.9	50.6	1.551	193.9	6.55	9.88	64.7	309.2	625	934.2	1.252
6	9.21	1.89	136.1	51.0	1.573	196.7	7.57	11.13	84.3	332.0	750	1082	1.450
7	10.43	1.89	137.3	51.3	1.593	199.1	8.59	12.37	106.3	356.7	875	1231	1.650
8	11.66	1.89	138.6	51.7	1.616	202.0	9.62	13.63	131.1	384.8	1000	1385	1.857
9	12.87	1.89	139.8	52.0	1.638	204.8	10.64	14.88	158.3	415.1	1125	1540	2.064
10	14.10	1.89	141.0	52.3	1.657	207.1	11.66	16.13	188.1	447.5	1250	1698	2.276

Column (8) is a summary of the load current and the shunt field circuit current in order to find the armature circuit current I_a. This column is the sum of columns (1) and (6).

Column (9) is the armature circuit voltage drop from Fig. 7-9 using the Forgue's method curve. The straight line portion of this curve has the equation $E_a = 1.224I_a + 1.86$ V. The armature may be seen to have a copper resistance of 1.224 Ω *and* a brush voltage drop of 1.86 V. This voltage drop is the drop due to all the resistive effects of the armature circuit. The data and the curve show that the commutating field voltage drop had been included. If this voltage were separate, it would have to be separately accounted for. In any experiment the data must show what was and what was not measured. Note that column (9) voltage plus the 125 V line voltage very closely equals the column (4) voltage, thus confirming the estimate of column (3): $125 + 16.13 \cong 141.0$ V.

Column (10) was treated as a $P = EI$ type loss since the total armature current was available at column (8) and the total voltage drop at column (9). An alternative approach is to square the current in column (8), determine the resistance of the circuit from Fig. 7-9, and then calculate the loss as an I^2R loss. The two methods are equivalent.

Column (11) is the summary of the rotational, shunt field, and armature circuit power losses. This is the simple sum of columns (5), (7), and (10).

Column (12) is the actual output power of the machine at an adjusted 125 V level. Here the $P = IE$ power law is used by multiplying 125 V times column (1) current.

Column (13) is the input power in watts. This is the sum of the output power and the total losses.

Column (14) is simply the input power of column (13) converted to horsepower by dividing watts by 746 W/hp.

Figure 7-10 is a plot of these results for the 1500 rpm and 125 V adjusted situation. It simply shows that if, for example, 875 W is delivered to a load (any load) the input power must be 1231 W or 1.650 hp. A similar procedure can be used to develop curves at various rotative speeds. If this is done, then by adjusting the output to 125 V and reading load current and speed the power input may be determined by interpolation between curves.

The same type of procedure will serve to calibrate a motor, except that the relation of output power = input $- \Sigma$ losses must be used. In addition, the total armature circuit voltage will be less than line voltage rather than more. If sufficient voltage range is taken on the rotational loss tests, the unit may be calibrated as either a motor or a generator.

The procedure described here has been dynamometer verified at Waterbury State Technical College. The actual data in Example 7-10 are part of a detailed experiment performed to verify Forgue's method.

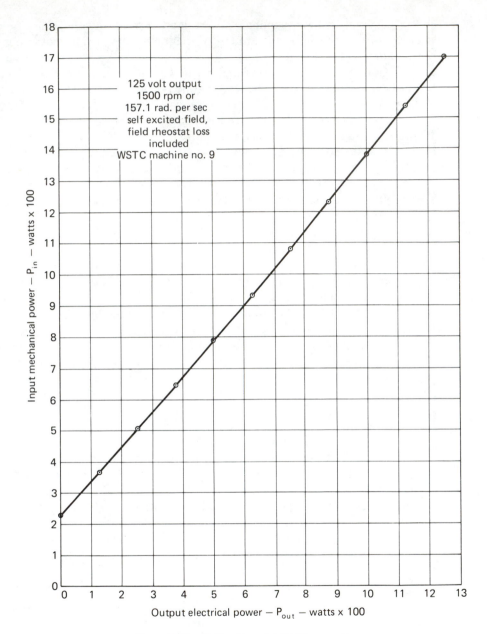

Figure 7-10. Calibration of 1.5-kW generator.

7-7 CONSIDERATIONS IN CALIBRATING A MACHINE

Each different calibration of a motor or generator raises specific questions that must be considered. Section 7-6 discusses a plain shunt machine. A series machine or a compound machine is calibrated with slightly different procedures. There are also modifications to shunt machine procedures that are perfectly acceptable.

It has been established that in large machines such as locomotive traction motors, there is a definite speed effect in the Forgue's method test. This is because a traction motor operates from standstill up to over 2000 rpm (209 rad/sec). In a large machine the copper resistance is comparatively smaller and the inductive effect of coil polarity reversals during commutation is comparatively larger. In this case, the increase in total armature circuit voltage drop from a low speed of 450 rpm (47 rad/sec) to a high speed of 2085 rpm (218 rad/sec) is of the same order as the voltage drop at the 450 rpm (47 rad/sec) situation. This held true for both the General Electric model 752E4 and the Electromotive Diesel (EMD) model D77 motors.

When calibrating a series machine it is necessary to use separately excited fields. It is more convenient to plot and use the rotational losses with the field current as the abcissa rather than armature circuit voltage. The same may be advantageous in shunt machines if the full range of field currents is recorded and used. In a compound wound machine, it may be more convenient to use combined field ampere turns as an abcissa since this is a true measure of the field conditions. It would be necessary here to know the separate shunt and series field currents and their individual number of winding turns so that the proper total flux could be assigned whether cumulatively or differentially connected.

With the techniques herein there is no need to apply any "stray load" corrections. The tests with the GE 752E4 motor connected shaft-to-shaft with the EMD D77 motor gave very satisfying confirmation of these methods. The two machines were connected in a Kapp back-to-back loop. Here one unit acted as a motor and the other as a generator at the same line voltage. The summary of both machine losses corresponded to the input power to within two or three parts per thousand over a very wide range of conditions.

7-8 DIRECT-CURRENT MOTOR EFFICIENCY USING CALIBRATED GENERATOR

An example of a direct-current motor characteristic curve set is shown in Fig. 7-11. This example was developed by use of a calibrated ac generator or alternator. When performed correctly the results are nearly identical to those derived from dynamometer tests.

In the situation shown in Fig. 7-11 the same direct-current machine was used in four different configurations: plain shunt, cumulative compound, differential compound, and series connections. The shunt-field rheostat adjustment was set to achieve 1800 rpm (188.5 rad per sec) at the rated 1.341 hp (1 kW) output and was left at the same point for the two different compound connection tests.

The series-field was severely bypassed in the compound connection runs because the specific machine tested had a large-series field for plain-series operation. As a result, the 120 turn series-field was bypassed with about an 0.08 ohm shunt copper connection. The effect approximately matched an 8 or 9 turn series field. This resulted in borderline speed instability when differential compound connected. Also, a perceptibly larger speed drop compared to the plain-shunt configuration was shown with load when in cumulative-compound connection.

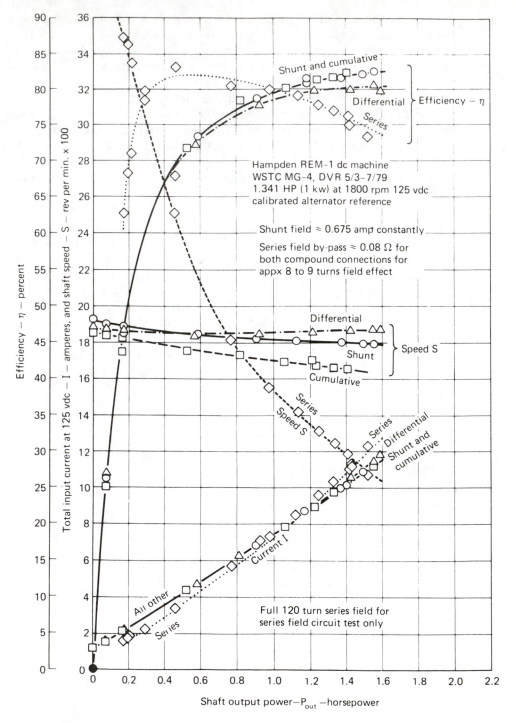

Graph axes and labels:

Left vertical axis: Efficiency — η — percent (90, 85, 80, 75, 70, 65, 60, 55, 50, 45, 40, 35, 30, 25, 20, 15, 10, 5, 0)

Second vertical axis: Total input current at 125 vdc — I — amperes, and shaft speed — S — rev per min. × 100 (36, 34, 32, 30, 28, 26, 24, 22, 20, 18, 16, 14, 12, 10, 8, 6, 4, 2, 0)

Horizontal axis: Shaft output power — P_{out} — horsepower (0, 0.2, 0.4, 0.6, 0.8, 1.0, 1.2, 1.4, 1.6, 1.8, 2.0, 2.2)

Chart annotations:

Shunt and cumulative
Differential
Series
Efficiency — η

Hampden REM-1 dc machine
WSTC MG-4, DVR 5/3–7/79
1.341 HP (1 kw) at 1800 rpm 125 vdc
calibrated alternator reference

Shunt field ≈ 0.675 amp constantly

Series field by-pass ≈ 0.08 Ω for
both compound connections for
appx 8 to 9 turns field effect

Differential
Shunt
Cumulative
Speed S

Series
Speed S
Series
Differential
Shunt and cumulative

Series

All other
Current I
Series

Full 120 turn series field for
series field circuit test only

Figure 7-11. Direct-current motor characteristics, with various field connections.

The series-motor connection used the full series field and no shunt field. There is a whole family of intermediate connections with partial shunt and full or even partially bypassed cumulatively connected series fields but these are not shown here.

Observe that efficiency in plain shunt and the cumulative compound connection shown are nearly indistinguishable from each other. The differential compound shows perceptibly lower efficiency under the higher loads. In this case, the loss of efficiency is probably caused by armature reaction effects with the weaker total field strength. On the other hand, the series-motor connection in this case is more efficient at low loads and less efficient at high loads. This different efficiency curve shape is the result of low rotational losses at low loads because of the weak field. When the field becomes stronger at higher loads the rotational losses increase and thus reduce efficiency.

QUESTIONS

7-1. Define efficiency in broad terms.

7-2. Must the input of any device be smaller or larger than the output? Why?

7-3. What form of power data is most convenient to measure on a dc generator?

7-4. What form of power data is most convenient to measure on a dc motor?

7-5. What is the difficulty in directly measuring the output power of a motor or the input power of a generator?

7-6. Name the major categories of losses in a dc machine.

7-7. What are some components of the rotational losses?

7-8. What are some winding losses in a dc machine?

7-9. Why is the shunt field loss considered separately?

7-10. What is the loss relation at peak efficiency of a machine?

7-11. How is rotational loss determined?

7-12. How are winding resistances found?

7-13. What is the advantage of Forgue's method?

7-14. How may a motor or generator be calibrated?

7-15. Of what use is a calibrated dc machine?

PROBLEMS

7-1. A 5 hp shunt motor draws 34.6 A at 125 V under rated conditions. What is its efficiency?

7-2. A 7.50 kW shunt motor draws 33.8 A at 250 V under rated conditions. What is its efficiency?

7-3. A 20 hp dc motor has 89.3 percent efficiency at rated power. What are its total losses?

7-4. A 3.5 kW motor is 87.2 percent efficient at rated power. What is its input power?

7-5. A 10 hp motor has an input of 8.425 kW while its losses are 925 W. What is its efficiency?

7-6. A 2.24 kW rated motor has 630 W of total loss. What is its efficiency?

7-7. A dc generator is tested when unloaded and disconnected from its prime mover. It is operated at its rated field flux and speed, and the armature then uses 268 V and 0.93 A. What is its rotational loss?

7-8. A dc motor draws 33.8 A and its shunt field uses 1.35 A. If its armature circuit resistance is 0.385 Ω, what is its armature circuit power loss?

7-9. If the motor of Problem 7-8 runs on 250 V, what is its shunt field loss assuming that there is no shunt field rheostat?

7-10. Using the same motor as in Problems 7-8 and 7-9, if the shunt field has a resistance of 125 Ω but with a field rheostat adjusted to maintain the same 1.35 A shunt field current, what would be the power loss of the shunt field itself?

7-11. If the motor of Problems 7-8 and 7-9 has a rotational loss of 216 W and the armature and field losses from Problems 7-8 and 7-9, what is its efficiency?

7-12. (a) What is the maximum efficiency of the motor in Problems 7-8, 7-9, and 7-11?
(b) At what power output is this maximum efficiency developed?

7-13. If an experimental electric vehicle motor is to operate on 125 V dc and produce 20.5 hp at 90 percent efficiency, what is the highest armature circuit resistance that can be considered in the preliminary design? Assume that the armature current is the same as the line current, as a first approximation.

7-14. It is desired to run a rotational loss test on the motor of Problem 7-13 to find its P_{rot} at 2550 rpm, $I_l = 136$ A, and using an assumed $R_a = 0.046$ Ω. What armature voltage should be used?

7-15. From the information in Fig. 7-4, can a rotational loss test be run at 1000 rpm and an armature voltage of 110 V? State the reason briefly.

7-16. What is the resistance of a shunt field coil that is wound with 3350 ft of AWG-24 copper wire at 20°C? Calculate by basic wire resistivity and cross section, and verify by resistance in Table 7-3.

7-17. A shunt field coil is wound with 979 m of 0.50 mm diameter copper wire. Calculate its resistance by basic resistivity and area and verify in Table 7-4. Use a temperature of 20°C.

7-18. If the shunt field coils in Problems 7-16 and 7-17 are operated with a 50°C temperature rise, what is their resistance?

7-19. Using Fig. 7-9, find the armature circuit voltage drop for the generator shown with an armature current of 10 A. Use the Forgue's method curve. If the generator is to be operated at a nominal 125 V and 10 A output, what rotational power loss could be expected?

7-20. (a) Under the conditions of Problem 7-19, determine the total losses of the generator in Figs. 7-8 and 7-9.

(b) Determine what input power will be needed to drive the generator for the 125 V at 10 A output level.

(c) Compare the value found with Fig. 7-10.

Answers.   **7-1**, 86.2%; **7-2**, 88.8%; **7-3**, 1787 W; **7-4**, 4014 W; **7-5**, 89.0%; **7-6**, 78.0%; **7-7**, 249 W; **7-8**, 407 W; **7-9**, 338 W; **7-10**, 228 W; **7-11**, 88.6%; **7-12(a)**, 88.7%, **(b)** 8707 W or 11.7 hp; **7-13**, 0.0459 Ω; **7-14**, 119 V; **7-15**, No; **7-16(a)**, 86.0 Ω, **(b)** 86.0 Ω; **7-17(a)**, 86.0 Ω, **(b)** 86.0 Ω; **7-18**, 103 Ω; **7-19**, 52.3 W; **7-20(a)**, 448 W, **(b)** 1698 W, **(c)** 1700 W.

8

Basic Control Components

As we have seen up to this point in the text, there are many factors that must be considered in the design of machines. The production of motors and generators can be considered to fall into one of two general categories: 1) general purpose machines and 2) custom-built machines. Many motors and generators, usually of the smaller frame sizes, are designed and produced with general applications in mind. Others are "custom designed," satisfying very specific operating requirements. In either case, it is important to realize that the machine is an instrument of work. It performs a task. Just how well it performs this task depends on the quality of design, materials, and production. But that is not all: the machine must be controlled. The control circuit must be able to selectively start and stop, control speed/torque operating points, and/or reverse so that it satisfies the accomplishment of the task for which it was designed and selected. Ultimately, the goal is to control a motor or generator for effective task performance at high efficiency.

8-1 CONTROL

Control of electric machinery is accomplished by first evaluating what control parameters are required. As will be seen in Chapter 9, Section 1, there are innumerable questions to be answered so as to delineate what, specifically, is needed in the *control scheme*. Once the task of the control scheme is known, a control circuit can be designed and produced to control the machine itself. It can be said that the control scheme or system is the "brains" of the machine. It is responsible for starting, stopping, accelerating, retarding, and reversing the machine. It is also responsible for machine protection, a very important consideration.

A simple control schematic is shown in Fig. 8-1. A brief study of this figure reveals some symbols that represent electrical/electronic components which, when connected as shown, will control the behavior of the motor they are serving, and together their task will be accomplished. Figure 8-1a is what is called a *conventional control* scheme, in that it uses electromechanical components. Figure 8-1b can be considered to illustrate what will be called *semiconductor control*, in that it utilizes power semiconductors to perform the same control functions as in Fig. 8-1a.

The function of this control circuit is not the concern of this chapter and will be discussed in detail in Chapter 9. It is important, however, to recognize the symbols for various components and, in fact, to be familiar with the basic types and functional uses for which they are designed and suited. Toward this end, this chapter will introduce the student to basic characteristics and symbols of devices used in both conventional and semiconductor control of electric machines.

The simplest control for a motor is a plain on–off switch. This is rarely resorted to except in the very small appliance field. In an autombile a motor-driven accessory may be controlled with an on–off switch, but this is usually backed up with a fuse for motor protection. When a motor becomes large enough that an internal failure could result in a serious short circuit, and where the motor is a part of an expensive appliance, a *manual starter switch* is used. The fractional-horsepower manual starter switch combines the manually controlled switch contacts with a thermally operated circuit breaker. Electrical codes which are based on protection from fire hazard require this type of switch for small motors that are started automatically. A domestic built-in ventilation motor will be protected this way.

The industrial motor control field has evolved a standardized schematic symbol language. These symbols can be combined in convenient diagrams for the interconnections needed in wiring the device. The resulting diagram can be followed for its actual function, if sufficient familiarity and knowledge are at hand. A diagram form that is designed to allow easy understanding of the control circuit paths is known as a *schematic diagram* or *schematic*.

The schematic and the wiring diagram are equivalent but are not arranged in the same fashion, and are really intended to convey different information.

The symbols used in the diagrams and schematics vary according to the industry that is involved. It will be seen that the same device may have two or three different symbols. The electronic field does not use the same symbols as industrial controls, even though the same device is being used for the same function. These differences will be explored in this text, because no student really knows where his working career may lead. It is well to have some familiarity with both major symbol languages.

Returning to the simple on–off switch, it may be shown in a number of ways (see Fig. 8-2). Note that the symbols may be shown vertically or horizontally without change of meaning. Unfortunately, the industrial switch contact symbol is easily confused with the symbol for a capacitor. The capacitor symbol, when on an industrial diagram, has one plate shown distinctly curved and should always carry a C or CAP designation.

When a *starter switch* is shown, an accompanying symbol for the thermal

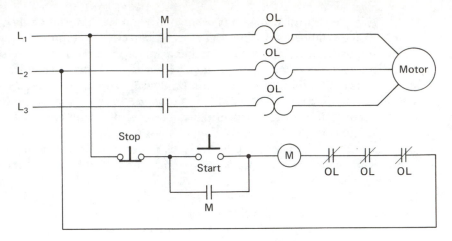

(a) Conventional control

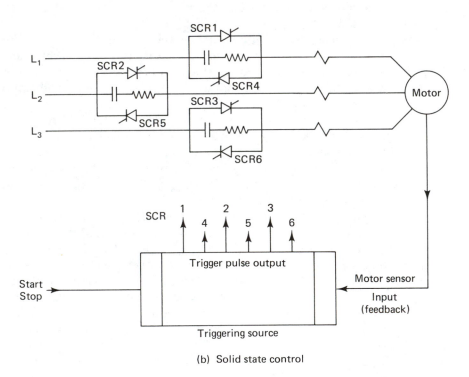

(b) Solid state control

Figure 8-1. Examples of conventional and solid state control.

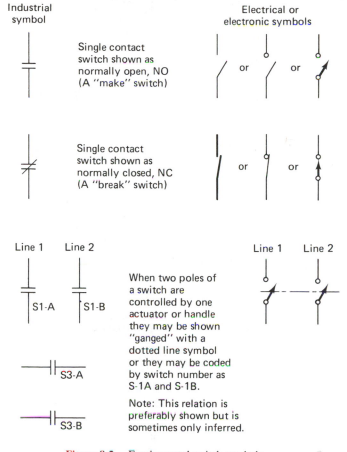

Figure 8-2. Fundamental switch symbols.

overload device is drawn close by the switch contact (see Fig. 8-3). This simple type of control is not used with motors larger than $\frac{3}{4}$ or 1 hp (1 kW). The larger motors need starting current control, which requires added components.

There are two basic families of components: *primary devices* and *pilot devices*. A primary device connects the load to the line or, in this case, the motor to the power source. The simple switches above are primary devices. Pilot devices control or modulate primary devices and do not themselves carry the main line current or voltage. These two families will be explored next.

8-1-1 Primary Control Devices. Primary devices include, but are not limited to, the following types of devices:

(1) Hand-operated *mechanical switches*. This may be the knife switch; a blade-like element moves between and is clamped by fixed spring jaws, or the contacts may be pressed together.

(2) *Rotary switches*: one or more metal segments are rotated into or away from contact with spring-loaded fixed contacts. These are frequently called *drum switches*

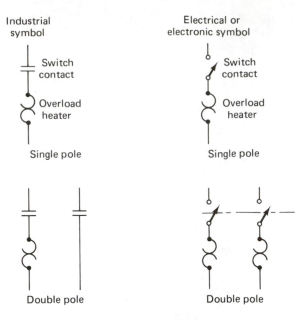

Industrial symbol

Switch contact

Overload heater

Single pole

Double pole

Electrical or electronic symbol

Switch contact

Overload heater

Single pole

Double pole

Note: The overload heater element that trips the release of the contacts may be incorrectly shown

as  rather than

The first symbol is more correctly used to show an element that burns out with overload, thus requiring replacement. The second is intended to show an element that can be manually reset. These switch elements are intended for repeated manual operation. A pure circuit breaker is different, as will be shown.

Figure 8-3. Manual starter switch symbols.

because of their roughly cylindrical construction. Rotary switches may have their fixed contacts arranged like the numbers on a clock face. These contacts are connected by a moving contact arm. The result is called a *face plate rotary switch* to differentiate it from the drum rotary (see Fig. 8-4).

(3) *Magnetic contactors* are electromagnetically actuated spring-loaded contacts, usually of at least two circuits and frequently as many as four or five separate circuits. A contactor requires some form of pilot device for its actuation. This is really a special power-handling form of a relay. A relay is defined as an electromagnetically operated switch. In the motor control field, if the device is used for opening or closing the main power lines, it is a contactor. Conversely, if it is a pilot device used to control the operation of a contactor, it is a relay. The two different functions may be performed by the same device, or, stated another way, the fields of application overlap. Figure 8-5 shows a representative magnetic contactor and its symbol.

Note: Any reasonable
number of contacts
may be used.
Off position may not
have contact button.

Electrical or
electronic symbols

Or

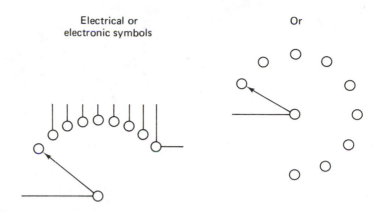

Figure 8-4. Face plate rotary switch symbols.

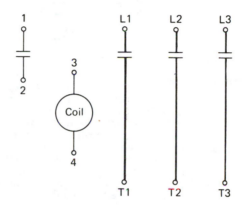

There may be two, three and more main contacts

which are normally open ⊥ or normally closed ⊥⊥

or some of both types.
The smaller contacts at 1 and 2 are auxiliary
contacts which may be used to keep the
actuating coil in operation after outside
initiation. These are "holding" contacts in
this sense. The coil may be identified by agreed
initials as MF or MR, or other, and may be
shown away from direct association with the
contacts.

Figure 8-5. Magnetic contactor symbols.

(4) A *starter switch* may be manually or *magnetically operated*. In either case,
a starter switch is a combination of a contactor and appropriately matched thermal
overload devices packaged in a coordinated and standardized assembly. Figure
8-6 shows a magnetically operated starter switch that combines the features of a
magnetically operated contactor with matching thermal overload devices. Note
that in Fig. 8-6 all three main load-carrying lines have thermal overload heaters.

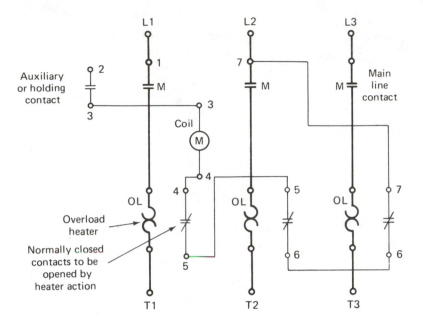

Typical three line magnetic starter switch for ac or dc
service, $L_3 - T_3$ would not be used for dc or single
phase ac. Note that if any line overheats due to high
current its associated normally closed contacts will
open and break the coil-holding current. The external
pushbutton actuation is not shown. This is really a
magnetic contactor with matching associated overload
protection. A manual starter will have mechanical rather
than electromagnetic actuation and mechanical overload
release.

Figure 8-6. Magnetic starter switch symbols.

Many existing three-line controls have only two thermally protected lines, but these
are fast being replaced owing to local electrical codes or to OSHA (Occupational
Safety and Health Act) regulations. Note especially that any *one* of the overload
(OL) heater elements can serve to break the normally closed auxiliary contact that
is associated with it. Since all the OL contacts are in series, the opening of any
one of them causes the contactor coil to relax and the contactor to "drop out."
The OL *contacts* do not carry motor current. The *heater elements* do carry motor
current.

A manual starter switch of similar size will be actuated by a mechanical
motion. This is usually a large mechanically linked push button. When closed,
the circuit is latched and held. Either the operation of the off button or the effect
of one of the overload heaters will serve to mechanically unlatch the mechanism
and allow the main contacts to drop out.

The overload heater symbols are the same as are used in the fractional-
horsepower manual starter switch, but their actual function is not quite the same.

In the fractional-horsepower situation, the line circuit was broken by the action of the heater releasing the latch of the main contacts. In a magnetic starter switch the OL heater usually serves to melt a capsule of low-temperature eutectic solder, which allows a ratchet wheel to rotate. The rotation releases the small auxiliary contacts, which open the coil circuit. The advantage here is that the thermal heater strip is readily changed to cover a wide range of trip-out currents. At the same time, the eutectic solder pot will refreeze and solidify after a minute or so and allow operation without changing any part. This type of mechanism reliably holds its calibration over a long period of time.

(5) *Current-limiting resistors* are primary devices by virtue of handling the main armature circuit current. They must be of appropriate resistance and wattage for the task. The resistors may be inserted or removed from the circuit by contactors or by a face plate rotary switch.

(6) Protective devices such as *fuses* or *circuit breakers* are used to break the current flow to the motor if a trouble develops that results in excessive current flow. Larger motors usually require both types of protection since each has different characteristics.

A fuse has very small mass and will heat to its melting point very rapidly if a higher-than-rated current is flowing. Conversely, a circuit breaker or related overload device has a greater mass of material and will react more slowly than a fuse. The result is that the circuit breaker is used to protect against a service overload and will open the circuit with a prolonged current of around 125 percent of normal. This value is, of course, determined by the particular circumstances. A typical starting current might be as much as 150 percent or even 200 percent of running current. The few seconds of this current level that are needed for starting will not "blow" or release the 115 or 125 percent breaker.

On the other hand, a fuse that is calibrated for a margin above the normal starting current will still blow out quickly if a serious short circuit is encountered.

In this way the long-term damage to insulation that is the result of operating a normal motor at a steady overload is limited by the circuit breaker or the thermal overload devices in the typical magnetic starter switch. At the same time, the fuses will prevent serious or even disastrous damage from a breakdown or internal short circuit.

Some installations have heat sensors in the windings to detect a heat rise from loss of ventilation. If the heat-sensitive unit is arranged to open the main lines, it is a primary device. This type is represented by the typical "Klixon" circuit breaker, which is so widely used in fractional-horsepower motors. In this case the short circuit protection is the remote fuses in the household circuit. If the heat sensor in the motor is not a part of the primary windings, it is a pilot device and not a part of this discussion. Figure 8-7 shows circuit symbols for current-limiting resistors, circuit breakers, and fuses.

As in the case of the circuit breaker with magnetic trip coils, there are coil elements in series with the main load-carrying wire lines. If the coil actuates some device in the main lines, it is part of a primary device. If its action results in the operation of a device that is not a part of the main lines, it is then a pilot device.

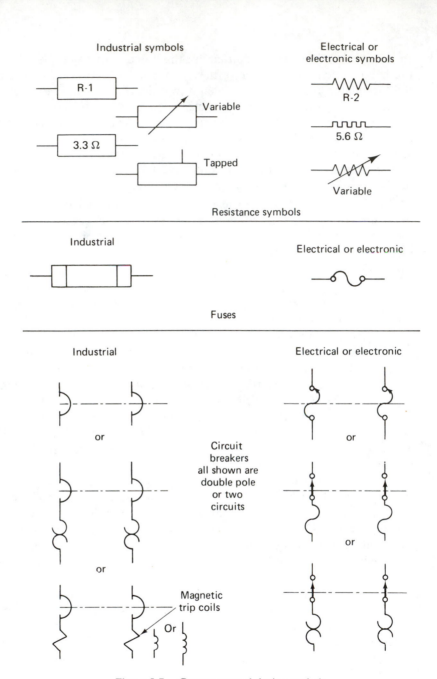

Figure 8-7. Current control device symbols.

Sometimes the distinction is not clear. However, the nomenclature is not as important as the understanding of the function of the device in question. Usually, if the function of a coil is to operate a control device, whether primary or pilot, the coil is shown as a circle, such as the symbol in Fig. 8-5. Exceptions would be the current coils in the magnetic trip coil version of a circuit breaker, as shown in

Fig. 8-7. In this case the coil is shown as one cyclic zigzag or perhaps two loops of a coil symbol.

8-1-2 Pilot Control Devices.
Pilot devices include but are not limited to the following types of devices:

(1) *Pushbutton* switches. The first link in the chain of control elements is frequently a pushbutton that can be forceably or gently jabbed with a finger. This device simply closes or opens a contact against the force of a light spring. As soon as the force of the finger is removed, the internal spring returns the movable element to its original position. This is said to be *momentary* action. If the button is mechanically linked to another, as in a start–stop combination, so that when pressed it stays in the on position, it is said to be *maintained*. Sometimes a maintained pushbutton set is two combined buttons mechanically linked. It can also be one button with a sequential on–off cyclic relation.

Most frequently a pushbutton is single and independent. Usually, the pushbutton has dual contacts so that it can be used either normally open (NO) or normally closed (NC), or even as both. When both contact sets are not actually installed, some mechanical provision is provided so that conversion can be easily accomplished. The pushbutton electrical connection is usually either a double make or double break. This means that two large fixed contacts are connected by a robust movable link. This link is electrically neutral until actuated. By using this construction, the movable link does not need a flexible pigtail connection. It also has two effective contact gaps so that it can make or break a reasonable voltage with a small motion; it is very rugged and long lived. These industrial pushbutton units should not be confused with such toy-like apparatus as a doorbell pushbutton (see Fig. 8-8).

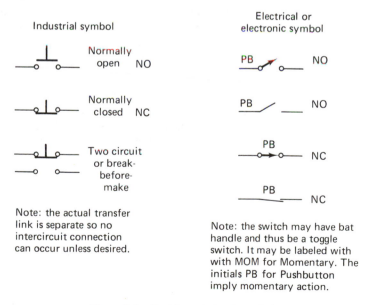

Figure 8-8. Pushbutton circuit symbols.

Processes or states of being are used to operate switch contacts in many ways as pilot devices. Fluid level, pressure, temperature, fluid or gas flow, location or proximity, such as machine tool table travel position or crane hook position, or a safety condition, such as requiring both hands to be clear of a press, can be used to actuate switches. Again there are many, many variations.

(2) *Float switches* are operated by a fluid level in a tank or process channel. They may simply be operated by rising water level to turn on a sump pump. The actual switch can of course be normally open or normally closed, or have both types of contacts. The actuation may be with rising fluid level or dropping level, or both combined. See Fig. 8-9 for some symbol variations.

(3) *Pressure switches* may be actuated by increasing or decreasing pressure. The pressure involved may be low or high, and of course this would affect the construction. A typical version of a pressure-operated switch that many will recognize is the one that operates an air compressor motor in a gasoline station air supply. Another is the switch that operates the motor in a household water pressure system. There are many varieties. See Fig. 8-9 for symbols.

(4) *Temperature switches* may be actuated by increasing or decreasing temperature. There are many ranges and applications. Of course, either float, pressure, or temperature switches may have normally open, normally closed, or both types of contacts. These types of units could also be used as primary devices rather than pilot devices if the current to be switched was reasonable for the switch contacts. All these various types of switches are available in sealed enclosures for hazardous conditions.

(5) *Flow switches* may be actuated by increasing flow or by decreasing flow in relation to a set flow point. The fluid measured may be liquid or gaseous. The actual phenomenon being measured is frequently the difference between two pressures measured in specially shaped portions of the tubing or ducting, such as a venturi section. In this case, the switch actuation is by a change of the pressure differential. Since the basic phenomenon in question is flow, the switch is considered to be and is symbolically represented as a flow switch. Again see Fig. 8-9.

(6) Location, proximity, or position is generally detected by *limit switches*. It is sometimes necessary to actuate a switch within a travel location as small as 0.001 in. (or about 0.025 mm). In other situations, such as in crane hook travel, a location of plus or minus $\frac{1}{32}$ in. (or perhaps 1mm) is sufficient. For safety and reliability considerations, or for maximum accuracy for the given type of switch, or to accommodate a reasonable overtravel of the given motor-driven mechanism, it is sometimes necessary to use a switch in the opposite of its normal sense. For example, the circuit may require that the switch be opened at the end of the desired travel, but a normally closed switch may not allow sufficient overtravel. In this case a normally open switch may be held closed with a spring and be forced open against the spring at the desired limit point. This distinction of use can be shown by the industrial symbol used, whereas the electrical or electronic switch would require special notes. See Fig. 8-9, which shows this distinction.

(7) *Foot-operated* switches are a special form that is used in industrial control

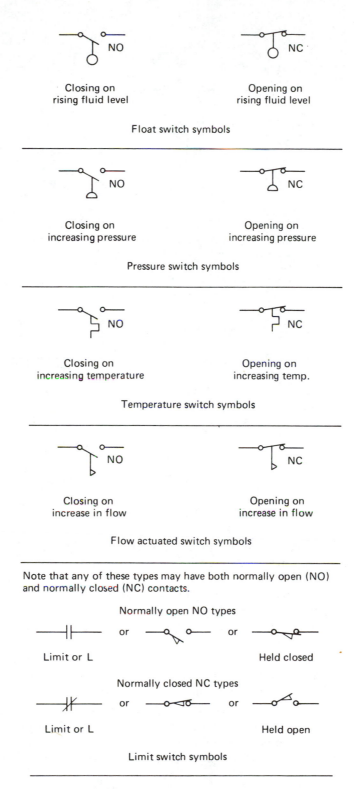

Closing on
rising fluid level

Opening on
rising fluid level

Float switch symbols

Closing on
increasing pressure

Opening on
increasing pressure

Pressure switch symbols

Closing on
increasing temperature

Opening on
increasing temp.

Temperature switch symbols

Closing on
increase in flow

Opening on
increase in flow

Flow actuated switch symbols

Note that any of these types may have both normally open (NO)
and normally closed (NC) contacts.

Normally open NO types

or   or

Limit or L

Held closed

Normally closed NC types

or   or

Limit or L

Held open

Limit switch symbols

Figure 8-9. Pilot device symbols.

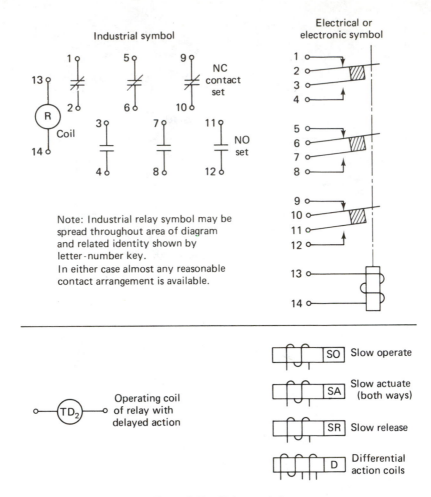

Note: Industrial relay symbol may be spread throughout area of diagram and related identity shown by letter-number key.
In either case almost any reasonable contact arrangement is available.

Figure 8-10. Relay symbols.

when both hands are occupied. The switch can be related to a pushbutton or a limit switch, but a special enclosure is required for reasonable life.

A good deal of space has been devoted to special forms of switches, but a fluent understanding of the symbols and their use in diagrams and schematics can help a great deal in reading and understanding motor control information. It is unfortunate, but many industrial motor control diagrams are designed to show an electrician how to connect the circuit and do not help in the understanding of the circuit involved. Unfortunately, many otherwise well-trained and skilled technicians do not know the special motor control industrial symbols and so cannot quickly translate information even with a solid electrical or electronic background.

(8) *Relays*, as mentioned when discussing contacts, are magnetically operated switches. The connotation here for "relay" is as a pilot device, although the same physical unit could be used as a primary contactor if the motor to be controlled were small enough that its current did not exceed the relay contact ratings. There

are all sorts of relays as far as size, shape, and appearance are concerned. Almost any conceivable contact arrangement has been or could be used. The real advantage from a motor control standpoint is that a circuit may be manipulated in many ways, with much interconnection over reasonable distances, without routing the high-current motor primary control cable from one control element to another.

Many varieties of relays are available from stock that have adjustable and reproducible operating parameters. Time delay in operation or release or both is readily obtainable. Pull in or drop out at particular desired current levels or voltage levels is also obtainable. The result is that many control functions, some of them not yet invented, can be readily accomplished by judicious use of relays. Contacts on a relay can serve to *interlock* functions so that prescribed sequences of operation are automatically obtained. On the other hand, control interlock functions are frequently designed so that undesired or unsafe operation sequences are "locked out" or unobtainable until the necessary conditions and sequences are in existence. This saves much operator training, much potential equipment damage is automatically avoided, and certain categories of accidents can be positively prevented. The safety and sequence aspects of a control can be literally designed in by judicious use of relays. A few of the many relay uses will be touched on when some of the specific controls are described. Figure 8-10 shows a few relay arrangements in the different symbol languages. The understanding of relay logic circuits requires that relay diagrams can be followed.

(9) There is a whole field of *sensors* or *pickups* or *transducers* that may be used as pilot devices. In general these terms mean a device that converts one phenomenon into another. For example, a fluid level may be sensed by a capacitor device; the fluid becomes the dielectric of the capacitor as it displaces the air or gaseous vapor in the tank. The resulting change in capacitance can be used in a tuned circuit to change a current flow and operate a relay that is sensitively adjusted. Here, one pilot device operates another. Radio signals can operate tuned circuits, which in turn operate relays, as in a garage door opener. Here again the field is literally wide open, and many new systems will be invented for specific conditions. From a motor control situation, it is necessary to know what phenomenon is being detected and at what level it is required to achieve a control function. With the desired phenomenon electrically detected, it is frequently necessary to operate some intermediate pilot device such as a relay. Once a relay is operated, the rest of the control can be quite conventional.

8-2 BASIC SEMICONDUCTOR CONTROL DEVICES

Control of electric motors has been changing in recent years due to technological advances in the design and fabrication of semiconductor devices. More and more, the control of motors is being accomplished utilizing two-, three-, four-, and five-layer semiconductor devices which can handle large amounts of power and be connected and packaged so as to take up less space and have improved reliability over the conventional primary devices. An important reason for improved reliability is that, unlike the conventional devices, solid state devices have no moving parts. Overall cost considerations (future maintenance, overall efficiency, etc.) may also favor solid state control.

Although solid state control is termed "state of the art" and is currently being implemented to control the smallest to the largest machines, it will take many years before solid state will completely replace all the conventional forms of control that are in place today. It is extremely important, however, that the rotating machinery student become familiar with the basic characteristics of the solid state devices which will undoubtedly be encountered routinely in the newer machine control design. It could be said that "as went the vacuum tube so goes the conventional control."

The scope of the material presented here is aimed at the technology student and will overview the devices at this level. The references should be consulted for those who require an in-depth study of these semiconductor devices.

8-2-1 Power Diodes. A diode is a two-terminal semiconductor device which ideally acts like a short circuit for current flow in one direction and an open circuit in the opposite direction. One terminal is connected to an N-type material, called the cathode, and the other to a P-type material, called the anode, as shown in Fig. 8-11a, along with its corresponding schematic symbol. Essentially a diode acts like a switch. The on or off condition of the diode switch is determined by the magnitude and polarity of voltage applied across the device. This condition is known as biasing. If the anode-to-cathode voltage is positive, see Fig. 8-11b, the diode is "turned on" and is *forward biased*. The diode conducts and has only a relatively small voltage drop across it, ideally zero volts while the forward current,

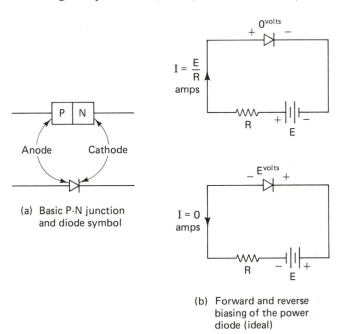

(a) Basic P-N junction and diode symbol

(b) Forward and reverse biasing of the power diode (ideal)

Figure 8-11. Ideal diode.

BASIC CONTROL COMPONENTS   CHAP. 8

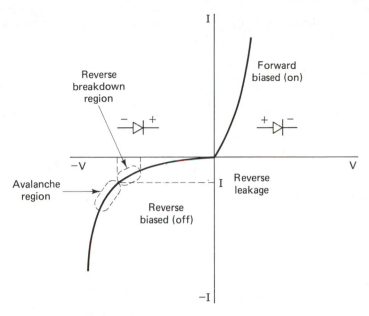

Figure 8-12. General forward and reverse characteristics of a diode switch.

I, is ideally E/R (see Fig. 8-11b). If the anode-to-cathode voltage is negative, the diode is "turned off" and is said to be *reverse biased*. The diode blocks current flow and has a large voltage drop ideally equal to the source voltage, E.

(1) *Characteristics and limits of operation.* For a given power diode, the two most critical ratings are the maximum steady state forward current and the maximum reverse breakdown voltage. These values are critical from a cost standpoint because a power diode must be operated close to these points, as the cost is proportional to these ratings. It would be expensive, say, to use a diode rated at 550 V reverse breakdown voltage in an application where it would only be subjected to a reverse voltage of 120 volts. Similarly, a diode should be operated close to its steady state forward current rating. For this reason, the diode is chosen and "matched" to the circuit application with the manufacturer's specifications of a particular diode clearly in mind.

Figure 8-12 shows a general behavioral characteristic of a diode. It should be noted that the first quadrant depicts the VI relation of the forward-biased (turned on) condition, and the reverse-biased (turned off) condition is noted in the third quadrant. Note that as the forward current increases, there is a corresponding, almost proportional, increase in the forward voltage drop. In the reverse direction it can be seen that as the voltage increases to the knee of the curve (reverse breakdown area), the value of reverse current can be appreciable. This current is called the reverse leakage current. Since the specifications of a given diode include the values of forward voltage drop and reverse leakage current, at rated forward and reverse operating points the designer can accommodate these variables in the application, and the power diode does an effective job.

When considering the implementation of power diodes in the switching control of motors or other inductive loads, it is necessary to consider the transients that occur in terms of the reverse breakdown voltage and forward current of the diode. As can be seen in Fig. 8-12, if the reverse breakdown voltage is exceeded (even momentarily) the reverse leakage current will rapidly begin to increase. This phenomenon is called the avalanche effect, and is caused by exceeding the reverse breakdown voltage. With this condition, the diode switch can obviously conduct appreciably, and the effectiveness of the diode as an open switch is thwarted. The reverse leakage current can increase to the point where, depending on time duration, the power rating of the diode is exceeded and burn out results. The concern of transients with regard to forward current is basically limited to the power rating of the diode and the VI forward characteristic.

(2) *Power diodes in series.* In some switching control circuits, very high voltages are encountered across diodes in the reverse direction either by design, as in high voltage dc control, or due to inductive kicks that occur due to the switching action itself. This condition, if only one diode is used, can cause avalanche and possible permanent damage or destruction of the diode if the specified reverse breakdown voltage is exceeded. To accommodate high reverse voltage from an economical and practical approach, two or more power diodes, usually of similar type and specification, are connected in series. In this case, the reverse breakdown voltage capability is essentially the sum of the reverse breakdown voltage of each diode connected in the series connection.

As can be expected, diodes with the same manufacturer's number and specification sheet do not have exactly the same characteristics and can vary because of slight production variations. If, for example, the peak reverse voltage of a particular diode is rated at 800 volts, the diodes with this specification will meet (or exceed) this rating. If the diodes are operated individually at this voltage, their respective reverse leakage currents will be different, as detailed in Fig. 8-13. The forward characteristics, generally, are very closely matched and can be considered the same. The forward voltage drops are about the same for the same forward current, which will occur when connected in series, and the forward current rating is the same for both units.

Figure 8-13 illustrates two diodes of the same manufacturer and number connected in series with a reverse voltage applied. Since the reverse leakage current must be the same due to the series connection, it is noted that the reverse voltage drops are unequal due to the differences in reverse characteristics, but the voltage rating of the diode pair will be larger than that of one diode alone, which is the obvious advantage of the series connection. It would be ideal if each diode had exactly the same reverse characteristic so that they each would have one-half of the impressed voltage E across them and, consequently, each could operate at the optimum point; i.e. close to the maximum reverse voltage rating. This is not usually the case, so it ends up that one or both of the diodes are not being operated efficiently.

Although it is possible to purchase "matched" diodes from manufacturers (i.e. diodes with the same VI characteristics), the cost is prohibitive from both a production and a maintenance perspective. Fortunately, the problem can be han-

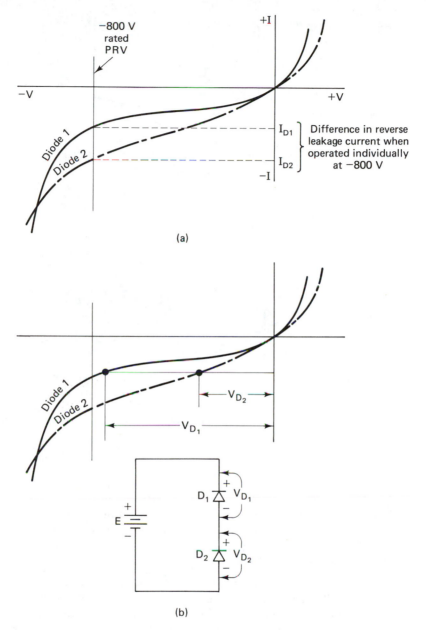

Figure 8-13. Differences in series diodes.

dled with a practical approach by shunting each diode with a resistor to balance out the voltage division.   Again, ideally the resistor values could be individually selected for each diode circuit so as to optimize the voltage across each diode, but usually the resistances are of the same value so as to "average" the voltage division. A trade off between the ideal and the practical is achieved, and production cost and maintenance headaches are minimized.

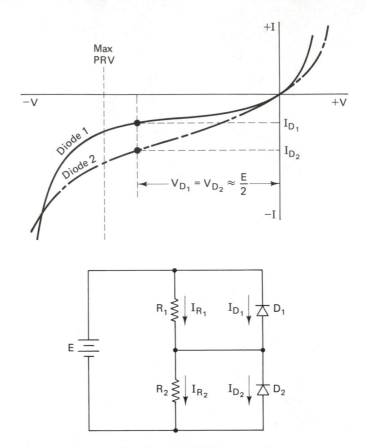

Figure 8-14. Balancing PRV in series diodes.

Figure 8-14 illustrates how added resistors benefit the reverse voltage division problem. Resistors R_1 and R_2 must be of high ohmic value so as not to undermine the off condition of the circuit and yet balance out the different dynamic resistances of the individual diodes so that D_1 and D_2 operate as nearly as possible with $E/2$ volts across them.

(3) *Power diodes in parallel.* Just as it is necessary to group connect diodes in series to handle high reverse voltage, it is also necessary to be able to accommodate high load currents when the diode is switched on in the forward direction. When the rating of one diode is not adequate to safely pass the forward current, two or more power diodes are connected in parallel to share the forward current equally.

As stated before, there are slight variations in the actual characteristics of individual diodes of the same number due to slight production/material variations. To insure that each diode carries the same forward current, a very small resistance is placed in series with each diode to limit the current (see Fig. 8-15).

With this parallel configuration, peak reverse voltage is essentially the same across each diode, and the rating of one diode is the limiting factor for all branches.

8-2-2 The Transistor.

A transistor is a three-terminal semiconductor device which can be used as a switch or as a variable resistance. The discussion here will be restricted to its use as a switch. Unlike the diode whose on or off state simply depends on the forward or reverse biasing of its simple P-N junction, the transistor has a third semiconductor layer, as shown in Fig. 8-16, and may have two variations in layering. The names associated with the two types of layering are analogous to the construction, namely the P-N-P and the N-P-N transistor. Note that the terminals of the device are called the collector, emitter, and base. The schematic symbols are clearly different in that the arrow of the base-emitter is pointing toward the base in the P-N-P and away from it in the N-P-N transistor. (As will be noted, this schematic difference will help to associate the base-emitter junction to that of a P-N diode. This junction will be forward biased and allow base current to flow when the broad end of the arrow is positive with respect to the other end and will be reverse biased, disallowing base current to flow, when the broad end of the arrow is negative with respect to the other.)

Figure 8-17a shows the basic collector characteristics of an N-P-N transistor. (The characteristics and operation of a P-N-P transistor are the same except that polarities are reversed.) As can be seen, the horizontal axis is the collector-to-emitter voltage, V_{CE}, and the vertical axis is the collector current, I_C. Note from this figure that higher values of I_C can be attained for the same value of V_{CE}, simply by increasing the value of the base current, I_B.

Transistor action is, in simple terms, the control of currents and voltages in one junction by the currents and voltages in another junction. The collector current and collector-to-emitter voltage in a transistor are controlled by the base current and the base-to-emitter voltage. External circuitry must be added to the transistor to attain this control, as shown in Fig. 8-17b. Since the base and emitter are one P-N junction, it acts much like a diode P-N junction.

With the switch open in Fig. 8-17b, there is no voltage applied to the base

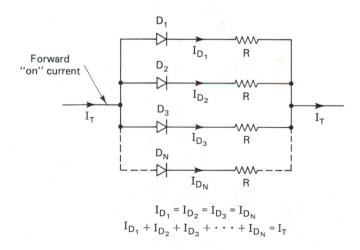

$$I_{D_1} = I_{D_2} = I_{D_3} = I_{D_N}$$
$$I_{D_1} + I_{D_2} + I_{D_3} + \cdots + I_{D_N} = I_T$$

Figure 8-15. Diodes in parallel.

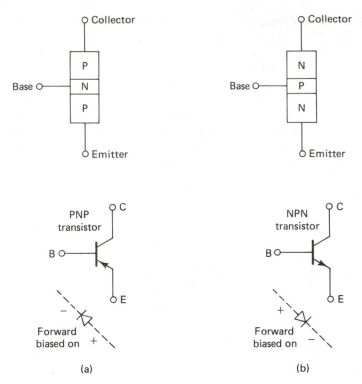

Note: The diodes shown are *not* part of
the schematic symbol but can be
associated with biasing the base-
emitter junction.

Figure 8-16.   Transistor nomenclature.

and no base current flows; $I_B = 0$.   Note in Fig. 8-17a that when $I_B = 0$, $V_{CE} = V_{CC}$ and $I_C = 0$.   When the switch is closed, the base-to-emitter junction will become forward biased and have a relatively small voltage drop similar to that of the diode previously discussed.   The voltage across R_B will determine the value of I_B, and the corresponding value of I_C will increase from zero, according to the collector characteristic curve for a given transistor type.   It should be noted here that for power transistors, typical values of base current are in the 100 to 500 milliampere region, and this relatively small current controls 5 to 10 amperes in the collector circuit.   It is very important that the maximum power rating of the transistor not be exceeded.   A typical maximum power curve is incorporated in Fig. 8-17a.   Note that because the base current is usually so small in comparison to the collector current, the power dissipated is essentially the product of the collector-to-emitter voltage and the collector current.   As is true for all semiconductor power devices, heat-sinking and sometimes forced cooling are used to remove heat from the device.

Figure 8-17c incorporates a basic approach to controlling the transistor as a switch.   As indicated in the figure, two voltage sources are available to be switched

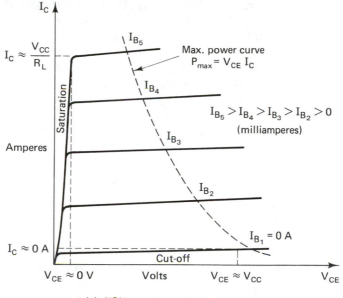

$$I_c \approx \frac{V_{CC}}{R_L}$$

$$I_{B_5} > I_{B_4} > I_{B_3} > I_{B_2} > 0$$
(milliamperes)

$I_{B_1} = 0$ A

$I_c \approx 0$ A

Max. power curve $P_{max} = V_{CE} I_C$

Saturation

Amperes

Cut-off

$V_{CE} \approx 0$ V   Volts $V_{CE} \approx V_{CC}$

(a) NPN transistor collector characteristics

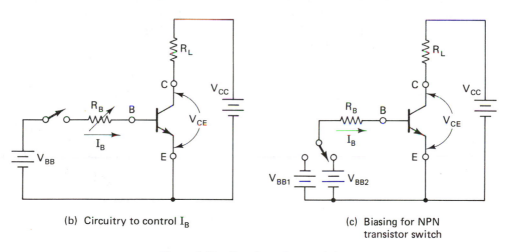

(b) Circuitry to control I_B

(c) Biasing for NPN transistor switch

Figure 8-17. Transistor characteristics.

into the base circuit of the transistor: V_{BB1} and V_{BB2} are supplies of opposite polarity. If V_{BB1} is switched in, the base-to-emitter junction will be reverse biased, and the base current will surely be zero. Under this condition of reverse bias, the transistor is said to be in the cut-off region (see Fig. 8-17a) and $V_{CE} = V_{CC}$. Essentially no current will flow through load resistor, R_L, and the transistor is fully turned off. If the switch is thrown so as to connect V_{BB2}, the base-to-emitter junction will be forward biased, and a high base current will flow, equal to I_{B5} in Fig. 8-17a. When this value of base current flows, the transistor is said to be in the saturated region and fully turned on. Under this condition, $V_{CE} = 0$ and the full-load current will flow equal to V_{CC}/R_L. When used as a switch, then, the transistor is switched between the cut-off and saturation regions to insure true on and off behavior.

With the constant improvement in techniques of semiconductor design and production, the new power transistors are capable of handling more and more power. A major advantage of using the transistor, besides reduced size and increased reliability due to no moving parts, is the fact that switching time is many times faster than in conventional electromechanical switches. For this reason alone, many times transistors are implemented in the control of machines.

8-2-3 Silicon-Controlled Rectifier (SCR)

(1) *Characteristics and operation.* Another solid state control device frequently employed as a switch is the SCR (Silicon-Controlled Rectifier). The SCR is a three-terminal device with four semiconductor layers, as shown in Fig. 8-18a, and with a schematic symbol, as shown in Fig. 8-18b. Note that the terminals are designated as anode (A), cathode (K), and control gate (G).

The SCR is very similar to a power diode under reverse-bias conditions. That is, when a positive voltage is applied to the cathode and a negative applied to the anode (refer to Fig. 8-19), just as a diode, the SCR acts like an open switch under these reverse-bias conditions and has similar ratings. Similar precautions must be exercised in terms of reverse leakage current and reverse breakdown voltage, as is seen in quadrant III of Fig. 8-19.

When forward biased (or anode positive with respect to the cathode, as in quadrant I), however, the SCR will also block the forward current until a forward-bias voltage is applied of such a magnitude (V_{BO}) that "breakover" occurs and the SCR conducts, or a pulse of current is applied to the gate causing the SCR to be "triggered" to the on state, or by a combination of the two. When the SCR conducts, either by the application of its particular breakover voltage or by triggering with a gate current of steep wavefront (usually a combination of the two), the SCR conducts current from anode to cathode much like a forward-biased diode

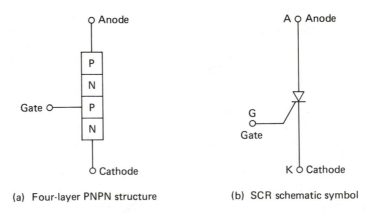

(a) Four-layer PNPN structure   (b) SCR schematic symbol

Figure 8-18. SCR nomenclature.

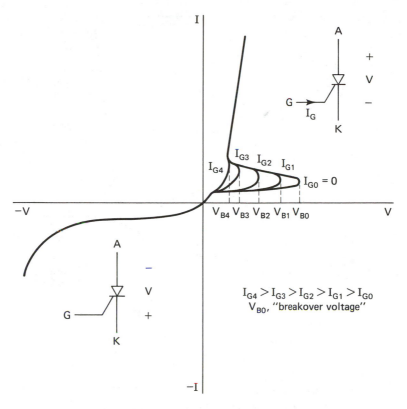

Figure 8-19. VI characteristics of the SCR.

and is said to be turned on and in the high conduction region. See Fig. 8-20. It then acts like a closed switch, with a relatively small forward-voltage drop.

An important and unique characteristic of the SCR is that it will continually act like a closed switch even if the gate current is removed under forward-biased conditions. Forward-conduction current flow can be stopped only by reducing the anode-to-cathode forward voltage to zero or by reducing the forward anode-to-cathode current to zero, as discussed in a later section.

The SCR has an advantage over power transistors in that a very small signal of gate current can trigger the conduction from anode to cathode of a very large current, and this trigger current does not need to be maintained. As previously noted, a few hundred milliamps of base current is typically needed to sustain 5 to 10 amperes of collector-to-emitter current for a power transistor, and once this base current is removed, the transistor will be switched off. The SCR, on the other hand, can conduct many times that amount from anode to cathode (about 2 000 amperes or more) with as little as 300 milliamps of gate current injected for just a short time duration in the microsecond range. With the continuing improvements being made in the fabrication of these devices, it may prove to be true that "the sky's the limit" in terms of power capability and switching rates of future models.

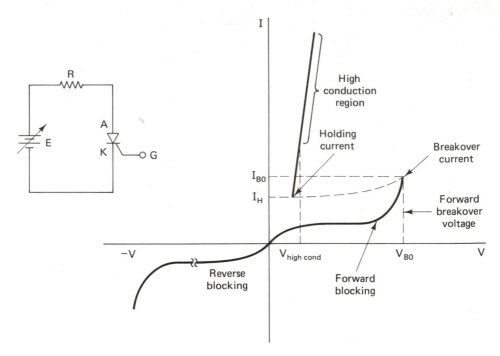

Figure 8-20. SCR forward characteristics.

(2) *SCR turn-on considerations.* As noted, the reverse VI characteristics of the SCR are typical of a reverse-biased diode.

The forward characteristics are such as to require a closer look at the critical areas of operation so that the SCR can be turned on (triggered on) consistently.

As can be observed in Fig. 8-20, there are several points of interest in the forward VI characteristics of the SCR. As the voltage, *E*, in Fig. 8-20 is slowly increased along the forward blocking region, the anode-to-cathode voltage will equal *E* until a point is reached where breakover occurs. Before this breakover voltage is reached, the SCR blocks current in the forward direction, has a high resistance, and is in the off state. When the breakover point is reached, the SCR "breaks over" to the high conduction region where it conducts heavy current with very little resistance and is in the on state. As can be seen, at the same time breakover occurs, the voltage across the SCR quickly changes to a low value, V_{HC}. The current from anode to cathode is limited by the resistor, *R*, and is approximately equal to *E/R*.

The high conduction region has an upper limit of current which is restricted by the power rating of the particular SCR and a lower limit, called the holding current, I_H, which is the minimum current needed to sustain heavy forward conduction. If the forward current drops below the specified value of holding current, I_H, the SCR will revert back to the forward-blocking region and thus act as an open switch, as before.

Using forward voltages alone to cause high-forward conduction and to turn on the SCR is seldom done and would eliminate the need for the control gate.

Injecting a current, I_G, into the gate will cause the SCR to breakover to the high conduction region as a function of two important variables: (1) the magnitude of the gate current, and (2) the phase angle when the gate current is applied when used in AC circuits. In Fig. 8-19, as noted, it is apparent that as the gate current is increased from $I_{G0} = 0$ through subsequently larger values (I_{G1}, I_{G2}, and I_{G3} to I_{G4}) of gate current, smaller and smaller values of breakover voltages are realized for a given SCR.

(3) *SCR turn-off considerations (commutation)*. As noted previously, if the forward current drops below the value of the holding current, the SCR will revert to its forward-blocking state. Because of the physical internal properties of the SCR, the forward current must drop below I_H for a specific time before the forward-blocking characteristics overcome forward conduction. This time varies from SCR to SCR and depends on the cross sectional area and other subtle properties internal to the device.

Turn off of the SCR, called commutation, can be accomplished as illustrated in Fig. 8-21. As can be seen in Fig. 8-21, the forward current can be made to drop below I_H by a) opening a switch connected in series with the SCR and thereby causing zero current, or b) closing a switch that is connected in parallel with the SCR, thereby causing zero current by diverting it around the SCR, or c) applying a reverse voltage across the anode to cathode and forcing the SCR to operate in the third quadrant.

The first two methods are not practical in high power applications and are used only rarely in low-power dc circuits. Commutation of the SCR is generally accomplished by the third method of applying a reverse voltage from anode to cathode. There are several types of commutation circuits, and the student is referred to the references for a detailed design perspective of the various types. One of the more simple types, the class C commutation circuit, is presented here to illustrate how an SCR can be added to a circuit to control another SCR's commutation.

Figure 8-22 depicts a class C commutation circuit and the sequence of action which occurs. Figure 8-22a shows the circuit with SCR1 and SCR2 both in the

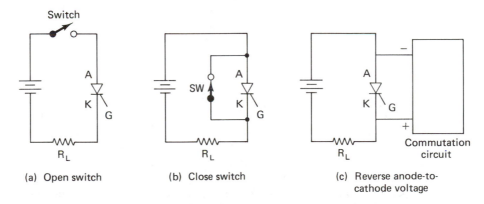

Figure 8-21. SCR turn-off circuits.

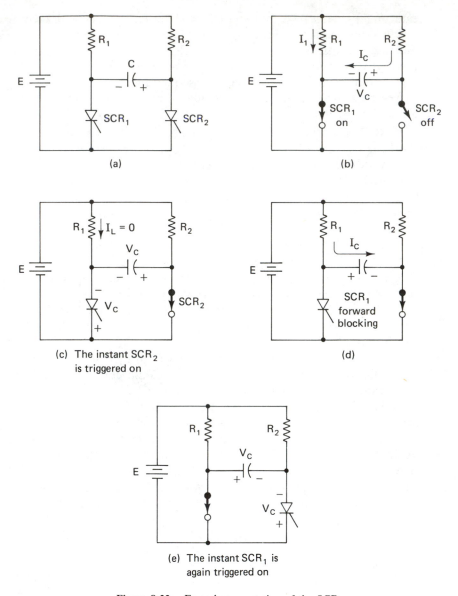

(a)

(b)

(c) The instant SCR$_2$
is triggered on

(d)

(e) The instant SCR$_1$ is
again triggered on

Figure 8-22. Forced commutation of the SCR.

forward-blocking mode of operation. When a pulse is applied to the gate of SCR1, it turns on and goes into high conduction and acts like a closed switch, as shown in Fig. 8-22b. At this instant, SCR2 is still in the forward-blocking region and is effectively an open switch. The capacitor, C, charges up to E volts through the resistor, R_2. At a later predetermined time, a turn-on pulse is applied to the gate of SCR2. It turns on and, at the same instant, the capacitor voltage V_C is applied across SCR1 due to the closed switch action of SCR2. Note that this voltage, V_C, commutates a reverse voltage across SCR1 and turns it off, as shown in Fig.

8-22c. Now with SCR1 off and SCR2 on, the capacitor will charge up through R_1, as shown in Fig. 8-22d. Finally, another pulse is applied to the gate of SCR1 which triggers it to the on condition. Now with SCR1 acting as a closed switch, the capacitor voltage, V_C, is applied across SCR2 in the reverse direction, as shown in Fig. 8-22d, and SCR2 is turned off, completing the cycle. This sequence of events is repetitious and periodic and yields a "chopped dc" voltage across R_1. The frequency of the chopped voltage is determined by the pulse rates to the subject SCRs and the RC time constants of the circuit. More or less average voltage can be controlled depending on how long SCR1 is on.

It should be noted that the foregoing discussion involves switching a dc voltage on and off to vary the average dc voltage to R_1 and, ultimately, to vary the energy to R_1. When varying the energy to a load with an ac input voltage, the commutation process is less of a problem because the second half-cycle of the input voltage is always negative and thus turns off the SCR during that time. Consider the circuit of Fig. 8-23. The input voltage is sinusoidal, and the SCR is placed in series with

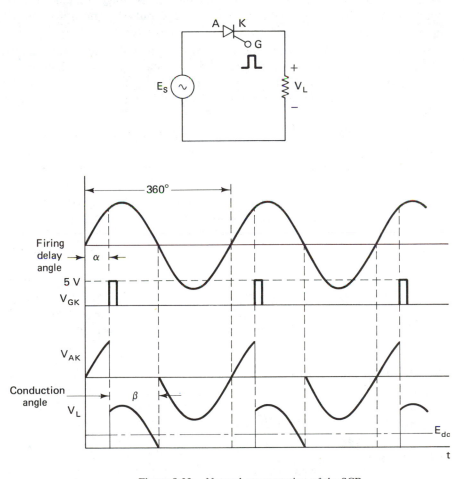

Figure 8-23. Natural commutation of the SCR.

the load.   To get the most possible energy to the load, the SCR could be triggered on just as the first half-cycle begins so that the whole first half-cycle of voltage appears across the load.   The second half-cycle impresses a reverse voltage across the SCR and, in so doing, turns off the SCR.   This type of circuit can be called naturally commutating because no external circuitry is required to turn off the SCR.   Note that the SCR is used as a rectifying device, and the load voltage, V_L, is dc.   The average value of this load voltage may be varied from the maximum, as mentioned above, to lower values.   Lower values of average load voltage can be obtained by delaying the turn on of the SCR until some time into the first half of the cycle.   The number of degrees of the cycle that pass before the SCR is triggered on is called the delay angle or firing delay angle, α.   The number of degrees of the cycle during which the SCR is on is called the conduction angle, β. In the circuit shown, if the gate trigger pulse is applied to the gate of the SCR at intervals as shown, the SCR will turn on with the trigger pulse and then off when the source voltage goes negative during the second half-cycle.   The voltage wave-forms are shown in Fig. 8-23.   Note that the load voltage appears at the load as soon as the SCR is fired with the gate voltage pulse.   The load voltage would appear sooner if the gate pulse came sooner and later if fired later.

(4) *Series-run SCRs*.   When the input voltage is higher than the voltage rating of the individual SCR, the SCRs must be connected in series.   Due to variations in production, the characteristics are not perfectly matched, so, as with the power diodes, resistors are connected in parallel with the SCRs so as to equally share the voltage.

(5) *SCRs in parallel*.   SCRs are connected in parallel if the current required exceeds the rating of one SCR.   As discussed with diodes, small resistances are connected in series with the SCRs in order to insure equal sharing of the current through them because of slightly different characteristics.

(6) *The snubber circuit*.   One important manufacturer's specification of an SCR is the *dv/dt* rating.   This rating indicates how fast the voltage across the SCR can change per unit time.   If the applied voltage changes faster than this rating, the particular SCR will turn on without a gate trigger pulse.   To minimize the rate of change of voltage per unit time across the SCR, a capacitor and series resistor are placed in parallel with the SCR, as shown in Fig. 8-24.   Since the voltage across a capacitor cannot change instantaneously, the *RC* snubber circuit "snubs" the fast voltage changes and can be designed for any condition simply by adjusting the *RC* time constant.   The snubber absorbs fast changes in voltage and allows the use of more inexpensive SCRs with lower *dv/dt* ratings.

8-2-4   The DIAC and TRIAC.   The DIAC, sometimes referred to as a bilateral trigger diode, is a two-terminal, five-layer semiconductor device which exhibits characteristics as shown in Fig. 8-25.   Note that as the voltage across its terminals increases, the DIAC remains in the open or off state until the voltage reaches the breakover voltage.   When the breakover voltage (V_{BO}) is reached, the DIAC conducts and is in the closed or on mode of operation.   If the voltage across it increases in the negative direction, the same action takes place, and the DIAC

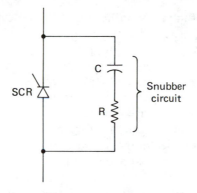

Figure 8-24. Snubber voltage suppression.

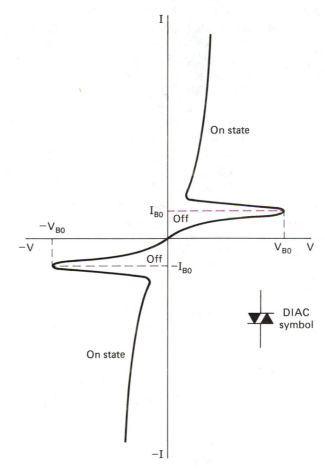

Figure 8-25. DIAC characteristics.

conducts current in the opposite direction. The DIAC is basically used to trigger other thyristors, especially the TRIAC.

The TRIAC is a three-terminal, five-layer semiconductor device which is similar to an SCR in operation except that it conducts current in both directions when appropriately triggered by a positive or negative gate pulse.

The bidirectional triode thyristor, the formal name for the TRIAC, acts very much like two SCR devices connected as shown in Fig. 8-26, and exhibits characteristics very similar to the forward-biased SCR (refer to Fig. 8-20). Since the two SCRs are connected anode to cathode in Fig. 8-26 and have a common gate connection, the polarity applied to the main terminals of the TRIAC, MT1, and MT2 is inconsequential. Note that if MT1 is positive with respect to MT2, or vice versa, one SCR or the other will be forward biased and an appropriate gate pulse will turn on the particular forward-biased SCR and cause conduction. Bidirectional flow may be controlled and thus it is possible to control both half-cycles of an ac sine wave and still maintain an ac signal. (Remember that the SCR is basically a rectifying dc device and that it can only control the positive half-cycle of a sine wave while blocking the negative half-cycle.) By varying the firing delay angle, greater or lesser output RMS voltage results. Figure 8-27 shows the load voltage waveform delivered to a resistive load with the delay angle at 90°. Note that by reducing the delay angle to 0°, full voltage would be delivered to the load and, conversely, if the delay angle was 135°, even less voltage is delivered. This technique of varying load voltage, using SCRs or TRIACs, is called phase control and is widely used in the control of motors.

There are four possible conditions for triggering the TRIAC; i.e. applying the trigger signal between the gate and MT1. These conditions are 1) when MT2 is positive with respect to MT1 and the gate pulse is positive, the TRIAC is operating in the first quadrant (I^+); 2) if MT2 is positive with respect to MT1 and the gate is negative, the TRIAC also operates in the first quadrant (I^-); 3) when MT1 is positive with respect to MT2 and the gate is positive, the TRIAC is operating in the third quadrant (III^+); 4) when MT1 is positive with respect to MT2 and the gate is negative, the TRIAC is also operating in the third quadrant (III^-). The

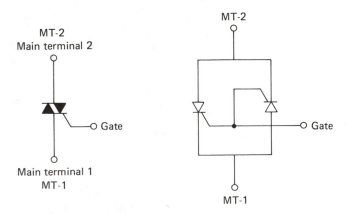

Figure 8-26. The TRIAC and its SCR equivalent circuit.

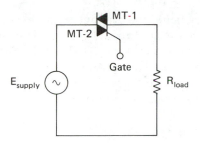

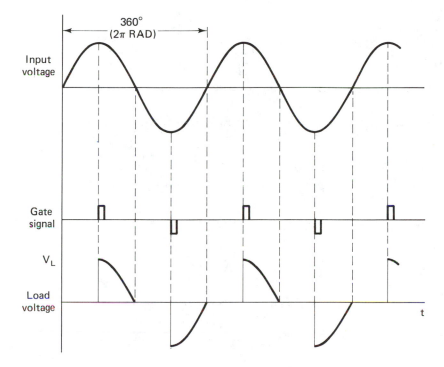

Figure 8-27. TRIAC phase control.

TRIAC usually is operated in the I^+ and III^- modes because it is most sensitive and responds quickly under these bias and gate conditions. The I^- mode is less sensitive and the III^+ is much less sensitive and is not normally used.

8-2-5 Diodes as Rectifiers. Since diodes are unidirectional devices (passing current only in one direction), they are used to convert ac waveforms into dc. When a diode is forward biased, it is a closed switch, and when reversed biased, it is an open switch.

When diodes are used to convert ac into dc, the circuit is called a rectifier or converter; the two terms are used synonymously. A converter's dc output voltage has the characteristic of either being half-wave rectified or full-wave rectified, the principal differences being the amount of dc voltage output, the frequency, and peak-to-peak voltage of the "ripple" that results from the conversion process.

A typical half-wave rectifier is shown in Fig. 8-28a. A two-to-one step-down transformer transforms a 60 hertz, 60 volt peak sine wave down to a 30 volt peak sine wave and feeds a diode in series with a 3 ohm load resistor. Figure 8-28b displays the transformer secondary voltage, e_s, as it varies with time. The frequency

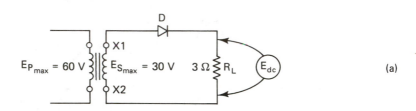

(a)

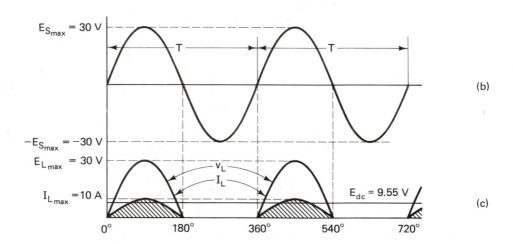

(b)

(c)

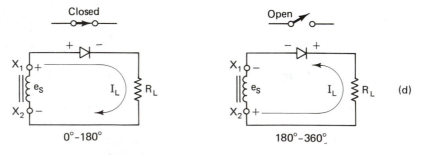

Figure 8-28. Half-wave rectification.

of the sine wave is the inverse of time period, T, so the time it takes for one full cycle can be calculated:

$$f = \frac{1}{T}$$

Eq. (8-1)

$$T = \frac{1}{f}$$

$$T = \frac{1}{60 \text{ Hz}}$$

$$T = 16.67 \text{ msec}$$

The time period, T, represents the time it takes the input signal to complete one full cycle of 360°. After this time, the sine wave continues to repeat itself exactly until the signal is removed.

Figure 8-28c depicts the load voltage and current that results from half-wave rectification. It is clear that the load resistance only receives the first full half of the sine wave; the second half is blocked. This is due to the switching action of the diode, D_1. As the voltage begins to increase at $t = 0$ (or 0°), diode D_1 becomes forward biased and acts as a closed switch. (It will be assumed in all discussions that the diode has no forward voltage across it, but in actuality it has about one volt across it when forward biased.) This forward-biased condition continues as long as the top of the secondary transformer winding, X_1, is positive with respect to the bottom of the winding, X_2, which, as can be seen in Fig. 8-28b, is true until 180°. At this time, $T/2 = 180°$. The secondary voltage switch's polarity and the X_1 terminal of the transformer winding go negative with respect to the X_2 terminal, thus causing a reverse-bias condition across the diode which then acts as an open switch, as shown in Fig. 8-28d. With the "open switch" reverse-biased condition between 180° and 360°, no current flows through the load resistance and therefore the load voltage is zero during the second half-cycle. It is obvious, then, that this simple rectifier circuit has converted an ac sine wave input appearing at the secondary of the transformer into a pulsating dc voltage (and current) appearing at the load.

If a dc voltmeter is connected across the load resistance, it will not read the pulsating value of voltage. Rather, a dc voltmeter always reads the *average* dc voltage it sees, E_{dc}. It can be shown through calculus that the average voltage, E_{dc}, that a dc voltmeter will read across the load resistance for the half-wave rectifier is:

$$E_{dc} = \frac{E_{L \text{ max}}}{\pi}$$

or

$$E_{dc} = 0.318 \, E_{L \text{ max}}$$

Eq. (8-2)

For the circuit of Fig. 8-28 then, the dc load voltage that a dc voltmeter will read is:

$$E_{dc} = \frac{E_{L\ max}}{\pi} = \frac{30}{\pi}$$

$$E_{dc} = 9.55 \text{ volts}$$

The dc current that a dc ammeter will read is:

$$I_{dc} = \frac{E_{dc}}{R} = \frac{9.55}{3}$$

$$I_{dc} = 3.18 \text{ amps}$$

Even though a dc instrument reads a steady value of voltage, it is important to remember that the load voltage varies with time. This variation is called the ripple voltage, and in this case, the peak-to-peak value is $E_{L\ max} = 30$ volts. We can consider the ripple voltage to vary around the average voltage, E_{dc}, with a specific frequency called the ripple frequency. In this half-wave circuit, the ripple frequency is the same as the line frequency, namely 60 hertz.

The half-wave rectifier is not used in power applications because of the relatively low dc voltage obtained, high peak-to-peak ripple voltage, and relatively low ripple frequency, but serves to introduce us to the single-phase, full-wave bridge rectifier of Fig. 8-29a. Note that this full wave rectifier has three additional diodes but has the same input secondary voltage and works into the same load resistance of 3 ohms.

At $t = 0$, the X_1 terminal of the transformer begins to go positive with respect to the X_2 terminal and remains positive for the first half-cycle. As seen in Fig. 8-29b, this forward biases diodes D_1 and D_2 while reverse biasing D_3 and D_4. During the first half-cycle, then, D_1 and D_2 act as closed switches and D_3 and D_4 as open switches. Load current then flows down through R_L for the first half-cycle or between 0° and 180° via D_1 and D_2, as noted in the figure. During the second half-cycle, the polarity changes and the X_2 terminal of the transformer is positive with respect to the X_1 terminal, as noted in Fig. 8-29c. D_1 and D_2 become reverse biased and open, and D_3 and D_4 become forward biased and act as closed switches. Load current then flows as indicated through diodes D_3 and D_4 and again the current flows down through the 3 ohm load resistor. In effect, the full-wave bridge action takes the bottom half of the sine wave and rectifies it up into the first quadrant so that the load resistance sees the voltage as shown in Fig. 8-29e. It can be seen that the secondary 30 volt peak sine wave (Fig. 8-29d) has been rectified and the full wave has been delivered to the load with the same polarity, as in Fig. 8-29e.

Again, a dc voltmeter will read the average dc voltage, E_{dc}, and it can be shown that:

$$E_{dc} = \frac{2}{\pi} E_{L\ max}$$

$$\boxed{E_{dc} = 0.637\ E_{L\ max}}$$

Eq. (8-3)

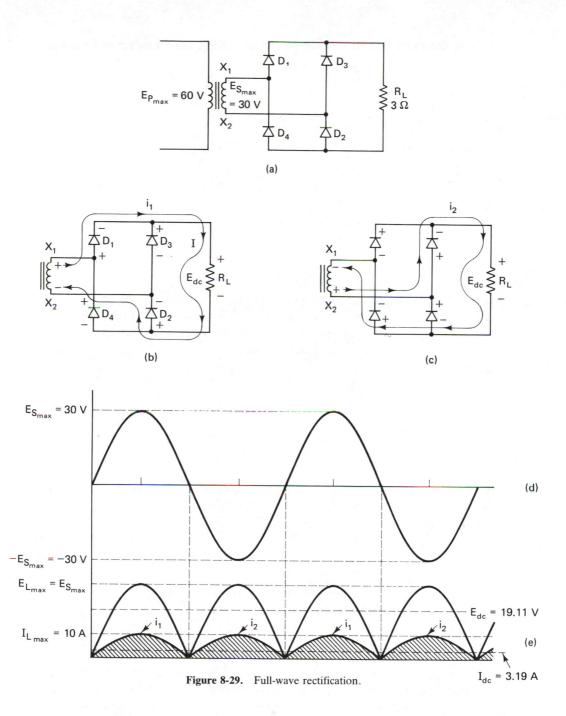

Figure 8-29. Full-wave rectification.

In the example of Fig. 8-29 the voltage that a dc voltmeter will read is:

$$E_{dc} = 0.637 \, E_{L \, max}$$

$$E_{dc} = 0.637 \, (30)$$

$$E_{dc} = 19.11 \text{ volts}$$

and a dc ammeter will read:

$$I_{dc} = \frac{E_{dc}}{R_L}$$

$$= \frac{19.11}{3}$$

$$I_{dc} = 6.37 \text{ amps}$$

The average dc current delivered to the load is 6.37 amps, as calculated, but notice that diode D_1 and D_2 conduct only for the first half of the input cycle, i_1, and D_3 and D_4 conduct only for the second half-cycle, i_2. For this reason, the average current carried by each diode is only one half of the average load current:

$$I_{D_1,D_2} = \frac{I_{dc}}{2} = I_{D_3,D_4} \qquad \text{Eq. (8-4)}$$

for D_1

$$I_{D_1} = \frac{6.37}{2} = 3.19 \text{ A} = I_{D_2,D_3,D_4}$$

The peak inverse voltage rating of each diode must also be considered. Figure 8-30 shows the full-wave bridge with diodes D_1 and D_2 conducting during the first half-cycle as closed switches. When, at 90° into the first half-cycle, the maximum voltage is delivered to the load, so too is it individually delivered across diodes D_3 and D_4. (Use Kirchhoff's voltage law to prove this.) It can be concluded, then, that the *PIV* rating of each diode must exceed $E_{L \, max}$.

$$PIV \, (D_1, D_2, D_3, D_4) > E_{L \, max} \qquad \text{Eq. (8-5)}$$

It is clear when comparing the dc voltages for the half- and full-wave converters that the full-wave bridge rectifier delivers twice as much dc voltage for the same ac input voltage. The ripple voltage of the full-wave bridge rectifier has a peak-to-peak value of $E_{L \, max}$, which is the same as the half-wave rectifier. Notice, however, that the waveform of the ripple voltage is now repeating twice for each time period of the input sine wave. This higher (2X) ripple frequency will allow more effective inductive filtering.

It should be noted here that the load for both rectifiers has been a pure resistance, and, for this reason, the current to the load has been in phase with the

BASIC CONTROL COMPONENTS   CHAP. 8

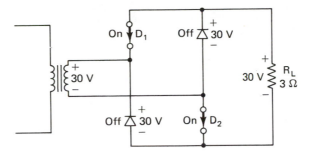

Figure 8-30. Full-wave circuit conditions to evaluate PIV.

load voltage for every instant in time and follows the shape of the pulsating sinusoid, as seen in Fig. 8-28c and Fig. 8-29e. The torque of a dc motor is directly proportional to the armature current and field flux. If the current of the armature or the field pulsates sinusoidally, then there will be sinusoidal pulsations of torque, and this is not desirable. To accommodate the need for a constant current, an inductive filter is used to smooth out the current waveform. By putting a large inductance in series with the resistor, called a smoothing inductor, the load current can essentially be made to follow a straight line. The larger the inductance used, the more constant the current will be, even with a fluctuating voltage. It is also true that the current will be smoother with higher ripple frequency. The student is encouraged to review the energy storage/release characteristics of the inductor, if necessary, for detailed proof of why this happens.

Figure 8-31 shows the single-phase, full-wave bridge with a smoothing inductor in series with the same 3 ohm load resistance and the resulting voltage and current waveforms delivered to the load. During the first half-cycle, diodes D_1 and D_2 still conduct, but notice that the current is flat and that during the second half-cycle, when diodes D_3 and D_4 conduct, the same is true. The smoothing inductor is a very important component when a constant current is needed.

It can be said that the half-wave and full-wave single phase rectifier circuits are "one-pulse" and "two-pulse" converters. The half-wave rectifier converts one cycle of the incoming 60 Hz sinusoid to one pulse of dc, as in Fig. 8-28. Similarly, the full-wave rectifier converts one cycle of the same incoming sinusoid to two pulses of dc, as in Fig. 8-29. It is noted that the amount of average voltage, E_{dc}, at the load increases as the number of pulses increase per cycle of the input sinusoid.

Consider the circuit of Fig. 8-32a. A Y-connected, 4-wire, three-phase (3ϕ) supply is connected as shown. Diodes D_1, D_2, and D_3 are each placed in series with their respective output phase (this could be the secondary of a 3ϕ transformer), and they in turn feed the same load as before consisting of a 3Ω resistor in series with a large smoothing inductance. Figure 8-32b represents the line-to-neutral sinusoidal 60 Hz, 3ϕ voltage output of the supply with a peak value, $E_{max} = 100$ volts, and each phase displaced from one another with the standard 120°. Notice also that the phases cross each other at $\frac{1}{2} E_{max}$ or 50 volts.

The resulting voltage delivered to the load is shown in Fig. 8-32c. In order to facilitate understanding of how the load voltage waveform is developed, it is

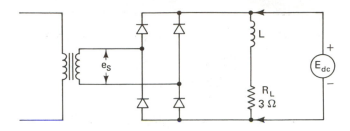

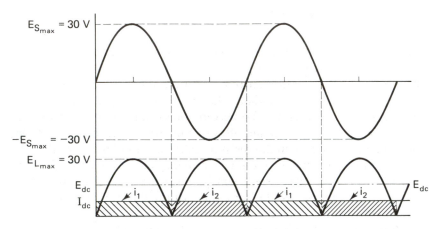

Figure 8-31. Single-phase, full-wave bridge rectifier with inductive filtering for current smoothing.

important to remember that the waveforms representing the 3ϕ line-to-neutral input voltage in Fig. 8-32b represent how each phase voltage varies with time. Thus, for any instant in time, the value of voltage for each phase can be evaluated, and it can be determined how the diodes are biased at that particular instant of time.

Note in Fig. 8-32c that five points in time, ① through ⑤, have been chosen at points where the load voltage reaches maximums and minimums, E_{max} and E_{min}, respectively. In Figure 8-33, ① through ⑤ represent the circuit conditions at these respective points in time and serve to illustrate the dynamics of the circuit as time passes.

At time $t =$ ① in Fig. 8-32b, note that phase 1 is +100 volts; phase 2 and phase 3 are each negative, at −50 volts. These values are shown in Fig. 8-33①, and it can be seen that D_1 is forward biased and closed, and D_2 and D_3 are reverse biased with 150 volts across each. Since diode D_1 is forward biased and acting as a closed switch, the +100 volts appear at the load. The other two voltages are blocked due to the reverse biasing.

Between time $t =$ ① and $t =$ ②, notice that phase 1 is always more positive

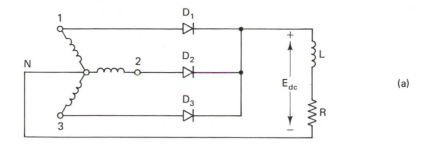

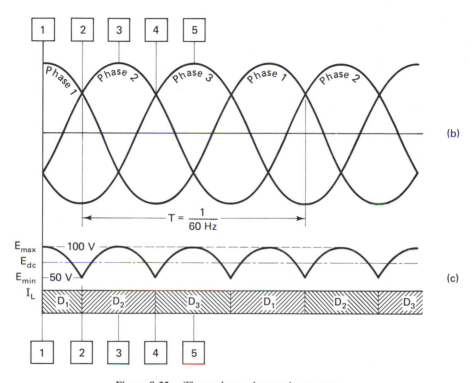

Figure 8-32. Three-phase, three-pulse converter.

than phase 2 and 3 so that it remains forward biased. Phase 1 voltage appears at the load until time $t = \boxed{2}$, as noted in Fig. 8-32c.

At time $t = \boxed{2}$ in Fig. 8-32b, notice that phase 1 voltage equals phase 2 voltage (both are $+50$ volts). Phase 3 voltage has reached its negative peak voltage, $-E_{max} = 100$ volts. At this point, diodes D_1 and D_2 are both forward biased, and $+50$ volts appears at the load while diode D_3 is reverse biased with 150 volts across it. Notice here that this condition only occurs at this *instant*, and this point, $t = \boxed{2}$ in Fig. 8-33$\boxed{2}$, only represents a transition point, where conduction is transferring from diode D_1 to diode D_2.

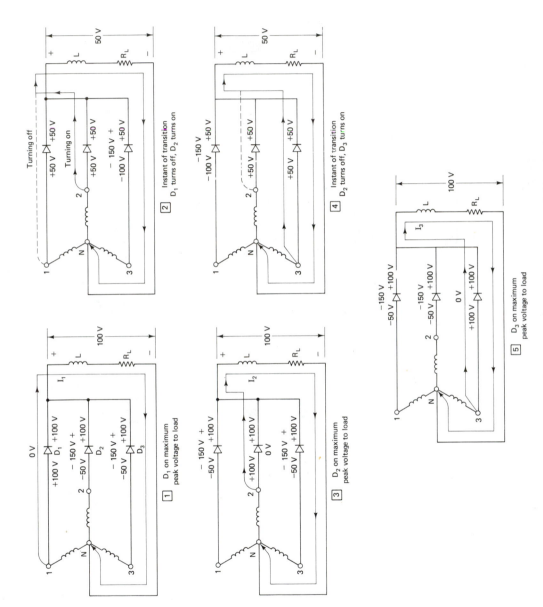

Figure 8-33. Three-phase, three-pulse converter switching circuit.

Between points $t = \boxed{2}$ and $t = \boxed{3}$, phase 2 voltage continues to increase toward $E_{max} = +100$ volts while phase 1 voltage goes negative and phase 3 voltage begins to increase but remains negative, as seen in Fig. 8-32b.

At time $t = \boxed{3}$, phase 2 reaches $E_{max} = +100$ volts, and both phases 1 and 3 cross at $-\frac{1}{2} E_{max} = -50$ volts. The circuit conditions at this instant are shown in Fig. 8-33$\boxed{3}$, and the maximum voltage is again delivered to the load due to the forward-biased, closed condition of diode D_2. Both diodes D_1 and D_3 are heavily reverse biased with -150 volts, which is actually the peak inverse voltage that each diode is subjected to throughout the entire switching sequence.

Between points $t = \boxed{3}$ and $t = \boxed{4}$, phase 2 voltage remains positive with respect to the other phases, and diode D_2 continues to act as a forward-biased closed switch, passing this voltage to the load as seen in Fig. 8-32c, while diodes D_1 and D_3 continue to block their respective voltages.

At time $t = \boxed{4}$, as shown in Figs. 8-32b and 8-33$\boxed{4}$, a point of transition is again encountered where phase 2 and phase 3 voltages are equal to $+50$ volts, while phase 1 voltage has reached its negative peak of -100 volts. This point represents where diode D_2 is becoming reverse-biased open and diode D_3 becomes forward-biased closed. This can be clearly seen at the instant just after point $t = \boxed{4}$ on Fig. 8-32b. Notice that as time increases from $t = \boxed{4}$, phase 3 definitely becomes more positive than phase 2, and diode D_2 is quickly reverse biased and turned off, while D_3 is forward biased on.

Between points $t = \boxed{4}$ and $t = \boxed{5}$, diode D_3 is forward biased on while diodes D_1 and D_2 are reverse biased off so that phase 3 voltage appears at the load for this interval while the other phase voltages are blocked.

At time $t = \boxed{5}$ of Fig. 8-32b, phase 3 reaches $E_{max} = +100$ volts, and this voltage appears at the load as shown in Fig. 8-32c. Phase 3 voltage continues to appear at the load until the next transition point.

This sequence continues to repeat itself and the student is encouraged to analyze the remaining maximum/minimum points in time in a similar fashion for Figures 8-32b and 8-32c to develop this type of approach in analysis and to verify the remainder of the load voltages shown.

Referring again to Fig. 8-32c, it is clear that a pulsating dc voltage appears at the load because of the switching action of the diodes and the 3ϕ sinusoidal input voltage. Note that for each time period of the incoming signal, three dc pulses appear in the ripple of the load voltage. This circuit is called a three-phase, three-pulse converter and the ripple frequency, f_R, for this circuit is:

$$f_R = 3 f_{in}$$

Eq. (8-6)

The advantages of this 3ϕ, three-pulse converter over the single-phase, one-pulse or two-pulse converters are: 1) more average voltage, E_{dc}, delivered to the load for the same peak value of the input sinusoid; 2) less peak-to-peak ripple voltage, and 3) a higher ripple frequency (three times the input frequency), which makes inductive filtering more effective and, consequently, the load current smooth.

It can be shown that the constant average voltage, E_{dc}, appearing at the load for a 3ϕ, 3-pulse converter is:

$$E_{dc} = \frac{3\sqrt{3}E_{max}}{2\pi}$$

$$\boxed{E_{dc} = 0.827\ E_{max}}$$

Eq. (8-7)

It is also true that the peak-to-peak ripple voltage $E_{R_{p\text{-}p}}$ is less than in the 1ϕ, two-pulse converter and has been shown for the 3ϕ, 3-pulse rectifier to be:

$$E_{R_{p\text{-}p}} = \frac{E_{max}}{2}$$

$$\boxed{E_{R_{p\text{-}p}} = 0.5\ E_{max}}$$

Eq. (8-8)

Applying these relationships to the circuit of Fig. 8-32 we find:

$$E_{dc} = 0.827\ E_{max}$$
$$= 0.827\ (100)$$
$$E_{dc} = 82.7 \text{ volts,}$$
$$f_R = 3f_{in}$$
$$= 3\ (60)$$
$$f_R = 180 \text{ Hz}$$

and

$$E_{R_{p\text{-}p}} = 0.5\ E_{max}$$
$$= 0.5\ (100)$$
$$E_{R_{p\text{-}p}} = 50 \text{ volts}$$

The performance of the 3ϕ, 3-pulse converter is clearly superior over the 1ϕ, 1-pulse and 2-pulse converters, but in many applications, the 1ϕ, 2-pulse system is used when a 3ϕ supply is not available or when the 1ϕ, 2-pulse converter characteristics are within acceptable limits for a given application. Generally, a 1ϕ converter is used for lower power loads and the 3ϕ system for loads requiring high power or more stringent ripple characteristics.

If the 3ϕ converter is required, a 3ϕ, 6-pulse converter is used. A 6-pulse converter is used instead of the 3-pulse because it is the most economical and has superior characteristics, as will be shown. The 3ϕ, 6-pulse converter is used in most large industrial converter installations and is simliar to the 3-pulse except that it has three additional diodes and is fed from a 3ϕ, 3-wire, Y-connected source, as shown in Fig. 8-34a, with line (-to-neutral) 3ϕ voltage, as shown in Fig. 8-34b.

Note that the neutral of the 3ϕ supply of Fig. 8-34a is floating, and there is

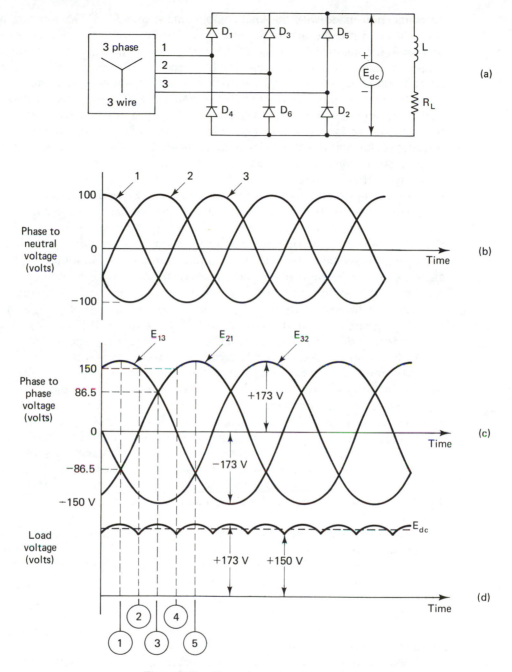

Figure 8-34. Three-phase, six-pulse converter.

no connection made with the load components L and R_L. The working voltage that will be impressed across the load is really the line-to-line voltage of the 3ϕ source, so the line-to-line voltage of the supply corresponding to the line-to-neutral supply voltage is depicted in Fig. 8-34c. Since the maximum value of line-to-neutral voltage is 100 volts, the maximum value of line-to-line voltage, $E_{\max}$, is $\sqrt{3}$ (100) or 173 volts. It is important to remember that these waveforms graphically represent how the voltage varies with time. It is possible, then, to determine the instantaneous value of each line-to-line voltage at any specified instant in time. Because of this, it is possible to analyze the somewhat cumbersome switching action of the 6-pulse converter by using a dc approach.

Five points in time, ① through ⑤, are indicated in Figs. 8-34c and 8-34d. Corresponding circuit conditions for each point are shown in Fig. 8-35, ① through ⑤. At each point, the value for each line-to-line voltage is determined from the waveforms of Fig. 8-34c and shown as a dc voltage in Fig. 8-35. This will simplify the development of the output voltage waveform, as shown in Fig. 8-34d. Also, when negative voltages are encountered in conjunction with double subscript notation, it should be remembered that the negative sign is associated with the first number of the double subscript. For example, when $E_{21} = -86.5$ volts, point 2 is negative with respect to point 1. With this in mind, note that the connection points for phases 1, 2, and 3 are marked on the circuits of Fig. 8-35. Batteries are connected to points 1–2, 2–3, and 3–1 to reflect the value of instantaneous line-to-line voltage, along with the proper instantaneous polarity.

At time $t = ①$, it can be seen from Fig. 8-34c that the values of instantaneous line-to-line voltages are: $E_{13} = E_{\max} = 173$ volts, $E_{21} = E_{32} = -E_{\max}/2 = -86.5$ volts. Figure 8-35① shows the condition of the 6-pulse network with these voltages installed as batteries. Diodes D_1 and D_2 are forward biased due to the 173 volts and act as closed switches. The remaining diodes are reverse biased with the reverse voltage across each as indicated. (Kirchhoff's voltage law should be used to prove these values.) Because of the closed condition of diodes D_1 and D_2, the 173 volts $= E_{\max}$ is transferred to the load as noted on the output voltage waveform for $t = ①$ in Fig. 8-34d with the polarity indicated, and current passes down through the load.

As time passes between $t = ①$ and $t = ②$, E_{13} decreases to 150 volts while E_{21} increases to 0 V and E_{32} decreases to -150 V. When evaluating the status of biasing of the diodes during this time period, it will be found that diodes D_1 and D_2 continue to conduct and E_{13} is passed to the output as indicated in Fig. 8-34d, and current flows down through the load.

At time $t = ②$, a point of transition occurs as can be seen in Fig. 8-35②. With E_{13} at $+150$ volts, E_{32} at -150 volts, and E_{21} at 0 volts, both D_1 and D_3 conduct *for this instant only*, and D_2 continues to conduct as before. Note, however, that an instant later E_{32} becomes greater in magnitude with respect to E_{13}, and diode D_1 then becomes reverse biased and D_3 takes over full conduction. At the exact point of transition, $t = ②$, the voltage across the load is 150 volts, as seen in Fig. 8-34d, and current flows down through the load.

At $t = ③$, it can again be seen in Fig. 8-35③ that diodes D_3 and D_2 continue to conduct with E_{32} at -173 volts. Note here that the load voltage is again *positive*,

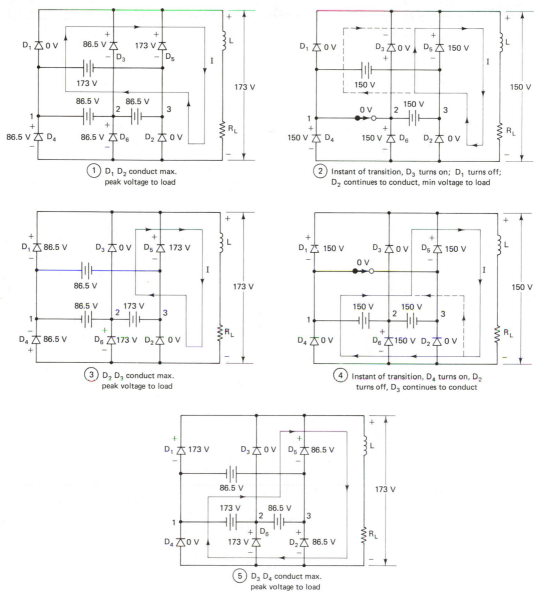

Figure 8-35. Three-phase, six-pulse converter switching sequence.

and current continues to flow down through the load. In effect, the voltage passed to the load is not E_{32} or -173 volts. Since the output voltage remains positive, the switching action of the diodes actually passes $E_{23} = +173$ volts to the load (and continues to do so between $t = ②$ and $t = ④$).

At time $t = ④$, another point of transition occurs exactly like that of time $t = ②$ except that, as can be seen in Fig. 8-35④, diode D_4 takes over conduction from diode D_2, and diode D_3 continues to conduct, passing $+150$ volts to the load. Current continues to flow down through the load.

At time $t = $ ⑤, E_{21} reaches $E_{max} = +173$ volts while E_{32} and E_{13} cross each other at $-86.5 = -E_{max}/2$. Diodes D_3 and D_4 continue to conduct, and $E_{21} = +173$ volts is passed on to the load. Load current continues to flow down through the load.

As the line-to-line voltages of the 3ϕ supply continue to vary with time, the output voltage will continue to follow the pattern of the first five time segments discussed here with the same results. The load voltage will cycle between $E_{max} = +173$ and $+150$ volts. It is observed that the peak values of each voltage pass to the output but always in a positive sense and with the sequence as shown in Fig. 8-34d, i.e., beginning from the origin, E_{13}, E_{23}, E_{21}, E_{31}, E_{32}, E_{12}, E_{13}, etc. Notice that E_{23}, E_{31}, and E_{12} have, in effect, been inverted or flipped up into the first quadrant. By observing this result we can see that positive maximum values of line-to-line voltage pass to the load, and the negative maximum values, due to the switching action of the diodes, are inverted and pass to the load as positive voltages.

Figure 8-36a depicts the as-supplied line-to-line voltages and also the inverse of each. It is then easy to observe and remember that the load voltage is continually positive with six pulses of ripple output resulting from one time period of the input line-to-line voltage. For a 60 hertz, 3ϕ supply, then, the ripple frequency, f_R, is six times that of the supply frequency, f_S. For a 3ϕ, 6-pulse network, then:

$$\boxed{f_R = 6\,f_S}$$ Eq. (8-9)

and the ripple frequency here is:

$$f_R = 6\,(60)$$

$$f_R = 360 \text{ Hz}$$

This relatively high ripple frequency allows much more effective inductive filtering and classifies the 3ϕ, 6-pulse converter superior to those previously discussed.

The peak-to-peak ripple voltage which results from the 6-pulse converter also has been greatly improved over the 3-pulse network. As was noted in Fig. 8-32c, for the 3-pulse converter, each pulse duration of output voltage was 120°, which resulted in a relatively high peak-to-peak ripple voltage. Notice, however, that the 6-pulse converter produces 6 pulses of ripple voltage, each with a duration of only 60°, as noted in Figures 8-34d and 8-36. The natural switching action of the diodes resulting from the 3ϕ supply line-to-line voltage variation results in the conduction of each line-to-line voltage of Fig. 8-36 for the duration of 60° to 120°. Thus, as can be seen, E_{13} is passed to the load between 0° and 60°, then E_{23} takes over between 60° and 120° and so on. The reader will note that these conduction angles correspond to the points of transition where diodes are switching, as previously discussed. In effect, then, the line-to-line voltages (and inverses: E_{23}, E_{31}, and E_{12}) seem to have their most positive portions clipped off from 60° to 120° and

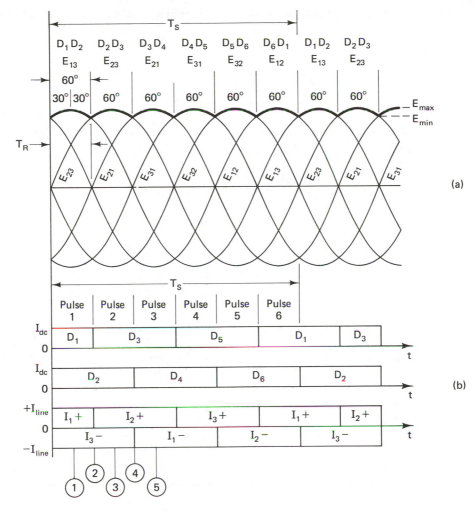

Figure 8-36. Six-pulse output voltages and diode currents.

passed on to the load. The load, in turn, sees a dc voltage that pulsates between $+150$ volts, E_{min}, and $+173$ volts, E_{max}.

The peak-to-peak ripple voltage has also been reduced and it is obvious that the dc load voltage is more constant than the 3-pulse network. Because of this, the average dc voltage which a dc voltmeter will read at the load is greater and can be shown to be:

$$E_{dc} = \frac{3}{\pi} E_{max}$$

$$\boxed{E_{dc} = 0.995\ E_{max}}$$

Eq. (8-10)

Furthermore, it can also be shown that the peak-to-peak ripple voltage is:

$$E_{Rp\text{-}p} = E_{max} (1 - cos\ 30°)$$

$$\boxed{E_{Rp\text{-}p} = 0.134\ E_{max}}$$ Eq. (8-11)

For the 6-pulse converter of Fig. 8-34a, the dc voltmeter connected across the load will read:

$$E_{dc} = 0.955\ E_{max}$$

$$= 0.955\ (173)$$

$$E_{dc} = 165.2\ \text{volts}$$

The peak-to-peak ripple voltage will be:

$$E_{Rp\text{-}p} = 0.134\ E_{max}$$

$$= 0.134\ (173)$$

$$E_{Rp\text{-}p} = 23.2\ \text{volts}$$

The currents of this 6-pulse network can be developed using the dc approach, as used in Fig. 8-35, and are shown in Fig. 8-36b. Referring to Fig. 8-35, and recalling that each line-to-line voltage passes to the load between 60° and 120°, it can be seen that only two diodes conduct per pulse. Note in Fig. 8-36b that during the first pulse diodes D_1 and D_2 conduct; during the second pulse D_2 and D_3 conduct, etc. As this is viewed, notice that each diode conducts for two pulses or 120°. Each diode, therefore, conducts for only one-third of the time (one-third of each period of the source, T_s) and, therefore, carries only one-third of the average dc current, I_{dc}. Since the diodes conduct only in one direction, the current associated with each diode is positive, as shown in Fig. 8-36b.

The line currents I_1, I_2, and I_3 are themselves ac in nature but are not continuous and certainly not sinusoidal. Note in Fig. 8-36b that the line currents are rectangular ac waveforms which pulsate between $+I_{line}$ and $-I_{line}$ and are "active" for 240° of the source period. Line current I_2, for example, is positive for 120° beginning at ②, then is idle for 60°, then goes negative for 120°, and again remains idle for 60°. When a line current is positive, current flows out of the line to the load, and when negative, it flows back into the line from the load, completing the path for current. In Fig. 8-35①, the line current leaves terminal 1, goes through the load, and returns into terminal 3. Line current I_1, therefore, is active and positive, and line current I_3 is active and negative. Compare other points in time in the figures to verify the line currents.

As can be seen, the 3ɸ, 6-pulse converter is an excellent network for conversion of ac to dc power. It is economical with low peak-to-peak ripple voltage; it has relatively high ripple frequency (which enhances effective inductive filtering), and it converts 95.5 percent of the peak ac sine wave, E_{max}, to the average dc voltage, E_{dc}. It should be noted that because of diode response times and other factors, the converter has certain frequency limitations and generally is operated in the 60–400 hertz region, although higher frequencies are sometimes used.

8-2-6 SCR Voltage Control. The 3φ, 6-pulse diode converter discussed in the previous section produces high quality dc voltage, but the output voltage, E_{dc}, is not variable in a practical sense because the 3φ supply voltage is usually fixed. When the control of a dc motor is considered, it is important to be able to vary the dc voltage which supplies the armature and/or field windings. The SCR can be effectively utilized in place of the diodes of Fig. 8-34 so that the average dc output voltage can be varied. A 3φ, 6-pulse thyristor converter is the same circuit as Fig. 8-34 except that the diodes are replaced with SCRs, as shown in Fig. 8-37. The same line-to-line voltages of the 3φ supply are used to power this thyristor converter as in Fig. 8-36a.

The SCR, it will be remembered, is also a unidirectional device which acts very much like the diode except that conduction in the forward direction must be triggered by a voltage pulse applied between the gate (+) and the cathode (−). Once conduction is initiated by the trigger pulse, the SCR will continue to conduct until the forward current drops below the holding current, I_H, or until the SCR becomes reverse biased. When the SCR is reverse biased, it shuts off and has been commutated, exactly like the diode.

In the 3φ, 6-pulse diode converter, all of the diode switching resulted naturally. Points of transition occurred every 60° (refer to Fig. 8-36a) and due solely to the variations of the line-to-line voltages. In the 3φ, 6-pulse thyristor converter, the same result would occur except that the SCR will not conduct without a properly timed trigger pulse. It follows, then, that the SCRs (thyristors) must be triggered with the same sequence with which the diodes conducted in Fig. 8-36b and at 60° intervals. The thyristors will be naturally commutated, as were the diodes, and so it is true that each thyristor will conduct for intervals of 120± just as the diodes did.

The points of transition are the points in time where the thyristors T_1 through T_6 must be fired sequentially, every 60°, to duplicate the output waveform of the 3φ, 6-pulse diode converter. The points of transition will be used as the reference, θ_R, for the firing sequence where $\theta_R = 0°$.

By delaying the firing of the thyristors by a firing delay angle, α, the average dc voltage can be reduced from the maximum (0.995 E_{max}) down through 0 volts to a maximum *negative* voltage (−0.955 E_{max}). It can be proven mathematically that, for the 3φ, 6-pulse thyristor converter:

$$E_{dc} = 0.955\ E_{max}\ cos\ \alpha$$

Eq. (8-12)

It is clear, then, that when $\alpha = \theta_R = 0°$, the maximum E_{dc} is obtained (*cos* 0° = +1.0), and as α is increased beyond 0°, the value of E_{dc} will decrease until α = 90°, at which time *cos* α = 0 and $E_{dc} = 0$. Going beyond α = 90° yields negative values of *cos* α and, of course, E_{dc} is also negative until a negative maximum of *cos* 180° = −1.0 or $E_{dc} = -0.955\ E_{max}$ occurs.

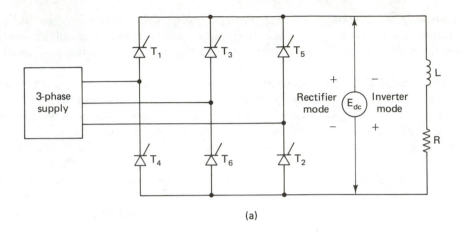

(a)

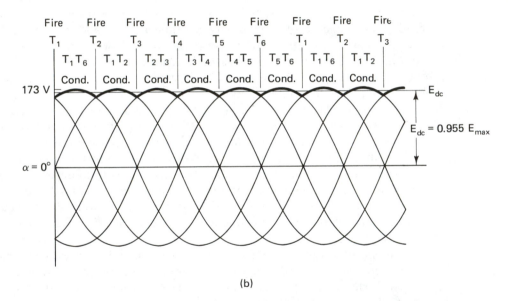

(b)

Figure 8-37. Three-phase, six-pulse thyristor converter.

      The 3φ, 6-pulse thyristor converter can operate in two modes of operation depending on the firing delay angle. Between 0° and 90°, the converter is said to be operating as a *rectifier* because it delivers a positive output dc voltage and load currents. Between 90° and 180° it operates as an *inverter* which, by definition, converts dc power to ac power. As will be seen, the inverter mode of operation requires a dc voltage source of proper polarity to be part of the "load" that will actually deliver power back to the ac source.

      Figure 8-37a shows the circuit configuration of the 3φ, 6-pulse thyristor converter in the rectifying mode with a load resistance, R, in series with a very large smoothing inductance, L. With the firing delay angle at $\alpha = 0°$, the waveform of Fig. 8-37b results (which is the same as that of the 3-phase, 6-pulse diode

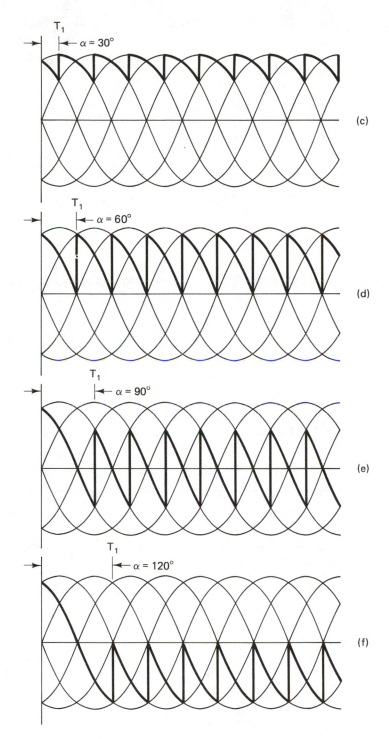

Figure 8-37. (continued)

converter of Figs. 8-34d and 8-36a). Note that the thyristor firing order is sequentially T_1 through T_6 with 60° between each firing. The triggering source (not shown) must be precise as it generates the triggering pulses to the gates of the SCRs to maintain a reliable dc output.

Figures 8-37c and 8-37d show the resulting output voltage waveforms for firing delay angles of 30° and 60° respectively. It is noticed that the average dc voltages, E_{dc}, of each are, respectively, 143.1 volts and 82.6 volts, based on E_{max} = 173 volts. Again, by varying the firing delay angle, α, it is clearly seen that E_{dc} is controllable. The ripple frequency, F_R, does not change but the value of peak-to-peak ripple voltage $E_{R_{p\text{-}p}}$, increases as E_{dc} is lowered.

A critical point occurs at α = 90°, as shown in Fig. 8-37e, where E_{dc} = 0. The output waveform consists of what appears to be a triangular, purely ac waveform with a peak-to-peak ripple voltage, $E_{R_{p\text{-}p}} = E_{max}$ = 173 volts. This represents the maximum peak-to-peak voltage possible for the properly fired 3ϕ, 6-pulse thyristor converter.

Beyond a firing delay angle of 90°, the output waveforms and E_{dc} are negative, as can be seen in Fig. 8-37f, with α = 120°. The average dc voltage can be calculated:

$$E_{dc} = 0.955 \, E_{max} \, cos \, \alpha$$

$$E_{dc} = 0.955 \, (173) \, cos \, 120°$$

$$E_{dc} = -82.6 \text{ volts}$$

Since a negative voltage is produced at the output of this converter while in the inverter mode ($+E_{dc}$ is now at the bottom), it is clear that current will tend to flow *up through* the load. (When a source is supplying power, the current leaves the positive terminal and returns to the negative terminal.) In this case, however, the current cannot flow out of the positive or into the negative because of the thyristors. The SCRs will not allow current to flow from cathode to anode in a reverse direction, yet Ohm's law dictates that a current must flow up through the load because of the presence of E_{dc} across the load. To accommodate this apparent contradiction, a voltage source *must be* placed in series with the load resistance and inductance while the converter is in the inverter mode. This voltage source must also be connected in series-opposing polarity with respect to E_{dc}. If it is not, a reverse breakdown condition could easily exist.

Figure 8-38 depicts the same 6-pulse thyristor converter as in Fig. 8-37 operating in the inverter mode. It is drawn inverted so that $+E_{dc}$ is at the top of the load for convenience and with an added "active" load, E_c, in series with R and L. Note that the polarity of E_c is opposing that E_{dc}, as it should be. (Remember also that if the firing delay angle is reduced to less than 90°, the converter would revert to the rectification mode and produce an output voltage of opposite polarity.)

If E_{dc} is greater in value than E_c, current would have to flow from the 3-phase supply up through the thyristor (T_4, T_6, or T_2) and leave the positive of E_{dc}. Again, this cannot occur because the thyristors will not pass current in the negative direction from cathode to anode. E_c, therefore, must be equal to or greater than E_{dc} under normal inverter operation for current to flow.

With the assumption that $E_c \geq E_{dc}$, the circuit of Fig. 8-38 can be replaced with an equivalent circuit of Fig. 8-39. For discussion purposes, assume that E_{dc} is a constant value because of a fixed firing delay angle and E_c is a voltage of slightly greater value than E_{dc}. Since E_c is greater than E_{dc}, current I_{dc} will flow out of the positive terminal of E_c and enter the positive terminal of E_{dc}, as shown in Fig. 8-39. This is permissible because the current will flow through the SCRs from anode to cathode, and E_{dc} is forward biasing all of the thyristors, as can be observed in Fig. 8-38. In effect, E_c is the active "source," even though it is part of the load circuit, because it is delivering current to E_{dc}. E_c is supplying power to the terminals of E_{dc}, or, in general terms, the "load" is supplying power to the

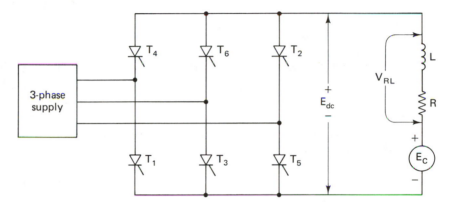

Figure 8-38. Three-phase, six-pulse thyristor converter–inverter mode with "active" load.

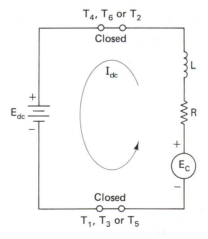

Figure 8-39. Equivalent circuit of three-phase, six-pulse converter–inverter mode.

"source." Assuming that the smoothing inductance, L, is very large with an ideal dc resistance of zero ohms, the current, I_{dc}, can be shown to closely approximate:

$$I_{dc} = \frac{E_c - E_{dc}}{R}$$

Eq. (8-13)

It follows, then, that the power "injected" into the source, E_{dc}, by the active "load," E_c, is

$$P_I = E_{dc}I_{dc}$$

Eq. (8-14)

This power, P_I, is received by the 3ϕ source through the thyristors as their constant firing sequence continues at a constant firing delay angle, α. It is absorbed into the 3-phase sinusoidal source and, if a transformer is being used, is reflected into the source of the primary windings where it can be effectively utilized somewhere at another load. (It is assumed that the 3ϕ source is part of a very firm power grid so that the sinusoidal nature of the waveform will not be distorted.)

SCRs can also be utilized in circuitry when the control of ac motors is considered. While the speed and torque of dc motors are controlled by controlling the armature voltage and field current, ac motors are controlled by varying the voltage and frequency supplied to the motor.

8-2-7 The Cycloconverter.

A cycloconverter is a circuit which uses SCRs to change the frequency of a source to a lower frequency. Consider the 3ϕ cycloconverter of Figure 8-40a consisting of six SCRs connected as shown and supplied by a 60 Hz frequency source. SCRs T_1, T_2, and T_3 are connected in phases 1, 2, and 3 respectively and when properly triggered will supply a positive voltage to the load. These are called the "positive group." SCRs T_4, T_5, and T_6 are each connected in parallel with the positive group as shown and, when properly triggered, will supply a negative voltage to the load. These are called the "negative group."

The operation of the cycloconverter of Fig. 8-40a can be easily understood by referring back to the 3ϕ, 3-pulse converter of Fig. 8-32. Remember that diodes D_1, D_2, and D_3 are switched on and commutated off naturally due to the variations in the phase-to-neutral voltages, and they produce the 3-pulse positive dc output voltage shown in Fig. 8-32c. If, on the other hand, the diodes of Fig. 8-32a are turned around and connected so that the diodes' cathodes are pointing toward the source, a negative dc output voltage will result, with an average dc voltage supplied to the load of $-E_{dc}$. The ripple voltage thus produced will follow the negative portions of the phase-to-neutral voltage between $-\frac{1}{2}E_{max}$ and $-E_{max}$. Notice that the SCRs of Fig. 8-40a are connected such that both conditions are possible.

If the positive group of SCRs is fired sequentially with a firing delay angle of $\alpha = 0°$ while the negative group is held inactive, they can be made to act as simple diodes, and the resulting output waveform will duplicate that of Fig. 8-32c. Similarly, if the positive group is held inactive by not applying firing pulses to the gates and the negative group is triggered with a firing delay angle of $\alpha = 0°$, a

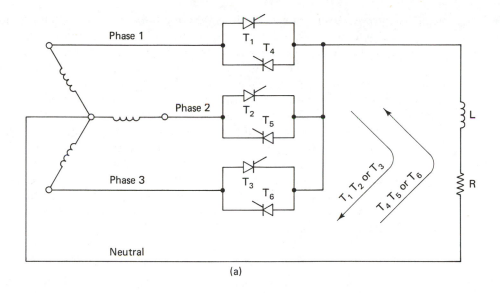

(a)

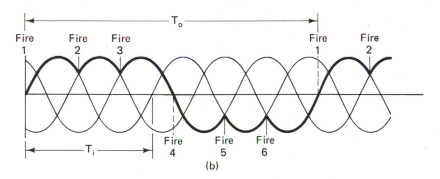

(b)

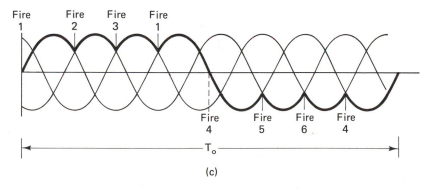

(c)

Figure 8-40. Basic cycloconverter and output waveforms.

negative output waveform will result as described above. (Note that the negative groups' cathodes are pointing toward the source.) By alternating the firing of the positive group and the negative group equally, an output waveform is produced which is ac in nature. The output has a lower basic frequency than the source frequency from which it is derived and, it will be noted, the output voltage is single phase in nature. Of course, the triggering source must be reliable and is often microprocessor-based with a closed loop control scheme.

Figure 8-40b shows an example of an output waveform which is produced by alternately triggering T_1, T_2, and T_3 in succession and then triggering T_4, T_5, and T_6 with the firing points as shown. It is observed that the time period for one cycle of the output waveform, T_o, is much longer than for the input cycle, T_i, of 60 hertz. Note that as the output waveform completes one cycle, the input has cycled 2-1/3 times, or 840° as noted. The frequency of the output can be calculated for the figure of 8-40b as follows:

$$T_o = \frac{d_o}{360} T_i \qquad \text{Eq. (8-15)}$$

$$f_o = \frac{1}{T_o} \qquad \text{Eq. (8-16)}$$

where

T_o = time period of the output (sec)
T_i = time period of the input (sec)
d_o = number of degrees the output takes to complete one full cycle based on the input cycle

For the figure of 8-40b then:

$$T_i = \frac{1}{f_i} = \frac{1}{60 \text{ Hz}} = 16.67 \text{ msec}$$

and

$$d_o = 840°$$

$$T_o = \frac{d_o}{360} T_i$$

$$T_o = \frac{840}{360} (16.67)$$

$$T_o = 38.9 \text{ msec}$$

$$f_o = \frac{1}{T_o} = \frac{1}{38.9 \text{ msec}}$$

$$f_o = 25.7 \text{ Hz} \cong 26 \text{ hertz}$$

It is clearly seen that with the implementation of the cycloconverter a 3ϕ, 60 hertz power signal can be converted to a single phase, 26 hertz power signal.

The output frequency can be varied from the 26 hertz down through lower frequencies to 0 hertz. This can be accomplished by firing the positive and negative groups equally for longer periods of time, which ultimately results in larger values of T_o. Figure 8-40c serves to illustrate how this is achieved. The positive group is now triggered with one additional pulse as is the negative group, i.e., T_1, T_2, T_3, and T_1 and then T_4, T_5, T_6, and T_4. By extending the "on time" of both groups, very low frequencies are obviously obtainable. The reader is encouraged to determine the output frequency for the conditions of Fig. 8-40c.

The cycloconverter described here serves to present the basic function of the cycloconverter in terms of basic frequency reduction. It should also be noted that by varying the firing delay angle to the SCRs, the RMS value of output voltage can also be varied simultaneously. To control these output variables, complex closed-loop computer controlled firing techniques are generally used with great success.

QUESTIONS

8-1. Name two basic considerations that affect the relation between a dc motor and the power supply that drives it.

8-2. What is the distinction between a pilot device and a primary control device?

8-3. What is meant by the duty cycle of a motor?

8-4. What is the function of a limit switch?

8-5. What type of motor-driven device might require position holding after the motor is shut off?

8-6. What is the difference between a magnetic contactor and a relay?

8-7. What is the difference between a magnetic contactor and a magnetic starting switch?

8-8. What is the distinction between a normally open and a normally closed contact on a control element?

8-9. What action is performed by a holding contact in a relay or contactor?

8-10. What is the broad overall meaning of the terms sensor, pickup, and transducer? Note that these terms are frequently used interchangeably.

8-11. What limits the range of dc motor speed control when it is performed by varying shunt field current?

8-12. What is the economic disadvantage of speed control by means of series armature resistance?

8-13. Under what conditions does a diode function as a closed switch? As an open switch?

8-14. What ratings are critical in a power diode? Why?

8-15. Why are power diodes sometimes connected in series? In parallel?

8-16. Draw a half-wave rectified diode bridge converter being fed from the secondary of a transformer. Briefly describe how it operates.

8-17. Repeat Question 8-16 for a full-wave rectified bridge converter.

8-18. Draw a PNP transistor and indicate polarities at all connection points so that it acts as an open switch. As a closed switch.

8-19. What is the purpose of a 3-phase, 3-pulse converter? How is it different from a full-wave bridge rectifier converter?

8-20. What is the advantage of using a 3-phase, 6-pulse converter instead of the 3-phase, 3-pulse system?

8-21. Compare in detail the characteristics of a diode and an SCR and note any similarities and differences.

8-22. How is an SCR turned on? Off?

8-23. What is a snubber circuit and why is it used?

8-24. Draw the symbol for a TRIAC and discuss how it operates. How is it similar to an SCR?

8-25. Discuss the operation of a 3-phase, 6-pulse thyristor converter, and detail the main advantage of using such a system in the rectifier mode.

8-26. Describe how and why the 3-phase, 6-pulse thyristor can be used as an inverter. What is the function of an inverter?

8-27. Describe a cycloconverter and its function.

9

Direct-Current Motor Control

It was shown in Chapter 6 that any substantially sized dc motor needs to have its armature current controlled at a reasonable level when starting. A starting resistance can be placed in series with the armature to limit the starting current to an appropriate level. As the motor builds up speed, it can be gradually reduced to zero ohms as the motor attains rated speed. This reduction of armature starting resistance can be accomplished using conventional switching relays and power resistors.

Chapter 6 also showed that to reverse the rotation of a dc motor, the polarity of the armature must be reversed in relation to the field, or vice versa. This, too, can be accomplished using conventional, electromechanical switching relays which physically change connections and therefore, polarity to one or the other.

As noted in Chapter 8, however, the dramatic improvement in power semiconductors has made the control of dc machines possible and more efficient with their use in what is called solid state motor control. By utilizing power semiconductors and the basic circuit configurations as introduced in Chapter 8, it is possible to start, stop, increase or decrease speed, and reverse a dc motor. These functions can be executed with a very fine degree of control with the added benefit of increased system efficiency and reliability.

Only basic conventional and solid state control circuits are presented in this chapter, but they will serve as the foundation for further study of the variations and other control concepts that the student will undoubtedly encounter in the field.

The listing below of some of the considerations that should be studied in order to determine the scope of a motor control scheme will help put the problem of motor control selection in its proper perspective. It is assumed that the motor choice has been considered based on Chapters 6 and 10.

9-1 MOTOR LOAD CHARACTERISTIC LIST

Any motor control selection must consider the following items.

9-1-1 Power Supply. At the risk of being obvious, it is necessary that the motor and its power supply have the same or nearly the same voltage rating. The power supply must make sufficient current available for steady state and peak current requirements. The interconnecting wiring must have adequate cross section and safe insulation.

9-1-2 Starting Requirement Considerations.

(1) Will the load be started by an *operator* or will some *automatic device* be used?

(2) If an operator is used, will he or she be a skilled person with specific training or must the control be *foolproof*, even "idiot proof"?

(3) If manually operated, is a *pushbutton* control desired or will a *rotary lever* that requires some judgment be used?

(4) Will a special control *location* be required for operator safety?

(5) If an automatic starting control is to be used, will the beginning of the start cycle be based upon *standard pilot devices,* such as a float switch, thermostat, pressure switch, or timing device?

(6) Will it be necessary to synchronize the starting with another operation or to avoid some interfering position condition so that some form of *interlocking* device is required?

(7) Is the starting requirement so unusual that a new type of control needs to be developed?

(8) What is the starting *duty cycle*? Will it be frequent or infrequent?

(9) Are there significant *starting torque* requirements, or does the load have *high inertia*? Note that either of these conditions would prolong the starting cycle.

(10) Will the starting *cycle time* be critical?

(11) Is starting *smoothness* or the *acceleration* profile a critical feature for satisfactory operation?

(12) When starting, is the machine always *loaded*, or always *unloaded*, or are the starting load conditions of a *random* nature?

(13) Does the motor require small increments of motion or *jogging*?

9-1-3 Stopping Requirement Considerations.

(1) Is the machine to be manually stopped by an *operator* or by an *automatic device* or both?

(2) Do the same operator *skill requirements* exist, or must the machine be stopped by a totally unskilled person?

(3) Will *limit switches* for temperature, pressure, or position be used, or will fixed intermediate *location stops* be required as in an elevator?

(4) Is there a different requirement for a *normal stop* versus an *emergency* stop?

(5) Will the driven device have characteristics requiring the motor to be stopped by some *braking means* or by a definite reverse torque? (Reversal here is called *plugging*.)

(6) Does the motor require *position holding* after it is stopped?

9-1-4 Reversing Requirement Considerations.

(1) Is motor reversing required *regularly*, occasionally, or never?

(2) Might reversing be required in an emergency even though not a normal sequence? Might reversing be required during an occasional loading or unloading cycle, as in a process machine, but not during normal operation?

(3) If *plugging* is defined as controlled reversing for maximum deceleration, might this type of reversing be required during a normal or emergency stop even though actual rotation in reverse is never foreseen?

(4) Is the normal reverse cycle of operation essentially *similar* to forward motion, or are there specific and unsymmetrical requirements to meet in one direction and not the other? For example, might one way require small increments of motion or *jogging* and the other not? View jogging as a small motion rather than an actual number of degrees of position.

9-1-5 Normal Running Requirement Considerations.

(1) Is the running duty cycle essentially *continuous* or is it of *short time* only? An extreme of this is an automotive starter motor, which has a duty cycle of from a few seconds to as much as 1 minute. Does the duty cycle have a regular relation that is repetitive so that advantage may be taken of a smaller motor that can meet the average load, but not a steady dose of the peak load?

(2) Is at least a part of the regular cycle such that the load *drives the motor* as in a descending loaded crane or hoist? This is called an *overhauling* load.

9-1-6 Speed Control Requirement Considerations.

(1) Does the load have essentially a *constant speed* requirement such that an initial matching or calibrating adjustment is sufficient?

(2) Is there a *precise speed requirement* that may, for example, require a closed-loop control system?

(3) Are there regularly *selected speed settings* such that fixed adjustments may be cyclically selected?

(4) Is an *adjustable speed* required, as in a changing machine process or a variable speed vehicle?

(5) If an adjustable speed, is the adjustment to be *manual* or *programmed* in an automatic way?

(6) Can the *speed vary with the load*, or must a relatively constant speed be maintained as load torque changes?

(7) Is there a requirement for *progressive slowdown*, as in a vehicle or a hoist condition? Need this slowdown be programmed?

(8) Are *combinations* of items 1 through 7 required?

9-1-7 Safety and Environmental Requirement Considerations.

(1) Is there a *hazardous* or *explosive atmosphere* around the load and/or control that requires explosion-proof features?

(2) Are there special *sanitation* and *cleaning* requirements, as in food processing or operating room environments?

(3) Is provision required for *operator incapacitation* to result in an immediate and perhaps controlled stop sequence? This is sometimes known as a *dead man* control.

(4) Are potential *short-circuit currents* suitably protected by fuses, and are *overload currents* provided for by suitable circuit breaking devices? Note that one or the other or both may be required depending on safety, legal, or device protection requirements.

(5) Is suitable *low-voltage protection* provided? When there is low-voltage dropout, this protection should require manual resetting to prevent inadvertent starts.

(6) Are suitable and appropriate *mechanical protections* provided for over-travel limit stops, *overspeed* cutouts, and automatic electromechanical *brakes*? Note that an overspeed protection is required for an unloaded series motor or for an unloaded shunt motor that has lost its field circuit current.

(7) Is mechanical *overtemperature protection* to be provided that may be needed for blocked ventilation even when electrical and performance parameters are normal?

The preceding lists, although extensive, are by no means all-inclusive. It should serve to alert the motor student to some of the factors that must be considered in a motor and motor control installation design. Fortunately, many installations are comparatively simple and can be reasonably accommodated with standard packaged devices. The more complicated automatic machine control circuits may require sufficient custom designing that, even with a combination of standard component elements and pilot devices, the control will exceed the cost and space requirements of the motor.

The control of motors is a very extensive and specialized field and cannot be comprehensively covered in a text such as this. As a result, only general types of control schemes will be presented with the hope that the student will be encouraged to study the various control references in detail.

It will be seen in these references that the control of motors is evolving dramatically and that solid state control has clearly overtaken the slower and more cumbersome conventional, electromechanical relay-type of control. The student, however, cannot dismiss the conventional concepts or devices because systems already in place generally are conventional types and must be maintained.

9-2 MANUAL DIRECT-CURRENT MOTOR STARTERS

The overall problem of starting a large dc motor was outlined in Sect. 6-9 along with a worked example to show how the current-limiting resistance values are determined. There are basically two ways in which these resistances can be switched in or out of a control circuit to perform their desired function: manually or automatically.

Manual switching of current-limiting resistances in series with the armature of the dc motor is rarely used in modern control design. In the past, two-point, three-point, and four-point manual starters were used to start dc motors by manually moving a lever which would sequentially reduce the current-limiting resistance in the armature circuit until full speed was reached.

9-3 AUTOMATIC DIRECT-CURRENT MOTOR STARTERS

The control of dc motors today is accomplished by automatic means. The concept of starting the dc motor, however, has not changed. It is still absolutely imperative that the starting armature current be restricted and controlled at a safe level for the reasons noted in Chapter 6. Any starting sequence that is initiated either by pushing a button or by the closure of some other form of pilot device must have automatic features which progressively remove the starting resistance until full-line voltage is applied to the armature circuit. Even the simplest control must automatically reset when the stop button is pushed or when another pilot device calls for a shutdown.

All automatic control circuits are either open or closed loop in function. The distinction between the two concepts is determined by whether operation of the motor is independent of its own output or at least partially dependent on its output. In open loop, there are no pilot control devices to detect mechanical output of the motor. In closed-loop control, pilot control devices detect shaft speed, for example, and feed this information back to the control circuit where adjustments are automatically made within the circuit to control the speed at a desired set point.

9-3-1 Definite Time Acceleration dc Starters.
This class of starters uses various forms of time-delay relays to achieve a time sequence in removing increments of the current-limiting resistance. The desired time delay is achieved by using one of a number of well-recognized variations:

(1) time-delay relays,
(2) time-delay contactors,
(3) time-delay relays that are deenergized after their period of use, and a
(4) motor-driven or mechanical escapement timer

Type 3 will be described in detail since it uses many interesting features and has a number of advantages. Consult one of the references for similar detail on the other types of definite time-delay relays.

Figure 9-1 shows a representative definite time acceleration dc starter of type

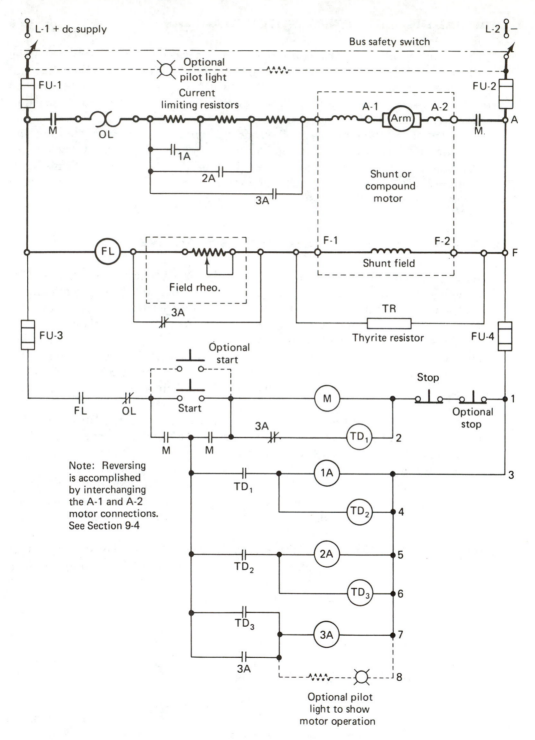

Figure 9-1. Definite time acceleration dc starter using adjustable time-delay relays.

3. This schematic should be followed in detail since it has features typical of many specific controls. As alternatives present themselves, they will be mentioned, since not all generically similar controls are identical. Local requirements will determine specific differences on this or any other control system.

This starter operates as follows:

(1) Power is established in the circuit by closing the bus safety switch. If this switch is overhead at stepladder height, as is frequently the case, another isolation switch must be established at the machine in order to shut off the field circuit. If the control location is remote from the motor, a pilot light is usually added to show a live circuit. This is shown in dotted lines. Note the distinctive symbol used in an industrial diagram. If this light is colored, the color is indicated by a letter code (G for green, and so on). When these switches are closed, the shunt field circuit is live, and the field loss relay coil FL will pull in and close the normally open FL contacts on control line 1. Only then is the starting control circuit live. Conversely, any open circuit anywhere in the shunt field line will cause the FL contacts to open and shut down the motor.

(2) The start button is momentarily pressed, which actuates the main start switch coil M on line 1. Coil M closes four sets of contacts, two main contacts in the armature circuit and two auxiliary contacts, that establish a path around the start button and hold coil M in its active state. Note that any reasonable number of start buttons may be placed in *parallel* with the main start button. All will perform the same function. The closing of the major M contacts places the armature across the line, and the motor starts and begins to accelerate. The armature current is limited to its design value by the full starting resistance. At the same time, the coil of the first time-delay relay TD_1 is also actuated. This is shown on line 2.

(3) After a time delay that has been preadjusted to match the acceleration rate of the motor, relay coil TD_1 overcomes its built-in holding effect and closes the TD_1 contacts on line 3. This contact closure simultaneously activates the coils of relay or contactor 1A on line 3 and time-delay relay coil TD_2 on line 4. Contactor 1A operates instantly and closes its 1A contacts, which short circuit a predetermined portion of the current-limiting starting resistors in the armature circuit. The motor current rises again and acceleration continues.

(4) After another preadjusted time, the time-delay relay coil TD_2 actuates its contacts. These can be seen on line 5. Contactor 2A then actuates, and time-delay relay TD_3 on line 6 starts its timed cycle. When contactor 2A closes its contacts, the second portion of the armature resistance is shorted out, thus continuing the acceleration cycle. It may seem redundant to have relay TD_1 operate contactor 1A and TD_2 operate 2A, and so on. This duality is used on large motors so that smaller and less costly time-delay relays may be used. Then the contactors, which need not be an adjustable delay tape, are chosen for their contact current-carrying capacity. On motors of around 5 hp or less, the two functions are reasonably combined. In addition, there may be any reasonable number of acceleration steps. It can be seen that control circuit lines 3 and 4 are exactly repeated on lines 5 and 6; they could, of course, be repeated again if needed.

(5) The final time-delay relay, TD_3 in the case shown, closes after its proper

delay and actuates contactor coil 3A, which is shown on line 7. This last contactor has a main set of contacts and three auxiliary sets, of which two are normally closed. The main contact set of 3A shorts out the last of the starting resistance and places the armature circuit finally across the full-line voltage. At the same time, three other actions take place. The other but normally smaller open contacts of $3A_1$, shown between lines 7 and 8, close and hold contactor 3A in its actuated state. A normally closed set of 3A contacts in control circuit line 2 opens and allows time-delay relay TD_1 to drop out. This lets 1A and TD_2 drop out, which in turn lets 2A and TD_3 drop out. Contactor 3A is now held by its own contacts, between lines 7 and 8. All the rest of the starting sequence relays are now inactive and do not consume wattage. In fact, only the main contactor M, the field loss relay FL, and the final 3A contactor are active and drawing coil current. The last set of 3A contacts is another normally closed set, which is connected to bypass the shunt field rheostat. It opens when 3A actuates and lets the field current take over and accelerate the motor to its final speed. By this means, the first three steps of acceleration are taken with the full field, and thereby the greatest possible back emf is generated. Finally, a pilot light, which is shown on line 8, can be connected if desired to show normal motor operation. This may be desired at a remote start or stop control station if the motor operation is not visible.

Normal operation can be stopped at any time by pressing the stop button, which is shown on line 1. Note that any additional stop locations which are desired for safety or convenience can be wired in *series* with the basic stop push button. A number of emergency or safety stop conditions are built in, as described below:

(1) An overload on the motor that draws a continued higher-than-acceptable current will "cook off" the overload thermal element OL in the armature circuit. This thermal element will release the normally closed OL contacts in control circuit line 1. The effect of opening the OL contacts is exactly the same as pressing the stop button. The OL element and OL contacts are physically a part of the main contactor assembly M. They must be manually reset after a shutdown. As in a normal stop cycle, everything resets to the off position.

(2) Any form of open circuit in the shunt field circuit will cause a potentially dangerous overspeed. This loss of field current will cause the FL field loss relay to drop out and open its FL contacts. This action in turn causes a normal stop sequence. Had this been a series motor with no shunt field circuit, a centrifugal overspeed switch would have been connected to the armature shaft. An overspeed switch, which is usually labeled OS, would have a *normally closed* set of OS contacts in the same place in the circuit as the normally open FL contacts. Remember that the FL contacts closed only after field current was initiated.

(3) A severe short circuit in the armature or field circuit will immediately open fuse FU-1 or FU-2, or both. This shuts down the whole circuit, preventing further damage or perhaps fire.

(4) Short circuits in the control circuit that might cause breakdown of control components would not necessarily draw enough current to open fuse FU-1 or FU-2. Lower-rating fuses FU-3 and FU-4 protect the small but costly relays.

The final component is the Thyrite resistor, which is connected around the shunt field in larger-sized motors. This resistor is normally coded as TR in these circuit diagrams. Since the shunt field has many winding turns and is intimately coupled with the field magnetic core, it is highly inductive. This inductance means that an inductively generated arc would be present at whatever contacts actually served to break the field circuit when the motor is shut down. In this case, the bus safety switch or the other protective motor isolation switch would have a short life without the Thyrite resistor. Thyrite has the characteristic of having low resistance under a high voltage and high resistance under normal operating voltage. Thus, the TR resistor does not draw appreciable current normally, but will quickly and effectively absorb the high inductively generated voltage when the switch is opened.

This whole control circuit is typical of four or five different ways of achieving a definite time acceleration of a dc motor. All related circuits share the fault that they do not recognize cycle-to-cycle variations in the starting and acceleration time requirements. Some types can be adjusted fairly readily, as can the one shown and described above. The ease of adjustment depends on which of the varieties of time-delay relays is chosen. In the case above, each step is separately adjustable and can be lengthened or shortened to fit the average starting cycle. Simpler circuits may not share this adjustability. A study of the references is recommended if a problem is encountered, since this chapter is only a brief overview of the subject.

The major disadvantages of the *definite time acceleration* type of motor starters can be overcome if the circuit operation can be made to be a function of the armature circuit current. The sensing of a load-related variable such as armature current and the use by some means of that variable makes the next class of starters much more flexible in operation.

The family of related starter designs are closed-loop controls in the sense that the input is at least partly related to the output. The input current is sensed in various fashions, and control sequence is made dependent upon that input current.

9-3-2 Current-Limit Acceleration Starters. These starters are so named to differentiate them from the previously described definite time acceleration starters. There are again four specific varieties:

(1) *Counter emf relays* or speed-limit method. In this type of control, voltage-sensitive relays are connected across the armature. As the proportion of line voltage shifts from the current-limiting resistors to the armature, these relays cut in at progressively higher voltages.

(2) *Holding coil relay* method. By the use of relays with two opposing windings, a decrease of armature current may be sensed and used to close relays. The two-winding or differential relay, in this sense, has a multiturn voltage winding in the normal sense and a relatively few turns of much heavier conductor for the current winding. They are connected so that the normal winding will pull the relay closed or actuate it. The current winding with its few turns is connected in series with the armature current in a direction such that its magnetic flux opposes the voltage

winding. Initially, the current winding has the greater ampere turns and overrules the voltage winding, thus keeping the relay in its normal or unactivated position.

(3) *Series relay starters* are another type where the relay coils carry the armature current. These relays have single coils, but with few turns of heavy wire and relatively strong mechanical springs to hold them unactivated. Due to small air gaps and low inductance, they are very fast acting. They are connected in this type of starter so that they close or actuate in milliseconds when they first "see" the high starting current. As the current decreases, they are forced back to their unactuated position. When this takes place, their normally closed contacts cause the operation of contactors, which sequentially short out parts of the current-limiting resistors. The contactors are normally connected to operate at the same time as the current relays, but have a much slower action owing to their high inductance coils.

(4) *Holding coil and voltage drop starters* combine the actions of items 1 and 2 but have a few advantages. The full armature current does not pass through the relay coils, so smaller units can be used and very large motors may be controlled.

A current-limit acceleration starter using holding coil relays is shown in Fig. 9-2. The type of holding coil or relay used here has two voltage coils connected in opposition. The coils do not have the same number of turns or the same wire size. The coil having the larger magnetic effect is used at a smaller voltage to create a greater magnetic force to hold the relay open. The action of this starter is as follows:

(1) Power is established in the circuit by closing the bus safety switch.

(2) Momentarily depressing the start button actuates the main contactor M, closing both its main and auxiliary contacts. The motor then starts and begins to develop a back emf as its speed builds up. When M closes, the greater part of the line voltage appears across the current-limiting resistors. This voltage is sensed by the holding coils of the various differential relays. At the same time, the actuating coils of the various differential relays are also excited. Because of the design of the opposing coils, the HC coils initially overcome the A coils and all the relays are held in their normal position.

(3) As the motor speed builds up, the voltage across the starting resistors decreases since the back emf in the armature is then a part of the total voltage. One HC coil loses its dominant force first and surrenders to the 1A coil so that contacts 1A are closed. This shorts out the first part of the starting resistance, and the starting current again builds up.

(4) As the speed continues to build up, the available voltage across the current-limiting resistance continues to decrease. Coil 2HC loses its control of the position of differential relay 2 and coil 2A assumes dominance, thus closing relay contacts 2A and repeating the process of step 3. The starting current again increases.

(5) As the back emf continues to increase, coil 3A overcomes the sagging force of coil 3HC, and differential relay 3A is closed. This completes the process of shorting out the starting resistance.

DIRECT-CURRENT MOTOR CONTROL   CHAP. 9

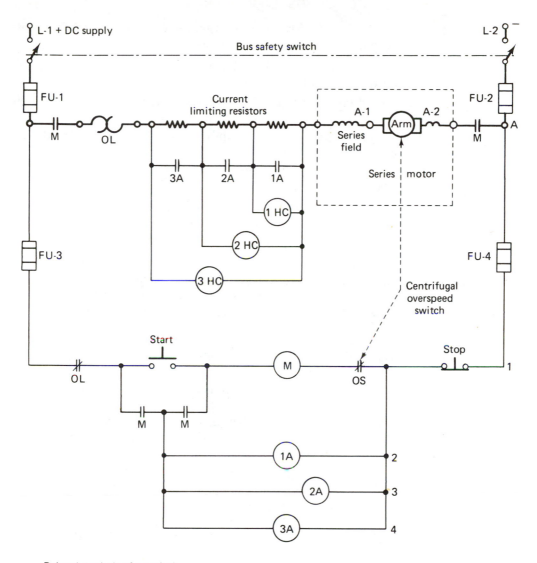

Related symbols of a typical
holding coil or differential relay

^{1A} ⊣⊢ Contacts

(1 HC) Holding coil holds contacts
in normal state

(1A) Actuating coil of same relay
actuates contacts after overcoming holding coil force

Note: Reversing
is accomplished
by interchanging
the A-1 and A-2
motor connections.
See Section 9-4

Figure 9-2. Current limit acceleration dc starter.

Some of the same features for stopping the motor are present as in the definite time acceleration control:

(1) A normal stop is achieved by pressing the stop button, which breaks the circuit across control line 1 and allows all relays to drop out.

(2) An overload condition is dealt with by the OL heat-sensitive element in the armature circuit.

(3) Short circuits in the armature circuit will melt fuse FU-1 or FU-2 and open the whole circuit.

(4) Again, as before, any short circuit in the control part of the circuit will open fuse FU-3 or FU-4.

(5) If, for any reason, a series motor loses its load, it will overspeed dangerously or at least mechanically expensively. A centrifugally operated overspeed switch will open the OS contacts in this situation.

This type of control requires that the relay coils be matched to the situation. After this match is designed and specified, further adjustment is difficult and limited in scope. However, the circuit will properly accelerate the motor in the best time relation for any load the motor can handle from minimum to maximum.

9-4 REVERSING CONTROL OF DIRECT-CURRENT MOTORS

Various means of reversing the armature circuits are used in automatic dc starters. In any case, the *whole armature circuit* must be reversed as a connected group, including commutating fields and compensating windings if they are present. The series field, if it is present, is the only part of the armature circuit that remains unreversed.

The fundamental use of reversing contacts or switches is shown in Fig. 9-3. The symmetrical rectangular array of two forward or F contacts and two reverse or R contacts surrounding the armature circuit is frequently used in control diagrams. Again, if commutating fields or compensating windings are present, they must be between the A-1 and A-2 contacts along with the armature itself.

9-4-1 Pushbutton Control with Reverse. Various types of reversing circuits are regularly used. Figure 9-4 shows a normal definite time-delay starter circuit arranged for pushbutton-controlled reversing. This circuit combines the functions of the TD time-delay relays and the A contactors of Fig. 9-1. Obviously, the same type of reversing control could have been added to Fig. 9-1, since once the main contactor M is closed the reversing part is fixed in position.

The specific reversing controller shown in Fig. 9-4 works as follows:

(1) The bus safety switch must be closed first as in all normal controls.

(2) Whichever direction is chosen, its appropriate button is pressed. In this case, it is pressed *forward*. First, this breaks any possible connection to the reverse contactor coil R, and then makes connection to the forward contactor coil F. All F contacts are then closed, thus selecting the armature circuit. This is line L-1

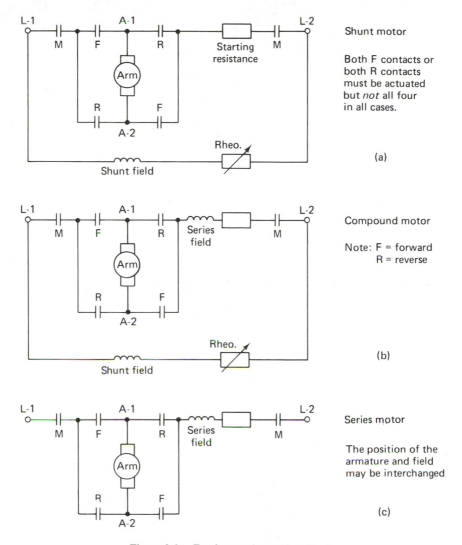

Figure 9-3. Fundamental reversing circuits.

through the starting resistors to armature A-1, then out armature A-2 to line L-2. As soon as the F contactor closes, its auxiliary F contacts hold the circuit past the forward button and through the F coil. At the same time, the auxiliary F contact in control line 2 closes the circuit to the main contactor coil M. This, in turn, closes the M contacts in the armature circuit, and the motor starts with the full starting resistor in series with the armature. The M auxiliary contacts close the circuit for the first time-delay relay coil TD_1.

(3) After relay TD_1 actuates, the first stage of the starting resistance is cut out by the TD_1 contacts. The starting sequence proceeds in the normal fashion of a definite time acceleration controller.

(4) Stopping the motor is accomplished by pressing the stop button in the normal

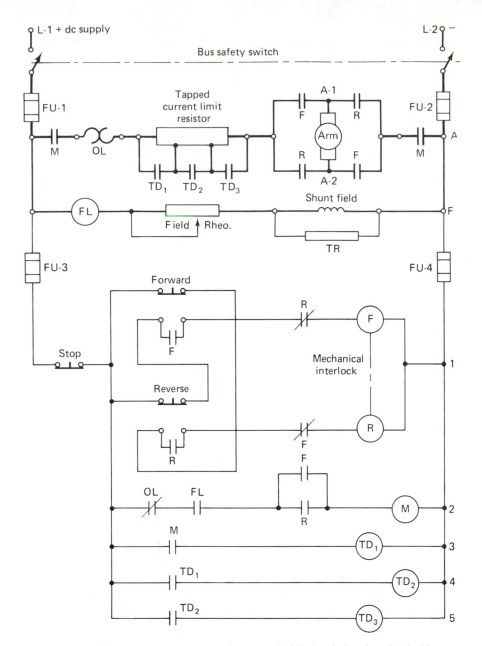

Figure 9-4. Direct-current motor reversing control with electrical and mechanical interlocking.

manner. The F contactor drops out, thus releasing the M contactor and in turn the various TD relay–contactors. The complete circuit is reset. In this case the motor will coast to a stop in its own good time depending upon its speed and the inertia of its load.

(5) Normally, the reverse operation is achieved by simply pressing the reverse

pushbutton. However, if the motor is locked in forward operation, nothing happens. This hold-forward action is due to the operation of the auxiliary F contacts.

(6) When the circuit is released and passive, pressing the *reverse* button will first break the line to the F contactor and then activate the reverse contactor coil R. The reverse connections are set up in the armature circuit, and the auxiliary R contacts activate the main contactor coil M. The reverse start then proceeds exactly as a forward start, with the progressive reduction of starting resistance until full reverse operation is achieved.

This could be a compound motor if its armature circuit was connected according to Fig. 9-3b, or a series motor if connected according to Fig. 9-3c. In this last case, there would be no shunt field circuit.

This particular control has no provision for rapid slowdown and reverse. It can be "beaten" to a certain extent, and in service, an operator would soon find out its features. As a result, when rapid cyclic reversals are required, some further control provisions are normally provided. Returning to this control and other similar simple reversing schemes, the rapid reverse problem is as follows: if an operator were to first press the stop and then immediately press reverse, the control would pull in and operate its M contactor and start its sequence. The problem is that the rapidly rotating armature is still generating a back emf. This back emf is now, after reversal, in a direction that aids current flow instead of opposing it. As a result, more current will flow than during a normal start. Sometimes nearly twice a normal start current, or 200 to 250 percent of normal running current, will flow. If the control is a definite time duration type and the load is of high inertia, the control may begin to switch out resistance before it is due. A current-limiting starter circuit will hold its maximum resistance setting until the current drops to design conditions. This will not take place until the motor has come to a stop and started in the opposite direction. Since this whole process is erratic and extremely hard on the motor and the control system, a better means is required when rapid controlled reverse is required.

9-5 PLUGGING REVERSE CONTROL

When a process that is being performed by a dc motor requires repeated and controlled rapid reversal of direction, a *plugging reverse* control is used. Plugging in this sense means a reversal of armature circuit polarity combined with sufficient armature circuit *added* resistance that the current is held to a predetermined maximum. At the moment of reversal, the torque direction is reversed, even though the rotation direction remains the same. The rotational speed decreases to zero and instantly begins acceleration in the reverse direction. At this point, the reverse acceleration procedure is normal. Acceleration continues until the full reverse rotational speed is reached.

A modification of the same circuit will produce a *plug stop* control. This control will apply full reverse torque to the motor until it is brought to a standstill. At the moment the rotation actually reverses, the circuit is broken by a *plugging switch* or *zero speed switch*.

9-5-1 Typical Plugging Reverse Controller. The operation of a typical plugging reverse controller is shown in Fig. 9-5 and may be described step by step:

(1) The initial start either forward or reverse is directly analogous to the control in Fig. 9-4 with a few exceptions:

    (a) This motor control has only two normal starting resistor steps controlled by time-delay relays TD_1 and TD_2.

    (b) The third step of resistance, which is controlled by the plugging contactor, P, is normally held closed when starting from a standstill. The voltage drop across the plugging forward, PF, or plugging reverse, PR, relay coils, as the case may be, is normally enough to pull one of them in and operate the P contactor.

    (c) Figure 9-5 shows a series motor, but a shunt or compound motor could have been controlled in a similar fashion.

(2) If the reverse button is pressed during normal forward operation, a plugging reverse sequence is initiated. Similar but mirror-imaged action is obtained if the forward button is pressed while running reversed. When the reverse button is pressed, the forward contactor coil circuit is opened and the F contacts revert to their normally open position. Further motion of the reverse button closes the normally open reverse contacts. This action causes the reverse contactor to pick up and hold as soon as the forward contactor drops out and closes the normally closed F interlock contacts. The R contact on line 3 closes immediately after the F contact on the same line opens. This causes the main contactor M to reclose, but not until the TD relays have dropped out. This time, however, the F contacts around the armature are open and the R contacts are closed. The armature circuit then sees a complete reverse of polarity, and reverse torque is immediately built up. In this case, the PF relay has dropped out and the P contactor has opened, thus placing additional series resistance in the armature circuit and preventing damaging currents from flowing. At this moment there is the substantial but designed-limit current flowing through all the tapped starting resistors. Thus, the polarity across the portion of the resistor that is to be bridged by the P contacts is as shown in Fig. 9-5. At the same time, the motor is still rotating at nearly its full forward speed. The field still has its original polarity and flux direction, and thus the armature has the polarity shown on the schematic. Since the R contacts are now closed, these two polarities are opposed, and there is insufficient voltage to close the PR relay.

(3) As the speed decreases under the opposed torque, the generated voltage in the armature decreases, and the voltage across the PR relay coil increases until at about zero speed the PR relay closes. The actual closure point is not critical as long as the current is held to desired limits. These limits are determined by the voltage drop in the resistor, the remaining emf in the armature, and the pull in voltage of the PR coil.

(4) When the PR coil has closed its contacts, the P contactor closes and shorts out the plugging portion of the starting resistance.

(5) From this point on the operation of the starter is as a normal definite time acceleration starter.

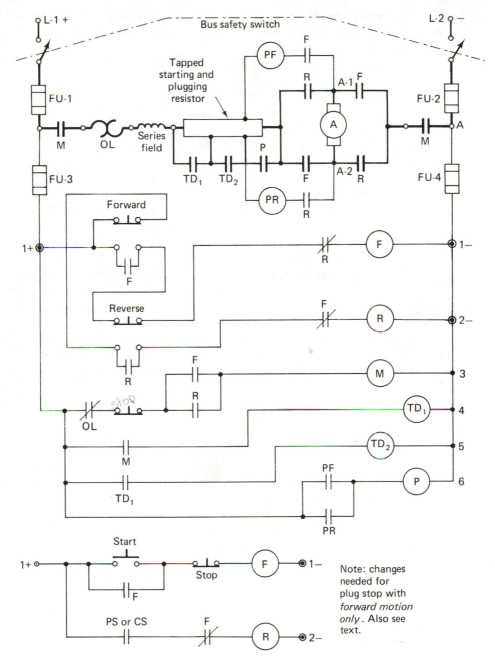

Figure 9-5. Direct-current plugging reverse control.

This could have been a current limit acceleration starter just as well if the control conformed to that type. The PR and PR relays are in fact of the nature of a current limit control.

An identical sequence takes place when the forward button is pressed while rotating in reverse. This control, as shown, is symmetrical.

A normal stop is realized when the stop button is pushed. The plugging feature is not available without further circuit features. It should also be noted that the forward and reverse pushbuttons could be *limit switches* so that the motor will automatically reverse after a particular amount of travel of its driven device or machine. In this case, either the forward or reverse limit switch would need a normally open pushbutton in parallel with its NO contacts in order to get the machine started.

If reverse action is only wanted in order to produce a plugging stop, the circuit modification shown in the bottom auxiliary view of Fig. 9-5 is required, along with the removal of the stop button on line 3. The operation of this control in a normal forward start is identical to the basic control of Fig. 9-5 just described.

This arrangement uses the full reverse circuit provisions and delivers reverse torque during deceleration, but does not produce reverse direction motion. If a plugging stop is wanted from either direction, further circuit complications are needed. In case this situation is wanted, consult reference (9).

Some situations call for a positive hold of position when stopped. If this is desired, it is easy to apply a spring-operated magnetically released brake that will be released when the start button (or forward or reverse button) is pressed. The final sequential step of a plug stop is the opening of the CS or PS switch, and these switches can be provided with additional normally open contacts, which can open the brake solenoid circuit. The brake will then be applied and held with an adjustable mechanical spring.

Any plugging reverse or plugging stop circuit requires an additional current-limiting starting resistor. This resistor can be determined by an adaptation of Eqs. (6-18) and (6-3). In this case, since the back emf or E_c is now in the *same* direction as the line voltage, Eq. (6-18) modifies by changing the minus to a plus and is then

$$R_{sp} = \left(\frac{V_l + E_c}{I_r \times M} \right) - R_a \qquad \text{Eq. (9-1)}$$

Here the terms are the same as in the original Eq. (6-18), which was in turn derived from Eq. (6-17).

When the motor is operating at its rated speed with no starting resistance in series with its armature, the back emf or E_c can be found from a modification of Eq. (6-3), where E_c is simply transposed out. This yields

$$E_c = V_a - I_a R_a$$

Since there is no appreciable external resistance other than the armature circuit itself, $V_a = V_{\text{line}}$. Note that V_a includes any series field.

$$E_c = V_l - I_a R_a \qquad \text{Eq. (9-2)}$$

Therefore, to find the required added starting resistance for a plugging situation, we find the E_c that will exist just before plugging from Eq. (9-2), then the total R_{sp} that will be required from Eq. (9-1), and finally the *added* R_{sp} over the original R_s that was needed for a normal start by simple subtraction.

9-6   RETARDATION AND STOPPING

There are many situations when it is desired to bring a dc motor and its driven load to a controlled reduction of speed. The cases where the torque required to do this is in the opposite sense from the normal driving torque are those under consideration here. This is to differentiate from speed control that results from more or less torque in the same directional sense. Speed control will be the subject of the next section of this chapter. Here we are concerned with the case of the "overhauling" load. Examples are the lowering of a loaded elevator, mine hoist, or crane hook load. In these cases, if the motor is shut off, the load will continue to accelerate downward and perhaps reach a runaway speed. Similarly, if an electrically driven vehicle such as a train is descending a long downhill grade, the whole vehicle will gain speed dangerously unless a retarding torque is applied. The obvious and historic method is to apply brakes and dissipate the energy as heat.

There are limitations, however, since no brake system yet built can control a loaded train on a mountain grade. Either stops are made to cool the brakes or the whole descent is made at sufficiently slow speed that the brake shoes can dissipate the heat. The problem then resolves itself into a need to controllably *reverse* the torque on the motor.

If the motor is converted so that it operates as a *generator*, it will convert shaft energy to electrical energy. The electrical energy can be dissipated remotely in a number of ways. Meanwhile, the motor can absorb as much or more mechanical energy input as a generator than it could deliver as a motor.

The problems of electric retardation are solved in two principal ways:

(1) *Dynamic braking*: the motor is converted to a generator by circuit arrangement, and the energy it produces is dissipated in a resistive element as heat.

(2) *Regenerative braking*: the motor also acts as a generator but, by careful arrangement, the generated electrical energy is fed back into the bus lines to become useful and recoverable electric power.

Dynamic braking is simple and does not require voltage adjustment in the usual case. Regenerative braking involves careful adjustment, but is much more efficient since energy is returned.

At the present time, diesel electric locomotives that are used in mountain-grade service use dynamic braking. The enormous power developed by their motors when serving as generators is connected to physically large but low-resistance elements on the roof of the locomotives. These resistances are blower cooled. The power to operate the large blowers could also be drawn from the traction motors. This scheme works beautifully since there is no brake-shoe wear and wheel damage. Furthermore, the train rides safely and smoothly without the surging and bucking that may result from heavy braking.

Regenerative braking was formerly practiced on a few of the major electrified railroads, notably the then Chicago Milwaukee and Saint Paul Railroad. Their locomotives were switched from motor mode to generator mode at the top of a mountain grade. This proposition is clearly so attractive and energy saving that more use may be expected in the future, especially since the world use of energy is overhauling world availability.

The original railroad use of regenerative braking was limited to nearly constant speeds since no variable elements were provided. Refinements of this scheme should obviously be developed for battery electric vehicle use. Here the braking energy generated can be pumped back into the batteries.

9-6-1 Dynamic Braking. The distinction as to the type of generative braking chosen depends upon a number of elements. Usually, if braking time is short, dynamic braking is used because of simplicity. However, if an overhauling load must be controlled for a long period, or if economy of operation is required, then regenerative braking will be used.

The following sequence must be performed in dynamic braking control operation:

(1) The armature circuit is opened from the supply source by opening the main contactor. The shunt field is left connected and live, or, if originally a series motor, a separate shunt field is placed across the line.

(2) The motor is now operating as a separately excited shunt generator and will produce voltage according to its built-in characteristics, its rotational speed, and the field magnetic flux.

(3) The armature is connected across a resistance that is usually a part of the current-limiting starting resistance. This causes armature-generated current to circulate in the resistance. The power developed will be related to the current and the resistance by $P = I^2R$. This power is directly related to the torque required to turn the motor, which is now a generator. The torque absorbed can be adjusted by the field flux or by the resistance.

(4) The opposed load will adjust its speed until the power required is matched by the power delivered to the armature circuit resistance. A descending crane load may be made to move as moderately as desired.

With this type of control, the load *cannot* be brought to a stop if it is of an overhauling nature since, as the motor slows down, the generated voltage decreases and thus the current is less, so the absorbed power is less. To stop the dynamically braked motor requires either the application of a mechanical brake or actual reverse operation of the motor. If the load is of an inertial nature, rather than overhauling, dynamic braking *will* result in a rapid stop.

The setting of the braking resistance may be so low that it is a severe but temporary overload on the motor. As soon as a reduction of speed is achieved, the motor–generator voltage decreases in proportion, and the current, and thereby the wattage, is brought under control.

Returning to the case of a modern diesel electric locomotive with dynamic braking, some further circuit complication is required. Here, or in any other series motor situation, it is usual to provide separate shunt field windings or to separately excite the series fields with a low voltage but high current source. Some locomotives reset their various fields all in series, since there are at least four motors in any one locomotive unit. This requires the current of only one field, but the voltage drop of four (or more). The power is easily supplied by the diesel-driven generator with the diesel nearly at idle power. The voltage is adjusted by the main generator excitation. Meanwhile, the motors are working vigorously as high output generators feeding kilowatts or even megawatts to the large resistor grids on the roofs. A simple adjustment of the field flux determines the output of the generators and thus the power taken from the moving train. A loaded freight train is easily taken down a winding continental divide grade with the diesels working lightly and the wheel brakes released. A torrent of heated air is released from the resistor grids, but nothing is being overheated and nothing is being worn out in the process.

It is worthwhile pondering for a bit that no known nor yet foreseen process can convert that heat energy back to a usable liquid fuel. We will see that *regenerative* braking can, in effect, produce a useful return by feeding power back into the lines or into batteries.

Cranes, hoists, and elevators are either dynamically braked or, in more sophisticated installations, regeneratively braked, and have been for many years.

9-6-2 Regenerative Braking. This type of motor braking is surprisingly simple in concept. In the case of a shunt motor, with its field flux strength being nearly independent of the motor load and speed, regenerative braking can take place with no circuit change whatever. The motor operates normally with its back emf nearly as high as the line voltage. Since the back emf is directly proportional to the rotative speed, any increase of speed due to an overhauling load will cause the back emf to exceed the line voltage. This is all that is required to convert a motor to a generator. The higher back emf then causes the current to reverse its direction and flow back into the source of supply. The current that flows turns out to be nearly directly related to speed. At the same time, the power being taken from the motor (now generator) shaft is the voltage times the current.

The speed of a descending load can be controlled over a wide range by varying the field excitation current. However, situations that use regenerative braking are not normally self-stopping, and an actual stop either requires plugging or the use of mechanical brakes.

In the case of series motors, they are frequently supplied with separately controllable shunt field windings so that controllable regenerative braking can be used. Here the problem resolves itself into changing from series motor operation to shunt generator operation. Since this involves some shuffling around of field connections, the problem comes with placing the machine back on the line with the generated voltage being required to be close to the line voltage. In the original railroad installations, engineers were strictly required to shift from motor action to generator action in a narrow range of speeds. This was required because, when going back on the line as a shunt generator, the locomotive's motor–generator was being paralleled with the powerhouse generators. Large differences in voltage result in severely large currents. Since the field adjustments were fixed, the locomotive-generated voltage was directly related to train speed. The inertia of a moving railroad train is so enormous that no quick changes in speed are possible.

Regenerative braking is then ridiculously simple when going from shunt motor action to shunt generator action. Complications are seen to arise when shifting over from series motor action to shunt generator action. No specific circuits will be shown, but the various references treat this in some detail. Reference (11) is a vaulable source.

9-6-3 Electric Brakes. The final means of retardation and stopping of a motor is through mechanical brakes. Electrically operated mechanical brakes are usually *electromagnetically released* and *magnetically applied* by adjustable spring mechanisms.

One type of brake circuit places the brake magnet coil in series with a separate normally open contact on the main M relay or contactor. This releases the brake and holds it off while the M contactor is active. Since the brake is not, in this type, used to perform the whole stopping task, it is combined with a plug stop circuit. Here a plug stop (PS) or centrifugal stop (CS) switch can be provided with additional normally open contacts that can be used for holding contacts for the brake circuit. In this case, the M contacts drop out when the stop button is released, but the brake solenoid is not released until the motor has come to a stop or nearly to a stop. If the brake were to do the whole job, there would be considerable wear and adjustment problems, or a substantially larger brake would be used. Standardized product lines of electrically released motor brakes are available.

There are frequent situations where, because the load inertia is small, the electromagnetic brake is entirely adequate to perform the complete retardation and holding job.

The simplest variety is the *series brake*. Here a magnetic brake with an actuating coil that is comparable to a series field is simply made a part of the armature circuit. Since this type of actuation coil has relatively few turns, it has a low time constant for brake release. When the motor circuit main contactor M is actuated, it simultaneously releases the brake and starts the motor. The normal accelerating circuit need not be changed. When the motor is shut off and the M contacts open, the brake is automatically applied. This is particularly useful on a hoist, for example. There is a safety advantage also since no added circuit complication is required. The usual design proportioning will release the brake at about 40 percent of full-load current. Application will automatically take place if the motor current drops below around 10 percent of full-load current. This means that a runaway due to loss of load is automatically protected.

A shunt motor or a compound motor most often will use a *shunt brake*. The shunt or compound motor may have been chosen in the first place because of periods of light load or no load. This would drop the armature current sufficiently to apply a series brake when it was not wanted. The shunt brake with its fine-wire multiturn release coil is highly inductive and has a relatively long time constant.

As a result of this time constant, a shunt brake is usually connected so that its brake coil is in series with a resistance, and the group is connected across the line. The resistance is ordinarily such that it will take about one-half the voltage drop and the brake coil the other half. This brake coil will then have less turns and use more current. It will, however, have about one-quarter the inductance that a full-voltage coil could have. The added series resistance further reduces the time constant. If the resulting brake coil and resistance combination is not then of sufficiently short time constant, a further change is made. This time the starting circuit will be so arranged that the resistance itself will be shorted out for the first starting time delay period. This will apply the full-line voltage to a half-voltage coil and further decrease its time of operation. Then, as the various relays operate, the resistance short will be opened so that the coil will not overheat. There are further varieties, but this is sufficient to show that brakes can be added to most controls. One precaution necessary for a shunt brake is that its parallel circuit to the armature be opened by one of the two M contacts. This prevents the brake coil from acting as a dynamic brake load resistance. Had it remained in parallel with the armature, the generator action of the disconnected armature might well have kept the brake released when its application was desired.

Electromagnetically released brakes are very safe devices. They do not release inadvertently and allow a mechanism to drift or to run away. On the contrary, they will immediately grab and bring the mechanism to an emergency stop if the power fails or if a fuse or breaker goes out. A runaway elevator, hoist, or crane could be a real nightmare hazard if power were lost without such automatic power-loss safety features. A suspended load is always preferable to a dropped load.

9-7 JOGGING

The National Electric Manufacturers Association (NEMA) defines jogging as "the quickly repeated closing of a circuit in order to start a motor from rest for the purpose of accomplishing *small* movements of the driven machine." There are many circumstances where it is desired to produce a small motion on a motor-driven device, such as to complete a traversing operation or to "inch" a hoist or crane hook into a desired position. This may be wanted when unloaded to attach a sling or progressively check out the safety and balance of a load. Or a suspended load may be deposited very gently while mechanics or riggers position it by hand or by rigging lines.

Any of these circumstances calls for jogging of a motor. This requires a button control that will initiate a slow-speed motion and yet not the whole starting cycle. The circuit problem then comes down to a pushbutton operation that will close the M contacts without actuating the holding contacts or any of the time-delay acceleration features of a circuit. When the button is released, all should stop, as nearly instantaneously as possible. Figure 9-6 shows a portion of a reversible jogging circuit that might be used in a circuit such as shown in Fig. 9-4.

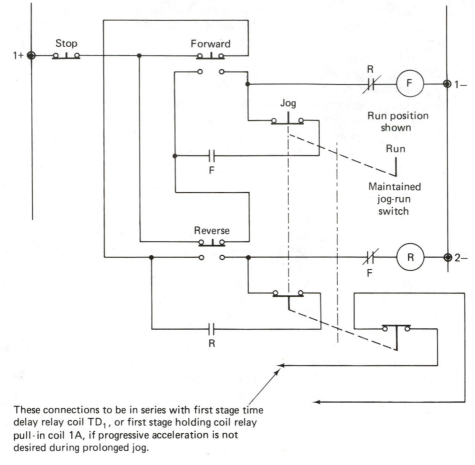

These connections to be in series with first stage time
delay relay coil TD_1, or first stage holding coil relay
pull-in coil 1A, if progressive acceleration is not
desired during prolonged jog.

Figure 9-6. Partial reversible jogging circuit.

Note that five pushbuttons are now provided: stop, jog, run, forward, and reverse. The jog–run switch is a maintained pair such that if jog is depressed it stays and forces run back out. On the other hand, if run is depressed it stays and forces jog back out. When run is selected, the control acts just as it did under Fig. 9-4. However, when jog is selected the following occurs:

(1) Pressing either forward or reverse will start operation at the lowest speed or with the maximum acceleration resistance in series with the armature. Since the circuit to the first step delayed action relay is *opened*, the motor does not finish its acceleration cycle but continues in a stabilized low speed.

(2) When the forward button is released, operation ceases immediately. If a magnetic brake has been provided, it will immediately apply and stop all rotation.

(3) Reverse operation is equal and opposite in this control. Jogging, however, may be applied to a unidirectional control if such is desired.

Sometimes a hoist is operated by jogging only in that no holding contact is placed across the "up" or "down" button. Thus, an operator must hold the button depressed or all will stop. This feature is desirable and may even be required in some hoists. For example, if an operator walks along with the dangling pendant control of an overhead hoist as it is traversing a load, he is apt to watch the load rather than his footing. If he trips over a foreign object and loses control, the hoist traverse and/or raise or lower must stop. If not, the stage is set for a potentially serious accident.

This mix of jog or run, coast or stop, accelerate or hold low speed, forward or reverse, and so on, can be anything whatever that is reasonable and safe, depending on the specific requirements of the task. All these features have been discussed step by step, but so far steady state speed control has not been introduced. This will follow in the next section.

9-8 DIRECT-CURRENT MOTOR SPEED CONTROL

Once a dc motor is set in motion and has finished its timed or load-related acceleration, the question of how fast or slow it is desired to run arises. Many relatively complicated motor and control systems run at a fixed speed once started on the way. However, a great many situations require that the speed be varied. Speed variations may be required more or less continuously, as in a vehicle in traffic. Conversely, speed variations may be needed only in an overall adjustment sense. There are many, many combinations of conditions.

There are four basic means of controlling a dc motor speed, and each means has a different applicable range of effectiveness:

(1) *Field control* or, more specifically, field magnetic flux control.

(2) *Armature resistance control* or control of available armature voltage by series resistance.

(3) *Series and shunt armature resistance control*, which uses resistance both in series with and in shunt with the armature. There is added complication and increased losses, but it has some desirable features.

(4) *Armature voltage control*, which, as it implies, uses a controlled voltage source for the armature. This may be in a separately excited shunt motor or a series motor. Normally, a bus supply is a relatively fixed voltage, but when it is designed to be variable, many advantages can be obtained.

9-8-1 Basic Causes and Limits of Speed. These four different types will be shown to affect each type of motor differently. It is first necessary to discuss the basic causes and limitations of rotative speed. A dc motor of any type will rise to an equilibrium speed where the current that is drawn from the lines will supply just sufficient torque for the conditions. The current is regulated by the difference between the line voltage and the back emf or countervoltage that is generated within the rotating armature. The back emf, in turn, obeys exactly the same voltage-generation formula that controls a generator. It was shown in Chapter 2 that the generated voltage is

$$E_c = \frac{\Phi ZSP}{60a} \times 10^{-8} \text{ V}$$

and that, for SI units,

$$E_c = \frac{\phi Z \omega P}{a} \text{ volts}$$

In either case, Z is the number of conductors per armature, P the number of poles, and a the number of parallel paths in the winding. All are fixed by the machine construction and are not adjustable in service. Similarly, in the English unit equation, the quantities 10^{-8} and 60 are constants of proportion necessary to make the units of the equation true, and cannot be varied. As a result, the two generation formulas become

(from Chapter 6) $E_c = K\Phi S$ Eq. (6-6$_E$)

and $E_c = k\phi \omega$ Eq. (6-6$_{SI}$)

In words, any specific machine whose winding arrangements and physical construction have been fixed will generate a voltage that is directly proportional to its total magnetic flux and its rotational speed. In this case, this voltage E is considered to be the back emf or countervoltage and will be identified as E_c, so that $E = E_c$.

During motor action, E_c must be less than the terminal voltage. It will adjust automatically. This can be seen by combining Eqs. (6-3), (6-6$_E$), and (6-6$_{SI}$):

$$\boxed{E_c = V_t - I_a R_a = K\Phi S}$$ Eq. (9-3$_E$)

and $$\left\langle E_c = V_t - I_a R_a = k\phi \omega \right\rangle$$ Eq. (9-3$_{SI}$)

If these equations are transposed and solved for the speed term, we find from Eq. (6-7$_E$) modified that

$$\boxed{S = \frac{E_c}{K\Phi} = \frac{V_l - I_a R_a}{K\Phi}}$$ Eq. (9-4$_E$)

Equation (6-7$_{SI}$) modified yields

$$\left\langle \omega = \frac{E_c}{k\phi} = \frac{V_l - I_a R_a}{k\phi} \right\rangle$$ Eq. (9-4$_{SI}$)

Note that $V_l = V_t$.

From these relations we can conclude that the steady-state rotation can be changed by changing either the countervoltage E_c or the flux The flux can be changed within the limits of the particular machine as a m. The low limit of flux is really the residual magnetism that the structure will ho. However, if the field is weakened too much, the commutation suffers and the motor becomes unstable. The countervoltage term is controlled *both* by the line voltage V_l and the current being drawn by the motor. The current required is almost directly related to the load torque required.

9-8-2 Specific Speed Relationships.

The speed, torque, and power developed by dc motors of shunt, series, and compound types were explored in some detail in Chapter 6, especially in Sects. 6-4 through 6-7. These areas should be reviewed so that the four basic control means that are normally used can be better understood.

The schematics and formulas shown in Figs. 9-7 through 9-9 will serve to summarize the information in usable form.

Chapter 6 and its worked examples show it is not always necessary to actually evaluate the Φ or ϕ terms. It is usually possible by working from a situation where the currents, voltages, and speeds are known to predict the effect of a change caused by an armature circuit or field circuit resistance without actual flux determination.

If the speed adjustment is made by changing only the voltage across the armature circuit terminals, and if the load is such that the armature current remains constant, the torque remains constant. At the same time, the gross armature power varies almost directly with speed. This is the situation in circuits 4, 5, 8, and 12 if the shunt fields are not changed at the same time.

On the other hand, if the speed adjustment is made by changing only the field flux and leaving the armature circuit voltage as it was, and if the load is again such that the armature current is unchanged, the developed power remains constant. In this case the torque varies about inversely as the speed varies. This condition holds in circuits 1 and 9. The other circuits are a mixture.

Note that in any circuit the *gross* developed motor *power* is the product of the countervoltage developed and the armature current. This is true from the basic power law of $P = IE$, but here $P_d = I_a E_c$ from Eq. (6-5). The actual shaft power delivered to the load is this gross armature power minus the rotational loss under the speed and flux conditions. This is then

$$\boxed{P_{\text{out}} = I_a E_c - P_{\text{rot}}}$$ Eq. (9-5)

Chapter 7 has shown that rotational losses require detailed measurement. Similarly, E_c is not a simple measurement. However, the various formulations given in circuits 1 through 12 in Figs. 9-7 through 9-9 show the E_c situation as a part of the speed formula. For example, in circuit 10 of Fig. 9-9, $E_c = V_l - I_a R_{se\ R} - I_a(R_a + R_{se})$. In words, E_c, the countervoltage in this particular case, is equal to the line voltage V_l minus some IR terms. These are, of

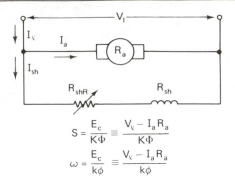

$$S = \frac{E_c}{K\Phi} \equiv \frac{V_\ell - I_a R_a}{K\Phi}$$

$$\omega = \frac{E_c}{k\phi} \equiv \frac{V_\ell - I_a R_a}{k\phi}$$

Φ or ϕ changes along saturation curve as R_{shR} changes

$$I_a = I_\ell - I_{sh} \qquad I_{sh} = \frac{V_\ell}{R_{sh} + R_{shR}}$$

Increasing R_{shR} increases speed
- Decreasing torque, constant power as speed increases

1 Field resistor control

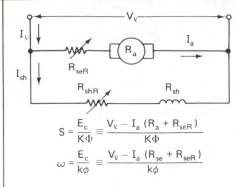

$$S = \frac{E_c}{K\Phi} \equiv \frac{V_\ell - I_a (R_a + R_{seR})}{K\Phi}$$

$$\omega = \frac{E_c}{k\phi} \equiv \frac{V_\ell - I_a (R_{se} + R_{seR})}{k\phi}$$

Φ or ϕ changes as in circuit 1

$$I_a = I_\ell - I_{sh} \qquad I_{sh} = \frac{V_\ell}{R_{sh} + R_{shR}}$$

Increasing R_{shR} increases speed
Increasing R_{seR} reduces speed

2 Shunt field and series armature res. control

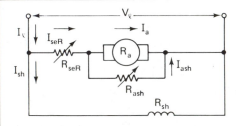

$$S = \frac{E_c}{K\Phi} \equiv \frac{V_\ell - I_{seR} R_{seR} - I_a R_a}{K\Phi}$$

$$\omega = \frac{E_c}{k\phi} \equiv \frac{V_\ell - I_{seR} R_{seR} - I_a R_a}{k\phi}$$

Φ or ϕ fixed point on saturation curve

$$I_{seR} = I_\ell - I_{sh} : I_{sh} = \frac{V_\ell}{R_{sh}} ; \ I_a = I_{seR} - \left[\frac{V_\ell - I_{seR} R_{seR}'}{R_{ash}} \right]$$

Increasing R_{seR} reduces speed
Decreasing R_{ash} reduces speed

3 Series and shunt armature res. control

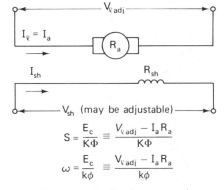

$$S = \frac{E_c}{K\Phi} \equiv \frac{V_{\ell \, adj} - I_a R_a}{K\Phi}$$

$$\omega = \frac{E_c}{k\phi} \equiv \frac{V_{\ell \, adj} - I_a R_a}{k\phi}$$

Φ or ϕ may be fixed or separately adjustable on saturation curve

$$I_\ell = I_a \qquad I_{sh} = V_{sh}/R_{sh}$$

Increasing $V_{\ell \, adj}$ increases speed

Increasing V_{sh} reduces speed

- Constant torque, power increases with speed

4 Line voltage control

Note: in all cases, R_a incudes commutating field and compensating winding, gross $P_a \cong E_c I_a$; net $P_a \cong E_c I_a - P_{rot}$, both in watts

Figure 9-7. Shunt motor speed control.

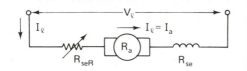

$$S = \frac{E_c}{K\Phi} \equiv \frac{V_\ell - I_\ell R_{seR} - I_\ell(R_a + R_{se})}{K\Phi; \text{ or } KK'I_\ell}$$

$$\omega = \frac{E_c}{k\phi} \equiv \frac{V_\ell - I_\ell R_{seR} - I_\ell(R_a + R_{se})}{k\phi; \text{ or } kk'I_\ell}$$

Φ or ϕ reduces along saturation curve as I_ℓ reduces

$$I_\ell = I_{seR} = I_a = I_{se}$$

Compare with 8 where: $V_\ell - I_\ell R_{seR} = V_{\ell adj}$

Increasing R_{seR} reduces speed

- Constant torque for same I_ℓ with power increasing with speed

5 Series resistance voltage control

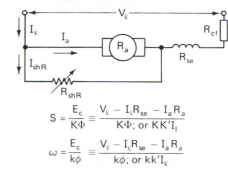

$$S = \frac{E_c}{K\Phi} \equiv \frac{V_\ell - I_\ell R_{se} - I_a R_a}{K\Phi; \text{ or } KK'I_l}$$

$$\omega = \frac{E_c}{k\phi} \equiv \frac{V_\ell - I_\ell R_{se} - I_a R_a}{k\phi; \text{ or } kk'I_\ell}$$

Φ or ϕ changes as in circuit 5

$$I_a = I_\ell - \left(\frac{V_\ell - I_\ell R_{se}}{R_{shR}}\right)$$

Note: R_{cl} may be added to limit I_ℓ and considered part of R_{se}

Reducing R_{shR} reduces speed

6 Armature shunt res. current control

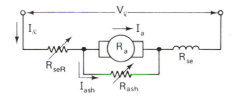

$$S = \frac{E_c}{K\Phi} \equiv \frac{V_\ell - I_\ell(R_{seR} + R_{se}) - I_a R_a}{K\Phi; \text{ or } KK'I_\ell}$$

$$\omega = \frac{E_c}{k\phi} \equiv \frac{V_\ell - I_\ell(R_{seR} + R_{se}) - I_a R_a}{k\phi; \text{ or } kk'I_\ell}$$

Φ or ϕ changes as in circuit 5

$$I_a = I_\ell - \left[\frac{V_\ell - I_\ell(R_{seR} - R_{se})}{R_{ash}}\right]$$

Increasing R_{seR} reduces speed
reducing R_{ash} reduces speed

7 Series and shunt armature res. control

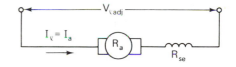

$$S = \frac{E_c}{K\Phi} \equiv \frac{V_{\ell adj} - I_a R_a - I_a R_{se}}{K\Phi; \text{ or } KK'I_{ae}}$$

$$\omega = \frac{E_c}{k\phi} \equiv \frac{V_{\ell adj} - I_a R_a - I_a R_{se}}{k\phi; \text{ or } kk'I_a}$$

Φ or ϕ changes as in circuit 5

$$I_\ell = I_a = I_{se}$$

Compare with 5 where: $V_{\ell adj} = V_\ell - I_\ell R_{seR}$

Increasing V_ℓ increases speed

- Torque and power as in circuit 5

8 Line voltage control

Note: in all cases, R_a includes commutating field and compensating winding gross $P_a \cong E_c I_a$; net $P_a \cong E_c I_a - P_{rot}$, both in watts

Figure 9-8. Series motor speed control.

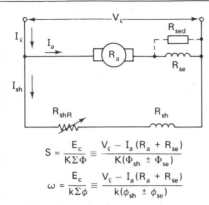

$$S = \frac{E_c}{K\Sigma\Phi} \equiv \frac{V_\ell - I_a(R_a + R_{se})}{K(\Phi_{sh} \pm \Phi_{se})}$$

$$\omega = \frac{E_c}{k\Sigma\phi} \equiv \frac{V_\ell - I_a(R_a + R_{se})}{k(\phi_{sh} \pm \phi_{se})}$$

$\Sigma\Phi$ or $\Sigma\phi$ varies along saturation curve according to connection and relative ampere turns

$$I_a = I_\ell - I_{sh}; \quad I_{sh} = \frac{V_\ell}{R_{sh} + R_{shR}}$$

Increasing R_{shR} increases speed
● constant power, reduced torque as speed increases

9 Field res. control

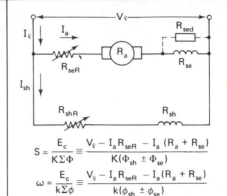

$$S = \frac{E_c}{K\Sigma\Phi} \equiv \frac{V_\ell - I_a R_{seR} - I_a(R_a + R_{se})}{K(\Phi_{sh} \pm \Phi_{se})}$$

$$\omega = \frac{E_c}{k\Sigma\phi} \equiv \frac{V_\ell - I_a R_{seR} - I_a(R_a + R_{se})}{k(\phi_{sh} \pm \phi_{se})}$$

$\Sigma\Phi$ or $\Sigma\phi$ changes as in circuit 9

$$I_a = I_\ell - I_{sh} \qquad I_{sh} = \frac{V_\ell}{R_{sh} + R_{shR}}$$

Increasing R_{seR} reduces speed
Increasing R_{shR} increases speed

10 Shunt field and series armature res. control

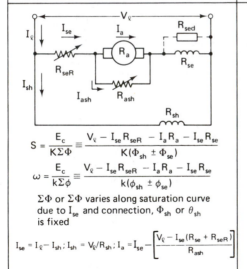

$$S = \frac{E_c}{K\Sigma\Phi} \equiv \frac{V_\ell - I_{se} R_{seR} - I_a R_a - I_{se} R_{se}}{K(\Phi_{sh} \pm \Phi_{se})}$$

$$\omega = \frac{E_c}{k\Sigma\phi} \equiv \frac{V_\ell - I_{se} R_{seR} - I_a R_a - I_{se} R_{se}}{k(\phi_{sh} \pm \phi_{se})}$$

$\Sigma\Phi$ or $\Sigma\Phi$ varies along saturation curve due to I_{se} and connection, Φ_{sh} or θ_{sh} is fixed

$$I_{se} = I_\ell - I_{sh}; \quad I_{sh} = V_\ell/R_{sh}; \quad I_a = I_{se} - \left[\frac{V_\ell - I_{se}(R_{se} + R_{seR})}{R_{ash}}\right]$$

Increasing R_{seR} reduces speed
reducing R_{ash} reduces speed

11 Series and shunt armature res. control

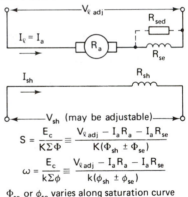

$$S = \frac{E_c}{K\Sigma\Phi} \equiv \frac{V_{\ell\,adj} - I_a R_a - I_a R_{se}}{K(\Phi_{sh} \pm \Phi_{se})}$$

$$\omega = \frac{E_c}{k\Sigma\phi} \equiv \frac{V_{\ell\,adj} - I_a R_a - I_a R_{se}}{k(\phi_{sh} \pm \phi_{se})}$$

Φ_{se} or ϕ_{se} varies along saturation curve according to I_a and connection

Φ_{sh} or ϕ_{sh} May be varied separately

$$I_a = I_\ell = I_{se} \text{ (If no } R_{sed})$$

Increasing $V_{\ell\,adj}$ increases speed
Increasing V_{sh} reduces speed

● Constant torque with same I_ℓ and I_{sh} Power increases with speed

12 Line voltage control

Note: in all cases, R_a includes commutating field and compensating winding, R_{sed} may be used to limit I_{se} especially with differential field connection, + is cumulative, − is differential compound versions of ± signs, gross $P_a \cong E_c I_a$; net $P_a \cong E_c I_a - P_{rot}$, both in watts

Figure 9-9. Compound motor speed control.

course, voltages from $E = IR$. These terms are the armature current I_a times the resistance of the series resistor $R_{se\ R}$ and the armature current I_a times the sum of the armature circuit resistance R_a and the resistance of the series field R_{se}. Most of these terms can be measured on an actual machine. The R_a is best found by Forgue's method, although, as Chapter 7 shows, a lesser measurement can be used.

Note that E_c should be a brush-to-brush voltage and not include the voltage drop across the commutating fields. This is an acceptable error, however, because other more significant errors exist, especially so since this relation *does not* take into account the field demagnetization effects of armature reaction. Also, the voltage across the commutating fields is usually very small.

In addition, it should be remembered that the R_{se} term should be the paralleled resistance of the series field and the series field shunt diverter $R_{se\ d}$, if one exists. There usually will be some form of diverter resistance unless the motor was specifically wound for its job.

9-8-3   Mechanisms Used in Speed Control. Many methods are used to vary the resistances that are used in motor speed control. Armature series resistance control, as used in circuits 2, 3, 5, 7, 10, and 11, is frequently part of the current-limiting acceleration resistances.

Armature shunt resistances, as in circuits 3, 6, 7, and 11, are high-wattage units and will be step controlled by contactors that short out portions of the resistance. These are part of a dynamic brake system.

Shunt field resistances, in circuits 1, 2, 9, and 10, are usually rheostats since they only carry the relatively smaller shunt field currents.

For specific circuits, consult the references. Note that with the growing awareness of energy shortages, some of the more inefficient means of speed control will be less attractive in the future. Any armature circuit resistance that is used for speed control represents a substantial loss. A series armature circuit resistance that drops the motor terminal voltage to one-half of the line voltage will consume as much energy as the motor itself. Thus, an 80 percent efficient motor running on one-half voltage is at least as poor as $80 \times 0.5 = 40$ percent efficient. Furthermore, under these conditions an armature circuit resistance must be very much larger than it would need to be for starting service.

9-8-4   Solid State dc Motor Speed Control. As already noted, the speed of a dc motor can be varied by varying the voltage to the field windings or by varying the voltage to the armature. By varying the voltage to the field while supplying constant voltage to the armature, the speed may only be varied above the rated speed. This will undermine the output torque, causing it to decrease as the speed increases. On the other hand, when the field voltage (and current) is held constant and the armature voltage is varied, the speed of the dc motor can be varied from some fixed value above zero speed right up to rated speed with constant torque, as shown in Fig. 9-7(4).

Figures 9-7 through 9-9 show various methods of speed control used in the conventional approach. In most of these cases, speed control is achieved by varying power resistors which, in effect, vary either the field voltage, or armature voltage, or both. Solid state control accomplishes the same goal electronically and, as an added result, has no power resistors to dissipate and waste power. The power losses external to the motor itself are restricted to the losses in the solid state and peripheral support circuits, namely the triggering and the commutation circuits.

Figure 9-10 illustrates a basic dc-to-dc chopper circuit which is used to obtain a variable dc voltage from a constant dc supply bus voltage, or where a dc supply voltage is a result of rectified ac voltage. The operation of the chopper is straight-forward. The SCR is switched on by a trigger pulse applied to the gate from the trigger circuit and stays on for a predetermined time (t_{on}). A commutating circuit will turn off and open the SCR at another predetermined time. The average value of the output voltage, V_L, will vary from zero to about E_S, depending on the on and off times of the SCR. Obviously, if the on time is zero, no voltage will be delivered to the load and no speed will result. If the off time is zero, a constant voltage of E_S will appear at the load and full speed will result. Intermediate values of on time will result in intermediate values of speed. The average dc voltage delivered to the load will be:

$$V_L = \frac{t_{on}}{T} E_S = t_{on} f E_S$$

Eq. (9-6)

The dc chopper can be effectively used in three standard modes of operation: 1) frequency modulation, 2) pulse-width modulation, or 3) a combination of 1) and 2).

A dc chopper is operating using frequency modulation when the on time t_{on} is held constant and the time period T is varied. Pulse width modulation occurs when the time period T is kept constant and the on time t_{on} is varied. In either case, separately or when used together, the net result is basically the same: the average voltage delivered to the load is variable. These techniques control the average voltage delivered to the load (and thus the speed of the motor) by varying the ratio of the on time to the off time. At low speeds the on time is much less than the off time. As the proportion of the on time to the off time is increased, the average voltage increases to the motor and so also does the motor's speed.

Consider the circuit of Fig. 9-11a where a source voltage, E_s, is placed in series with a switch and feeds a load consisting of a large inductance in series with an "active" load voltage, E_c, which is less than the supply voltage E_s. If this switch (which represents SCR1 in Fig. 9-10) is closed for a time, t_{on}, and then is opened for a time, t_{off}, a current will flow into battery E_c during the period t_{on}. In order to evaluate this current, it must be remembered that the voltage across and current through the inductor are related by the conventional equation:

$$e_L = L\frac{di}{dt} = L\frac{\Delta i}{\Delta t}$$

Eq. (9-7)

This equation in words states that the voltage induced across an inductor is directly proportional to the value of the inductance and the rate of change of current per

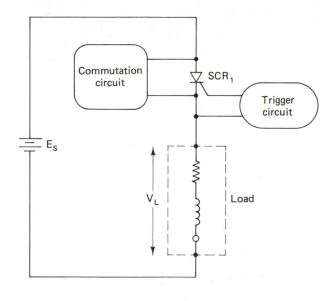

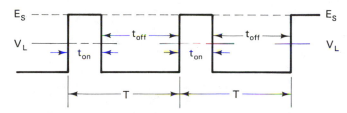

Figure 9-10. Basic dc chopper circuit and load voltage.

unit time through L. Since the voltage across the inductor of Fig. 9-11 is easily found to be $E_S - E_C$ volts (a constant), an equation needs to be found which will solve for current in terms of the voltage across the inductor. By the use of integration, the equation needed can be shown to be:

$$i_L = i_C = \frac{E_S - E_C}{L} t_{on}$$

Eq. (9-8)

It follows, then, that the value of current into E_C will increase at a constant rate between the time the switch is closed at $t = 0$ up until the time the switch is opened at $t = t_{on}$ as shown in higher values for a fixed value of E_S and E_C.

When the switch is opened at $t = t_{on}$, the current must continue to flow at the same value because the value of current through an inductor cannot change instantaneously. Furthermore, the voltage across the inductor must instantaneously change polarity and assume a very high value of voltage due to the very high value of the open switch resistance. This is because the inductor stored energy during the on time of the switch ($W_L = \frac{1}{2}LI^2$), and it must return the same amount of energy during the off time. This energy will be dissipated as heat and lost in

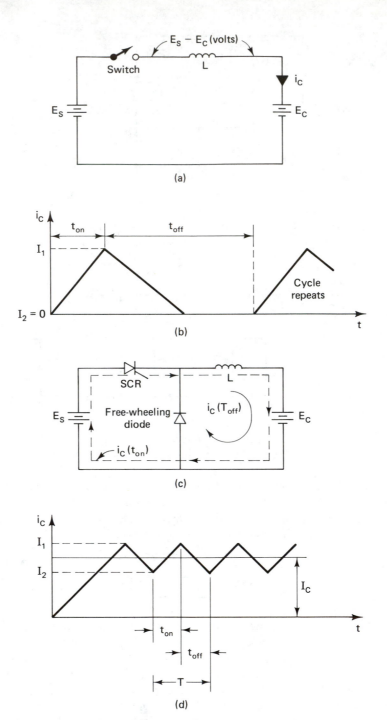

Figure 9-11.   The chopper and free-wheeling diode.

the form of an arc as the switch is opened and the current decays to zero during the off time. (Refer to Fig. 9-11b.) By the addition of one power diode to the circuit of Fig. 9-11a, commonly referred to as a "free-wheeling" diode, the circuit will behave much more efficiently and the motor will not experience voltage surges.

Figure 9-11c depicts the circuit with the free-wheeling diode and SCR in place. When the SCR is fired at $t = 0$, the current increases at a constant rate up to a value of I_1 at time $t = t_{on}$, at which time the SCR is commutated. During this time the source, E_S, is supplying power to the active load, E_C, the diode is reverse biased and energy is being stored in the inductor. (The current travels from $+E_S$ out through the loop and back to $-E_S$.) When the SCR is turned off at $t = t_{on}$ the same value of current, I_1, must continue to flow as the voltage across the inductor instantaneously changes polarity. The value of inductor voltage, instead of rising to a very high value as before, now rises to a value just slightly higher than E_C as the diode becomes forward biased and conducts. The current then coasts or "free wheels" down at a constant rate to zero amps as the inductor depletes its stored energy. With the addition of this free-wheeling diode, the stored energy of the inductor is absorbed by the active load, E_C, and not wasted. (The diode's power dissipation is negligible.)

It can be seen from the current waveform of Fig. 9-11b that the current is not continuous and can be rather choppy depending on the values of E_S, E_C, and L. If the load components of this circuit, L and E_C, in fact represent the inductance of the series field winding and counter emf respectively of a series dc motor, the torque will also be discontinuous and "jerky" as will the speed.

The current can be made to be continuous and approach a constant value by switching the SCR on and off much faster. Instead of letting the current fall to zero during the off time, the SCR is fired when the current falls to a value of, say, I_2. The current then will again increase to I_1 at which time the SCR is commutated and again falls to I_2 where the cycle is repeated continuously, as illustrated in Fig. 9-11d. The average dc current, I_C, that a dc ammeter would read I_C flowing into E_C can be shown to be:

$$I_C = \frac{I_1 + I_2}{2} \qquad \text{Eq. (9-9)}$$

It can be concluded that during the on time of the SCR, the source E_S delivers a pulse of current to the load and during the off time the decaying current is supplied by the inductor as it gives back the stored energy through the free-wheeling diode. Since the inductor does not dissipate any power, the average power absorbed by the active load, E_C, must be the same as the power delivered to the circuit by the source, E_S.

With the assumption that the switching rate (frequency) and value of inductance are both of sufficiently large value, it can be shown that:

$$I_S = I_C \frac{t_{on}}{T}$$

or

$$\boxed{I_S = t_{\mathrm{on}}\, f I_C}$$

Eq. (9-10)

From Eqs. 9-6 and 9-10 it is clear that the voltage applied to a motor and the current are functions of the on time, t_{on}, and the frequency or switching rate, f. The choice of using frequency modulation, pulse width modulation, or a combination of the two is closely tied with the type and complexity of triggering circuitry and associated feedback pilot control devices used.

Direct-current choppers, because of their high reliability, efficiency, and ability to handle large amounts of power, are generally used to power dc traction (series) motors. Trains, for example, use dc traction motors as do trolleys and some experimental vehicles. Speed control of the traction (series) motor was accomplished in the past by switching connections and high power resistors in and out to vary the motor's speed, which resulted in wasted power and jerky changes in speed. Increased efficiency and smooth speed control is achieved with chopper control.

The triggering and commutation of the SCR in the dc chopper must be considered at least briefly here because they are responsible for the controlled behavior of the motor to a large degree. As has been noted, commutation of the main SCR in many cases is actually a result of some other thyristor being triggered. The triggering circuitry, although beyond the scope of this text, is really in control of the control, so to speak.

In Fig. 9-2, for example, the dc motor was brought up to speed by sequentially (with relays) switching out portions of the total current limiting resistance so that a controlled, although not stepless, increase in motor speed resulted. Holding coil relays *sensed* the decreasing armature current as the speed increased and automatically reduced the starting resistance so that more line voltage was delivered to the motor, with a resulting increase in current and torque.

The same type of approach can be used in dc chopper speed control. The armature current is sensed by some device such as a small resistance, sometimes called a sensing resistance, placed in series with the armature. The proportional voltage drop developed across this sensing resistance resulting from the actual flow of armature current is fed back to the trigger circuit. The trigger circuit senses this feedback voltage and then adjusts the on time of the main SCR so that more voltage and current are available to the motor. This same trigger circuit can also be designed to accept actual shaft speed feedback from what is called an incremental shaft encoder, which is a disc attached concentrically to the shaft. As the shaft rotates, pulses of light are passed through slots of the rotating disc from a light source in the back of the disc. These pulses of light are then counted electronically by optoelectronic devices (solid state transistors and other semiconductor devices sensitive to light). The number of pulses per minute, being proportional to shaft speed, is converted to a voltage which is fed back to the triggering circuitry where, again, the on time of the main SCR is adjusted electronically. These closed-loop feedback techniques are complex and expensive but maintain a fine speed control. Simpler speed control may rely on the operator to supply feedback in terms of speed. By varying a potentiometer by hand, the operator may control the triggering source directly and adjust the machine's speed as required.

Although trains, trolleys, and some elevators are powered by the series dc motor, it is true that the dc shunt motor lends itself more handily to many other tasks. The series dc motor, with its high starting torque, is generally used with high inertial loads but in and of itself has poor speed regulation. The shunt motor, on the other hand, has excellent speed regulation and is sometimes called a constant speed machine, even though its speed does decrease slightly with load.

Referring back to Sect. 6-5, remember that the shunt motor usually has its field winding connected in parallel with the armature, but it can be fed by an independent source altogether, in which case the whole arrangement is called a separately excited dc shunt motor. Solid state speed control of the shunt motor is generally applied with the field separately excited.

Consider the circuit of Fig. 9-12, which represents a separately excited dc shunt motor. The armature circuit is shown with the armature resistance, R_a, and inductance, L_a, in series with a smoothing inductor, L_x. A generator, E_c, is shown in series with R_a, L_a, and L_x to represent the countervoltage that is generated within the armature of the motor as it spins. The field winding is also composed of the field resistance, R_f, and the inductance of the field, L_f.

Recalling equations (6-7) and (6-8):

$$\text{rpm} = S = \frac{V_a - I_a R_a}{K\Phi} = \frac{E_c}{K\Phi} \qquad \text{Eq. (6-7}_\text{E})$$

$$\text{radians per second} = \omega = \frac{V_a - I_a R_a}{k\phi} = \frac{E_c}{k\phi} \qquad \text{Eq. (6-7}_\text{SI})$$

and

$$T = C \Phi I_a \qquad \text{Eq. (6-8}_\text{E})$$

$$t = c \phi I_a \qquad \text{Eq. (6-8}_\text{SI})$$

Some observations can be made as to how speed and/or torque can be controlled in the dc machine.

It is obvious that the torque of a motor is directly proportional to two variables,

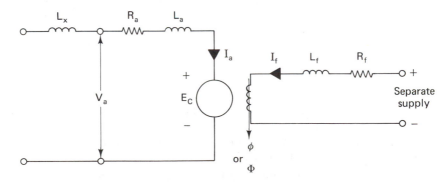

Figure 9-12. Separately excited dc shunt motor.

namely the field flux and the armature current. C (or c) in Eq. (6-8) is a constant, as already noted in Chapter 6. If the field flux is held constant (by maintaining a constant field current, I_f) then the torque produced will be dependent on the average value of the armature current, I_a. If *both* the field flux and armature current are each controlled at a constant value, the torque will be constant.

For the moment, consider that the field flux and the armature current are both fixed. The field current is flowing at rated amperes and the armature current is flowing at some value below the rated value. With this condition, the speed of the machine can be evaluated using Eq. (6-7). With I_a, R_a, K (or k), and Φ (or ϕ), all being constants, the speed of the machine becomes dependent on the value of voltage, V_a, applied to the armature. As the value of applied voltage to the armature is increased, the numerator of Eq. (6-7) increases in value and so then does the speed of the machine, while constant output torque is maintained. Reduction of the applied voltage, V_a, while keeping the field flux and armature current constant, results in lower speed while the same constant torque is maintained. This method of control can be referred to as a *variable speed*, *constant torque* control application. A possible use for this is where wire is wound around an object. Constant torque is required to maintain a constant tension on the wire being wound, but the speed is varied as the winding radius varies.

Another, perhaps more common, mode of operation occurs when the machine is run with constant speed and variable torque, as is generally used with pumps or conveyors. Again the field flux is held constant by maintaining rated field current with a separate source. Examination of Eq. (6-8) reveals that since C (or c) and Φ (or ϕ) are kept constant, the only way to vary the torque is by the variation of the armature current, I_a. To maintain constant speed with the given constants, the counter voltage E_c, of Eq. (6-8) must be held constant. Since $E_c = V_a - I_a R_a$, if I_a increases due to loading of the motor, V_a must be immediately increased an appropriate amount to keep E_c constant. This, in turn, will maintain a constant speed at the new torque output.

Finally, applications where variable speed and variable torque are required may be accommodated by the controlled variations of the applied armature voltage, V_a, and the armature current, I_a, simultaneously.

Although the modes of operation considered above may seem simple, the ability of the motor to perform as required obviously depends upon the control system. The block diagram of Fig. 9-13 shows the basic components of the control system comprised of power thyristors, a triggering source, and transducers. At the heart of the operation is the triggering source, which is really a microprocessor of some type. The microprocessor is programmed by the operator with the desired values of speed, torque, and/or other variables, called the controlled variable setpoints, which are within the safe operating range for the motor. Maximum limit values are also programmed into the processor for current, torque, speed, etc. for system protection. Finally, the transducers, which instantaneously measure actual speed, current, torque, etc., feed signals back to the processor. These signals are compared to the controlled variable setpoints by the electronics inside the processor. If there is a difference between, say, actual and set point speed, the firing delay angle, α, is electronically and automatically changed by the processor so that

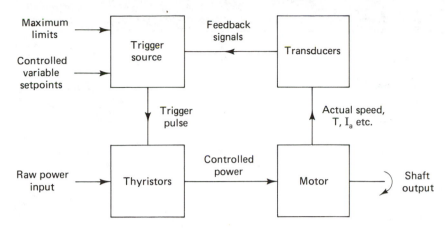

Figure 9-13. Block diagram of control.

the actual speed changes and equals that of the set point. If the speed increases beyond the maximum limit setting because of operator error or circuit malfunction, the motor is automatically shut down. The same sequence of events occurs for other monitored outputs (torque, current, etc.).

The three-phase, six-pulse thyristor converter is used for speed control and is shown feeding a separately excited shunt motor in Fig. 9-14. The armature circuit consists of the armature resistance, R_a, inductance, L_a, and a generator that represents the countervoltage produced by the generator action of the motor, E_c. Smoothing inductor, L_x, is included to smooth out the armature current, although if the inductance of the armature, L_a, is sufficiently large, L_x can be dispensed with. The field winding of Fig. 9-14 is fed from a full-wave diode bridge rectifier with a fixed dc voltage. The field current and the flux are therefore held constant.

When the start of the motor is called for (either by an operator or by some remote means, i.e., a computer), the microprocessor immediately begins to trigger the SCRs with a firing delay angle $\alpha = 90°$. This of course represents, from Eq. (8-12):

$$E_{dc} = 0.955 \, E_{max} \, cos \, \alpha$$

and here:

$$V_a = E_{dc} = 0.955 \, E_{max} \, cos \, 90°$$

therefore:

$$V_a = 0 \text{ volts}$$

It is obvious that at this starting point, the armature current is zero amperes because there is no voltage applied, and there is no countervoltage, E_c, because the armature is at a standstill. Sensing this through the feedback transducers, the triggering source reduces the firing delay angle, and the armature voltage, V_a, becomes positive and armature current flows. As this current flows into the armature, the shaft turns and the machine begins to accelerate. The triggering source, ever

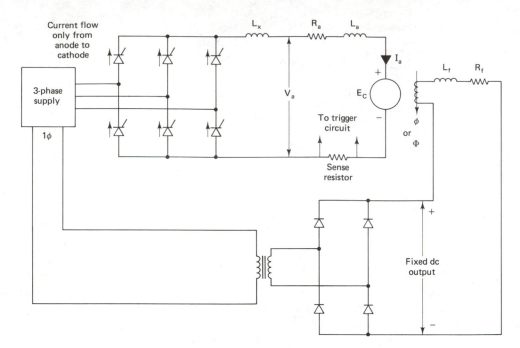

Figure 9-14. Three-phase, six-pulse motor control.

"mindful" of what the machine is doing, continues to lessen the firing angle as the machine approaches the setpoint (required) speed. Just how fast the firing delay angle is decreased depends on how much armature current is flowing. As the machine accelerates, the armature current is monitored by the small sensing resistor in series with the armature. If the armature current exceeds 150 percent of rated during acceleration, the triggering source *increases* until an acceptable I_a flows (below 150 percent of rated). As the motor comes up to rated speed, the firing delay angle approaches 20°. This is about the minimum delay angle for design limits because three-phase line fluctuations can cause intermittent problems if lower delay angles are used.

Once the machine is running at rated speed, the microprocessor triggering source maintains the required firing delay angle because the feedback and set point signals are closely in agreement. If a lower speed is called for, either by an operator or some remote source, the triggering source increases the firing delay angle, which decreases the converter's output, so that V_a becomes *less than* E_c. Since E_c is now greater than V_a, current will tend to flow out of E_c and into the converter, but this cannot happen because the current cannot flow into the cathodes of the SCRs. Consequently, there is no armature current flow, and the motor will slow down because no more torque exists. The rate at which it slows down depends on the inertia of the load and other factors such as friction.

As the speed drops, the motor is acting as a generator because its own voltage, E_c, is still greater than the converter's output, V_a. Since E_c decreases with shaft speed, a point is reached where E_c becomes less than V_a, and armature current again flows from the converter into the motor. At this point, the motor's torque is restored and it runs at the lower speed.

When the operator or remote source "calls" for an increase in speed from this lower-than-rated steady state speed, a similar process occurs. The "call" is really a reprogramming of the speed setpoint of the triggering source and immediately there is a difference noted between the feedback and setpoint signals. Now the triggering source begins to decrease the firing delay angle so that voltage, V_a, increases and, as a result, so does the shaft speed. This change continues until the set point and feedback signals are equal, at which time the firing delay angle is maintained.

As noted above, when a controlled reduction in speed of the motor is in progress, the motor *tends* to act as a generator. After all, when V_a is reduced below E_c, the motor no longer produces torque because the armature current stops flowing. At this point, however, the spinning load that has stored energy in the form of inertia during the motoring period gives this energy back by driving the shaft until the energy is depleted. This stored mechanical energy is not converted to electrical power as the shaft speed decreases but rather is dissipated mechanically in the form of friction, windage loss, and, possibly, in a few last strokes of useful work. The "bottom line" of this situation is really a loss of useful energy and, of course, a lowering of efficiency. Additionally, it may take an unacceptable amount of time for the shaft to "coast" to a lower speed.

Dynamic braking, of course, can be utilized to quickly slow the shaft's speed by first opening the converter circuit so that it no longer feeds the armature and then switching a resistance of appropriate value across the armature circuit so that current can flow from the armature as it acts as a generator. This method is often used to control the time it takes for speed changes but does not improve the efficiency of operation.

Regenerative braking offers improved efficiency with controlled reduction in speed although it is initially more costly, as will be seen presently.

Up to this point, the thyristor converter of Fig. 9-14 has been used in the rectifier (rectification) mode. That is, the output voltage, V_a, has been positive as a result of the triggering source controlling the firing delay angle betwen 90° down to about 20°. When the triggering source delays the firing angle beyond 90°, however, the polarity of V_a becomes negative and the converter is then said to be in the inverter mode, as discussed in Chapter 8. Merely reversing the polarity of the converter of Fig. 9-14, however, will not result in regeneration because current cannot flow into the cathodes of the SCRs of the converter. If, on the other hand, the armature's polarity is physically switched in conjunction with the converter being operated in the inverter mode, regeneration is possible, although the sequence must be carefully controlled and programmed into the triggering source. First, a call for speed reduction is initiated, and the triggering source, as usual, increases the firing delay angle so that V_a drops below E_c, at which point armature current flow stops. The triggering source receives this information because transducers are monitoring both I_a and E_c. The triggering source now delays the firing pulses beyond 90° so that V_a goes negative and is slightly less in magnitude than E_c. A high-speed power switch is then activated so that the connections to the armature are reversed and current flows from the armature into the three-phase source via the inverter. The energy absorbed by the converter quickly causes the speed to decrease, as does the generated voltage, E_c. When E_c becomes slightly less than V_a, the armature current again stops because current cannot flow out of the anode of the SCRs at the bottom of the converter, so the connections and circuit conditions must be restored as before by reversing the sequence. Although this scenario seems long and arduous, it can be implemented in well under one second, and *energy is returned to the line* from the load as it comes to a slower speed. Similar results can be achieved by reversing the polarity of the field windings instead of the armature using the same approach, but the time required for switching is increased due to the normally large inductance of the field windings. Armature reversal requires less time but a much larger switch because it must carry the large armature currents (sometimes greater than 1 000 amperes) and field reversal requires more time (seconds) but a relatively smaller switch. In either case, a switch is used which requires maintenance, and the response time may still not be fast enough.

It is possible to eliminate the switch and improve the response time considerably by considering the circuit of Fig. 9-15. Here is a circuit comprised of two converters connected in opposing parallel. Note that converter A can pass current out of the top and into the bottom while converter B can pass current out of the bottom and into the top. Both converters obviously require more triggering circuitry and support hardware, so the system is initially more costly.

The operation of the double converter speed control circuit is again centered in the triggering circuitry and involves the operation of one and only one converter at a time so that currents do not circulate between converters. Assume that the motor in Fig. 9-15 is running at some constant speed below rated speed, and develops a counter voltage, E_c, with positive polarity as indicated. Converter A is operating in the rectification mode and is supplying the armature voltage, V_a, and current I_a, while receiving triggering pulses at a firing delay angle, α_A. At the same time, converter B is not operating but the triggering source develops gate pulses anyway at a firing delay angle of α_B, which would result in the same value and polarity of V_a *if* converter B was operating. (Delay circuitry in the triggering source does not allow α_B pulses to reach converter B while converter A is receiving

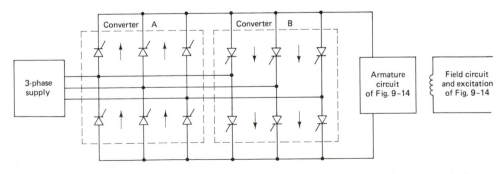

Figure 9-15. Double converter motor control.

DIRECT-CURRENT MOTOR CONTROL   CHAP. 9

α_A pulses.) Note that the polarity of V_a requires that converter B must be ready to operate with a *negative* voltage if called immediately into operation. Converter B then is always ready to take over with the same voltage level and polarity as converter A (and vice versa).

Suppose now that a call for a lower speed is initiated by the operator or remote control source. As soon as the speed setpoint is lowered, the triggering source reduces V_a so that it becomes slightly less than E_c by increasing the firing delay angle α_A. This causes the armature current to stop as in the previous single converter circuit. At this instant the motor acts as a generator, and the triggering source turns converter A off and converter B on. As noted above, the triggering source changes the angle of α_B any time α_A is changed, or vice versa, so that the same voltage and polarity of V_a are maintained and so converter B is activated with V_a slightly less than E_c. Because of this, converter B absorbs the energy generated by the armature and passes it to the source while α_B is decreased until the shaft speed matches the setpoint speed. Since α_A has been constantly adjusted by the triggering source, converter A is activated as converter B is turned off and α_A is decreased until V_A slightly exceeds E_c. The armature now receives power from converter A and the motor runs at the lower speed.

Increases of shaft speed are accomplished using converter A in the rectification mode just as in the single converter speed control circuit of Fig. 9-14. That is, converter A's output voltage is increased by decreasing the firing delay angle until the actual shaft speed matches that of the setpoint speed. Again, fixed-limit settings for overspeed, overcurrent, and so forth that are programmed into the triggering source provide protection and disallow damaging motor operation. As the motor increases in speed, it should be noted that converter B is on standby. The triggering source is generating, but not delivering, trigger pulses of delay angle, α_B, which would produce the same voltage and polarity produced by converter A.

Reversal of the dc motor requires that the polarity of either the armature or the field be reversed, but not both. The single converter speed control circuit of Fig. 9-14, as already seen, requires the switching of the polarity of either the armature or the field in order to control the decrease in speed of a motor. In order to reverse the direction of rotation, a slightly different sequence of operation is needed. First a decrease in speed is initiated and, as before, the polarity of the armature or field is switched, with the accompanying change in the firing delay angle so that the polarity of V_a is also switched. Once the regenerative action has resulted in a decrease of shaft speed (ideally to zero rpm), the firing delay angle must now be decreased so that V_a again becomes positive. The armature or field, however, is *not* switched back to the original configuration for this speed reversal condition. Now, as the armature voltage, V_a, is increased, the motor will accelerate in the opposite direction to that of the original direction as a result of an effective change in polarity of either the armature or field.

The double converter circuit of Fig. 9-15 is, of course, a faster responding alternative to that of Fig. 9-14 and can also be used for speed reversal. Note that both converter A and B of Fig. 9-15 can each be operated in either the rectification mode or the inverter mode. When one is operating in the rectification mode, the other is ready to operate as an inverter, as noted previously. It is obvious that converter A supplies a positive voltage when operating in the rectification mode, and it is also true that converter B supplies a negative armature voltage when operating in the rectification mode. With this capability, the reversal of polarity to the armature can be effected without any switching, per se. Thus, when reversal is called for, a speed reduction is initiated. Just as before, a controlled transition occurs such that converter B takes control of operation in the inverter mode, and regeneration occurs as the speed decreases rapidly. When a very low speed, ideally zero rpm, is in evidence via feedback, converter B is triggered into the rectification mode, and it supplies a negative voltage to the armature. This, of course, results in an opposite rotation due to the negative torque produced.

Although the reversal techniques described here seem straightforward, a complication does exist in reality. When in the reversal mode, it would be ideal to reduce the speed to zero rpm before producing countertorque. Under this condition, the converter voltage would be zero and intermittent currents can result (near zero rpm) which result in speed ripple and torque ripple. For this reason, a negative braking torque is applied before zero speed so that this need not be of concern.

QUESTIONS

9-1. What is the major operating difference between a manual and an automatic dc motor starter?

9-2. What does the term plugging mean?

9-3. What is an overhauling load?

9-4. What feature of a dc motor necessitates current-limiting resistors in a starter circuit?

9-5. What is the major fault of a definite time acceleration dc starter?

9-6. How does a current-limit acceleration starter overcome the fault problem of a definite-time acceleration starter?

9-7. When reversing a dc motor, what element of the motor circuit usually has its circuit sequence changed?

9-8. If the armature circuit connections are to be changed in a reversing control, what provision must be made for the commutating field if one is provided in the motor?

9-9. What added feature is necessary in a plugging reverse control?

9-10. What is the difference between dynamic braking and regenerative braking in a motor control?

9-11. What is meant by the term jogging?

9-12. What simple circuit provisions are necessary for a motor control to have a jogging feature?

9-13. Why is line voltage control of a dc motor more efficient than series armature resistance control?

9-14. What is a "chopper" and why is it used?

9-15. Briefly describe the three modes of operation in which a chopper may be effectively utilized.

9-16. Discuss the use of a smoothing inductor and free-wheeling diode in chopper control.

9-17. Draw a block diagram of a complete control system and detail how it functions.

9-18. Why are dc motors sometimes run as separately excited motors when using solid state control?

9-19. Draw a schematic diagram of a double-converter motor control and explain how it functions.

10

Direct-Current Motor and Generator Selection

Any motor or generator is rated by its *output power* capacity. For years this has been listed by horsepower in motors and kilowatts in generators. With the advent of SI measurements, *both* will be rated in kilowatts. This is an easy transition, since 0.746 kW equals 1 hp. Motor or generator selection must be far more comprehensive than just consideration of mechanical or electrical power delivered by the machine. Some of the factors that require consideration as to series, shunt, or compound type and that affect physical size are the following:

   (1) Shaft power in or out in horsepower or kilowatts.
   (2) Characteristics of driven load.
   (3) Rated speed in rpm or radians per second.
   (4) Frame size.
   (5) Speed classification.
   (6) Effect of duty cycle.
   (7) Ambient temperature.
   (8) Allowable temperature rise in the machine.
   (9) Voltage rating for input or output power.
 (10) Machine enclosure type and ambient conditions.
 (11) Maintenance requirements and accessibility.

10-1 SHAFT POWER

The electric machine shaft power should obviously be sufficient to carry the load that it is intended to drive. This is complicated by the fact that a motor or generator can briefly sustain a considerable overload. The full labeled rating may be such that the motor is rated as 10 hp at 1750 rpm continuous duty with a 50°C temperature rise. This size of motor might be able to produce 15 hp at around 1650 rpm but *not* continuously. The windings would overheat after 10 or 15 min.

The other condition is that the load may be far less for part of the time. This will be explored later under *duty cycle*. Remember that a motor is very inefficient at low loads. The motor should be chosen so that it runs at from 70 to 100 percent of its rated power most of the time.

In the future, when internationally standardized metric–unit–related motor and generator characteristics are finally widely accepted, the power rating will be in kilowatts rather than horsepower. The tentative kilowatt power rating that most closely corresponds to the 10 hp rating discussed above is 7.5 kW, but this is not expected to be a standard size rating: 5.6 kW or 8.0 kW would more likely be substituted for 10 hp.

Standardized power ratings are listed in Table 10-1. Note that the International Electrotechnical Commission (IEC) working with the International Organization for Standardization (ISO) has made recommendations for a wide, but so far incomplete, group of standardized sizes and dimensions. These ISO SC-2B power ratings, together with their successors, are shown as tentative only, so be aware and cautious.

The listing shown in Table 10-1 combines the long practiced U.S. horsepower ratings, as standardized by the National Electrical Manufacturers Association (NEMA), with the NEMA Guide for the Development of Metric Standards for Motors and Generators as revised in November 1980. Here NEMA is suggesting for its member subscribers in the industry their best interpretation as to the items and sizes to be chosen for standardization. This area is no longer "wide open," but is a long way from settling down to internationally accepted usage. The present NEMA Guide is just that—a guide. It is not yet a standard, but it does contain the latest information as of this writing.

10-2 CHARACTERISTICS OF DRIVEN LOAD

A load that might reasonably require 10 hp at 1750 rpm can have all sorts of different characteristics that must be considered. A blower has no particular starting torque requirements, but its torque needs build up as the square of the speed. A hoist or a vehicle may require far more torque to start it than it needs to keep moving. Another factor might be that periodic or cyclic plug braking and reversal may be required. The motor characteristics must be matched to the load, and it can be seen that power is only part of the story. The hard-starting load probably needs a series motor. The constant speed needs a shunt motor.

TABLE 10-1 MACHINE POWER RATINGS, REPRESENTATIVE DC AND AC NEMA-RECOMMENDED, PARTIAL COMPLIANCE WITH IEC

Existing NEMA std		Proposed NEMA guide all kilowatt						
hp	kW equiv.							
$\frac{1}{6}$	0.125			1.1	11.2		1120	11 200
$\frac{1}{4}$	0.187	0.125				125	1250	12 500
$\frac{1}{3}$	0.25		0.14				1400	14 000
$\frac{1}{2}$	0.375			1.6	16	160	1600	16 000
$\frac{3}{4}$	0.56	0.018					1800	18 000
1	0.75		0.20		20	200	2000	20 000
$1\frac{1}{2}$	1.12						2240	22 400
2	1.5	0.025		2.5	25	250	2500	25 000
3	2.25		0.28				2800	28 000
5	3.75				32	320	3200	32 000
$7\frac{1}{2}$	5.6	0.036					3600	36 000
10	7.5		0.40	4.0	40	400	4000	40 000
15	11.2						4500	45 000
20	15	0.050			50	500	5000	50 000
25	18.6		0.56	5.6			5600	56 000
30	22.5					600		
40	29.8				63		6300	63 000
50	37.5	0.071				710	7100	71 000
60	44.7		0.80	8.0	80		8000	80 000
75	56					850		
100	75	0.009					9000	90 000
125	93.2		0.10		100	1000	10 000	100 000
150	112							
200	150							
250	186							

10-3 SPEED RATING

The nameplate rating of rpm, or later in radians per second, is a steady-state design point only. A motor may be required to operate at any speed from "stop" up to its rated speed. Or it may be required to operate at its rated speed except for periodic excursions to about double the normal speed. The load may decrease suddenly, and yet constant or nearly constant speed will be required. This would rule out a series motor, for example.

Finally, the motor operation point and whether speeds above or below are required will then determine the base speed and the basic motor type.

Motor nameplate speeds in the standardized sizes are normally stated in ranges such as the following rpms: 3500, 2500, 1750, 1150, 850, 650, 500, 400, and 300. Up-to-date SI standardized speeds are not yet available.

Internationally standardized base speed ranges in radians per second are not yet agreed upon. However, NEMA is suggesting rpm ranges that closely correspond to those above. These are 3500, 1750, 1450, and 1150 rpm for dc generators and 3450, 2500, 1725, and 1140 rpm for small dc motors. Note that the generator speeds suggested are those that correspond to loaded ac induction motor speeds that would hold in direct-connected motor generator sets.

10-4 FRAME SIZE

Motor and generator frame sizes have long been standardized by the National Electrical Manufacturers Association (NEMA). These frame sizes may be visualized as diameter series with a few standardized lengths within each diameter. There is no one-for-one correspondence between output power and frame size since speed introduces such large torque and cooling variations. A typical family of horsepower, rpm, and frame size relations is shown in Table 10-2. This table is specifically related to General Electric drip-proof construction 60°C temperature rise motors, but is reasonably representative of quality construction in the industry. Note that a particular frame size may be used on as many as eight different power and rpm combinations. The 256A frame is used for 1 hp at 400 rpm on down to 10 hp at 2500 and 3500 rpm, with a complete blend of powers and speeds in between.

At this point, it is good to show the degrees of standardization that do exist in NEMA frame sizes. See Table 10-3 for some of this range. Note that all variations within a series are in extra length, with corresponding length change to mounting bolt spacing (see also Fig. 10-1).

NEMA frame sizes have evolved with major changes or additions to the specifications in 1953 and 1964. In general, a given frame number has the same

TABLE 10-2 REPRESENTATIVE HORSEPOWER RATINGS AND FRAME SIZES, DC MOTORS

| Hp | \multicolumn{9}{c}{Speed (rpm)} |
	3500	2500	1750	1150	850	650	500	400	300
$\frac{1}{2}$					187A	215A	216A	216A	
$\frac{3}{4}$				187A	215A	216A	218A	254A	
1			186A	187A	216A	218A	254A	256A	
$1\frac{1}{2}$	186A	186A	187A	215A	218A	254A	256A	284A	
2	186A	187A	215A	216A	254A	256A	284A	286A	324A
3	187A	215A	216A	218A	256A	284A	286A	324A	
5	216A	216A	218A	256A	286A	324A	326A		
$7\frac{1}{2}$	218A	218A	256A	286A	324A	326A			
10	256A	256A	284A	286A	326A				
15	284A	284A	286A	326A					
20	286A	286A	324A						
25		324A	326A						
30		326A							

TABLE 10-3 NEMA STANDARD FRAME DIMENSIONS FOR DC AND AC MOTORS AND GENERATORS

All dimensions in inches. Multiply by 25.4 for millimeters. See Figure 10-1 for location of dimensions.

NEMA frame	Shaft height D	Shaft Dia U	Shaft Hub length V	Key Over-all N-W	Key Square	Key Long	Mounting bolt location Width $2E$	Mounting bolt location Length $2F$	Mounting bolt location Dia H	Set back BA
42	2.62	0.375	—	1.12	0.050	flat	3.50	1.69	0.28 slot	2.06
48	3.00	0.500	—	1.12	0.050	flat	4.25	2.75	0.34 slot	2.50
56	3.50	0.625	—	1.88	0.187	1.38	4.88	3.00	0.34	2.75
56H	3.50	0.625	—	1.88	0.187	1.38	4.88	3.00	slots	2.75
56HZ	3.50	0.500	—	1.50	flat	flat	and	and	slots	and
		or		or	or	or				
		0.875		2.25	0.188	1.38	5.50	5.00	slots	2.25
143T	3.50	0.875	2.00	2.25	0.188	1.38	5.50	4.00	0.34	2.25
145T	3.50	0.875	2.00	2.25	0.188	1.38	5.50	5.00	0.34	2.25
182	4.50	0.875	2.00	2.25	0.188	1.38	7.50	4.50	0.41	2.75
184	4.50	0.875	2.00	2.25	0.188	1.38	7.50	5.50	0.41	2.75
182T	4.50	1.125	2.50	2.75	0.250	1.75	7.50	4.50	0.41	2.75
184T	4.50	1.125	2.50	2.75	0.250	1.75	7.50	5.50	0.41	2.75
213	5.25	1.125	2.75	3.00	0.250	2.00	8.50	5.50	0.41	3.50
215	5.25	1.125	2.75	3.00	0.250	2.00	8.50	7.00	0.41	3.50
213T	5.25	1.375	3.13	3.38	0.312	2.38	8.50	5.50	0.41	3.50
215T	5.25	1.375	3.13	3.38	0.312	2.38	8.50	7.00	0.41	3.50
254U	6.25	1.375	3.50	3.75	0.312	2.75	10.00	8.25	0.53	4.25
256U	6.25	1.375	3.50	3.75	0.312	2.75	10.00	10.00	0.53	4.25
254T	6.25	1.625	3.75	4.00	0.375	2.87	10.00	8.25	0.53	4.25
256T	6.25	1.625	3.75	4.00	0.375	2.87	10.00	10.00	0.53	4.25
284TS	7.00	1.625	3.00	3.25	0.375	1.88	11.00	9.50	0.53	4.75
286TS	7.00	1.625	3.00	3.25	0.375	1.88	11.00	11.00	0.53	4.75
284T	7.00	1.875	4.38	4.62	0.500	3.25	11.00	9.50	0.53	4.75
286T	7.00	1.875	4.38	4.62	0.500	3.25	11.00	11.00	0.53	4.75
324TS	8.00	1.875	3.50	3.75	0.500	2.00	12.50	10.50	0.66	5.25
326TS	8.00	1.875	3.50	3.75	0.500	2.00	12.50	12.00	0.66	5.25
324T	8.00	2.125	5.00	5.25	0.500	3.88	12.50	10.50	0.66	5.25
326T	8.00	2.125	5.00	5.25	0.500	3.88	12.50	12.00	0.66	5.25
364TS	9.00	1.875	3.50	3.75	0.500	2.00	14.00	11.25	0.69	5.88
365TS	9.00	1.875	3.50	3.75	0.500	2.00	14.00	12.25	0.69	5.88
364T	9.00	2.375	5.62	5.87	0.625	4.25	14.00	11.25	0.69	5.88
365T	9.00	2.375	5.62	5.87	0.625	4.25	14.00	12.25	0.69	5.88
404TS	10.00	2.125	4.00	4.25	0.500	2.75	16.00	12.25	0.81	6.62
405TS	10.00	2.125	4.00	4.25	0.500	2.75	16.00	13.75	0.81	6.62
404T	10.00	2.875	7.00	7.25	0.750	5.62	16.00	12.25	0.81	6.62
405T	10.00	2.875	7.00	7.25	0.750	5.62	16.00	13.75	0.81	6.62

TABLE 10-3 CONTINUED

NEMA frame	Shaft height D	Shaft Dia U	Shaft Hub length V	Key Over-all N-W	Key Square	Key Long	Mounting bolt location Width 2E	Mounting bolt location Length 2F	Mounting bolt location Dia H	Set back BA
444TS	11.00	2.375	4.50	4.75	0.625	3.00	18.00	14.50	0.81	7.50
445TS	11.00	2.375	4.50	4.75	0.625	3.00	18.00	16.50	0.81	7.50
447TS	11.00	2.375	4.50	4.75	0.625	3.00	18.00	20.00	0.81	7.50
444T	11.00	3.375	8.25	8.50	0.875	6.88	18.00	14.50	0.81	7.50
445T	11.00	3.375	8.25	8.50	0.875	6.88	18.00	16.50	0.81	7.50
447T	11.00	3.375	8.25	8.50	0.875	6.88	18.00	20.00	0.81	7.50

NOTES: NEMA letter designations after frame number

A Usually found on older dc machinery
C Face mount
H Increased 2F dimension
J Face mount
K Hub for sump pump mounting
M, N Flange mount for oil burners
T Integral hp motor dimensions as of 1964
U Integral hp motor dimensions as of 1953
Y Nonstandard mounting, see manufacturer
Z Nonstandard shaft or shaft extension

Frame numbers not listed here: 66, 203, 204, 224, 225, 254, and larger without suffix letters refer to NEMA frames before 1953. Differences are in shaft diameter and/or length.

Longer and/or shorter frame sizes within a given series are available. For example, a 216 frame size is 1.0 inch longer than 215.

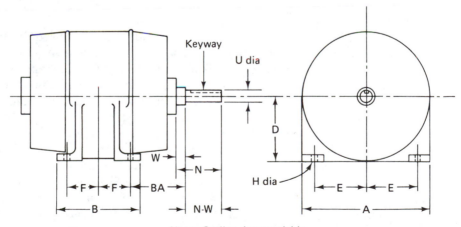

Note: Outline shape variable.
See Table 10.3 for sizes.

Figure 10-1. Standard motor dimensions of the NEMA.

overall size and mounting foot dimensions. However, as motor power outputs within a given size have continuously increased, the stresses on the shafts have also increased, requiring larger diameter shafts. For example, frame sizes 364 and 365 originally had 1.875 inch diameter shafts with an alternate increase to 2.125 in. and then with the 364T size to 2.375 in. Now the alternate 364 TS is available at 1.875 in., and the intermediate 364 U size of 2.125 in. is not so readily found. See ahead to Table 22-2 for clarification.

In spite of the best efforts of the NEMA, the various catalogs of current manufacturers occasionally disagree, so that Table 10-3 does not agree with all manufacturers at all locations. This area of standardization is important because of the long useful life of electric motors and generators and the machines that they are used on. The somewhat muddied area between frame sizes 56 and 143 T shows this. The 56 size was originally a single-phase ac appliance motor, while the 143 and 143 T are originally three-phase ac industrial motors. A 56H frame size will fit in either location as a replacement.

Direct current machines have been kept as close to NEMA frame standards as is practical, but they are sometimes found in *longer* standard frames to accommodate the commutator. For example, the 216 frame size is just one inch *longer* than a 215. Some units will have shafts on both ends; if so in larger sizes, one end will then be the larger, and the other the smaller, of the two major sizes for that basic frame.

The process of standardizing electric machine frame sizes for internationally recognized metric-dimensioned incremental sizes is slowly progressing. The IEC has issued suggested standards, and the NEMA has, in turn, issued to its members the recommended guide for newer designs. The trend is toward metric dimensions of existing frame sizes, with minor adjustments in mounting-hole dimensions and shaft diameters. In fact, Table 10-4 shows that NEMA is recommending sizes which are the nearest whole mm conversion of the inch dimensions of the regular NEMA frames shown in Table 10-3. However, the shaft heights are in accordance with ISO R496, with every third or every second ISO-recommended shaft height chosen.

Since there is much difference between various manufacturers' motors within a given NEMA frame at the present, and since only the shaft location and mounting feet dimensions are really critical, it is felt that this is a practical approach. Most manufacturers will be able to save most of their tooling, and with minor adjustments they will be able to market internationally acceptable electric machines.

Shaft diameters are not yet matched to suggested frame sizes. On the other hand, ISO recommended shaft sizes and tolerances will be adhered to. See Table 10-5 for probable shaft sizes and possible, but not yet probable, matching frame sizes by comparison with the inch sizes in Table 10-3 and Table 22-2.

There are many possible spacings for the mounting bolt holes—especially for the 2F dimension—but the likely ones are shown in Table 10-4.

At present, it can be seen that physical outline differences will not be drastic. For example a 10 hp, 1150 rpm, 286A, or 286 TS frame size motor is to be replaced by its nearest internationally standardized equivalent.

TABLE 10-4 NEMA-PROPOSED GUIDE FOR ISO-RELATED MOUNTING DIMENSIONS FOR AC AND DC MOTORS AND GENERATORS
All dimensions in millimeters. Refer to Figure 10-1.

New NEMA Frame	Shaft height D	Shaft U	V	$N\text{-}W$	Key sq.	Key lg.	Mounting bolt location Width $2E$	Length $2F$	Dia H	Set back BA	Nearest old NEMA frame
	66.7	9.525	28.6	—	flat		88.8	42.9	7.1	52.4	42
	76.2	12.700	38.1	—	flat		108.0	69.8	8.7	63.5	48
	76.2	12.700	38.1	—	flat		108.0	120.6	8.7	63.5	
	88.9	15.875	47.6	—	4.8	35.7	123.8	76.2	8.7	69.8	56
	88.9	15.875	47.6	—	4.8	35.7	123.8	127.0	8.7	69.8	56H
90D	90	Note: Shaft dimensions are					140	100	9.5	56	143T
90F	90	not yet chosen					140	125	9.5	56	145T
112C	112	Refer to Table 10-5 for range					190	114	12	70	182T
112E	112	of proposed sizes					190	140	12	70	184T
132D	113	Shaft heights 66.7,					216	140	12	89	213T
132F	132	76.2 & 88.9					216	178	12	89	215T
160E	160	are direct mm					254	210	14	108	254T
160G	160	conversions of NEMA frames					254	254	14	108	256T
180E	180	42, 48 & 56, as no ISO R496					279	241	14	121	284T
180G	180	sizes are proposed below					279	279	14	121	286T
200E	200	90 mm.					318	267	18.5	133	324T
200G	200						318	305	18.5	133	326T
225E	225	All other dimensions					356	286	18.5	149	364T
225F	225	are nearst mm					356	311	18.5	149	366T
250E	250	conversion of previous					406	311	23	168	404T
250F	250	NEMA dimensions.					406	349	23	168	405T
280E	280						457	368	23	190	444T
280F	280						457	419	23	190	445T
315	315						508	—	27	217	—

(1) Ten hp is equivalent to 7.5 kW, but from Table 10-1 the nearest kW power size is 8.0 kW.

(2) Rpm standards will be closely equivalent, since even in dc machines they are usually derived from ac induction motor speeds. Thus, 1150 rpm is in the loaded range for a six-pole induction motor when it is operating on 60 Hz ac power. A typical dc motor loaded speed can be as close as 1140 rpm or its radians per second equivalent.

(3) Motor sizes are manufactured in as small a frame size as they can reasonably be made with normal efficiencies and acceptable temperatures. An ISO equivalent motor might well be the 180G frame size from Table 10-4. The choice will depend upon acceptable temperature and efficiency, but a shaft height greater than 180 mm is not likely. Since 180 mm is only 2.2 mm larger than 7 in. (actually 7.087 in.), it would be prudent to design a new machine to have a 180-mm shaft height space. An extra 0.087 in. (2.2 mm) shim could well be provided, since at present

TABLE 10-5 NEMA PROPOSED METRIC SHAFT SIZES IN PARTIAL COMPLIANCE WITH ISO STANDARDS

All dimensions in millimeters except where noted.
Column headings U, etc., refer to Figure 10-1.

Shaft					Keyseat			Key	
Shaft dia. U	Equiv. inch dia.	Nearest NEMA inch	Shaft extension N-W	Usable length V	Width S	Bottom to far side R	Length ES	Wide	High
19	0.748		38	32	6	15.5	32	6	6
		0.875							
24	0.945		48	42	8	20	42	8	7
28	1.102		56	50	8	24	50	8	7
		1.125							
32	1.260		64	58	10	27	58	10	8
		1.375							
38	1.496		76	70	10	33	70	10	8
		1.625							
42	1.654		84	78	12	37	78	12	8
		1.875							
48	1.890		96	90	14	42.5	90	14	9
50	1.969		100	94	14	44.5	94	14	9
		2.125							
55	2.165		110	104	16	49	104	16	10
60	2.362		120	114	18	53	114	18	11
		2.375							
65	2.559		130	124	18	58	124	18	11
70	2.756		140	134	20	62.5	134	20	12
		2.875							
75	2.953		150	144	20	67.5	144	20	12
80	3.150		160	154	22	71	154	22	14
85	3.346		170	164	22	76	164	22	14
		3.375							
90	3.543		180	174	25	81	174	25	14
95	3.740		190	184	25	86	184	25	14
100	3.937		200	194	28	90	194	28	16
110	4.331		220	214	28	100	214	28	16
120	4.724		240	234	32	109	234	32	18
130	5.118		260	254	32	119	254	32	18

an existing NEMA frame would probably be used initially. At replacement, perhaps 10 or 20 years later, or when the 180 F motor or its equivalent is manufactured, it could be dropped into the same space. If a 160-mm shaft height replacement were to be used, an additional 20 mm spacer would be easy to provide. If this were a geared drive machine, it would be wise to provide the 180 mm shaft height and shim, because if only the basic seven inches were provided (177.8 mm), a 180 mm unit would not fit at a future time.

(4) The shaft diameter is a little less sure, but from Tables 10-3 and 10-5 | probably be 42 or 48 mm. Possibly a shaft as small as 38 mm would be | a 160-mm shaft height. It is relatively easy matter to bore or to bush a p | a gear. If a pulley, it is a low cost replacement, especially if it is one | proprietary brand conical wedging hubs where only the insert need be replaced.

(5) The longitudinal spacing (2F) of the mounting bolt holes from Table 10-4 would be 279 mm, which is very close to the original 11.00 in. Actually 279 mm is 10.984 in., so the nomimal difference of 0.016 in., or about $\frac{1}{64}$ in., can probably be ignored when the bolt-to-bolt hole clearance is considered. The lateral spacing (2E) happens to be the same in this case. As a result, a 180G motor can be directly substituted for a NEMA 286 frame, and the bolts will drop in. This is especially so if a preferred size 12-mm-diameter bolt is chosen. Since this is 0.472 in. in diameter, there is additional clearance over the original 0.5 in. preferred size.

This whole subject can be seen to be under very careful consideration, and if sufficient thought is used at the outset of a new design, most motor interchange problems will be seen to be quite simple.

10-5 SPEED CLASSIFICATION

Motors have become classified by their speed-load characteristics. Some classifications fit dc or ac types, but only those applicable to dc types are named:

(1) *Constant speed* is a motor whose speed varies only a small amount from no load to full load. This amount is understood to be 20 percent or less, but this is not officially adopted. Shunt motors and some compound motors can meet this specification.

(2) *Varying-speed* motors change more than 20 percent from no load to full load. This type is represented by series and some compound motors.

(3) *Adjustable-speed* motors can have their base speed changed over a wide range, but will hold that speed within approximately 20 percent from no load to full load. A shunt and some compound motors meet this requirement with proper control.

(4) *Adjustable varying speed* motors change their speed widely as their load changes, but can have their speed at any particular load controlled. Here again series and some compound types fit this requirement.

(5) *Reversing motors* is a classification that can be met by all dc types. This becomes more complicated with ac types.

10-6 EFFECT OF DUTY CYCLE

One of the most important considerations for motor selection is that of the range of its operating loads or its duty cycle. If the motor is substantially too large, it will operate at a low efficiency. This is wasteful and also means that a larger and therefore more expensive machine is unnecessarily chosen. Conversely, if the motor is too small in size, and therefore too low in power handling ability, it will run at a higher than normal temperature if it is able to handle the task at all.

Since every 10°C temperature rise in the windings of a machine will cut the useful life of the insulation in half, this factor must be carefully watched. A motor must operate at or below its maximum winding temperature if a reasonable life is to be expected.

A steady load situation can then be easily determined, and the proper-sized motor can be easily chosen. It is the situation where the load varies cyclically and where periods of rest are intermixed that poses the challenge for proper motor rating choice. The heating of a motor is not necessarily limited by a short but heavy peak load. The root-mean-squared or rms values of all the cyclic portions of the load must be accounted for.

If the various load powers and times are gathered in a usual form of square root of the sum of the squares fashion, the rms power emerges. Rest periods or times with the motor shut off are counted as one-third as long as they actually are. This is because the motor is not as well cooled when not rotating. Equation (10-1) can be used for either English or SI units as long as they are consistent within the formula. The time periods can be either in seconds or minutes as long as they are consistent.

$$\text{rms power} = \sqrt{\frac{(p_1^2 t_1) + (p_2^2 t_2) + (p_n^2 t_n)}{t_1 + t_2 + t_n + (t_r/3)}} \qquad \text{Eq. (10-1)}$$

where P is power in horsepower or kilowatts, and t is time, usually in minutes.

Example 10-1 will show this use.

Example 10-1. A test motor was connected to a variable-cycle machine load for a period of 1 hr. The motor operated at 11.0 hp (8.21 kW) for 4 min, 1.25 hp (0.933 kW) for 6 min, 6.1 hp (4.55 kW) for 11 min, and was off for a period of 9 min. The cycle repeated on this basis. Calculate the rated machine horsepower (or kilowatts) required.

Solution: An inspection shows that the time periods total 30 min and then repeat, so a 30-min calculation period is chosen. Using Eq. (10-1), the data are applied as shown. Note that, if the rest period had been running at no load, the full running time would be used rather than the 9-min rest divided by a factor of 3.

$$\text{rms hp} = \sqrt{\frac{[(11.0)^2 \times 4] + [(1.25)^2 \times 6] + [(6.1)^2 \times 11]}{4 + 6 + 11 + (9/3)}}$$

$$= \sqrt{\frac{(121.00 \times 4) + (1.5625 \times 6) + (37.21 \times 11)}{24}}$$

$$= \sqrt{\frac{484.00 + 9.375 + 409.31}{24}} = \sqrt{\frac{902.68}{24}} = \sqrt{37.612}$$

$$= 6.133$$

From Table 10-1, a 7.5 hp motor would be chosen as the *next larger* standard NEMA power rating. It would be running at $(11/7.5) \times 100 = 146.7$ percent of rated load for 4 min, 16.6 percent of rated load for 6 min, and 81.3 percent of rated load for 11 min. A 5 hp motor would *average* over 20 percent overload and thus be doomed to a short life even if it could carry the 11 hp peak briefly. In SI units the problem becomes:

$$\text{rms kW} = \sqrt{\frac{[(8.21)^2 \times 4] + [(0.933)^2 \times 6] + [(4.55)^2 \times 11]}{4 + 6 + 11 + (9/3)}}$$

$$= \sqrt{\frac{(67.404 \times 4) + (0.870\ 49 \times 6) + (20.703 \times 11)}{24}}$$

$$= \sqrt{\frac{296.62 + 5.2229 + 227.73}{24}} = \sqrt{\frac{502.57}{24}} = \sqrt{20.940}$$

$$= 4.576$$

As a check, 4.576 kW/0.746 kW per hp = 6.134 hp.

In this case, a new standard 4.0 kW motor would already be significantly overloaded. Specific design details and tests would be needed here. A 5.6 kW motor would be a natural choice.

This example has its highest power held for an interval of only 4 min. Since this time would not be sufficient to bring the motor to an equilibrium temperature, the windings would not reach their highest temperature. This rms averaging takes advantage of that fact. If each time were 10 times as long and in the same proportion, the method would be *invalid*. Forty minutes at a high overload would bring the windings to a dangerous temperature. As a result, the motor would have to be sufficiently large to carry the maximum load without its being seriously overtaxed. It would at least require a 10 hp (or a new standard 8.0 kW) motor to carry the assumed 40 min interval at 11 hp (or 8.21 kW).

Conversely, if the operating periods were much shorter (say by a factor of 0.1 or even 0.01 times), the problem complicates in another way. In the short-interval case the starting current and plug stop current, if used, become significant heating factors. The analysis here goes beyond the scope of this book. However, it can be understood that constant start and stop and frequent speed change, if required, become heating factors due to higher current in the armature circuit.

10-7 AMBIENT TEMPERATURE EFFECTS

In many kinds of device usage, the term "ambient temperature" means the temperature of the medium used for cooling. The vast majority of electric motors are air cooled, so that ambient or surrounding air temperature determines the effectiveness of the motor or generator cooling. The temperature that limits motor life is the temperature of the armature and field winding wires. Here the problem is not so much the copper as it is the insulation effectiveness. High temperatures gradually oxidize and carbonize the insulation materials, which reduces their insulation ability. As the materials slowly change, they gradually shift from good to poor insulators and finally to partial conductors. Ultimately, the former insulation will allow a short-circuit current to flow where it did not flow before. This short-circuit may only partially cut out a turn or so of winding. When this progressive increase of conductivity reaches the situation where a substantial current can flow, failure follows rapidly. The usual course is that a local part of the windings will melt owing to the effect of this short circuit. At the very least, the winding is destroyed. The failure may well result in a progressive failure to other previously unaffected windings. The failure may result in a serious fire since most insulations will burn when sufficiently heated. This is the process that keeps the motor replacement market and the motor rebuilders in business.

The ambient temperature is the base temperature from which all internal heating starts. Obviously, a high ambient temperature location will reduce the capacity of a motor, since it will result in abnormal temperatures within an otherwise normal motor. Ambient temperatures of up to 40°C may be perfectly acceptable even though above a normal room temperature of about 20°C. If the motor is associated with a high-temperature device, such as an oven or furnace, its ambient temperature might well be in the 140°F or 60°C range. This is already approaching some insulation limiting temperatures. Here special insulations must be employed or the motor will progressively fail without even operating. Special cooling air ducts that operate continuously are one alternative. Each dubious situation must be explored for the design alternatives that may be available.

10-8 ALLOWABLE TEMPERATURE RISE

Motors and generators normally carry allowable temperature rise data on their nameplates. Typically this will be 50°C. It may be 60°C or more with heat-resistant insulations. Even higher temperatures are allowed with inorganic insulations.

Since it is really the interior winding hot spot temperature that limits insulation life, the combination of ambient temperature plus temperature rise is the true criterion.

Insulation materials are at present standardized for temperature rating under NEMA agreements. It is not known what classification forms will be internationally agreed upon under the IEC or ISO, but it is believed that a similar arrangement will be adopted. The similarity will be based upon the very real temperature limitations of commercially available insulation materials. This type of limitation is international in nature.

Present temperature limitations are summarized in Table 10-6. The older insulation category known as class O, or organic, is of less importance because this represents dry, unvarnished, or unimpregnated organic materials, which are not used in this fashion now.

The listing of temperature-rise data in relation to 20°C ambient is unofficial, but it is included to show what can be expected if the motor is favorably located.

TABLE 10-6 INSULATION MATERIAL TEMPERATURE LIMITS

Partial list of materials	Class	Allowable temp. rise for assumed 40°C ambient air temp. (Nameplate)	Reasonable temp. rise if ambient air temp. is 20°C room temp.	Probable winding surface temp. under max. cond.	Hottest spot in winding max. limit temp., °C
Organic insulation without impregnation or binder liquid	O	50	70	75	90
Impregnated class O materials, enamel insulations, cellulose materials, molded cellulose, phenolics, resins	A	65	85	90	105
Inorganic minerals with some organic class A binder	B	90	110	115	130
Epoxy enamels modified polyester–epoxy materials, fiber glasses, Teflon	F	115	135	140	155
Inorganic materials with silicone binders, silicone rubbers	H	140	160	165	180
Mica, porcelain, glass, quartz in pure form	C	No assigned limits			

Similarly, the 15°C difference between the surface temperature of a coil and its hottest spot is conventional rather than specific. This is based on average differences when measurements are made with thermocouples that are buried in the windings.

The nameplate allowable-temperature-rise data appear in the second column.

10-9 VOLTAGE AND CURRENT RATINGS

There are millions of small dc machines that operate on 6 and 12 V. The 12 V range actually includes up to 14.7 V, since it includes the whole range of automotive auxiliary devices that operate on 12 V or less from battery sources and up to 14.7 when the battery charging system is in operation. This voltage is determined by the electrochemical requirements of the normal lead acid storage batteries. For similar reasons, there is a whole range of dc machines that operate on the normal 28.5 V of an aircraft dc electrical system.

The NEMA has long adopted standard voltage ratings for pure dc-operated machinery, as follows:

(1) For generators, 125, 250, 275, and 600 V.

(2) For motors, 120, 240, and 550 V.

The differences between the generators and motors are to allow for the nominal voltage drops in connecting wiring. Industrial systems have operated on 120–125 V and 240–250 V, while transportation traction or rapid transit systems operate on 550–600 V.

Now the trend is toward industrial control systems that take advantage of the inherent flexibility of dc motors, but that use ac as a primary source of power. The dc machines then operate on rectified and sometimes filtered ac, perhaps modified by chopped or pulsing voltage supplies. This trend has required dc motors with laminated field structures, which frequently results in machines with different voltage ratings for their armature circuit and their shunt field circuit.

The NEMA is now recommending in its guide that small dc motors, or motors with the shaft height D of under 90 mm, be made to operate on armature voltages of 75, 90, or 150 V when the primary ac supply is single phase. At the same time,

TABLE 10-7 FULL-LOAD CURRENTS IN AMPERES, DC MOTORS

kW rating	hp NEMA	120 V load current	Inferred efficiency at 120 V	240 V load current	Inferred efficiency at 240 V
0.10		2.1	40.0	1.1	40.0
	$\frac{1}{6}$	2.3	45.5	1.2	45.5
0.14		2.5	48.0	1.3	48.0
	$\frac{1}{4}$	2.9	53.6	1.5	53.6
0.20		3.1	54.3	1.6	54.3
	$\frac{1}{3}$	3.6	57.6	1.8	57.6
0.28		4.0	58.4	2.0	58.4
	$\frac{1}{2}$	5.2	59.8	2.6	59.8
0.40		5.5	60.5	2.8	60.5
0.56	$\frac{3}{4}$	7.4	63.0	3.7	63.0
	1	9.4	66.1	4.7	66.1
0.80		10.0	66.6	5.0	66.6
1.12	$1\frac{1}{2}$	13.2	70.6	6.6	70.6
	2	17.	73.1	8.5	73.1
1.6		18.2	73.3	9.1	73.3
	3	25	76.4	12.2	76.4
2.5		26.6	78.2	13.3	78.2
	5	40	77.7	20	77.7
4.0		42.7	78.1	21.3	78.1
5.6	$7\frac{1}{2}$	58	80.4	58	80.4
	10	76	81.8	38	81.8
8.0		81.2	82.1	40.6	82.1
11.2	15	110	84.8	55	84.8

NOTE: Basis from NEC 430–147, all kW power data derived.

the fields are suggested to operate on 50 or 100 V. Sizes larger than 400 watt or 0.4 kW (or around $\frac{1}{2}$ hp or larger), when operated on single-phase ac sources, are expected to use 90 or 180 V for the armature and 50, 100, or 200 V for the fields.

When these same small dc machines are operated from a rectified three-phase source, the armature dc supply is then 240 V, and the fields will operate on 100, 150, or 240 volts.

Medium-size motors, or those larger than 90 mm shaft height, will use slightly higher voltages when operating on rectified ac control systems. Armature voltages are then 90 or 180 V, and fields are again at 50, 100, or 200 V.

When medium-size motors are operated from true dc lines, the old value of 240 V dc is suggested for armature circuits for machines of up to 200 kW or over 250 hp. These machines are to have field circuits of 100, 150, or 240 V, depending on whether an electronic control or straight dc line voltage is used. When the higher 500 or 550 V dc is used for armature circuits in the size range of 5.6 kW to 1000 kW, then field voltages are expected to be 240 or 300 V.

Still larger dc motors are expected to operate on dc supplies with either 500 or 700 V being chosen, depending on size. Obviously, the picture is no longer simple, nor are the voltage ranges fully agreed upon.

Each motor input at a particular line voltage will then result in a specific line current under rated load. This current must be sufficiently larger than the volts times current equivalent of the output power to allow for normal machine losses. As a result, the National Electrical Code 430–147 recognizes certain volts, power, and current relations that are the basis of Table 10-7.

It can be seen from the various current ratings that with appropriate safety factors, breakers, fuses, and contactors can be selected.

TABLE 10-8 STANDARD NEMA RATINGS OF DC CONTACTORS

Contactor size	8-hour open rating amperes	Power rating					
		120 V		240 V		550 V	
		kW	hp	kW	hp	kW	hp
00	8*	0.56	3/4	1.12	$1\frac{1}{2}$		
0	17*	1.12	2	2.5	3		
1	25	1.6	3	4.0	5		
2	50	4.0	5	8.0	10	20	20
3	100	8.0	10	20	25	40	50
4	150	11.2	20	25	40	63	75
5	300	25	40	63	75	125	150
6	600	63	75	125	150	250	300
7	900	80	110	160	225	400	450
8	1350	125	175	250	350	600	700
9	2500	250	300	500	600	1250	1200

*Contactor sizes 00 and 0 are not NEMA-rated for dc service. Values are estimated.

NOTE: kW-rated motors are estimates that stay within current ratings for reasonable efficiencies. Contactor ratings are for open locations. For enclosed control situations, consult manufacturer's ratings.

Contactors are sized by NEMA rating, as shown in Table 10-8.   There are no known IEC or ISO agreed equivalents for overall design purposes when working in SI units at this time.   However, an ampere is an ampere in either English or SI units, and at this time control systems can be developed around NEMA-rated components.

10-10   MACHINE ENCLOSURE TYPES

A dc or ac motor or generator may be physically packaged in a number of standardized ways.   The NEMA defines a group of motor enclosure types, which have long been honored by the electrical industry.   There will presumably be some such similar IEC or ISO agreed types in due time.   The probability is that some of the types will be combined in order to simplify manufacture, inventory, and sales.

The motor types begin with the simplest and least costly and work on up to the most elaborate.   The simple open *drip-proof* frame was referenced in Table 10-2.   More elaborate enclosures usually require larger frame sizes for a given power since ventilation is restricted.   Consult specific manufacturers' lists for specific design problems.

The enclosure types as rated by the NEMA are as follows:

(1) *Open enclosure.*   An open end frame structure that permits maximum air circulation for ventilation.   This construction usually is such as to prevent dropped objects from contacting electrically live or moving parts.

(2) *Drip-proof enclosure.*   An enclosure so constructed that liquid or solid particles that fall at not greater than 15° from vertical will not enter the enclosure either directly or while running off the surface.

(3) *Splash-proof enclosure.*   Carries the drip-proof situation further so that particles arriving at up to 100° from the vertical will not penetrate inside.   Note that each increase in protection usually reduces ventilation.

(4) *Guarded enclosure.*   This enclosure is so arranged that no accidental or intentional object can penetrate.   Specifically, a $\frac{1}{2}$ in.-diameter rod must not be able to penetrate screens or guards.

(5) *Weatherproof enclosure.*   A variation of the drip-proof and splash-proof design that prevents blowing rain, snow, or dust from contacting electrical parts.

(6) *Totally enclosed enclosure.*   Closed and/or covered but not necessarily airtight enclosure.

(7) *Explosion-proof enclosure.*   Designed to contain an inside explosion and/or to prevent ignition of specified gases or vapors surrounding the motor.   This may be accomplished by specified screens and/or flame traps, as in a marine gasoline engine carburetor air opening.   It does not specifically mean total enclosure, although it may be so.

(8) *Dust-ignition-proof enclosure.*   Totally enclosed and constructed to exclude entrance of ignitable dusts or dusts that would build up and affect performance.

(9) *Waterproof enclosure.*   Designed so that its total enclosure may be sprayed by the stream from a hose without detrimental effect.   Shaft leakage is allowed if

it is drained away in specified fashion. This enclosure is used in dairy and other food-processing machinery where daily cleaning and even sterilization take place.

Manufacturers may not necessarily use these terms in their nomenclature but will reference them. A totally enclosed fan-cooled motor (TEFC) may be specified to meet various of these NEMA classifications, for example. Again, consult manufacturers' literature for specific problems.

10-11 MAINTENANCE AND ACCESSIBILITY

A motor or generator installation should always take into account the need for easy access to inspect, maintain, repair, or replace the unit. With dc machines, the commutator and brushes should be regularly checked for arcing and mechanical wear. The wear products, that are primarily finely powdered brush particles, should be periodically cleaned out for maximum life. This dust, when mixed with excess shaft bearing oil or grease, will cling tenaciously to any surface. It is conductive so that it may lead to flash over and damage if not removed.

Modern bearings are very long lived but need periodic checking. A normal well-designed ball bearing may go from 5000 to 10 000 hours without service. However, it will eventually need regreasing. If properly done, another 5000 or 10 000 trouble-free hours may be expected. However, if not attended to, the bearings will eventually fail and perhaps spoil the motor windings owing to progressive overloading.

Bearing life is greatly affected by misalignment during installation or by too tight belt tension. Winding temperature and therefore winding life expectancy are greatly affected by accumulated dust and grime, since this prevents effective cooling. These problems are interrelated, since a major cause of electrical failure is *overlubrication* of the bearings. The manufacturer's recommendations should be closely followed for best results.

The machine operator should be aware of abnormal sounds, vibrations, smells, and excess surface temperature. Each abnormality can usually be traced to improper maintenance and, if detected soon enough, can be cleared up in a few minutes. Ignored symptoms have a way of resulting in serious and expensive breakdown. The resemblance to human preventive medicine and checkup is quite valid.

QUESTIONS

10-1. What is the probable difficulty if a motor is loaded substantially beyond its nameplate power rating but is at a power level where it appears to run satisfactorily?

10-2. What is the major reason for not using a readily available motor of the proper voltage rating, but one that is rated for substantially more power than is required to operate the load?

10-3. Is it possible that a motor of adequate power to run a load may not be able to start the same load?   Explain.

10-4. What is the special value of a series motor?

10-5. By what means is it possible to operate a load in a satisfactory manner if its speed requirement is between normally rated motor speeds?

10-6. What is the most general meaning of motor frame size number?

10-7. What is the major physical difference between sizes 215A and 216A?

10-8. What is the interpretation of the term constant speed in a dc motor?

10-9. What is meant by duty cycle?

10-10. What is meant by the term ambient temperature and how does it affect a motor selection?

10-11. What is meant by a class B insulation?

10-12. Why does a more fully enclosed motor frame enclosure sometimes require a larger frame size for the same power rating?

10-13. Name some typical maintenance operations that may be needed on dc motors and generators.

PROBLEMS

10-1. A load is calculated to require 2.27 hp.
(a) What NEMA power rating is required?
(b) NEMA suggested kW rating is required?

10-2. The motor of Problem 10-1 is to be used at 1750 rpm.   What NEMA frame size is available?

10-3. If a motor selection requires a 3 hp, 2500 rpm 215A frame-size motor, what shaft diameters are available?

10-4. What would be the shaft height above the mounting surface for the 215A frame size motor?

10-5. (a) What are the lateral distances between mounting holes for a standard 284A frame size motor?
(b) What is the longitudinal distance between foot mounting holes?
(c) What is the standard mounting hole diameter?

10-6. A new material-handling apparatus has its power requirements tested by measuring input kilowatts, and then converting this figure to the probable mechanical power by use of a calibration curve such as Fig. 7-10 that was prepared for the specific motor.   The power and time data are as follows: 3.7 hp for 3.5 min; 6.5 hp for 8.25 min; 1.7 hp for 11.33 min; 4.1 hp for 2.67 min; and off for the remainder of a 20-min cycle.
(a) What horsepower motor is required?
(b) What kilowatt-rated motor is required if tentative SI motor ratings are used?

10-7. A NEMA standard motor is chosen with a standard shaft size of 1.875 in. It is desired to bore the driven coupling to the next larger size in the IEC proposed SI shaft sizes, and then to use a sleeve to reduce to 1.875 in. The coupling bore should then be what size in millimeters?

10-8. The NEMA standard motor in Problem 10-7 has a shaft height D of 7 in. It is proposed to machine the mounting surface of the driven machine to accommodate the next larger shaft mounting height in the proposed SI machine shaft heights, and then to shim up the NEMA motor. What shaft height space should be provided in millimeters?

10-9. If the motor chosen for Problem 10-6 is to be operated on 240 V dc, what will be its approximate average current?

10-10. If the motor in Problems 10-6 and 10-9 is to draw about 20 A on 240 V, what NEMA-size contactor could be used?

Answers. **10-1(a)** 3 hp, **(b)** 2.5 kW; **10-2**, size 216A; **10-3(a)** 1.375 in., **(b)** 1.125 in.; **10-4**, 5.25 in.; **10-5(a)** 11.0 in., **(b)** 9.5 in., **(c)** 0.53 in.; **10-6(a)** 5 hp, **(b)** nearest kW rating is 4.0 kW; **10-7**, 48 mm; **10-8**, 180 mm; **10-9**, 20A; **10-10**, size 1.

11

Alternating-Current Dynamos

Alternating-current dynamos are visually different from dc machines of comparable sizes. The difference in appearance has a fundamental basis because the machines are "inside out" in relation to dc machines. Most ac machines have the armature in the fixed or stator position and the field in the moving rotor position. This arrangement is the natural order of things for a few very good reasons:

(1) The high voltage, high current, and therefore high power-handling element is the armature on any ac or dc rotating electrical machine. Armature coils are therefore larger than field coils.

(2) Since no alternate switching of coil polarities is needed on an ac machine, no commutator function is needed. Thus, the high power windings may be made stationary for direct connection. The universal motor is an exception to this condition.

(3) The field structure and coils are not ordinarily required to handle more than a fraction of the total power. Thus, their rotating electrical connection may be made smaller. Since no polarity switching is required, collector rings are usually used.

(4) The armature and field coils are both placed in slots in the punched magnetic structure, but the stationary armature structure can be conveniently made with deeper slots to handle the required larger coils.

(5) It is easier to cool the stator than the rotor, which is an advantage of the normal ac construction.

342

11-1 PHYSICAL CONSTRUCTION OF ALTERNATING-CURRENT MACHINES

Except for specialized types, such as the universal ac–dc motor, which appears much like a dc series motor, almost all ac motors and generators are built to take advantage of the natural relation of having the armature fixed and surrounding the field, and the field moving and inside of the armature.

11-1-1 Fixed Armature or Stator.
This fixed and outside armature has a complete ring of teeth and slots on its inner face. In the usual machine, all the slots are filled with similar and symmetrical coils. As a result, it may not be at all obvious how many poles or phases are present in the winding. In the field rotor, the construction may be much like that of a dc armature with a complete circular magnetic structure having a continuous group of slots and teeth on the outer surface. Again, these slots are filled with similar and symmetrical coils, and it is not readily apparent how many poles or phases are present in the machine. With *salient pole* field construction, the number of poles is visible, as in a dc machine.

If the winding slot sides are parallel to each other in a single slot, which is a frequent construction, it may be seen from Fig. 11-1 that the stator tooth structure becomes stronger as it grows deeper. On the other hand, Fig. 11-1 shows that a rotor tooth becomes weaker as it grows deeper. This tooth structure advantage for the stator is used in the ac stator.

One may already wonder if parallel side slots need be used. They are not

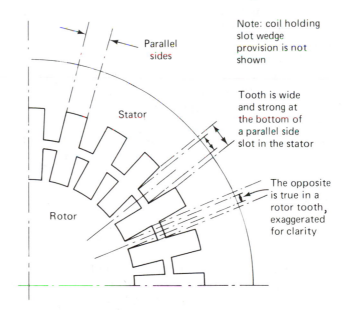

Figure 11-1. Typical magnetic lamination slot structure in ac dynamo.

necessary on small sizes. However, on large sizes, where the coils are wound with large cross-sectional wire and where the insulation must be most carefully distributed, the parallel side slot is required. Since large coils are preformed, bound with insulation, and impregnated with varnish and baked, they cannot readily change shape during installation in the magnetic core.

Smaller ac machines are wound with loose coils of round wire, which may be slipped down into the slots turn by turn during winding or installation. In this fashion almost any slot shape may be used. Full use of the slot cross section seems to require parallel side slots in large sizes. In any shaped slot, some provision must be made to capture and hold the windings in place. As a result, the slot will have some provision for a covering wedge, even if parallel sided.

In the ac machine stator, the current is continually varying at the frequency repetition rate. The resulting magnetic flux then varies cyclically, and there are hysteresis and eddy current losses in the magnetic structure. Minimizing the losses requires the use of a laminated magnetic structure. The structure is built up of thin plates of silicon steel alloy that are readily punched to shape in press dies built for the task. The punched stator laminations usually cover a full circle in small- and medium-sized machines. There are proprietary families of progressively larger sizes that each manufacturer standardizes. Since punch dies are expensive, only a few different numbers of slots and teeth are provided for a basic size. As will be seen later, these slot numbers determine the number of coils that can be accommodated.

The larger-sized machines are built up with laminations in segments of reasonable sizes. The size depends on the available stock width and press die sizes.

Lamination stock thickness is dictated by eddy current loss considerations and by material-handling considerations. The thin lamination has less eddy current loss, but becomes difficult to handle and the teeth will bend too easily. A stock thickness of about 0.014 in. (0.35 mm) has long been used for 60 Hz ac machines. The lamination stock size under SI-agreed material sizes will perhaps stay the same, since 0.35 mm is a tentative second preferred-size thickness. If not, perhaps either 0.3 or 0.4 mm will be used, since the same economic and frequency consideration will hold regardless of measurement units.

Numbers of slots are standardized around 36, 48, 60, and 72 slots, and so on, for good mechanical reasons, which will become more apparent when actual windings are discussed.

11-1-2 Rotating Field Structure. The mechanical construction of the rest of an ac motor or generator follows closely that described in Chapter 3 for dc machines, except for the lack of a commutator. With synchronous alternators and synchronous motors, slip rings, which are used to carry dc power into and out of a rotating field, are used in a similar location to a commutator.

A slip ring is a copper alloy ring that is insulated from the rotor shaft and connected to the rotor windings. A carbon brush is supported in a brush carrier rigging to complete the connection. Since there is no requirement for a particular internal resistance to aid commutation, the slip ring brush is harder and denser than a commutator brush. It has a lower voltage drop and is therefore responsible for less power loss than its dc counterpart.

When the rotor windings carry three-phase ac power, three slip rings are used. In some cases of larger synchronous motors, multiple windings are used and five or more slip rings may be present.

The high-power armature windings are placed on the stator structure, which has relatively larger winding space. An ac machine can usually be smaller overall than its dc counterpart with the same power rating. The lack of a commutator also contributes to size reduction. Later chapters will show that ac machine sizes are also standardized in NEMA frame sizes and tentative SI frame sizes. However, the ac machine averages about 50 percent more power within the same frame size.

11-2 ALTERNATOR WINDINGS

The types of windings used in ac machinery are closely related to dc windings. Both lap and wave windings are used, but lap is much more common owing to shorter coil connections. See Chapter 3 and Figs. 3-8 and 3-9 for lap and wave-winding descriptions. In single-phase machines a winding form called concentric coil is used, not because of any circuit advantage, but because it lends itself to prepared coil skeins that can be put in place rapidly. Here the economics of construction dominate. More on single-phase machines later.

Since three-phase machines require three identical groups of windings spaced 120 electrical degrees apart, and since poles must exist in pairs, certain rules affect windings and magnetic structure slot spacing.

Electrical degrees refers to the cyclic angle of the repeating sine wave, where one full cycle is 360 electrical degrees. Since opposite magnetic poles produce opposite voltages in a moving coil to pole situation, the maximum voltage difference along a sine wave is found 180 electrical degrees apart. This 180 electrical-degree shift takes place between each successive field pole. The total electrical degrees in a 360 mechanical-degree rotation is then simply 180 times the number of poles or

$$\text{total electrical degrees in one revolution} = 180 \, P \qquad \text{Eq. (11-1)}$$

where P is the number of poles.

A three-phase, four-pole machine will be used as an example. Assume that 36 stator slots are available in the stator lamination stack. If so, there are 36/4 = 9 slots per pole position and 9/3 = 3 slots per phase per pole. These 36 slots would allow twice as many, or six, slots per phase per pole for a two-pole machine, and, similarly, two slots per pole per phase for a six-pole machine.

11-2-1 Chording of Windings. One design factor is *chording* of the pole windings. If, on the 36-slot, four-pole machine, an individual coil enters slot 1 and comes back in slot 10, it will have spanned 90 mechanical degrees of the stator circular structure. Since there are four poles by definition in this case, 90 mechanical degrees is 180 electrical degrees, from Eq. (11-1). Thus, the two sides of the coil are in the same relative position on the adjacent north and south pole positions. This is a *full-pitch* coil construction (see Fig. 11-2 for a representation of this condition).

The more usual ac machine coil will cover less of the periphery of the machine and is then said to be *fractional pitch*. A typical situation might have a coil enter slot 1 and leave slot 7. This then covers six out of a possible nine slot pitches, and is a 6/9 or 66.7 percent pitch. The majority of ac machine coils are of fractional-pitch type, for which there are a few important advantages.

(1) The ends of the coils are shorter, which means less copper loss due to less total length.

(2) The end coils can be formed more compactly. The end bells will need less winding space, resulting in a shorter unit.

(3) There is a distinct reduction in machine harmonics due to cancellation of

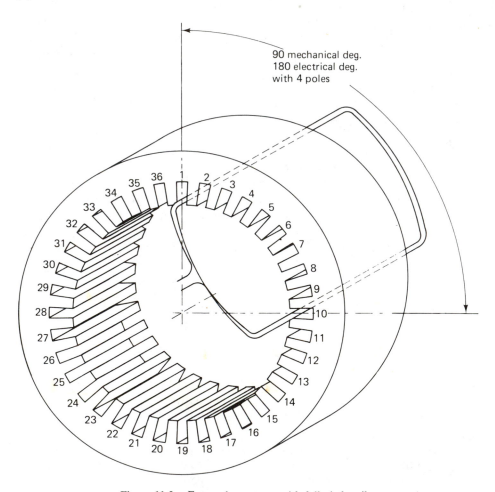

Figure 11-2. Four-pole ac stator with full pitch coils.

higher harmonics. Since all ac equipment is designed to operate on a pure sine wave, the generation of harmonics is to be avoided. This is especially so when the factor that achieves it is otherwise desirable.

Again, in the 36-slot situation, and with either full- or fractional-pitch coils, all 36 coils are alike.

11-2-2 Coil Group Connections. Figure 11-3 shows how the coils are laid into the slots. The vast majority of lap- or wave-wound machines use this double-layer winding arrangement. This is very similar to the manner of winding a dc armature. The interconnection of the coils will result, in this 36-coil situation, in 12 groups of three coils per group. Each group is then involved with one phase and one pole. Since there are four poles in this simple but real situation, there are four coil groups in each phase. This is the usual situation, even when more slots and coils are used.

A 72-slot, six-pole machine when wound for three phase would have 72/6 = 12 slots per pole and 12/3 = 4 slots per phase per pole. Here the coils would be connected in groups of four, and there would be six of these four-coil groups per phase.

Many varieties of coil group connections are possible, but only a relatively few are used today. In a three-phase machine, the coil groups per phase are collected for all poles, and this larger grouping is usually divided into two parts. On the 36-slot machine, two three-coil groups are permanently connected. There are then two of these six-coil collections per phase. If they are series connected, the motor or generator will be set for operation on the higher of its two rated

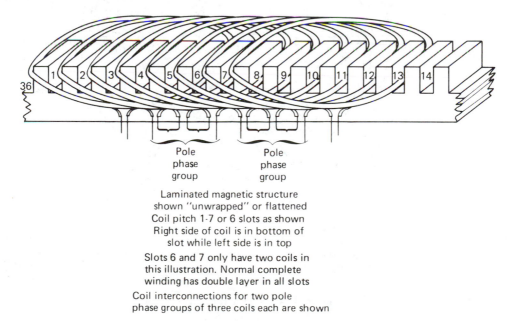

Pole phase group   Pole phase group

Laminated magnetic structure
shown "unwrapped" or flattened
Coil pitch 1-7 or 6 slots as shown
Right side of coil is in bottom of
slot while left side is in top

Slots 6 and 7 only have two coils in
this illustration. Normal complete
winding has double layer in all slots

Coil interconnections for two pole
phase groups of three coils each are shown

Figure 11-3. Double-layer coils in ac stator.

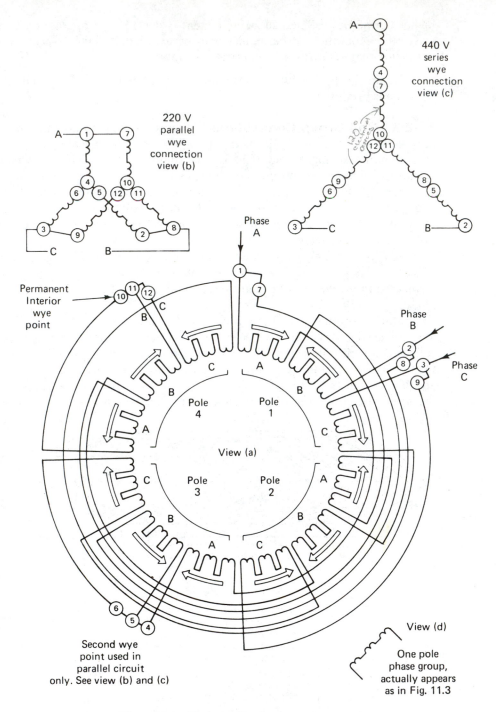

220 V parallel wye connection view (b)

440 V series wye connection view (c)

120 electrical degrees

Phase A

Permanent Interior wye point

Phase B

Phase C

Pole 4

Pole 1

View (a)

Pole 3

Pole 2

Second wye point used in parallel circuit only. See view (b) and (c)

View (d)

One pole phase group, actually appears as in Fig. 11.3

Figure 11-4. Typical coil and pole group connections.

voltages. If parallel connected, the lower of two rated voltages may be accommodated. In this way a motor or generator may be operated on either 110 or 220 V, or perhaps 220 or 440V, and so on. Great installation flexibility is inherently obtained.

As these coil groups are gathered together, the direction of winding or connection on opposite (north or south) poles must be opposite. Thus, each adjacent series connected per phase per pole group must be reversed for proper polarity (see Fig. 11-4 for typical coil interconnection).

A three-phase machine, when gathered into phase coil groups, is then connected either in wye or in delta, and also in series or parallel, as shown for wye connections in Fig. 11-4. Three-phase motor stator coil group ends are normally numbered from 1 to 9, as shown, and the points 10, 11, and 12 are normally "buried" unless specially needed.

11-2-3 Winding Distribution. A final definition is necessary before the specific calculation of machine voltage ratings is attempted. Since coils are usually laid as shown in Fig. 11-3, they are seen to be spaced uniformly around the periphery of the machine stator. Returning to the 36-coil, four-pole situation, we can see that in this specific case any one pole has three phase groups of three series-connected coils per pole. The voltages generated in coils of a single-phase group of three coils are not simply additive. Since each coil is not swept or cut by the same intensity of magnetic flux at the same time, they are not in the same time–phase relation even though they are a part of the same phase winding. The individual coil voltages must then be combined as phasors in a manner that will be shown. All these factors, which make up a multiphase, two-level, chorded or whole-pitch distributed winding, are applicable for a variety of ac generators and motors, both large and small.

11-3 SYNCHRONOUS ALTERNATOR

The first type of machine to be treated is the *synchronous alternator*, which is the basic ac generator. It is called synchronous because its generated frequency is directly related to its number of armature and field poles and to its rotative speed. From Chapter 2, we find that an individual coil of winding generates a full cycle of ac voltage each time it is swept by a pair of magnetic poles. At this point it does not matter if this ac is a sine wave or not, although it is a great disadvantage to an ac machine if its final output voltage and thus its current are not sine waves. The generated frequency is converted from cycles per pole pair to a machine basis by the following relation:

$$f = \frac{PS}{120}$$

where

f = frequency in hertz
P = number of poles
S = speed in rpm

This relationship is of *fundamental* importance and can be easily followed if it is constructed from its basics.

(1) One full cycle of alternating current is developed for each pair of magnetic poles swept by a winding: cycle/2 poles.

(2) There are a fixed number of poles in a full circle of construction or one revolution: poles/rev.

Note that there must be an even-integer number of poles, as in a dc machine, since a north pole cannot exist without a south pole.

(3) The rotative speed is measured in revolutions per minute: rev/min.

(4) There is 1 min for each 60 sec: min/60 sec.

Gathering and canceling, we have

$$\frac{\text{cycle}}{2 \ \text{poles}} \times P \frac{\text{poles}}{\text{rev}} \times S \frac{\text{rev}}{\text{min}} \times \frac{\text{min}}{60 \ \text{sec}} = \boxed{f = \frac{PS}{120} \frac{\text{cycles}}{\text{sec}}} \qquad \text{Eq. (11-2}_{\text{E}}\text{)}$$

Note that cycles/sec = hertz (Hz) in English or SI units.

If the rotative speed is given in radians per second or ω, then

(1) Cycle/2 poles and poles/rev are as before, but speed is expressed as rad/sec.

(2) A mechanical revolution is composed of 2π radians of angle: rev/2π rad.

Again, gathering and canceling,

$$\frac{\text{cycle}}{2 \ \text{poles}} \times P \frac{\text{poles}}{\text{rev}} \times \frac{\omega \ \text{rad}}{\text{sec}} \times \frac{\text{rev}}{2\pi \ \text{rad}} = \left\langle f = \frac{P\omega}{4\pi} \frac{\text{cycles}}{\text{sec}} \right\rangle \qquad \text{Eq. (11-2}_{\text{SI}}\text{)}$$

Note that, if machine speed is given or needed in S rpm, rather than ω rad/sec, Eq. (11-2$_{\text{E}}$) must be used even if other considerations are in SI units.

There may then be seen to be fixed relationships between cycles per revolution and number of poles in the windings. This relationship is shown in Eq. (11-1). There are also only a few recognized and used ac power frequencies. These are 25, 50, 60, and 400 Hz, with 60 Hz by far the most common in the United States; 400 Hz is used almost exclusively for aircraft ac power because it allows small high-speed machines, which also require less magnetic structure size and weight.

From Eq. (11-2) pole–speed–frequency relations, we may conveniently construct Tables 11-1 and 11-2, which show the built-in relationships in synchronous ac machines. Note that the highest speed available at 60 Hz is 3600 rpm for a two-pole machine.

Note that Table 11-2 lists speeds as multiples of π. This was done to save long repeating decimals where possible.

A few comparisons from these tables are in order. Note from Table 11-1 that frequency conversions are possible with directly connected machines. A 12-pole, 60 Hz machine turning at 600 rpm will directly connect to a 10-pole, 50 Hz

TABLE 11-1 FREQUENCY–POLE–SPEED RELATION

Freq. (Hz)	Speed in S (rpm) for Various Numbers of Poles									
	2	4	6	8	10	12	14	16	20	40
25	1 500	750	500	375	300	250	214.29	187.5	150	75
50	3 000	1 500	1000	750	600	500	428.57	375	300	150
60	3 600	1 800	1200	900	720	600	514.28	450	360	180
400	24 000	12 000	8000	6000	4800	4000	3428.57	3000	2400	1200

From Eq. (11-2$_E$), $f = PS/120$.

TABLE 11-2 FREQUENCY–POLE–SPEED RELATION

Freq. (Hz)	Speed in ω (rad/sec) for Various Numbers of Poles									
	2	4	6	8	10	12	16	20	40	
25	50π	25π	16.667π	12.5π	10π	8.333π	6.25π	5π	2.5π	
50	100π	50π	33.333π	25π	20π	16.667π	12.5π	10π	5π	
60	120π	60π	40π	30π	24π	20π	15π	12π	6π	
400	800π	400π	266.67π	200π	160π	133.33π	100π	80π	40π	

From Eq. (11-2$_{SI}$), $f = P\omega/4\pi$.

machine turning at the same 600 rpm. This is in fact how intersystem power conversions were performed. Similarly, a six-pole synchronous 60 Hz motor can directly connect to a 40-pole synchronous alternator and produce 400 Hz when both machines are turning at 1200 rpm. This is how accurate aircraft 400 Hz power is generated for use in a factory that has only 60 Hz power available. If inexact frequency is acceptable, many factories generate the 400 Hz power needed for testing by connecting a two-pole induction motor, which we shall see runs *below* 3600 rpm on 60 Hz, to a 14-pole nominally 400 Hz synchronous alternator. We can see that almost any prime-mover inherent rotative speed can be matched by a combination of pole and frequency combination.

A worked example is shown to clarify the relationships.

Example 11-1. A large hydroelectric power plant is under consideration. Its hydraulic head or water level difference above and below the dam and its total power requirement dictate that its water turbine or "runner" must turn at from 137.00 rpm (14.347 rad/sec) to 140.00 rpm (14.661 rad/sec) to reach peak efficiency; 60 Hz power is required.

(a) How many poles must a direct-connected alternator have?
(b) What rotational speed must be used?

Solution:

(a) Using Eq. (11-2$_E$), transpose for finding P and investigate what pole numbers would result using the boundary speeds specified.

$$f = \frac{PS}{120} \quad \text{so} \quad P = \frac{120f}{S}$$

Since f is specified as 60 Hz,

$$P = \frac{120(60)}{137} = 52.5 \text{ poles}$$

and

$$P = \frac{120(60)}{140} = 51.4 \text{ poles}$$

But poles can only exist in integer *pairs*, so in this case $P = 52$ poles. (b) Using 52 poles and Eq. (11-2$_E$) transposed for S, we find that

$$f = \frac{PS}{120} \quad \text{so} \quad S = \frac{120f}{P}$$

$$= \frac{120(60)}{52} = 138.46 \text{ rpm}$$

Note that a broader choice of speed might well have been possible, but the alternator can be built with only one number of poles. If the boundaries had been wider, the *even* pole number nearest the middle of the range would be chosen.

(a) Using SI quantities and the same limitations, we have $f = P\omega/4\pi$. Then

$$P = \frac{4\pi f}{\omega} = \frac{4\pi 60}{14.347} = 52.5 \text{ poles}$$

or

$$P = \frac{4\pi 60}{14.661} = 51.4 \text{ poles}$$

Poles can only exist in *even integer* numbers, so again we must have $P = 52$ poles.

(b) Using the 52 poles and Eq. (11-2$_{SI}$) transposed for ω, we have

$$f = \frac{P\omega}{4\pi} \quad \text{or} \quad \omega = \frac{4\pi f}{P} = \frac{4\pi 60}{52}$$

$$= 14.500 \, \frac{\text{rad}}{\text{sec}}$$

Checking with rpm $\times$ 0.104 72 = rad/sec, we have 138.46 $\times$ 0.104 72 = 14.500 rad/sec.

Note that poles are the same for either units since they are a numeric rather than unit situation. However, the rpm and radians/second are not the same number but a different unit expression of the same absolute value.

ALTERNATING-CURRENT DYNAMOS   CHAP. 11

QUESTIONS

11-1. What is meant by saying that most ac machines are "inside out" in relation to dc machines?

11-2. Why is a commutator not necessary in usual ac motors and generators?

11-3. What is meant by the terms rotor and stator as distinct from the terms armature and field?

11-4. What is meant by chorded windings?

11-5. There is a hierarchy of windings, starting with an individual armature winding coil and ending with a completely connected armature circuit. What part of this succession is a pole–phase group?

11-6. If each pole–phase group of coils is gathered into total phase groups, all poles of a particular phase are interconnected. In what manner can this phase interconnection be performed so that two different voltage levels can be accommodated?

11-7. In addition to voltage-level considerations, what other major precaution must be taken in connecting a phase group of coils that connects various poles?

11-8. Why is a normal ac generator called a synchronous alternator?

PROBLEMS

11-1. How many electrical degrees are passed in one revolution of a six-pole synchronous alternator?

11-2. How many cycles of alternating current are generated in one revolution of a 14-pole synchronous alternator?

11-3. If a four-pole, three-phase winding is placed in a stator that has 48 slots, (a) how many slots are there per phase? and (b) how many slots per pole per phase?

11-4. What frequency is generated by a six-pole alternator that rotates at 1200 rpm?

11-5. What frequency is generated by a 10-pole alternator that rotates at 62.83 rad/sec?

11-6. Using Table 11-1, what frequency is developed by an alternator that has 12 poles and rotates at 4000 rpm?

11-7. Using Table 11-2, what frequency is developed by a six-pole machine rotating at 125.66 rad/sec?

11-8. A large diesel engine is to be used as a prime mover in a standby or emergency power plant. Its normal rated speed is 440 rpm, and it can be adjusted to operate a small range above or below this point.
(a) How many poles should be specified in a directly coupled alternator?
(b) What operating speed should be used to produce 60 Hz?

11-9. The same type of situation as in Problem 11-8 is to be solved in SI units. The diesel engine normally runs as 32.2 rad/sec and 25 Hz is desired.

(a) How many poles are required in a closely matched directly coupled synchronous alternator?

(b) What operating speed should be specified?

Answers. **11-1**, 1080 electrical degrees; **11-2**, 7 cycles; **11-3(a)**, 16 slots per phase, **(b)** 4 slots per pole per phase; **11-4**, 60 Hz; **11-5**, 50 Hz; **11-6**, 400 Hz; **11-7**, 60 Hz; **11-8(a)**, 16 poles, **(b)** 450 rpm; **11-9(a)**, 10 poles, **(b)** 31.42 rad/sec.

$$f = \frac{Poles \; speed \; rpm}{120}$$

$$60 = \frac{4/5}{120}$$

$$S = 1,800 \, rpm$$

3,600
1,800 1,780 } synchronous
1,200 1,600 } motor

12

The Synchronous Alternator

Voltage is generated in a synchronous alternator in the same manner as described in Chapter 2 for dc generators. The magnetic flux term of Φ lines/pole or ϕ webers/pole is exactly the same as before.

12-1 VOLTAGE RELATIONS

In calculations to determine the required ampere turns per field pole, the same procedures are used that are described in Example 3-1. This procedure holds except that the structural layout would be different. In an ac machine the same division of the circuit through the center of the poles would be followed. However, the actual pole geometry would not be so obvious unless a salient pole field rotor construction were used. Salient here means individual and separate in the sense of the dc machine field poles. With salient poles, an arbitrary division of 360/number of poles would be considered. Here the poles are attached to the rotor core, and the air gap is at the outer end of the poles instead of the inner end. The outer main frame is toothed on its inner face to accommodate the stator or armature windings. A nonsalient or cylindrical pole field rotor will have a toothed structure similar to the dc armature, and its diameter will fill the space inside the rotor except for the air gap. In this case the pole locations will be just as definite as in salient pole construction but not as readily apparent to the eye.

If a special interest is developed in ac machine magnetic structure calculations, the same references as in Chapter 3 may be consulted, except that reference (12) has a separate ac machines volume dating from 1961.

It should also be realized that the same type of magnetic materials are used and that therefore the same limitations for actual flux density in B lines/in.^{2}, or β Wb/m^2, would hold for an ac machine structure.

12-1-1 Basic Voltage Generation Formula.

Again referring back to Faraday's basic law, as quantized by Neumann, the basic equation for the *average* voltage generated in a single turn of a coil winding is, from Eq. (2-1$_E$),

$$E_{av} = \frac{\Phi}{t} \times 10^{-8} \frac{\text{volt}}{\text{turn}}$$

or, from Eq. (2-1$_{SI}$),

$$E_{av} = \frac{\phi}{t} \frac{\text{volt}}{\text{turn}}$$

The units in each case are the same as in Chapter 2. Note that the average voltage term will need modification for use in a sine-wave condition.

If these equations of per turn of coil winding are modified to a per coil situation in a basic two-pole machine, Eqs. (2-5) hold; these are, from Eq. (2-5$_E$),

$$e_{av/coil} = 4\Phi Ns \times 10^{-8} \frac{\text{volt}}{\text{coil}}$$

and in SI, from Eq. (2-5$_{SI}$),

$$e_{av/coil} = 0.636\,62\ \phi N\omega \frac{\text{volt}}{\text{coil}}$$

Again the units are the same as in Chapter 2 using s or ω for rotative speed. Note that s in rps is used rather than S in rpm in Eq. (2-5$_E$).

For convenience, since we are working with a generator and since the resulting frequency of the ac voltage is a basic and critical parameter, we can substitute f in hertz for s in revolutions per second on a one-for-one basis. This is true for a two-pole machine, since one revolution will produce one full cycle. Similarly, in SI units, f can be substituted for 2π rad/sec.

In addition, "per coil" is not particularly convenient, but a per phase per pole coil group is, since this is normally a permanently connected group of windings. Therefore, N turns per coil is modified to become Nn in either set of units, where n is the number of coils per phase per pole. An *intermediate* formula then is

$$E_{av/pp} = 4\Phi Nnf \times 10^{-8} \text{ V in English units}$$

$$= 4\phi Nnf \text{ volts in SI units}$$

Note that 0.636 62 had come from $4/2\pi$, but in going from ω to f we have multiplied by 2π and so the factor 4 has returned.

This was represented above as an intermediate step. It is so because the *average* voltage is not wanted, but the *effective* ac voltage of a sine wave is desired. In a sine wave the average voltage is $2/\pi$ or 0.636 62 times the maximum or peak value of the sine wave. The desired *rms* (root mean squared) or *effective* voltage is $1/\sqrt{2}$, or 0.707 11 times the peak voltage. Therefore, the desired ratio of rms to average is 0.707 11/0.636 62 or $(1/\sqrt{2})(\pi/2) = 1.1107$. Reference (13) has a simple integral calculus proof of this.

We must then convert the above intermediate formulas by a factor of 1.1107 and by two *additional* factors.

12-2 WINDING PITCH

A *fractional-pitch double-layer* winding coil does not have the same voltage being generated at the same time in each side of the coil. The coil winding distribution and the flux intensity distribution are both arranged so that a sine-wave voltage results from a pole–phase group of windings. The individual coil side voltages must then be combined as phasors. The pitch factor is the ratio of the voltage generated by a fractional-pitch coil to the voltage generated by a full-pitch coil and is always less than 1. This is trigonometrically the sine of one half the coil span angle in electrical degrees. This becomes, for *both* unit systems,

$$k_p = \sin (p/2)$$

Eq. (12-1)

where

k_p = pitch factor (dimensionless) but ≤ 1

p = coil span in electrical degrees where full pitch is 180 electrical degrees

Since coil spans can only be constructed by integer numbers of lamination slots, and only a relatively few slots and resulting slot pitches are practical, k_p may be conveniently tabulated. Coil spans below two-thirds of the available slots are not normally used since they are of no advantage.

Table 12-1 contains all likely pitch factors for ac motors and generators that have an integer number of slots per pole per phase. The process used to construct the table is explained in Example 12-1.

> **Example 12-1.** Find the pitch factor to be used in calculations involving a six-pole, three-phase alternating current generator. It has a total of 54 winding slots in its stator, and the coils span seven slots.
>
> *Solution:* The 54 total slots divided by six poles gives 54/6 = 9 slots per pole. Nine slots per pole divided by three phases yields 9/3 = 3 slots per pole per phase. With nine slots per pole and a coil span of seven slots, the fractional pitch is 7/9. In electrical degrees, this is 180(7/9) = 140° = p. Using Eq. (12-1), which is $k_p = \sin (p/2)$, we have sin 140/2 = sin 70° = 0.939 69. If a full-pitch coil voltage is unity then this generator has a pitch factor of k_p = 0.939 69.

This can be checked in Table 12-1. The problem is usually in the understanding of what fractional pitch actually exists. If a coil enters one slot and leaves the core at the eighth slot away, counting the entering slot, it then has a coil span of seven slots. The first or entering slot must be considered as the zero slot.

TABLE 12-1 PITCH FACTOR K_p FOR ALL POSSIBLE SLOT COMBINATIONS FOR THREE-PHASE ALTERNATORS HAVING 3 TO 15 SLOTS PER POLE

Slots per pole	Slots per pole per phase	Full pitch 180°	14/15 168°	11/12 165°	8/9 160°	13/15 156°	5/6 or 10/12 150°	12/15 144°	7/9 140°	9/12 135°	11/15 132°	10/15, 9/12 6/9, 4/6 or 2/3 120°
							Fractional pitch or slots used per slots per pole					
3	1	1.000										0.866 03
6	2	1.000					0.965 93					0.866 03
9	3	1.000			0.984 81				0.939 69			0.866 03
12	4	1.000		0.991 44			0.965 93			0.923 88		0.866 03
15	5	1.000	0.994 52			0.978 15		0.951 06			0.913 55	0.866 03

Eq. (12-1): $k_p = \sin(p/2)$.

For a two-phase situation, the value of k_p will have to be calculated since the common values are not charted.

12-3 WINDING DISTRIBUTION

When the several coils in a pole group are connected in series, their individual coil voltages are not directly additive unless two or more coils lie in the same slot. This *distribution factor* is related to but not identical to the pitch factor just described. The distribution factor is determined by the phase angle differences due to the individual coil placement, so its formulation is also based on phasor summation. The distribution factor is really related to the number of slots per pole per phase (n) and to the number of electrical degrees between these slots (α).

The relationship of the phasor voltage combination involved is shown in Fig. 12-1, where four slots per pole per phase are involved. The segments labeled coil voltage (E_c) are proportional to the individual coil voltages whether or not they are chorded or are of fractional pitch. The long phasor (E_{pg}) is the desired voltage for a phase group of coils. The relationship desired is then

$$k_d = \frac{E_{pg}}{E_c} = \frac{\text{phasor sum of coil voltage per phase}}{\text{arithmetic sum of coil voltages per phase}}$$

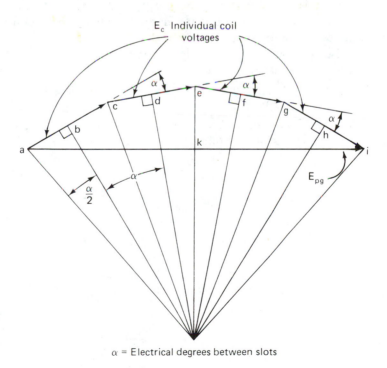

α = Electrical degrees between slots

Figure 12-1. Coil distribution factor.

TABLE 12-2 DISTRIBUTION FACTOR K_d FOR THREE-PHASE ALTERNATORS

Slots per pole	Slots per pole per phase (n)	Electrical degrees per slot (α)	Distribution factor (k_d)
3	1	60	1.000 00
6	2	30	0.965 93
9	3	20	0.959 80
12	4	15	0.957 66
15	5	12	0.956 68

Eq. (12-2): $K_d = \dfrac{\sin{(n\alpha/2)}}{n \sin{(\alpha/2)}}$.

which reduces to

$$k_d = \frac{\sin(n\alpha/2)}{n \sin{(\alpha/2)}}$$

Eq. (12-2)

Since *n* is slots per–pole phase, *n* can only be an integer such as 1, 2, 3, 4, 5, etc., unless *uneven* coil groups are chosen. Similarly, α in a three-phase situation can only be 180/3*n* or, in general, 180/slots per pole. Then, as in the case of k_p, k_d can also be tabulated (see Table 12-2).

The process used to determine the distribution factor of an ac generator winding can be followed in Example 12-2.

Example 12-2. An eight-pole three-phase alternator is wound on a 72-slot core. Find the distributions factor of the winding.

Solution:

$$\frac{72 \text{ slots}}{8 \text{ poles}} = 9 \text{ slots/pole}$$

$$\frac{9 \text{ slots/pole}}{3 \text{ phases}} = n = 3 \text{ slots/pole/phase}$$

Since there are 9 slots/pole and each pole is 180 electrical degrees, 180/9 = α = 20 electrical degrees between adjacent slots. Thus, α = 20° and *n* = 3. Using Eq. (12-2), where

$$k_d = \frac{\sin{(n\alpha/2)}}{n \sin{(\alpha/2)}}$$

THE SYNCHRONOUS ALTERNATOR   CHAP. 12

we have

$$k_d = \frac{\sin (3 \times 20°/2)}{3 \sin (20°/2)} = \frac{\sin 30°}{3 \sin 10°}$$

$$= \frac{0.5}{3(0.173\ 65)} = \frac{0.5}{0.520\ 94} = 0.959\ 80$$

k_d can be checked in Table 12-2.

12-4 COMPLETE POLE-PHASE GROUP VOLTAGE RELATIONS

The intermediate formulas in Section 12-1 can then be corrected to account for the full voltage developed in a per pole per phase group of coils. The individual units are as before. $E_{av/pp} = 4\Phi Nnf \times 10^{-8}$ V expands in English units to

$$\boxed{E_{gpp} = 4.4428\Phi Nnfk_pk_d \times 10^{-8} \text{ V}} \quad \text{Eq. (12-3}_\text{E}\text{)}$$

[handwritten annotations: Fixed values 4Nf×10⁻⁸ V; Eg = 4.4428 Φ; trans are frequency dependent]

and in SI units to

$$\left\langle E_{gpp} = 4.4428\phi Nnfk_pk_d \text{ V} \right\rangle \quad \text{Eq. (12-3}_\text{SI}\text{)}$$

Note that E_{gpp} means voltage generated per pole per phase. This voltage can then be directly expanded by however many pole–phase groups are connected in *series* to get the full voltage per phase. This is because each pole–phase group is in *time phase* with other *poles* of the *same* phase. A typical four-pole machine would have all four poles connected in series for its highest designed voltage. The other choice is usually two in parallel and two parallel groups in series for *one-half* the emf of the higher voltage connection.

Remember that the final generated voltage per phase must be corrected for the type of three-phase connection. If *delta* connected (Δ), the phase voltage is the line-to-line voltage. If *wye* connected (Y or "star"), the phase voltage must be multiplied by the square root of three ($\sqrt{3}$ or 1.732 05) to find the *line-to-line voltage*. Finally, many industrial and commercial locations will take advantage of this relation and use the individual phase voltages or *line to neutral* as 120 V, and use the resulting wye-connected *line-to-line* voltage of 208 V as a low- to medium-power three-phase combination. This 208/120 V common combination takes advantage of the $\sqrt{3}$ factor.

A final word of caution is in order since Eq. (12-3$_\text{E}$) appears in many references as $E_p = 4.44\Phi N_p fk_pk_d \times 10^{-8}$ V. This is *equivalent* except for the roundoff to three significant figures in the 4.44 factor. This really resulted from traditional slide rule accuracy and instrument accuracy. If three significant figures is sufficient or if instrumentation does not allow more, use 4.44 and round off k_p and k_d to the nearest three significant figures. Also, E_p and N_p are *per phase* quantities. N_p then equals Nn and $E_p = E_{gpp}$ times any series-connected multiples that may exist. Again use care and thought and the formula form that is most useful.

Example 12-3. A small three-phase synchronous alternator can be connected in series wye for nominal 440 V line-to-line loads or parallel wye for nominal 220 V line-to-line loads. It has a total of 36 slots and is connected as a four-pole machine. Each of its 36 coils has 15 turns and is formed to span 8 slots (from 1 to 9). The field coils are adjusted to produce a Φ of 600 000 lines/pole ($\phi = 0.006$ Wb/pole). What is its line-to-line voltage at open circuit when connected parallel wye and turning at 1800 rpm (188.50 rad/sec)?

Solution: With 36 slots and 4 poles there are 36/4 = 9 *slots/pole*. Also, there are 9/3 = n = 3 *slots/pole/phase*. There are also 180/9 = α = 20 *electrical degrees/slot*, and with a span of 8 slots there are $8 \times 20 = 160$ *electrical degrees/ coil*. From Table 12-1 we find that $k_p = 0.984\,81$, and from Table 11-2 that $k_d = 0.959\,80$. Also given are turns per coil: $N = 15$.

A pole–phase group is made of three coils, $n = 3$, so $Nn = 15 \times 3$. For parallel-wye connections *each phase* will have *two* permanently connected pole–phase groups in series and two of these series-connected groups in parallel. The connection will resemble Fig. 11-4. As a result, Eq. (12-3$_E$) or (12-3$_{SI}$) voltages will be multiplied by 2 to get the *per phase* voltage. This in turn will be multiplied by $\sqrt{3}$ to get the final line-to-line voltage.

The frequency f is not given, but from Eq. (11-2$_E$) we have

$$f = \frac{PS}{120} = \frac{4(1800)}{120} = f = 60 \text{ Hz}$$

or from Eq. (11-2$_{SI}$),

$$f = \frac{P\omega}{4\pi} = \frac{4(188.50)}{4\pi} = f = 60 \text{ Hz}$$

From Eq. (12-3$_E$),

$$E_{gpp} = 4.4428\Phi Nnfk_pk_d \times 10^{-8} \text{ V}$$

$$= 4.4428(600\,000)15(3)60(0.984\,81)0.959\,80(10^{-8})$$

$$= 68.031 \text{ V}$$

per pole group. Then

$$E_{gpp}(2)\sqrt{3} = 68.031(2)\sqrt{3} = E_l = 235.67 \text{ V}$$

where 2 is the number of pole–phase groups.
In SI units, using Eq. (12-3$_{SI}$),

$$E_{gpp} = 4.4428(0.006)15(3)60(0.984\,81)0.959\,80$$

$$= 68.031 \text{ V}$$

per pole group.

THE SYNCHRONOUS ALTERNATOR   CHAP. 12

$$E_{gpp}(2)\sqrt{3} = 68.031(2)\sqrt{3} = E_1 = 235.67 \text{ V}$$

So the same line-to-line voltage will hold in either units, as it should.

QUESTIONS

12-1. What is the difference between a salient pole and a distributed pole field structure?

12-2. Why is rms voltage wanted rather than average voltage?

12-3. Describe the meaning of pitch factor k_p.

12-4. Why is k_p never larger than unity?

12-5. Describe the meaning of distribution factor k_d.

12-6. Why is distribution factor always less than unity if there is more than one coil and coils do not lie in the same slots?

12-7. Why is the per pole per phase coil group a convenient quantity?

PROBLEMS

12-1. What is the pitch factor of the winding coils in an eight-pole synchronous alternator that has 72 lamination slots and whose coils span six slots? Calculate and then verify from Table 12-1.

12-2. What is the distribution factor of a six-pole, three-phase alternator wound on a 72-slot core? Calculate and then verify from Table 12-2.

12-3. A three-phase synchronous alternator is designed to produce 60 Hz when operated at 1200 rpm. The stator structure has 90 slots, and the 90 coils have four turns each. The coils span 11 slots each. Determine (a) the number of poles required, (b) the pitch factor of the coils, and (c) the distribution factor of the coils.

12-4. The alternator of Problem 12-3 operates at a total flux per pole of $\Phi = 1\ 033\ 000$ lines. Determine (a) the generated voltage per pole per phase, (b) the generated voltage per phase if three pole groups are operated in series, and (c) the line-to-line generated voltage.

11-5. A three-phase alternator is designed to produce 400 Hz while running at 837.76 rad/sec. The stator has 54 slots and the 54 coils have two turns each. The coils span seven slots. Determine (a) the number of poles required, (b) the pitch factor of the coils, and (c) the distribution factor of the coils.

12-6. The alternator of Problem 12-5 operates at a total flux of $\phi = 0.001\ 42$ Wb/pole. Determine (a) the generated voltage per pole per phase, (b) the generated voltage per phase if the phases are arranged with their pole–phase groups in two parallel paths, and (c) the line-to-line generated voltage.

Answers. **12-1**, 0.866 03; **12-2**, 0.957 66; **12-3(a)** 6 poles, **(b)** 0.913 55, **(c)** 0.956 68; **12-4(a)**, 48.14 V, **(b)** 144.4 V, **(c)** 250.2 V; **12-5(a)** 6 poles, **(b)** 0.939 69, **(c)** 0.959 80; **12-6(a)** 13.66 V, **(b)** 40.98 V, **(c)** 70.97 V.

13

Synchronous Alternator Regulation

When a synchronous alternator is in operation, a number of conditions have to be satisfied:

(1) The alternator must be connected to a prime mover and driven at its synchronous speed so that the proper ac frequency may be delivered. This is usually an exact requirement rather than merely approximate.

(2) The alternator must be properly synchronized before it is paralleled with any other alternators on the bus line. The first two items are much simplified in the relatively rare situation where an alternator operates by itself rather than in parallel with others.

(3) The voltage that the alternator is to deliver must be properly set by adjustment of the rotating field excitation current. The field excitation is direct current to produce a steady magnetic field flux.

When there is no electrical load on the alternator, its generated voltage per phase E_{gp} and its terminal voltage per phase V_p are the same. The use of E for an internal voltage and V for external is general. As with a dc generator, the terminal voltage is reduced by the IR drop through the winding resistance. This effect is similar but not quite identical to the dc situation, because the dc resistance of the windings does not account for all the voltage drop, as will be explained later.

The terminal voltage is also affected by the armature winding inductive reactance. Reactance and resistance effects are combined as phasors. As with any ac situation, the reactive voltage drop is 90 electrical degrees out of phase with the resistance drop.

The terminal voltage is also affected by the armature reaction, which is the result of the stator ampere turns acting across the main field. This effect is more complex than that of a dc armature, because the armature reaction acts to increase the apparent winding inductance; together the two effects can be quite large. The armature reaction has a variable effect since, depending upon the *load* power factor, it can act to demagnetize the field or to increase the field magnetization.

As a result, the voltage regulation of an alternator is both variable and large. The same alternator may display substantial voltage drop or significant voltage rise, depending upon its load power factor. There is, as yet, no self-regulating synchronous alternator internal connection. In the usual situation, voltage is controlled by external regulating circuits that vary the rotating field winding current so that a constant voltage is delivered to the load.

13-1 REPRESENTATIVE ALTERNATOR PERFORMANCE

Before discussing the individual effects of various power factor loads, it is reasonable to consider typical alternator performance. Figure 13-1 shows alternator voltage regulation for representative load power factors with the same field excitation current. Figure 13-2 shows the same alternator and power factors, but with the no-load voltage adjusted to different values for each load power factor so that the full-load voltages coincide. Although these curves appear superficially similar to Fig. 4-8, the two situations are not comparable. The curves in Fig. 4-8 were

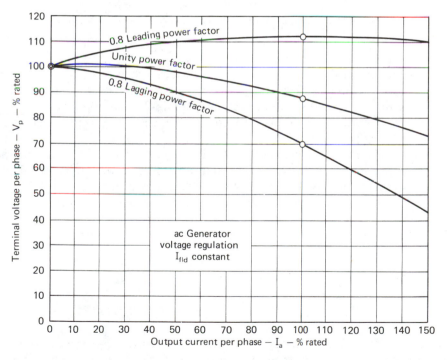

Figure 13-1. Typical terminal volts versus output current for various load power factors, I_{fld} constant.

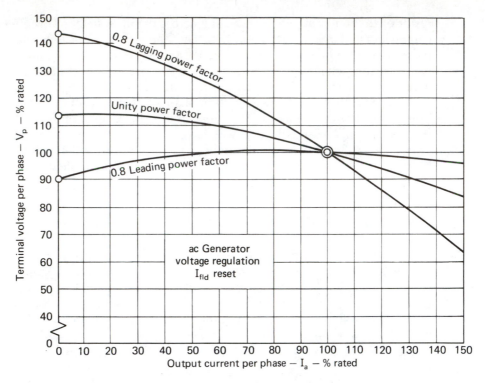

Figure 13-2. Typical terminal volts versus output current for various load power factors, I_{fld} reset.

developed by using different field connections, whereas the situations shown in Figs. 13-1 and 13-2 have only the one rotating field circuit. Figure 13-1 shows the results for a *constant* field flux. Figure 13-2 shows the results for constant flux with a particular load power factor, but each power factor requires its particular *different* field current.

These two figures are plotted in the *per unit* manner. This method does not specify the rated capacity of the unit or its normal output values. However, if the alternator in Figs. 13-1 or 13-2 was, for example, rated at 250 kW, three phase at 440 V line-to-line and 80 percent power factor (PF), and was wye connected, then the per phase voltage at full load would be $440/\sqrt{3} = 254$ V, and this value would be 1.0 or 100 percent. The rated or 100 percent current *per phase* would then be

$$\frac{250 \text{ kW}(1000 \text{ W})}{(3) \text{ kW}(0.8 \text{ PF})} = 104 \ 167 \text{ VA/phase}$$

$$\frac{104 \ 167 \text{ VA}}{254 \text{ V}} = 410 \text{ A/phase}$$

Here 410 A is 100 percent of the output current per phase. A similar approach can be used for any type of device and enables comparison between similar units of different rated capacities. A base of 1 or 100 percent can be used.

It is worthwhile noting that the conversion from three-phase ratings to individual voltage and current is usually confusing to a beginner. The one cardinal relation to remember that holds for wye or delta connections is that *phase* voltage times *phase* current times *three* always equals total *three-phase* volt-amperes in a balanced system.

$$V_p \times I_p \times 3 = \text{volt-amperes, three phase}$$ Eq. (13-1)

Similarly, for power

$$V_p \times I_p \times 3 \times \text{PF} = \text{watts, three phase}$$ Eq. (13-2)

From these fundamental conditions, wye or delta can be accommodated remembering that in balanced *wye*

$$I_{\text{line}} = I_{\text{phase}} \quad \text{or} \quad I_l = I_p$$

and in balanced *delta*,

$$V_{\text{line-to-line}} = V_{\text{phase}} \quad \text{or} \quad V_l = V_p$$

Every other condition requires $\sqrt{3}$. In balanced *wye*,

$$V_{\text{phase}} = V_{\text{line-to-neutral}} = \frac{V_{\text{line-to-line}}}{\sqrt{3}} = \frac{V_l}{\sqrt{3}}$$

or

$$V_p = V_{ln} = \frac{V_l}{\sqrt{3}}$$

In balanced *delta*,

$$I_{\text{phase}} = \frac{I_{\text{line}}}{\sqrt{3}} \quad \text{or} \quad I_p = \frac{I_l}{\sqrt{3}}$$

Rearrange these conditions to accommodate a specific situation, but use Eq. (13-1) to *check* a volt-ampere situation and Eq. (13-2) to check a power or wattage situation. Overall, in Y or Δ, volt-amperes = $\sqrt{3}\, V_l \times I_l$ and watts = $\sqrt{3}\, V_l \times I_l \times \text{PF}$ in balanced three phase.

13-2 ALTERNATOR PERCENTAGE REGULATION

The voltage regulation percentage change of an ac alternator is figured in the same manner as for a dc generator. Thus, Eq. (4-5) is valid except that the subscripts change for the ac situation.

$$\text{Percent voltage regulations} = \frac{E_{gp} - V_p}{V_p} \times 100$$ Eq. (13-3)

where

> E_{gp} = internal generated voltage *per phase* at *no* load
> V_p = terminal voltage *per phase* at *rated* load

Please note that while in service the field current in a synchronous alternator requires constant adjustment unless the load power factor happens to be the particular small leading power factor that results in zero percent regulation. For the generalized case of Fig. 13-1 or 13-2, this would be *about* 0.88 to 0.90 leading PF. Each specific model of alternator would have a different but related form of curves for its particular regulation. It will be shown that the regulation depends upon the armature circuit inductive reactance and the various modifications of the field magnetization due to the particular armature reaction. First, Example 13-1 will show typical regulation percentage measurements.

Example 13-1. A typical Y-connected, three-phase alternator is adjusted to its rated line-to-line voltage of 230.0 V while under its rated load and at 80 percent lagging power factor. The same field excitation current results in a no-load line-to-line voltage of 328.6 V. What is its regulation?

Solution: Using Eq. (13-3) and recognizing that the zero load voltage is the E_{gp}, or what will be called voltage generated per phase, we have for the three-phase condition

$$E_{gp} = \frac{328.6}{\sqrt{3}} = 189.7 \text{ V}$$

$$V_p = \frac{230.0}{\sqrt{3}} = 132.8 \text{ V}$$

and from Eq. (13-3),

$$\frac{189.7 - 132.8}{132.8} \times 100 = 42.8\% \text{ regulation}$$

Note that the results would have been the same if the voltages had been used as line-to-line rather than per phase. We shall see later that per phase is preferred for phasor calculations since it relates to individual winding measurements. This problem would also have produced the result if the per unit situation had been given and Fig. 13-2 used, as may be seen next, since the voltages chosen match the situation in the Fig. 13-2.

Example 13-2. The same three-phase alternator has a rated load voltage of 100 percent while carrying an 80 percent lagging power factor load at 100 percent current. If its no-load voltage is 142.8 percent of rated voltage, what is its regulation?

Solution:

$$\frac{142.8 - 100}{100} \times 100 = 42.8\% \text{ regulation}$$

In this case the voltage measurement is not specified as line-to-line or per phase since 100 percent or any other percent of line-to-line is also 100 percent or the same other percent of the per phase voltage. The per unit method is a convenient way of comparing machines that are rated under *different* conditions. It does, however, require an additional step if actual voltage measurements are supplied.

13-3 UNITY POWER FACTOR REGULATION

The voltage drop situation within the synchronous alternator under unity power factor conditions is shown in Fig. 13-3a. All the phasor relations in Fig. 13-3 are voltages either directly or because $IR = E$. Thus, some phasors are voltages such as E_{gp} or V_p and others are currents times impedances, which are voltages by Ohm's law. Hence, $I_a R_a$ is a *voltage*.

Under *unity* power factor conditions then, by definition, the V_p phasor or the terminal load voltage is *in phase* with the armature current I_a. The V_p phasor and the $I_a R_a$ phasor are then aligned, as shown in Fig. 13-3a.

The voltage drop across the inductive reactance X_a of the armature winding is, by definition, leading the current by 90° because the current *lags* the voltage by 90°. Therefore, in Fig. 13-3d the *voltage* $I_l X_a$, which is a part of the phasor $I_a X_s$, leads the *voltage* $I_a R_a$, because the $I_a R_a$ phasor is in phase with the *current* caused by the $I_a X_a$ *voltage*.

The other part of the $I_a X_s$ phasor can be seen, from Fig. 13-3d, to be the E_{ar} section. E_{ar} is the voltage due to *armature reaction*. Section 3-4 defines armature reaction. It is the same physical situation here. This cross-magnetizing effect is due to the load current in the armature windings, which produces ampere turns in the stator. The cross-magnetizing can be represented as a voltage E_{ar} that is considered to act 90° from the load current, and thus 90° from the $I_a R_a$ voltage.

The hypotenuse of the phasor triangle in Fig. 13-3d is the resultant *impedance* of the effective armature winding times the armature current, or $I_a Z_s$. The determination of the *synchronous impedance* Z_s of the armature winding will be discussed later. Note from Fig. 13-3a through 13-3c that the winding *impedance triangle* always has the same shape, but its position changes with power factor in relation to the V_p vector. The $I_a R_a$ phasor is much smaller in proportion than shown.

Returning to Fig. 13-3a, it can be seen that the generated voltage E_{gp} is subject to the voltage drop of $I_a Z_s$ *within* the armature winding. The remaining per phase voltage available at the terminals of the machine is then V_p. It is not quite as easy to determine the relationship between E_{gp} and V_p from $I_a Z_s$ as it is from the other two sides of the winding voltage phasor triangle.

Since $I_a R_a$ is in line with V_p for unity power factor by the terms at the beginning of this section, the base of the whole phasor relation is $V_p + I_a R_a$. The vertical leg of the triangle is $I_a X_s$. The final relationship between V_p and E_{gp} is then

$$E_{gp} = \sqrt{(V_p + I_a R_a)^2 + (I_a X_s)^2}$$ Eq. (13-4)

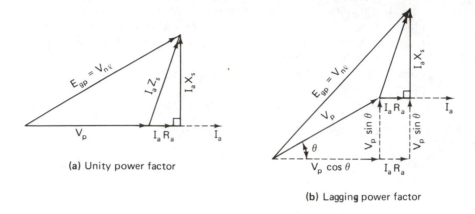

(a) Unity power factor

(b) Lagging power factor

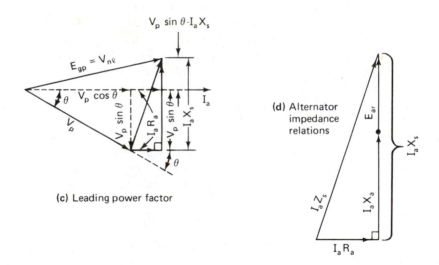

(c) Leading power factor

(d) Alternator impedance relations

Figure 13-3. Synchronous alternator voltage relations with various load power factors.

An alternative way of stating the *same* relationship is

$$E_{gp} = V_p + I_aR_a + jI_aX_s$$ Eq. (13-4)

The use of one or the other depends upon the methods that you have learned to use for phasor solutions; they are *equivalent*. However, with modern pocket calculators of scientific capability, the first form is very straightforward. Take your choice, but it is good to be familiar with both methods. In practice, the I_aX_s phasor is so much larger than the I_aR_a phasor that the effect of R_a is sometimes ignored.

13-4 LAGGING POWER FACTOR REGULATION

When the load power factor is of a lagging nature, as is the usual case, the relationship between V_p and E_{gp} is more involved. Consulting Fig. 13-3b, we can see that the I_aR_a phasor of the basic winding voltage phasor triangle is no longer in line with the V_p phasor. The angle θ is the power factor angle where the phasor I_aR_a *lags* the phasor V_p. V_p could have been drawn horizontally in a more conventional diagram, but the illustration chosen in Fig. 13-3b is easier to follow trigonometrically.

To see the relationship between V_p and E_{gp}, the following construction is made: the phasor $V_p \cos \theta$ is made parallel to I_aR_a, and phasor $V_p \sin \theta$ is made parallel to I_aX_s. A second I_aR_a is constructed parallel to the original I_aR_a such that it is an extension of $V_p \cos \theta$. A second $V_p \sin \theta$ is drawn parallel to the true $V_p \sin \theta$ such that it is in line with I_aX_s. The resulting figure, as shown in Fig. 13-3b, is easy to analyze. The small four-sided figure is a rectangular parallel-epiped, since the two I_aR_a sides are of identical length, as are the two $V_p \sin \vartheta$ sides. This figure is a rectangle because all four corners are right angles. This follows from the fact that the original I_aR_a phasor and I_aX_s phasors are 90° to each other by original definition.

It is easy to define E_{gp} by use of the Pythagorean theorem, since one side of the overall right triangle is $V_p \cos \theta + I_aR_a$ and the other orthogonal side is $V_p \sin \theta + I_aX_s$. E_{gp} is then seen to be

$$E_{gp} = \sqrt{(V_p \cos \theta + I_aR_a)^2 + (V_p \sin \theta + I_aX_s)^2} \qquad \text{Eq. (13-5)}$$

In a similar fashion, this is

$$E_{gp} = V_p \cos \theta + I_aR_a + j(V_p \sin \theta + I_aX_s) \qquad \text{Eq. (13-5)}$$

Study of the figure will show that, as the load power factor becomes progressively poorer, or as the phase angle θ becomes larger, E_{gp} becomes larger in relation to V_p. This process reaches its maximum at just short of $\theta = 90°$, when the I_aZ_s phasor lines up with V_p. In actual practice, R_a is very small in relation to X_s. As a result, the maximum E_{gp} is reached at very nearly $\theta = 90°$. The more usual power factor relationship is about as shown in Fig. 13-3b.

13-5 LEADING POWER FACTOR LOADS

When the load power factor is *leading*, the armature winding phasor triangle voltages swing the other way. See Fig. 13-3c, where the phasor I_aR_a *leads* the phasor V_p by the load power factor angle θ. Again, in drawing this figure the I_aR_a phasor has been drawn horizontally to simplify the trigonometric understanding. $V_p \cos \theta$ and $V_p \sin \theta$ are drawn in relation to the V_p phasor in a similar fashion to that of Fig. 13-3b. Again I_aR_a is extended in line with $V_p \cos \theta$. In this situation the

$V_p \sin \theta$ projection is a part of the $I_a X_s$ phasor. It may be shorter than $I_a X_s$, as shown in Fig. 13-3c, or it may be longer if θ is a larger angle. The sides of the triangle that are used in a Pythagorean solution for E_{gp} are $V_p \cos \theta + I_a R_a$ and $V_p \sin \theta - I_a X_s$. E_{gp} is then

$$\boxed{E_{gp} = \sqrt{(V_p \cos \theta + I_a R_a)^2 + (V_p \sin \theta - I_a X_s)^2}} \qquad \text{Eq. (13-6)}$$

In the other form, this is

$$\boxed{E_{gp} = V_p \cos \theta + I_a R_a + j(V_p \sin \theta - I_a X_s)} \qquad \text{Eq. (13-6)}$$

Note that as the power factor angle θ becomes larger, and the winding impedence triangle swings around, as a result, E_{gp} may actually be *smaller* than V_p. The fact that the terminal voltage V_p can be actually higher than the generated voltage E_{gp} is explained by the alignment of the $I_a Z_s$ phasor. There is one point where V_p and E_{gp} are the same. As a result, V_p, E_{gp}, and $I_a Z_s$ form an isosceles triangle. At this point, the synchronous alternator has a zero percent voltage regulation. This may be likened superficially to a flat compound dc generator, but the reasons for the zero voltage change are not the same.

13-6 LOAD POWER FACTOR RELATIONS SUMMARIZED

A glance at Eqs. (13-4), (13-5), and (13-6) shows that they are closely related and that they can be stated in a combined fashion that holds for *any* power factor:

$$\boxed{E_{gp} = \sqrt{(V_p \cos \theta + I_a R_a)^2 + (V_p \sin \theta \pm I_a X_s)^2}} \qquad \text{Eq. (13-7)}$$

or alternatively, in complex form,

$$\boxed{E_{gp} = V_p \cos \theta + I_a R_a + j(V_p \sin \theta \pm I_a X_s)} \qquad \text{Eq. (13-7)}$$

Using this equation for *unity* power factor $\cos \theta$ is *one*, so the term really becomes $V_p + I_a R_a$ as it was in Eq. (13-4). At unity power factor, $\sin \theta$ is *zero*, so $V_p \sin \theta$ drops out. Stated another way, Eq. (13-4) is merely a special case of Eq. (13-7) with $\cos \theta = 1$ and $\sin \theta = 0$. By comparing Eq. (13-7) to Eqs. (13-5) and (13-6), it can be seen that the only difference is the plus or minus sign in the $V_p \sin \theta \pm I_a X_s$ term. For *unity* and *lagging* power factor, the sign is *plus*. For *leading* power factor, the sign is *minus*. If these special conditions are remembered, Eq. (13-7) is all that is required, since Eqs. (13-5) and (13-6) are shown to be special varieties of Eq. (13-7). Worked examples will clarify this situation, but first it is necessary to define the $I_a R_a$ and $I_a X_s$ phasors so that their use may be understood. Realize that the relations shown in Eq. (13-7) and Fig. 13-3 may be plotted to scale with the same answer results.

13-7 WINDING RESISTANCE

The effective ac resistance of a particular winding is usually determined by using *direct current* and the voltmeter–ammeter method. A dc current is passed from terminal to terminal of either two of the three leads of a three-phase winding or a complete single-phase winding. If the voltage drop from terminal to terminal is recorded at the same time as the current used, the dc resistance is $R = E/I$. Multiple readings are averaged for accuracy. This resistance is usually that of *two* phases, since the neutral point of a wye winding is frequently not accessible. See Fig. 13-4a for a typical test circuit. In the less common delta-wound alternator,

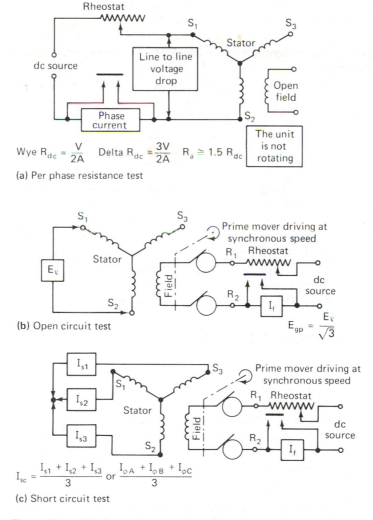

$$\text{Wye } R_{dc} = \frac{V}{2A} \quad \text{Delta } R_{dc} = \frac{3V}{2A} \quad R_a \cong 1.5 \, R_{dc}$$

(a) Per phase resistance test

(b) Open circuit test

$$E_{gp} = \frac{E_\ell}{\sqrt{3}}$$

$$I_{sc} = \frac{I_{s1} + I_{s2} + I_{s3}}{3} \quad \text{or} \quad \frac{I_{\phi A} + I_{\phi B} + I_{\phi C}}{3}$$

(c) Short circuit test

Figure 13-4. Winding resistance and synchronous impedance test circuits.

the windings are usually permanently spliced together at the delta points. The resulting per phase resistance of *one* phase in *wye* is

$$R_{dc} = \frac{\text{voltage drop}}{\text{current reading}} \left(\frac{1}{2}\right) = \frac{V}{2\ A}$$

Eq. (13-8)

R_{dc} for connected delta winding is

$$R_{dc} = \frac{3\ V}{2\ A}$$

Eq. (13-9)

Since there are harmonics generated in individual coils of the alternator windings, some high frequencies are present even though their combined effect in a phase is minimal. As a result of the fact that high-frequency currents travel near the surface of a conductor, all the cross-sectional area of the windings is not equally used by the current. For this and other reasons, the effective ac resistance is greater than the dc resistance. The factor of difference varies according to the base frequency of the alternator as well as to the winding configuration, and is from 1.2 to 1.8. Resistance of the winding to the passage of alternating current is then usually taken as 1.5 times the dc resistance for 60 Hz machines.

$$R_a = 1.5 R_{dc}$$

Eq. (13-10)

The R_a is usually a very low resistance, particularly on large machines where it is an extremely small part of an ohm. The error introduced into Eqs. (13-4) through (13-7) is minimal even with a two-significant-figure approximation of R_a. The ohmmeter is useless. Voltmeter–ammeter methods are usually used. The currents used are either up to the rated current of the machine or as high as can be reached, whichever is applicable. Really large alternators have winding currents in the thousands of amperes. This current level is not lightly achieved in a test rig. Accurate determination of R_a is more important in efficiency calculations.

13-8 SYNCHRONOUS IMPEDANCE

It has been shown that the phasor relations of Fig. 13-3(a), 13-3(b), and 13-3(c) can be used to determine the generated voltage per phase E_{gp} necessary to deliver a particular terminal voltage per phase V_p. The phasor relationship in Fig. 13-3(d) holds for any power factor situation. Note that all three sides of Fig. 13-3(d) contain the armature winding current I_a as a multiplier. If I_a is divided into each side, a similarly shaped *impedance triangle* results with R_a and X_s as the orthogonal sides and Z_s as the hypotenuse. These values are, respectively, as follows:

(1) R_a is the *effective* ac resistance of the winding *per phase* in whatever series and parallel coil combination is used.

(2) X_s is the effective combined winding *reactance* and reactance effect of the armature *reaction*. Note that these effects operate together and that X_s is the *synchronous reactance*, also a *per phase* value.

(3) Z_s is then the phasor sum of R_a and X_s, and is defined as the *synchronous impedance* of the armature winding on a *per phase* basis.

The determination of the values of the Z_s and the X_s components of this impedance triangle is known as the *synchronous impedance method*. This procedure is widely used and is almost universally recognized. The synchronous impedance method is actually simpler than other recognized methods, and when used with real equipment gives slightly pessimistic results. This means that an actual alternator will have slightly better voltage regulation (lower percentage regulation) than that calculated from its synchronous impedance measurements. All the other recognized methods also have limitations and are considered to be beyond the scope of this book.

Two specific tests are necessary to determine the Z_s or effective winding impedance. These are the *open-circuit test* and the *short-circuit test*. From these two tests and Ohm's law for ac circuits, the impedance is determined where

$$Z_s = \frac{E \text{ open-circuit per phase}}{I \text{ short circuit}} = \frac{E_{gp}}{I_{sc}} \qquad \text{Eq. (13-11)}$$

Specifically defined procedures are used to get mutually comparable values of E open circuit and I short circuit.

The open-circuit voltage test is comparable to the dc machine open-circuit saturation curve, as discussed in Chapter 4 and Fig. 4-1. The alternator is driven at its *synchronous speed* in either rpm or ω units. Field current is varied from a low value up to that sufficient for a voltage reasonably beyond the rated voltage. Data are recorded in suitable steps. If the alternator is a three-phase wye unit, as is usually the case, the line-to-line voltage E_1 is divided by the $\sqrt{3}$ to find the per phase voltage. This step is not applicable for a single-phase unit. However, since the resistance R_a, the reactance X_s, and the impedance Z_s are all per phase units, the data used to determine these values must be per phase also. Figure 13-4b shows a simplified test circuit, and Fig. 13-5 shows a typical open-circuit voltage curve. Note that the shape of the E_{gp} versus I_f curve is very similar to Fig. 4-1. The similarity is due to the same degree of saturation of the magnetic circuit in each case.

For the short-circuit test, the machine is shut down and reconnected as shown in Fig. 13-4c. Ammeter protection switches are not shown, since the test will controllably build up and then reduce current. The ammeters for large machines require heavy shunts or suitable current transformers, since these currents may be in the thousands of amperes in a large unit.

It is worth noting that it is not a simple thing to connect and drive a large alternator. Many large hydroelectric units are not rotated at all until they are built into their final location because of the magnitude and expense of a suitable prime-mover drive. In addition, a set of characteristic curves, such as Figs. 13-1

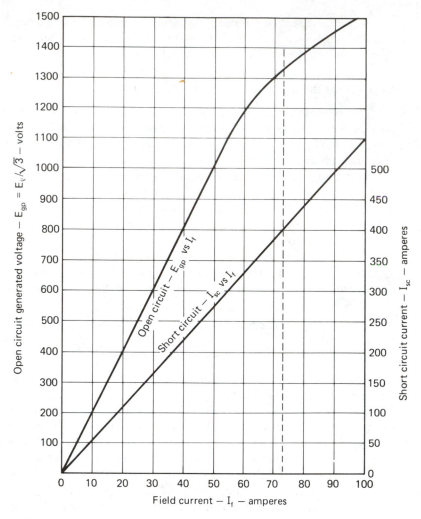

Figure 13-5. Alternator test curves to determine synchronous impedance Z_s.

or 13-2, requires *actual loading* of the unit. On a large alternator the only suitable load is the combined industrial, business, and domestic electric customers of a whole city. This difficulty of performing a full-scale test with a controllable load is the real reason for the synchronous impedance test. Much less power is consumed in the prime mover since only the internal losses of the machine must be overcome. No *external* power is delivered during either the open-circuit or the short-circuit test. As a result, only a small percentage of full-load torque is needed.

When a suitable version of Fig. 13-4c is connected, the alternator is brought up to its synchronous speed while short circuited through the ammeters. Initially, the field current I_f is very low or zero, so that no abnormal short-circuit currents are generated. Then, as the field current is increased, the short-circuit currents are watched and recorded as I_{sc} and its simultaneously corresponding I_f. A current of substantially beyond the normal rated current may be briefly reached with safety as long as the lines are not switched open. The data will show a straight line, as in Fig. 13-5, because the load is almost entirely reactive due to the high inductance and high armature reaction. As a result of the demagnetizing effect of the armature reaction, the machine is operating in the linear region of its saturation curve during this test.

This is not a disastrous short circuit in the sense of a circuit failure or incorrect connection. It is really a situation where only the normal load currents and reasonable overload currents are carefully and controllably applied.

Returning to Fig. 13-3b and 13-3d, it can be seen that if there is *no* external V_p, as in the case of a full short circuit, the E_{gp} phasor then coincides with the I_aZ_s phasor. Thus, the relation of Eq. (13-11) is applicable.

Since the currents and magnetic paths per phase are not truly identical, the I_{sc} value used is the average of the three ammeter readings:

$$I_{sc} = \frac{I_{s1} + I_{s2} + I_{s3}}{3} \qquad \text{Eq. (13-12)}$$

When the current I_{sc} is taken from the curve at the same I_f as that which produced the rated voltage per phase, it may be seen that the open-circuit voltage E_{gp} would have produced I_{sc} through the windings. This means that E_{gp} was dropped across the synchronous impedance, Z_s, when I_{sc} was measured, and Eq. (13-11) can be used. Example 13-3 will show this use.

Example 13-3. A 1000 kVA, 2300 V, three-phase wye-connected synchronous alternator is tested to determine its synchronous impedance. The dc resistance between two lines averages 0.412 Ω. The open-circuit voltage and short-circuit current are determined to have the relationships shown in Fig. 13-5. Find the values of R_a, Z_s, and X_s, assuming that R_a effective resistance is 1.5 times the dc resistance per Eq. (13-10).

Solution:

$$R_{dc} = \frac{0.412}{2} = 0.206 \ \Omega$$

$$R_a = 1.5 \times 0.206 = 0.309 \ \Omega$$

From Fig. 13-5 at the open circuit per phase voltage corresponding to the rated line to line voltage, find the corresponding field current. $E_{gp} = E_l/\sqrt{3} = 1328$ V. At $E_{gp} = 1328$ V, the corresponding field current $I_f = 73$ A. See the dotted line in Fig. 13-5. At the same I_f, the corresponding I_{sc} is 400 A from the curve. Using Eq. (13-11),

$$\frac{E_{gp}}{I_{sc}} = \frac{1328}{400} = Z_s = 3.32 \ \Omega$$

From the orthogonal relationship between R_a and Z_s, we may use the Pythagorean theorem:

$$X_s = \sqrt{Z_s^2 - R_a^2} = \sqrt{(3.32)^2 - (0.309)^2} = 3.31 \ \Omega$$

Note that X_s has very nearly the value of Z_s because of the comparatively small value of R_a. This impedance triangle relationship is reasonably typical. Probable error sources keep this from being any better than a three-significant-figure value. More on the errors later.

Example 13-4. With the same alternator described in Example 13-3, calculate the percent regulation for a full-load lagging power factor of 0.8.

Solution: The full-load line current I_l is the same as the current in the connected group of armature coils that comprise one phase, so $I_l = I_a$. In three-phase wye,

$$I_l = \frac{\text{kVA } (1000)}{\sqrt{3} \ V} = \frac{1\ 000\ 000}{\sqrt{3} \times 2300} = I_a = 251.0 \text{ A}$$

To use Eq. (13-7), we find the various values needed for substitution:

$$V_p = \frac{V_l}{\sqrt{3}} = 1328 \text{ V}$$

$$I_a R_a = 251.0 \times 0.309 = 77.6 \text{ V}$$

$$I_a X_s = 251.0 \times 3.31 = 831 \text{ V}$$

The power factor when expressed as a decimal is the cosine of the power factor angle. In this case it is a lagging angle.

$$\cos \theta = 0.8$$

so that

$$\theta = 36.870° \quad \text{and} \quad \sin \theta = 0.6$$

Note that this situation is the familiar 3–4–5 triangle for simplicity. It is rare that the power factor is an even value.

Using Eq. (13-7), we use for lagging PF

$$\begin{aligned}
E_{gp} &= \sqrt{(V_p \cos \theta + I_a R_a)^2 + (V_p \sin \theta + I_a X_s)^2} \\
&= \sqrt{[1328(0.8) + 77.6]^2 + [1328(0.6) + 831]^2} \\
&= \sqrt{(1140)^2 + (1628)^2} = \sqrt{1\ 299\ 600 + 2\ 699\ 700} \\
&= \sqrt{3\ 949\ 300} \\
&= 1987 \text{ V}
\end{aligned}$$

Using Eq. (13-3) to find the regulation,

$$\frac{E_{gp} - V_p}{V_p} \times 100 = \frac{1987 - 1328}{1328} \times 100 = 49.6\%$$

Thus, this particular machine must have its field set so that its no-load voltage would be 1987 V/phase, or $1987\sqrt{3} = 3441$ V line-to-line. This setting is necessary in order to deliver its rated 2300 V line-to-line at its rated current and 0.8 lagging power factor. This rather large regulation percentage is typical of the synchronous alternator.

The synchronous impedance method has a number of inherent errors that should be recognized for what effect they have. Even so, the method is reasonable and conservative. Consequently, the machine will always perform a bit better than the calculations. An inspection of Fig. 13-5 will show that the synchronous impedance was taken at the *rated* V_p voltage rather than the actual E_{gp}. Since the probable E_{gp} was not known at the beginning, it could not be used. However, it would be farther up the saturation curve, and the saturation curve *turns toward* the short-circuit curve. Thus, had the synchronous impedance been taken at an I_f corresponding to the actual E_{gp}, the ratio of E_{gp} to I_{sc} would be *smaller*. This way the Z_s would be smaller. For this reason some users prefer to calculate X_s and then arbitrarily multiply it by a factor of 0.75. Since the use of this 0.75 factor is judgmental, its use is not recommended unless the machine in question corresponds fairly closely with another where the 0.75 factor results in good experimental correlation. A beginning student does not have the experience needed to juggle factors that sometimes hold and sometimes do not. In addition, the E_{ar} voltage of Fig. 13-3(d) does not always exactly line up with the $I_a X_a$ phasor as shown. The synchronous impedance method is then viewed as a *useful approximation*. It is considered to be the simplest of the various approximations available, and therefore the only one to be considered at this level of study.

QUESTIONS

13-1. Why are Figs. 13-1 and 13-2 shown with per unit scales?

13-2. Can a synchronous alternator ever show a negative voltage regulation? If so, under what load conditions can negative regulation be expected?

13-3. What combination of voltage and current measurements can be used for three-phase volt-ampere calculations whether the unit is wye or delta connected?

13-4. What combination of voltage and current measurements can be used for three-phase wattage calculations whether the unit is wye or delta connected?

13-5. Can a synchronous alternator be connected and adjusted so that its voltage is inherently self-regulating as in a flat compound dc generator? State reasons.

13-6. What is the effect of stator winding impedance on the voltage regulation of an alternator?

13-7. What is the effect of armature reaction on the voltage regulation of an alternator?

13-8. Explain how Eq. (13-7) covers the total scope of Eqs. (13-4), (13-5), and (13-6).

13-9. Why is the simple dc resistance of the armature phase windings modified by a factor of 1.5 to achieve an effective ac resistance?

13-10. Why is the short-circuit test portion of the synchronous impedance test not damaging to the alternator?

13-11. Why is the synchronous impedance method of calculating alternator voltage regulation called a conservative method?

13-12. What are some of the difficulties of directly testing the voltage regulation characteristics of a really large synchronous alternator?

PROBLEMS

13-1. A wye-connected, three-phase ac generator is delivering power to a three-phase line. The line-to-line voltage is 460 V. The line currents are 7.73 A and the total wattage is 5.12 kW. What are the following: (a) phase voltage, (b) phase current, and (c) load power factor?

13-2. If the line-to-line voltage of the machine in Problem 13-1 rises to 618 V at no load with the same field excitation, what is its voltage regulation?

13-3. What is (a) the E_{gp} and (b) the voltage regulation of an alternator that has $R_a = 0.152 \ \Omega$ and $X_s = 9.33 \ \Omega$, and delivers 230 V line-to-line at 9.5 A per line? Use unity power factor.

13-4. (a) What E_{gp} and (b) what regulation will the alternator in Problem 13-3 display if the load is 83.30 percent power factor lagging and is set to deliver 230 V line-to-line at 8.86 A?

13-5. The alternator in Problems 13-3 and 13-4 operates at a 76.41 percent leading PF and delivers 9.05 A to each line. If it is adjusted to the same 230 V line-to-line while under load, what will be (a) its E_{gp} and (b) its percent voltage regulation?

13-6. A voltage regulation test is to be performed on a three-phase wye alternator. Its line-to-line voltage drop is taken with a dc supply to find its armature circuit resistance. The readings are dc voltage drop = 11.15 V and line current = 18.5 A.
(a) What is its per phase dc resistance?
(b) What effective armature circuit resistance should be used?

13-7. A synchronous impedance test is taken on the alternator of Problem 13-6. Under short-circuit conditions the currents in the three lines are 18.53, 19.08, and 18.41 A. What current should be assumed for the test?

13-8. During the synchronous impedance test in Problem 13-7, the resulting open-circuit voltage line-to-line is 240.3 V.   What is the synchronous impedance?

Answers.   **13-1(a)**, 265.6 V, **(b)** 7.73 A, **(c)** 83.1%; **13-2**, 34.3%; **13-3(a)** 160.9 V, **(b)** 21.13%; **13-4(a)** 192.1 V, **(b)** 44.67%; **13-5(a)** 102.9 V, **(b)** −22.5%; **13-6(a)** 0.301 Ω, **(b)** 0.452 Ω; **12-7**, 18.67 A; **13-8**, 7.43 Ω.

14

Ideal and Practical Transformers

The transformer is a simple, reliable, and efficient device for changing an ac voltage fron one level to another. The ratio of voltage change, α, can be almost any reasonable number either to step up or step down voltage.

Remember, Faraday found that if the number of magnetic lines linking a coil of wire were *changed* a voltage was *induced* in the coil. So far, we have confined that change of magnetic linkage to that caused by a physical motion between the coil and a relatively constant magnetic flux. The action is the *same* if there is no physical motion and if the amount of magnetic flux *varies*. We will find that almost the same rules hold and that our knowledge is already nearly sufficient. The change is in configuration and in varying flux rather than motion.

In fundamental studies of ac inductances in basic ac circuits courses it was shown that, when two coils of wire are *inductively coupled*, the magnetic flux that *passes through* one coil passes through the other to a greater or lesser degree. In other words, the *magnetic circuit* is common or largely common to both coils. If the magnetic flux is cyclically varying because the coil that creates it has a cyclically varying current in it, the magnetic flux linkage to the second coil is also cyclically changing. As a result, the varying flux generates a second varying voltage in the second coil. This second voltage is a *transformer voltage* and is said to be created by *transformer action*.

The availability and use of transformer action is one of the major reasons for commercial preference of alternating current for power use. Alternating current is also subject to power losses in transmission by the I^2R effect, or current in amperes squared times resistance in ohms. This product results in watts of *lost* power, I^2R = watts. If the current can be substantially reduced by *raising* the voltage, the I^2R losses can be drastically *reduced*. This is the major reason for long transmission lines using very high voltage levels. Note that very *little* power is lost in a transformer. If many megawatts of power are being transmitted or carried through many miles, the resistance is appreciable even in large cables, and transmitted energy losses can be extremely expensive. Transformers change voltage and inversely change current. They also may serve for electrical isolation and to change impedances.

By the use of transformers, the generated voltages, which may already be as high as 18 000 V line-to-line, are raised to very high levels. The transmission-line voltages in use vary with the date of installation, the power carried, and the distance, but are routinely in many hundreds of thousands of volts. Research has long been underway leading to practical technology for the 1 000 000 V level.

At the far end the transmission-line voltages are *reduced* by use of transformers to an urban distribution level. This is frequently as high as 18 000 V, although there is much use of levels around 4600 or 2300 V. Finally, at a commercial or factory customer substation, the voltage is transformed again to either 460 or 230 V line-to-line. All this is three-phase power. Household power is further reduced by local line-pole-mounted transformers to the 230 to 240 V level. Only a single phase is brought into any one house. Older installations use 115 to 120 V, single phase, but this is largely superseded by the three-wire single-phase distribution systems that have both 115 and 230 V (or 120 and 240 V). The higher 230 V level has made practical high-power electrical appliances, such as stoves, dryers, and water heaters.

Power transformers at any of the commercial or aeronautical line frequencies, or from 25 to 400 Hz, are invariably of iron-core construction. Air-core construction is used only for high-frequency work. A study of this type properly belongs in a communications course.

As in either dc or ac armatures, the magnetic iron core of a transformer is of laminated construction. For 60 Hz, laminations of or nearly 0.014 in. (or 0.35 mm) are widely used and for the same reasons as in motor structures.

The input ac connection is to the *primary coil*. This coil may be the high or the low voltage. If the input is the high-voltage end, the transformer is called a *step-down* transformer because the output is at a lower voltage. The step-down primary coil connections are conventionally labeled as H_1 and H_2 terminals for *high* voltage. In this case the *secondary coil* terminals are labeled as X_1 and X_2.

When operated in the reverse order, with the *input* or again *primary* coil using the low voltage, the unit is called a *step-up* transformer. Here the labeling is again X_1 and X_2 for the low-voltage coil and H_1 and H_2 for the high-voltage coil. Formerly, the coils were labeled P_1 and P_2 for the primary with S_1 and S_2 for the secondary connections. The advantage of the later form of labeling is that the terminal nomenclature is the *same* whichever way the individual transformer is used. Any transformer may be used as a step-up or step-down unit depending on the way it is connected. The necessary precautions are that the *insulation* be sufficient to cope with the high-voltage end and that the voltage per turn of coil not be beyond reason.

The primary and secondary coils are connected *magnetically* by the laminated iron core. They are *not* connected by electrical conductors. If the core forms a simple rectangle in construction, with the primary coil surrounding one leg and the secondary coil surrounding another parallel leg, it is known as a *core-type* transformer. Conversely, if the core forms a rectangular figure eight, with both coils concentrically mounted on the middle leg, it is known as a *shell-type* transformer. Figure 14-1a shows a core-type transformer, and Fig. 14-1b shows a shell-type transformer.

The two types are different in construction and, as might be expected, have somewhat different properties. The differences, as will be shown, are such that neither type has dominated the field. Both types are available.

In general, the core type has a longer mean length of core and a shorter mean length of coil turn. The core type also has a smaller cross section of iron and so will need a greater number of turns of wire, since, in general, not as high a flux may be reached in the core. However, the core type is better adapted for some high-voltage service since there is more room for insulation. The shell type has

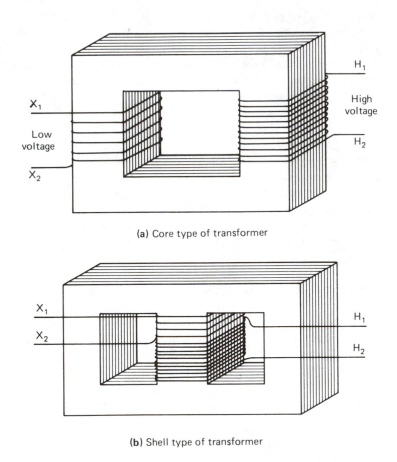

(a) Core type of transformer

(b) Shell type of transformer

Figure 14-1. Laminated core construction.

better provision for mechanically supporting and bracing the coils. This allows better resistance to the very high mechanical forces that develop during a high-current short circuit. Optimization of a transformer and choice of a specific core form are beyond the scope of this text.

Many proprietary forms have been developed for assembling the cores of transformers. These are better studied in specialist texts on transformers. However, many satisfactory forms of lamination stacks have been developed using simple rectangular sheared or punched plates of core steel. Higher production units use L-shaped or E-shaped punchings. Finally, the core may be wound of a continuous ribbon of a special high permeability steel. Reference (13) gives excellent treatment of core constructions.

With either of the transformer constructions, the coils are carefully wound, mechanically supported in bobbins or coil forms, and impregnated with insulation. The terminals are carefully supported and brought out to a terminal board or insulators.

The transformer core is assembled inside the coils and is mechanically clamped or otherwise supported in position. Clamping is necessary to keep the laminations from moving under cyclic magnetic forces. The electrical and magnetic assembly is then placed in a case except for the very smallest sizes. Very small units are sometimes of open construction with mounting brackets attached to the core. Medium and large units are fully cased. A surprising amount of structural support is used because of the mass of core and coil that must be mounted and supported. There is hardly any common device that is more dense than a transformer, since closely stacked steel laminations and closely wound copper coils are both dense materials.

Small and medium transformers are air cooled even if of cased construction. Larger sizes are filled with an insulating transformer oil in order to transmit the heat from the coils and core to the outside surface where air cooling is available. Still larger sizes have either cooling fins or actual separate oil-to-air radiators to enable the heat to be transmitted to the surrounding air. In the very largest sizes, the cooling oil is pump-circulated or the oil volume is cooled with circulating water in immersed cooling coils. Transformers are built in an astonishing range of sizes, from very tiny units used in transistorized communication devices that are of almost negligible weight up to monsters that weigh hundreds of tons (hundreds of metric tons). All operate on the same principle.

The theoretical operation of an ideal transformer will be discussed first, then real forms with loss problems, and finally practical simplified analysis methods will be shown.

14-1 IDEAL TRANSFORMER

If a transformer could be constructed with ideal properties it would need to have the following not quite attainable properties:

(1) All magnetic flux created by the primary coil would ideally *link* with the

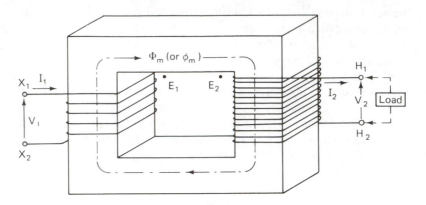

Figure 14-2. Ideal transformer.

secondary coil. There would be *no* leakage flux. This is nearly achieved in a carefully designed iron-core transformer.

(2) The primary and secondary coils would have *zero resistance*. Again, this is *nearly* achieved, but some resistance is present since the conductor cross section is limited.

A transformer having these ideal properties is shown in Figure 14-2. The operation of the ideal transformer is then as follows:

(1) With the incoming primary coil voltage momentarily *positive*, the direction of the primary current is as shown with the arrow I_1. This produces the magnetic flux Φ_m (or ϕ_m) in the direction shown. The subscript m signifies *mutual flux*. In an ideal transformer, that is the only flux present. We will need other subscripts later, so the m is useful.

(2) Since this flux Φ_m (or ϕ_m) is changing, it is inducing a voltage E_1, which opposes the applied voltage V_1. The *dot convention* shows that the induced voltage is positive at the top of the coil when the applied voltage is positive. This is in accordance with Lenz's law.

(3) At the *same time*, the magnetic flux is also inducing a voltage E_2 in the *secondary coil*.

(4) Again in accordance with Lenz's law, this voltage must be of such polarity that any current, I_2, that it produces must also oppose the mutual flux.

If these conditions are true, as they certainly are qualitatively even in a real transformer, then:

(1) If there is *no load* or the secondary circuit is open, $I_2 = 0$ A.

(2) Since the applied voltage V_1 is alternating in polarity, its resultant current I_1 is also alternating. The alternating I_1 produces the flux Φ_m (or ϕ_m), which also alternates at the same frequency. The alternating flux induces voltage E_1, which constantly opposes V_1, and induces voltages E_2. These are also alternating voltages

whose instantaneous polarities follow the dot convention shown.

(3) There is a small component of the I_1 current that remains, because I_1 is not completely canceled. Thus, E_1 does not quite equal V_1. This small component is called I_m or the *magnetizing current*. This is the current necessary to overcome the reluctance of the magnetic circuit. The reluctance is low in a good magnetic circuit, but it is not zero. This magnetizing current is the only current during no load.

(4) Figure 14-3a shows the relations of these various phasors under *no-load* conditions. I_m lags the primary voltage by 90° because the coil is assumed to be a pure inductance (no resistance). The flux Φ_m (or ϕ_m) is in phase with the current.

(5) The varying flux leads the voltage that it induces by 90°. Another way of stating this is that induced voltages E_1 and E_2 are lagging the flux by 90°. This brings E_1 180° out of phase with V_1, or E_1 *opposes* V_1.

This is entirely in accordance with an ideal inductance at this stage since the secondary coil is an open circuit and does not yet have any effect.

(6) Assume that an inductive or lagging load is connected to the secondary terminals (Fig. 14-3b). Note that a lagging load is by far the most usual situation for a power transformer and thus is most realistic. The lagging current I_2 can be seen from Fig. 14-3b to lag the secondary voltage E_2 by the power factor angle θ_2.

(7) Even though this is an assumed ideal transformer, it nonetheless has real properties. Its secondary current I_2 and secondary coil turns N_2 together produce a demagnetizing flux that is proportional to $I_2 N_2$ and which opposes Φ_m (or ϕ_m). This effect, if not compensated, would tend to *reduce* both E_2 and E_1 voltages.

(8) This assumed reduction of E_1 voltage causes the primary component of the load current I_1' to flow in the primary. $I_1' N_1 = I_2 N_2$, so that the same number of ampere turns of magnetization that was lost in step 7 is restored. The sinusoidal flux level in a transformer then remains nearly constant. This restores Φ_m (or ϕ_m) to its original value (in this ideal case). Note the symmetry between I_2 and θ_2 on the one hand and I_1' and θ_1' on the other, in Fig. 14-3b.

(9) Consulting Fig. 14-3c, it can be seen that now there are two components of the primary coil current, I_m and I_1'. The *phasor sum* of these currents is then I_1, which may be seen to have a different power factor angle θ_1. Thus, when an ideal transformer is being loaded, its primary current will assume a smaller lagging angle than its original $\theta = 90°$ when unloaded. Furthermore, its power factor angle θ_1 reflects the load power factor angle as one of its component parts. Therefore, the power factor angle θ_1 of the primary current is not the same as θ_2 of the load and, as a result, angle $\theta_1 > \theta_2$.

14-1-1 Transformation Ratio Relations. This effect of loading is roughly analogous to the loading of a dc shunt motor as described in Chapter 6. The back emf of generator action decreases as the speed decreases until enough additional current is drawn for the load to be supported. Here the presence of the I_2 phasor forces I_1' to grow.

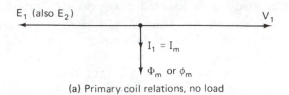

(a) Primary coil relations, no load

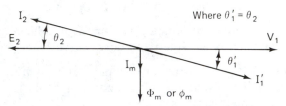

(b) Secondary coil relations, under load

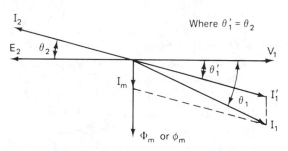

(c) Primary coil relations, under load

Figure 14-3. Ideal transformer phasor relations.

In turn, the phasor sum of I_1' and I_m grows until sufficient current is drawn to support the load and maintain the magnetizing ampere turns.

The idealized but nonetheless nearly realistic phasor relations described contain one very fundamental transformer relation, which is the equality between the secondary demagnetizing ampere turns and the primary magneto motive force ampere turns I_1N_1. This can be equated as shown:

$$\boxed{I_1'N_1 = I_2N_2}$$

Eq. (14-1)

This equation, when cross multiplied to produce a different form, shows another fundamental transformer relationship:

$$\boxed{\frac{I_2}{I_1'} = \frac{N_1}{N_2} = \alpha} \qquad \frac{V_1}{V_2} = \frac{N_1}{N_2}$$

Eq. 14-2

Here α is the *transformation ratio* or ratio of primary coil turns to secondary coil turns. This is popularly used as the *turns ratio*. The other elements of the equation were described and defined above.

The turns ratio is a fixed quantity depending on the actual number of turns in the winding coils as the transformer is wound and connected. It is not a constant in a fundamental sense, but is rather a built-in fixed ratio.

Examples will show the utility of this number α.

Example 14-1. The high-voltage coil of a transformer is wound with 700 turns of wire, and the low-voltage coil is wound with 292 turns. When used as a step-up transformer (the low-voltage coil used as the primary), the load current is 10.5 A. Find (a) the transformation ratio α, and (b) the I_1' current, which is the *load component* of the primary current.

Solution:
(a) Using Eq. (14.2),

$$\frac{N_1}{N_2} = \alpha = \frac{292}{700} = \alpha = 0.417$$

(b) Using Eq. (14-2), this time the current portion,

$$\frac{I_2}{I_1'} = \alpha \quad \text{or} \quad I_1' = \frac{I_2}{\alpha}$$

$$I_1' = \frac{10.5}{0.417} = 25.2 \text{ A}$$

In a real situation, the actual I_1 current will be a few percent larger than I_1'. This is in order to allow for the component I_m. There will also be added currents necessary to account for leakage of magnetic flux, because the full Φ_m (or ϕ_m) is not really all that is present. This simplification is sufficiently good for a first approximation. The high-voltage side always has the smaller current, and vice versa.

Example 14-2. Using the same transformer as in Example 14-1, calculate its transformation ratio when used as a step-down transformer.

Solution: Again use Eq. (14-2), but this time the high-voltage winding is the primary. N_1 turns is now 700.

$$\frac{N_1}{N_2} = \alpha = \frac{700}{292} = 2.40$$

Equation (14-2) then works both ways, since the subscript 1 means primary and subscript 2 means secondary, but the transformer can be used either for step-up or step-down functions. The transformation ratio α then has two different values for a given transformer, depending on how it is used. These two numbers are reciprocals. Thus, $1/0.417 = 2.40$ or $1/2.40 = 0.417$.

Since the step-up or step-down term refers to *voltage*, the transformation ratio or turns ratio α can also refer to voltages.

Referring back to the discussion of Eq. (2-1), it was stated that the voltage developed was seen to be directly proportional to the *rate of change* of the *linked* lines of force. In differential calculus this means that the voltage is proportional to $\frac{d\Phi}{dt}$ or to $\frac{d\phi}{dt}$. Avoiding the calculus, but remembering the progressive development of Eqs. (2-1) through (2-5), and by the same type of reasoning, we have

$$E_1 = N_1\frac{d\Phi_m}{dt} \times 10^{-8} \quad \text{and} \quad E_2 = N_2\frac{d\Phi_m}{dt} \times 10^{-8}$$

or in SI units

$$E_1 = N_1\frac{d\phi_m}{dt} \quad \text{and} \quad E_2 = N_2\frac{d\phi_m}{dt}$$

In either situation, if we divide the first equation by the second, the differential and constant terms cancel out:

$$E_1 = N_1\frac{d\Phi_m}{dt} \times 10^{-8} \bigg/ E_2 = N_2\frac{d\Phi_m}{dt} \times 10^{-8}$$

which then produces

$$\frac{E_1}{E_2} = \frac{N_1}{N_2}$$

Similarly, with the SI situation,

$$E = N_1\frac{d\phi_m}{dt} \bigg/ E_2 = N_2\frac{d\phi_m}{dt}$$

which also produces

$$\frac{E_1}{E_2} = \frac{N_1}{N_2}$$

Quite simply, the *induced* voltage ratio is proportional to the turns ratio and thus equal to α, since α has been defined as N_1/N_2. Thus

$$\alpha = \frac{N_1}{N_2} = \frac{E_1}{E_2}$$

From this and Eq. (14-2), we then can say

$$\alpha = \frac{N_1}{N_2} = \frac{E_1}{E_2} = \frac{I_2}{I_1'}$$

Finally, we may also consider that with an ideal transformer the induced voltage E_1 equals the applied voltage V_1. The delivered *output* voltage V_2 equals the induced secondary voltage E_2 so that one more term appears:

$$\alpha = \frac{N_1}{N_2} = \frac{E_1}{E_2} = \frac{I_2}{I_1'} = \frac{V_1}{V_2}$$

Eq. (14-3)

In a *practical* transformer it will be shown that the induced E_1 is *less* than V_1 by a small amount. The output voltage V_2 is less than E_2. The relations really are as shown below and are only approximate:

$$E_1 < V_1, \quad \text{but as an approximation } E_1 \cong V_1$$

$$E_2 > V_2, \quad \text{but as an approximation } E_2 \cong V_2$$

Since these inequalities are not in the same direction, we will subsequently find that the voltages are approximately related: $E_1/E_2 \cong V_1/V_2$. But this is still a useful working approximation.

Still in an ideal transformer situation, we may see from *part* of Eq. (14-3) that

$$\frac{E_1}{E_2} = \frac{I_2}{I_1'} \quad \text{so that} \quad E_1 I_1' = E_2 I_2$$

Eq. (14-4)

The conclusion from this equation is that in an ideal transformer the volt-amperes in the primary circuit equals the volt-amperes in the secondary circuit if we neglect I_m. If I_1' is much larger than I_m, we may say that $E_1 I_1 \cong E_2 I_2$.

Finally, for an ideal transformer with no flux losses, no I^2R losses, and negligible magnetizing current,

$$V_1 I_1 \cong V_2 I_2$$

Eq. (14-5)

There are a few more assumptions in Eq. (14-5), but the output power VI cos θ does approximately equal the input power. In a real transformer of *very large size*, the difference between the output power and the input power is around 1 percent. Even in small to medium sizes, less than a 10 percent loss is not unusual, so that the approximations in Eqs. (14-3), (14-4), and (14-5) are very useful. Notice that these equations are in volt-ampere terms. Transformers are rated in volt-amperes or kilovolt-amperes rather than watts. The *volts per turn* is a *constant* since each turn is cut by the same flux.

Example 14-3. A 4600 to 230 V, 60 Hz, 7.5 kVA transformer is rated at 2.6 volts per turn of its winding coils. Assume that it is an ideal transformer and calculate the following:

  (a) Step-up transformation ratio.
  (b) Step-down transformation ratio.
  (c) Total turns of the high voltage coil.
  (d) Total low-voltage turns.
  (e) Primary current as a step down.
  (f) Secondary current as a step down.

Solution:
  (a) As a step-up transformer, the primary voltage V_1 = 230 V. From Eq. (14-3), V_1/V_2 = 230/4600 = α = 0.05.
  (b) As a step-down transformer the primary voltage, V_1 = 4600 V, so V_1/V_2 = 4600/230 = α = 20.
  (c) At 2.6 V per turn rated, we can state this as 2.6(V/t). We want *turns* in the denominator, so we simply take the reciprocal ($1t/2.6$ V) and multiply by 4600 V. Note that a turn is given the unit t, but the *number* of turns is given the unit N by conventional use. *Volts cancel*:

  (d)
$$\frac{4600 \text{ V}t}{2.6 \text{ V}} = N_h = 1769 \text{ turns}$$

$$\frac{1t(230 \text{ V})}{2.6 \text{ V}} = N_x = 88.5 \text{ turns}$$

Actually, turns are usually made as integer numbers, although it is perfectly possible to bring a finish end of a coil 180° around the bobbin from the start.
  (e) As a step-down transformer, the 4600 V coil is the primary.

$$\frac{7.5 \text{ kVA}(1000 \text{ VA})}{(4600 \text{ V})\text{kVA}} = I_1 = 1.63 \text{ A}$$

  (f) 230 V is the secondary rating:

$$\frac{7500 \text{ VA}}{230 \text{ V}} = I_2 = 32.6 \text{ A}$$

14-1-2 General Transformer Equation. To develop the basic voltage relationships in a transformer on a basis that does not involve the turns ratio, we start with Eqs. (2-1$_E$) and (2-1$_{SI}$). These are:

$$E_{av} = \frac{\Phi}{t} \times 10^{-8} \frac{\text{volts}}{\text{turn}} \qquad \text{Eq. (2-1}_E\text{)}$$

$$E_{av} = \frac{\phi}{t} \quad \frac{\text{volts}}{\text{turn}} \qquad \text{Eq. (2-1}_{SI}\text{)}$$

These equations related a *steady value* of magnetic flux, Φ (or ϕ), depending on units, with a time to link that flux. As a result, the voltage produced was given as E_{av}, since nothing was assumed about the uniformity or lack thereof of the velocity required to traverse the flux in time t. If we now realize that the winding turn does not move, but that the flux changes sinusoidally and cyclically from maximum through zero and to an opposite polarity maximum, we can then develop the following:

$$E_{av} = 4\Phi_{pm}fN \times 10^{-8} \quad \text{and} \quad E_{av} = 4\phi_{pm}fN$$

Here the terms Φ_{pm} (and ϕ_{pm}) refer to the *peak mutual* flux in the appropriate units of lines or webers. The term 4 simply recognizes the fact that one full cycle of flux variation requires going from zero to maximum flux, or maximum to zero, a total of four times per cycle. The f is the cyclic frequency of flux variation, consistent with Chapter 2 as the dc voltage relations were developed. The N term introduces the number of turns per coil.

Since the voltage term wanted is the root mean squared rather than the average, the exact relation needed is 4.4428 instead of 4. This development is traced in Sect. 12-1. The equations then become

$$\boxed{E = 4.4428f\Phi_{pm}N \times 10^{-8}} \qquad \text{Eq. (14-6}_E\text{)}$$

and

$$E = 4.4428f\phi_{pm}N \qquad \text{Eq. (14-6}_{SI}\text{)}$$

where

Φ_{pm} = peak mutual flux in lines for English units
ϕ_{pm} = peak mutual flux in webers for SI units
N = the number of turns in coil under consideration; this may be N_1 for primary, N_2 for secondary, N_x for low-voltage coil, or N_h for high-voltage coil
f = cyclic frequency in hertz

These are the *general transformer equations* in the English and System International units. The peak flux terms have the same real limits as in other dc or ac machinery. This is a limitation of the magnetic material chosen. See Figs. 3-16 and 3-17 for magnetic steel magnetization curves in the two unit systems.

Quite obviously, as a transformer is made larger, it has room for a larger cross section core. This means that Φ_{pm} and ϕ_{pm} are related to the physical size of the unit in question. The higher the total flux, the less turns are needed to induce a given voltage. This pair of equations works both ways in that the voltage can be specified and the number of turns or the maximum flux can be determined. Conversely, if the flux and turns are specified, the induced voltage can be determined.

The peak magnetic flux that is chosen is normally not far around the knee of the curve (see Figs. 3-16 and 3-17), since a nonlinear change in flux with the primary coil current will introduce a nonlinear relation between applied voltage and induced voltage. This will result in appreciable harmonics in the transformer output.

As a result of being limited in maximum usable *flux*, a transformer is also limited in its usable frequency. Since the maximum permissible magnetic flux density may not be exceeded, a reduction in frequency *must* be accompanied by a reduction in the applied *voltage*.

The fact that we are still working with *ideal* transformer relations should be kept in mind while some illustrative examples are followed.

Example 14-4. A 4600 to 230 V, 60 Hz step-down transformer has core dimension of 3.00 by 4.40 in. (76.2 mm by 111.8 mm). A maximum flux density of 60 000 lines/in.^{2} (0.930 Wb/m^2) is to be used. Calculate the following, assuming 9 percent loss of area due to stacking factor of laminations: (a) primary turns required, (b) turns per volt, (c) secondary turns required, and (d) transformation ratio.

Solution:
(a) The flux units given are in B (or β) lines/in.^{2} (or webers/square meter) so that the core area must be determined and then the total flux Φ (or ϕ) calculated.

$$3 \text{ in. } (4.4 \text{ in.})0.91 = 12.0 \text{ in.}^2$$

$$12 \text{ in.}^2 \frac{60\,000 \text{ lines}}{\text{in.}^2} = \Phi = 720\,000 \text{ lines}$$

In SI remember that, although mechanical dimensions are normally given in *millimeters*, calculations are performed in *meters*.

$$0.0762 \text{ m}(0.1118 \text{ m})0.91 = 0.007\,752 \text{ m}^2$$

$$0.007\,752 \text{ m}^2 \frac{0.930\,00 \text{ Wb}}{\text{m}^2} = \phi = 0.0072 \text{ Wb}$$

Transposing the equations to solve for turns N, we have, for Eq. (14-6$_E$),

$$N = \frac{E \times 10^8}{4.4428 f \Phi_{pm}}$$

and for Eq. (14-6$_{SI}$),

$$N = \frac{E}{4.4428f\phi_{pm}}$$

Solving by substituting available data,

$$N_p = \frac{4600 \times 10^8}{4.4428(60)720\ 000} = \frac{4600}{4.4428(0.6)0.72}$$

$$= 2397 \text{ turns}$$

or in SI,

$$N_p = \frac{4600}{4.4428(60)0.0072} = 2397 \text{ turns}$$

The answers in the two unit approaches are the same, as of course they should be, since the input data had the same absolute value.

(b) 2397 t/4600 V = 0.521t/V.

(c) The secondary turns are found from the same turns per volt: 230 V (0.521 t/V) = N_s = 120 turns.

(d) The transformation ratio can be found from the ratio of voltages or the ratio of turns:

$$\frac{4600}{230} = \alpha = 20 \quad \text{or} \quad \frac{2397}{120} = \alpha \cong 20$$

If the turns calculation had used rounded-off values of flux, the results would have been exactly the same. With rounded-off values in the second case, α = 19.975, which is 20.0 to the nearest three significant figures.

The transformer specified is a fairly large one owing to its core size. A bit of thought will show that had its core area been less, say 0.1 times the area specified, the required turns would have been 10 times as much. This is so because 0.1 times the area would have meant 0.1 times the peak mutual flux. This in turn is reasonable, since maximum *flux density*, or *flux per unit area*, is an inherent quality of the magnetic steel.

Similarly, since the frequency f is also a linear component of the basic equations, Eqs. (14-6$_E$) or (14-6$_{SI}$), if the frequency is changed, the voltage must change in a given transformer. A transformer designed for a low frequency can be used at a high frequency. If the input voltage is not changed, the peak magnetic flux will be less. The other way around, if the *frequency* is *reduced*, the voltage *must also decrease*. This is so because the magnetic flux *cannot* substantially increase. Example (14-5) will show this.

Example 14-5. A 120 V to 27.5 V, 400 Hz step-down transformer is to be operated at 60 Hz. Find (a) the highest safe input voltage, and (b) the transformation ratio in both frequency situations.

Solution:

(a) Since the peak mutual flux cannot increase, the allowable voltage must decrease in proportion, because the *general transformer equation* must fit in either situation: $120(60/400) = E_{p60} = 18$ V.

(b) The 27.5 V secondary must decrease by the *same ratio* since the volts per turn are the same for both primary and secondary. Therefore, the transformation ratio α is the *same* in either case. Thus, using part of Eq. (14-3),

$$\alpha = \frac{V_1}{V_2} = \frac{120}{27.5} = 4.36$$

The general transformer equation does not specify the current involved. However, the magnetization of the core requires ampere turns through the coil. The amperes are the basic I_m or magnetization current of Fig. 14-3. This current is related to the lamination steel used and how far beyond the true linear portion of the magnetization curve the designer cares to go. Single-phase transformers will develop severe harmonics in the output voltage and current if the operating point of Φ_{pm} (or ϕ_{pm}) is too far up the curve. In three-phase transformer circuits, the third harmonic largely cancels. Therefore, higher magnetization can be used.

Working with the situation in Example 14-4, we find the following:

Example 14-6. The transformer in Example 14-4 is operated at no load or with an open secondary. The mean magnetic path length is 48.00 in. (1219 mm). Find the current required to maintain the core magnetization.

Solution: The specified magnetic flux densities in Example 14-4 are entered into Figs. 3-16 and 3-17 to find the required ampere turns per inch (or per meter) of the core length. For $B = 60\,000$ lines/in.^{2}, $H = 10.35$ AT/in.; for $\beta = 0.930$ Wb/m^2, $H = 420$ AT/m. The curves for annealed sheet silicon steel were used in each case. In English units, the required ampere turns is $10.35 \times 48 = 497$ AT. In SI, $405 \times 1.219 = 494$ turns. These figures should agree, but it is not possible to interpret the curves to sufficient accuracy.

The peak magnetization takes place during the *peak current*, but ac currents are rated in root mean squared (rms) values.

Since the number of ampere turns required is *about* 495 and the number of primary turns is 2397 (from Example 14-4a), 495 AT/2397 T $= 0.207$ peak amperes. Since the rms value is $1/\sqrt{2} = 0.707\,11$ times the peak, $0.707\,11 \times 0.207 = I_m = 0.146$ A.

Remember that this is still under ideal transformer conditions and there will be some loss. A transformer of this size would have a rating in the 15 to 20 kVA range, so that its rated I_p current would be in the 2 to 3 A range.

14-2 PRACTICAL TRANSFORMER CONDITIONS

All along in the discussion on ideal transformers it has been hinted that the ideal was not really achieved. One basic problem is that the inductive coupling between

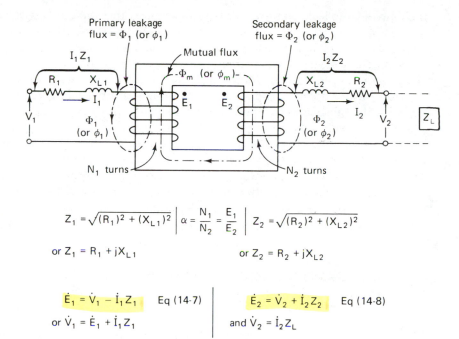

$$Z_1 = \sqrt{(R_1)^2 + (X_{L1})^2} \quad \bigg| \quad \alpha = \frac{N_1}{N_2} = \frac{E_1}{E_2} \quad \bigg| \quad Z_2 = \sqrt{(R_2)^2 + (X_{L2})^2}$$

or $Z_1 = R_1 + jX_{L1}$ \qquad\qquad\qquad or $Z_2 = R_2 + jX_{L2}$

$$\dot{E}_1 = \dot{V}_1 - \dot{I}_1 Z_1 \quad \text{Eq (14-7)} \qquad \dot{E}_2 = \dot{V}_2 + \dot{I}_2 Z_2 \quad \text{Eq (14-8)}$$

or $\dot{V}_1 = \dot{E}_1 + \dot{I}_1 Z_1$ \qquad\qquad and $\dot{V}_2 = \dot{I}_2 Z_L$

Figure 14-4. Practical transformer circuit relations.

the coil windings is not quite perfect. In spite of all precautions in the design of the magnetic circuit, there is some small magnetic flux leakage from each coil. Each coil has some resistance, which produces I^2R losses. Each coil has inductance by virtue of its construction so that the currents passing through the coils see IZ voltage drop through the impedance. A loaded practical transformer then corresponds to Fig. 14-4.

The new terms introduced in Fig. 14-4 over those in Fig. 14-2 are the following:

(1) The Φ_1 (or ϕ_1) primary leakage flux.
(2) Φ_2 (or ϕ_2) secondary leakage flux.
(3) R_1, primary coil resistance.
(4) X_{L1}, primary coil inductive leakage reactance due to leakage flux.
(5) R_2, secondary coil resistance.
(6) X_{L2}, secondary coil inductive leakage reactance due to leakage flux.

The R and X components are really combined in the same wire length of the coils. These are simply the real-world circuit values that must exist, since it is not possible to build coils around a magnetic core without some resistance and substantial inductance. These values are minimized by careful construction, as are the unavoidable leakage fluxes. One can see now that the shell construction will nearly eliminate the leakage fluxes. However, the coils will then be longer (have a longer mean coil length) and the resistance inevitably will be greater. As in any other practical device, the actual physical design is a compromise of many factors. Here again, improvements in the properties of lamination steel directly improve the transformer design. Higher attainable flux density would allow less winding turns for the same-sized core owing to a larger total flux in a smaller physical size. This effect would reduce the winding size and resistance.

Consulting Eq. (14-6) in its various versions will show that it is not easy to directly relate the flux terms from external voltage and current measurements. These equations relate the internally induced voltage and the number of turns in the windings. Possibly, neither value is known. However, from Fig. 14-4 and ordinary ac circuit theory, it can be seen that the following instantaneous relations are true if correct phasor relations are used:

$$\dot{E}_1 = \dot{V}_1 - \dot{I}_1 Z_1 \quad \text{or} \quad \dot{V}_1 = \dot{E}_1 + \dot{I}_1 Z_1 \qquad \text{Eq. (14-7)}$$

and
$$\dot{E}_2 = \dot{V}_2 + \dot{I}_2 Z_2 \qquad \text{Eq. (14-8)}$$

These relations can be expanded to the following instantaneous equations:

$$\dot{E}_1 = \dot{V}_1 - \dot{I}_1 \sqrt{(R_1)^2 + (X_{L1})^2} \qquad \text{Eq. (14-9)}$$

and
$$\dot{E}_2 = \dot{V}_2 + \dot{I}_2 \sqrt{(R_2)^2 + (X_{L2})^2} \qquad \text{Eq. (14-10)}$$

So far these sums do not accurately account for power factor effects. However, it can be seen that for a *loaded* practical transformer with lagging PF,

$$V_1 > E_1 \text{ and } E_2 > V_2 \qquad \text{Eq. (14-11)}$$

The simple algebraic addition or substraction of the IZ terms corresponds to the worst possible power factor, but is a *useful first approximation*:

Example 14-7. A 2300 V to 230 V step-down transformer is rated at 750 kVA and 60 Hz. Its windings have the following resistances and inductances: $R_1 = 0.093\ \Omega$, $X_{L1} = 0.280\ \Omega$, $R_2 = 0.000\ 93\ \Omega$, and $X_{L2} = 0.002\ 80\ \Omega$. The transformer is operating at rated load. Calculate the following:
   (a) Primary and secondary currents.
   (b) Primary and secondary winding impedances.
   (c) Primary and secondary winding voltage drops.
   (d) Primary and secondary induced volatages.
   (e) The transformation ratio (turns ratio).
   (f) Ratio of terminal voltages.

Solution:

(a) $I_1 = \dfrac{\text{kVA} (1000)}{V_1} = \dfrac{750\ 000}{2300} = 326.1\ \text{A}$

$I_2 = \dfrac{750\ 000}{230} = 3261\ \text{A}$

Note that I_1, when determined in this fashion, assumes no losses in the transformer.

(b) $Z_1 = \sqrt{(R_1)^2 + (X_{L1})^2} = \sqrt{(0.093)^2 + (0.280)^2} = 0.295\ \Omega$

$Z_2 = \sqrt{(R_2)^2 + (X_{L2})^2} = \sqrt{(0.000\ 93)^2 + (0.002\ 80)^2} = 0.002\ 95\ \Omega$

(c) $I_1 Z_1 = 326.1(0.295) = 96.2\ \text{V}$

$I_2 Z_2 = 3261(0.002\ 95) = 9.62\ \text{V}$

(d) Using Eqs. (14-7) and 14-8),

$$E_1 = V_1 - I_1 Z_1 = 2300 - 96.2 = 2204\ \text{V}$$

$$E_2 = V_2 + I_2 Z_2 = 230 + 9.62 = 239.6\ \text{V}$$

(e) Transformation ratio α was defined as the ratio of N_1/N_2 and E_1/E_2. The discussion under Eq. (14-3) showed that the ratio V_1/V_2 was only an approximation, but E_1/E_2 is more definitive.

$$\alpha = \frac{E_1}{E_2} = \frac{2204}{239.6} = 9.198$$

(f) $\dfrac{V_1}{V_2} = \dfrac{2300}{230} = 10$

Note that, to *achieve a terminal voltage ratio* of 10, a *transformation ratio* α of 9.198 appears to be necessary with this type of algebraic simplification. This simply means that a transformer actually has a *turns ratio* numerically *less* than the terminal voltage ratio. The amount of difference depends upon the winding impedance voltage drops of the transformer in question. Note that, if this *same* transformer were to operate as a 230/2300 V *step-up* unit, the same currents would hold for the same rating. This time the primary would be 230 V and 3261 A. It would be extremely unlikely that this particular large unit would be used as a step-up transformer since such high currents at the 230 V level are not usually obtainable. However, a transformer can be *used* either way within its ratings. The only real problem is that *either way* the internal impedances act to *drop* the available voltages. To achieve a 230/2300 V step up would require an α of *smaller* than 0.1, or in the region of 0.093.

Note also that the *same* kilovolt-amperes was assumed for both the input and output coils. This is *not true* since there are losses due to the internal resistances and magnetic losses. These losses will be determined in the next chapter.

Example 14-8. Using the same transformer as in Example 14-7 as a step-up unit with the rated kilovolt-amperes at the low-voltage winding, determine the probable secondary voltage.

Solution: The same low-voltage winding terminal voltage is now the input, and this is 230 V. The same low-voltage IZ voltage drop exists, but it is now the I_1Z_1 drop since the transformer is used the other way around. Thus, E_1, which is now the low-voltage side induced voltage, is from Eq. (14-7):

$$E_1 = V_1 - I_1Z_1 = 230 - 9.62 = 220.4 \text{ V}$$

This time the induced voltage ratio is the *reciprocal* of 9.198. Thus, $\alpha = 1/9.198 = 0.1087$. This is also the turns ratio and thus the E_1/E_2 ratio.

The E_2 or secondary induced voltage is then

$$E_2 = \frac{E_1}{\alpha} = \frac{220.4}{0.1087} = 2028 \text{ V}$$

Since the secondary is the high-voltage winding, Eq. (14-8) may be used, but it will need transposing.

$$E_2 = V_2 + I_2Z_2 \quad \text{or} \quad V_2 = E_2 - I_2Z_2$$

Thus, $2028 - 96.2 \text{ V} = V_2 = 1932 \text{ V}$

The same transformer used the other way around, or step up instead of step down, *does not* achieve the reciprocal voltage ratio when loaded. It can be used this way but 1932 V is a poor substitute for 2300 V. This unit with an α of 9.198 achieves a 10-to-1 terminal voltage ratio under step-down conditions at rated load. When used the other way as a step up, with the α now the reciprocal of the original, or $\alpha = 0.1087$, the resulting terminal voltage ratio is 230/1932 = 0.1190 instead of the hoped for 0.1. Comparable ratios of performance hold for any transformer under reversed conditions. As a result, many transformers have *tapped windings* or a choice of various turns ratios.

QUESTIONS

14-1. What is transformer action?

14-2. What advantageous use is made of transformers in long-distance electric power transmission?

14-3. What is the difference between the primary coil and the secondary coil of a transformer?

14-4. What is a step-down transformer as distinguished from a step-up transformer?

14-5. In what manner are the primary and secondary coils of a transformer mutually connected?

14-6. What is meant by leakage flux?

14-7. What is meant by mutual flux?

14-8. How is voltage induced in transformer coils?

14-9. What is the magnetizing current in a transformer?

14-10. Does the sinusoidal magnetic flux level vary with load or remain at the same level in a transformer? Why?

14-11. What is the meaning of transformation ratio?

14-12. State Eq. (14-3) in your own words.

14-13. Why is volts per turn a constant?

14-14. Why must a change in frequency be accompanied by a change of voltage in a transformer?

14-15. What two features distinguish a practical transformer from an ideal transformer?

14-16. Why should the peak magnetic flux density be not far beyond the linear portion of the transformer lamination material B-H curve?

PROBLEMS

14-1. A transformer has 120 primary turns and 720 secondary turns. If its load current is 0.833 A, what is its primary current load component?

14-2. What is the turns ratio of the transformer in Problem 14-1?

14-3. What would be the turns ratio of the transformer in Problem 14-1 if the 720-turn coil were used as the primary?

14-4. If an ideal transformer has a turns ratio of 10 and a primary line voltage of 230 V, what is its secondary voltage?

14-5. In an ideal transformer situation, if the output voltage is 120 V at 8.333 A and the input voltage is 240 V, what is the input current?

14-6. A 2300 to 230 V, 60 Hz, 2 kVA transformer is rated at 1.257 V/turn of its winding coils. Assume it is an ideal transformer and calculate:
(a) Step-down transformation ratio.
(b) Total turns of the high voltage coil.
(c) Total turns of the low voltage coil.

14-7. Using the transformer of Problem 14-6, what is its secondary current?

14-8. With the transformer in Problem 14-6, what would be its peak magnetic flux in the core (a) using English units and (b) using SI units?

14-9. The transformer of Problem 14-6 is to be considered as a practical transformer. Its windings have the following R and L: $R_1 = 9.1\ \Omega$, $X_{L1} = 28.4\ \Omega$, $R_2 = 0.091\ \Omega$, and $X_{L2} = 0.284\ \Omega$. If it is operating at rated load, calculate:
(a) Primary current.
(b) Primary winding voltage drop.
(c) Secondary winding voltage drop.

14-10. Again using the transformer of Problem 14-6, calculate as *first approxi-mations*:
  (a) Primary induced voltage.
  (b) Secondary induced voltage.
  (c) The transformation ratio.

Answers. **14-1**, 5.0 A; **14-2**, 0.1667; **14-3**, 6; **14-4**, 23 V; **14-5**, 4.167 A; **14-6(a)**, 10, **(b)** 1830 turns, **(c)** 183 turns; **14-7**, 8.696 A; **14-8(a)**, 0.4715×10^6 lines. **(b)** 0.004 715 Wb; **14-9(a)**, 0.8696 A, **(b)** 25.9 V, **(c)** 2.59 V; **14-10(a)**, 2274 V, **(b)** 232.6 V, **(c)** 9.78.

15

Transformer Equivalent Circuits

To perform practical measurements on a transformer that lead to the ability to predict performance, some further simplifications of the circuit are desirable. The ability to lump together the equivalent combined circuit of a primary and secondary is useful. This depends upon the concept of *reflected impedance*.

15-1 REFLECTED IMPEDANCE

Consulting Fig. 14-4 and using the nomenclature shown there, we may develop an impedance reflection scheme as follows:

(1) If the load impedance Z_L is removed so that the secondary is open circuit, $I_2 = 0$ and $Z_L = \infty$.

(2) Looking back into the transformer secondary terminals, the impedance is $Z_2 = V_2/I_2$.

(3) Looking into the primary side terminals, the impedance is $Z_1 = V_1/I_1'$.

(4) A change in the load current I_2 is reflected by a change in the primary current. Since the primary and secondary are thus related, we can use a single equivalent circuit with the secondary *reflected* to the primary. In this case, from Eq. (14-3), $V_1 = \alpha V_2$ and $I_1' = I_2$. Thus,

$$\frac{\alpha V_2}{I_2/\alpha} = Z_1 = \alpha^2 \frac{V_2}{I_2}$$

Since $Z_2 = V_2/I_2$, then Z_1 in turn is $Z_1 = \alpha^2 Z_2$ or

$$\alpha^2 = \frac{Z_1}{Z_2}$$ Eq. (15-1)

From Eq. (14-2) squared,

$$\alpha^2 = \left(\frac{N_1}{N_2}\right)^2$$ Eq. (15-2)

This assumes now that the secondary winding impedance is very small or negligible compared to the load impedance. The whole secondary impedance, load and all, has been transferred to the primary. The fact that the ratio of the input to output impedance is equal to the transformation ratio squared is of vital importance to the use of transformers in the communication industry. Here, in electrical power equipment, it is of great use in transformer performance calculations.

Primary and secondary impedances are related to Ohm's law so that the following holds:

(1) The high-voltage side has relatively *lower* current and thus, from $V/I = Z$, *higher* impedance.

(2) The low-voltage side has relatively *higher* current and thus, from $V/I = Z$, *lower* impedance.

(3) When reflected to the high-voltage side, impedance is higher, and vice versa.

Example 15-1. Use the same transformer as in Example 14-7 for comparison purposes. Here the turns ratio $\alpha = 9.198$ and the input voltage $V_1 = 2300$ V. A load of $Z_L = 0.070\,53\ \Omega$ impedance is connected to the secondary. Determine the following using impedance reflection:
   (a) Secondary voltage.
   (b) Secondary current.
   (c) Primary current.
   (d) Primary input impedance from part c and given primary volts.
   (e) Primary input impedance by impedance reflection.

Solution:
(a) $V_2 = V_1/\alpha = 2300/9.198 = 250.0$ V.
(b) $I_2 = V_2/Z_L = 250.0/0.070\,53 = 3545$ A.
(c) $I_1 = I_2/\alpha = 3545/9.198 = 385.4$ A.
(d) $Z_{1L} = V_1/I_1 = 2300/385.4 = 5.968\ \Omega$.
(e) $Z_{1L} = \alpha^2 Z_L = (9.198)^2(0.070\,53) = 5.967\ \Omega$.

Note that Z_L and Z_{1L} here are load and reflected load impedances and thus are *not* the same as Example 14-7 versions of Z_1 and Z_2, which are winding impedances. The results from Example 15-1 are not too close to those in Example 14-7 since no internal losses were assumed. However, if an α that corresponded to the desired loaded voltage ratio had been used, $\alpha = 10$, instead of the *presumed* actual $\alpha = 9.198$, the results would be quite reasonable.

Example 15.2. The transformer of Examples 14-7 and 15-1 is assumed to have a turns ratio $\alpha = 10$. Use the same load impedance and calculate the same quantities.

Solution:
(a) $V_2 = V_1/\alpha = 2300/10 = 230$ V.
(b) $I_2 = V_2/Z_L = 230/0.070\ 53 = 3261$ A.
(c) $I_1 = I_2/\alpha = 3261/10 = 326.1$ A.
(d) $Z_{1L} = V_1/I_1 = 2300/326.1 = 7.053\ \Omega$.
(e) $Z_{1L} = \alpha^2 Z_L = (10)^2(0.070\ 53) = 7.053\ \Omega$.

The results of steps a, b, and c are exactly the same as in Example 14-7 since the algebraic implications are the same. This input impedance is less realistic than that of Example 15-1, but that output voltage of 250V is also unrealistic. These are practical working relations and do not give theoretically accurate results, since neither the procedures in Examples 15-1 nor 15-2 yet take any account of losses in the transformer. Example 14-7 allowed for winding losses but not for magnetic core losses.

15-2 SIMPLIFIED EQUIVALENT CIRCUITS

To predict the *efficiency* of a transformer, certain equivalent circuit assumptions that involve the reflected impedance concept can be made. Consulting Fig. 15-1a, we can see that the secondary winding resistance and inductive reactance have been reflected back to the primary, as has the load. The resulting circuit has the primary, secondary, and magnetization circuits shown in series parallel. As is shown in Fig. 14-3c, the primary current is composed of the magnetization current I_m and the load component I_1'. R_m represents the in-phase component of the magnetization current; X_{Lm} represents the inductive reactance component of the transformer with an open secondary. Remember from Fig. 14-3c that this total current I_m is almost 90° lagging in respect to the V_1 voltage.

Figure 15-1a shows a reasonable circuit for a *loaded* transformer if I_1' is a finite current. If the transformer is *unloaded*, the I_1' is zero and the right branch does not affect the circuit.

In Fig. 15-1b, the R_m and X_{Lm} block has been shifted to the input voltage V_1 side of the R_1 and X_{L1} components. This would involve the magnetizing current being fed from the full V_1 voltage without the small reduction due to the I_1Z_1 drop. Since the I_m current is already very small in relation to I_1' and at a very substantial phasor angle to it, this does not appreciably affect the I_1' current. However, it does allow the primary resistance and reactance components and the *reflected*

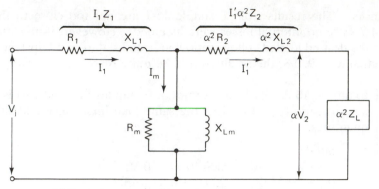

(a) Equivalent circuit of loaded power transformer

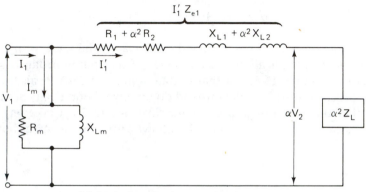

(b) Equivalent circuit approximation
with combined primary and reflected
secondary impedances

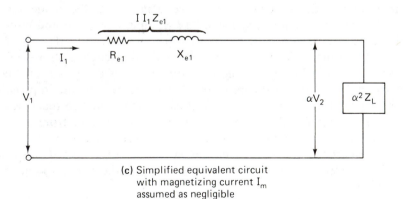

(c) Simplified equivalent circuit
with magnetizing current I_m
assumed as negligible

Figure 15-1. Practical power transformer equivalent circuit.

secondary resistance and reactance to be lumped or combined, as shown in Fig. 15-1b.

Then since the I_m current component is considered negligible when the transformer is loaded, the I_m branch can be discarded, leaving Fig. 15-1c as the final simplification. Here the following combinations have been made:

$$R_{e1} = R_1 + \alpha^2 R_2$$ Eq. (15-3)

$$X_{e1} = X_{L1} + \alpha^2 X_{L2}$$ Eq. (15-4)

which combine as phasors to become

$$Z_{e1} = \sqrt{(R_{e1})^2 + (X_{e1})^2}$$ Eq. (15-5)

or in complex form,

$$Z_{e1} = R_{e1} + jX_{e1}$$ Eq. (15-5)

Because of the assumption that I_m is minimal,

$$I_1 \cong I_1'$$

With these simplifications, we then combine the resistive and reactive components of the load:

$$I_1 = \frac{V_1}{\sqrt{(R_{e1} + \alpha^2 R_L)^2 + (X_{e1} \pm \alpha^2 X_L)^2}}$$ Eq. (15-6)

or in complex form this is

$$I_1 = \frac{V_1}{R_{e1} + \alpha^2 R_1 + j(X_{e1} \pm \alpha^2 X_L)}$$ Eq. (15-6)

The plus or minus sign in the last term of both forms of Eq. (15-6) allows for *load* power factor. The plus sign is used for lagging load power factor, the minus sign is for leading, and the $\alpha^2 X_L$ term drops out for unity power factor.

Example 15-3 compares this method to the others preceding it.

Example 15-3. Again using the transformer in Example 14-7, assume the turns ratio α to be 9.50, since an α of 9.198 did not assume proper phasor combination of the factors and $\alpha = 10$ does not allow for losses. Calculate the following:

(a) Internal resistance R_{e1} referred to the primary.
(b) Internal reactance X_{e1} referred to the primary.
(c) Internal impedance Z_{e1} referred to the primary.

(d) Secondary load impedance $Z_L = 0.070\ 53\ \Omega$ reflected to the primary. Assume that this load is entirely resistive.

(e) Primary load current at rated primary voltage.

Solution: Note that if the turns ratio *had not* been specified it would be necessary to assume $\alpha = V_1/V_2 = 2300/230 = 10$, since we cannot assume that the developed $\alpha = 9.198$ is available without repeating Example 14-7. Use $\alpha = 9.5$.

(a) $R_{e1} = R_1 + \alpha^2 R^2 = 0.093 + (9.5)^2(0.000\ 93) = 0.177\ \Omega$.

(b) $X_{e1} = X_{L1} + \alpha^2 X_{L2} = 0.280 + (9.5)^2(0.002\ 80) = 0.533\ \Omega$.

(c) $Z_{e1} = \sqrt{(R_{e1})^2 + (X_{e1})^2} = \sqrt{(0.177)^2 + (0.533)^2}$
$= \sqrt{0.3154} = 0.562\ \Omega$.

(d) $\alpha^2 Z_L = (9.5)^2(0.070\ 53) = 6.365\ \Omega$.

(e) In this unity power factor situation, there is no $\alpha^2 X_L$ term, since $\alpha^2 Z_L = \alpha^2 R_L$:

$$I_1 = \frac{V_1}{\sqrt{R_{e1} + \alpha^2 R_L)^2 + (X_{e1})^2}}$$

$$= \frac{2300}{\sqrt{(0.177 + 6.365)^2 + (0.533)^2}} = \frac{2300}{\sqrt{42.798 + 0.284}}$$

$$= \frac{2300}{\sqrt{43.081}} = \frac{2300}{6.563} = 350.4\ A$$

Notice with this approach to the calculation that the I_1 current is larger than the $I_1 = 326.1$ A of Example 14-7. The 326.1 A was simply determined from the 750 kVA rating divided by the 2300 V primary voltage. That process does not allow for losses. The process in Example 15-3 comes closer to the truth.

So far, in summary, the calculation process in Example 14-7 gives a first approximation of the primary and secondary voltage drops. The currents are good for the rated secondary, but only roughly correct for the primary.

The procedure in Example 15-2 gives a quick determination of *approximate* overall impedances and currents.

The last procedure in Example 15-3 using the equivalent circuit is closer to the truth, but it requires prior knowledge of the turns ratio. As in so many situations, use the procedures and equations that are applicable to the available data and that will produce the required accuracy. The last procedure, using Eqs. (15-3) through (15-6) and following Example 15-3, is particularly useful in finding the probable *input* current when an output *load situation* is known and when enough of the transformer parameters are known.

15-3 SECONDARY VOLTAGE PHASOR RELATIONS

Transformer impedances can be reflected either all to the primary or all to the secondary. To determine the secondary power factor and voltage regulation, it is desirable to reflect impedances to the secondary. Using reasoning that is consistant with Sects. 15-1 and 15-2, the impedance relationships become equivalent secondary relations:

$$R_{e2} = R_2 + \frac{R_1}{\alpha^2}$$

Eq. (15-7)

$$X_{e2} = X_{L2} + \frac{X_{L1}}{\alpha^2}$$

Eq. (15-8)

The equivalent impedance when reflected to the secondary is

$$Z_{e2} = \sqrt{(R_{e2})^2 + (X_{e2})^2}$$

Eq. (15-9)

or in complex form this is

$$Z_{e2} = R_{e2} + jX_{e2}$$

Eq. (15-9)

The space relation of these various impedances is shown in Fig. 15-2d. The phasors expressed in Eqs. (15-3), (15-4), and (15-5) could have been drawn in a similar fashion.

15-3-1 Unity Power Factor Voltage Relations.
When the phasors are combined in diagram form as in Fig. 15-2, the *voltage regulation* of a power transformer can be calculated. With a unity power factor load, Fig. 15-2a shows that the output current I_2 is drawn in line with the output terminal voltage V_2. Thus, the *voltage phasor* I_2R_{e2} is also in line with the output voltage V_2. Since the combined or reflected primary and secondary impedances must be accounted for, the I_2Z_{e2} phasor is placed as shown. Note from Fig. 15-2d that Z_{e2} is the combined impedance of *both* windings. The relationship between the input voltage V_1 and the output voltage V_2 is then found from the trigonometric relations of Fig. 15-2a:

$$\frac{V_1}{\alpha} = \sqrt{(V_2 + I_2R_{e2})^2 + (I_2X_{e2})^2}$$

Eq. (15-10)

It can be seen that this is very similar to Eq. (13-4), which was developed for the voltage regulation of a synchronous alternator under unity power factor conditions. As can be foreseen from the other power factor conditions in Fig. 15-2, a general similarity will follow.

15-3-2 Lagging Power Factor Voltage Relations.
When the *load* power factor is lagging, the load current I_2 lags behind the load voltage V_2 by the power factor angle θ, as shown in Fig. 15-2b. Since the current through the secondary winding is in phase with the load current, the voltage phasor I_2R_{e2} is shown parallel

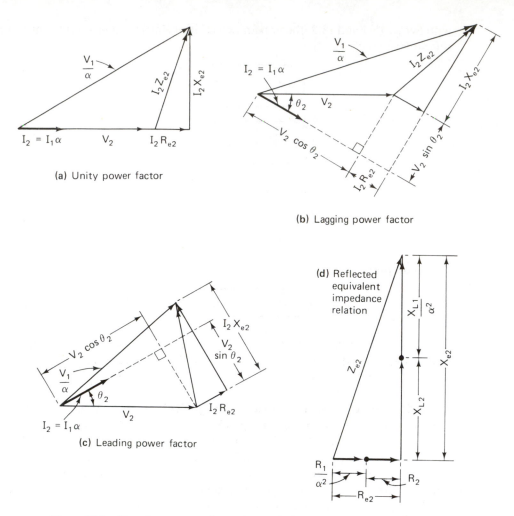

Figure 15-2. Transformer secondary voltage relations with various load power factors.

to the load current I_2. This establishes the equivalent impedance triangle relationship to the V_2 phasor. The trigonometric relationship can then be followed from Fig. 15-2b. Here the trigonometry is very similar to that in Fig. 13-3b and Eq. (13-5):

$$\frac{V_1}{a} = \sqrt{(V_2 \cos \theta_2 + I_2 R_{e2})^2 + (V_2 \sin \theta_2 + I_2 X_{e2})^2}$$

Eq. (15-11)

Note that this time the voltage phasor V_2 was drawn horizontally. This is the conventional practice in diagrams showing power factor angles. Figure 13-3b was drawn with the $I_a R_a$ phasor horizontal because it was felt that by doing so the trigonometric visualization was simplified. In this case, with some experience with this type of representation, the conventional diagram attitude has been followed.

410 TRANSFORMER EQUIVALENT CIRCUITS CHAP. 15

15-3-3 Leading Power Factor Voltage Relations. When the *load* power is a *leading* situation, the voltage phasor diagram assumes the relationship shown in Fig. 15-2c. Here, since the load current *leads* the load voltage, I_2 is shown leading V_2, and V_2 is still shown horizontally in a reference position. The trigonometric situation is related to Fig. 13-3c and to Eq. (13-6):

$$\frac{V_1}{\alpha} = \sqrt{(V_2 \cos \theta + I_2 R_{e2})^2 + (V_2 \sin \theta_2 - I_2 X_{e2})^2} \qquad \text{Eq. (15-12)}$$

The minus occurs in the last term because the $V_2 \sin \theta_2$ phasor sense is opposite to the $I_2 X_{e2}$ phasor. Note the similarity to Fig. 13-3c, where if the power factor angle, here θ_2, had been *larger* the θ_2 phasor might well be above the V_1/α phasor. Also, if θ_2 is sufficiently large an angle, the V_1/α phasor may become shorter than the V_2 phasor.

The relations in all three power factor situations involve the same reflected equivalent impedance triangle.

15-4 TRANSFORMER VOLTAGE REGULATION BY PHASOR RELATIONS

Again the similarity of transformer voltage regulation to synchronous alternator voltage regulations extends to a combined equation. Equations (15-10), (15-11), and (15-12) may be combined as shown:

$$\frac{V_1}{\alpha} = \sqrt{(V_2 \cos \theta_2 + I_2 R_{e2})^2 + (V_2 \sin \theta_2 \pm I_2 X_{e2})^2} \qquad \text{Eq. (15-13)}$$

Just as in similar situations before, this equation can be expressed in complex fashion:

$$\frac{V_1}{\alpha} = V_2 \cos \theta_2 + I_2 R_{e2} + j(V_2 \sin \theta_2 \pm I_2 X_{e2}) \qquad \text{Eq. (15-13)}$$

Here, as before in Eq. (13-7), the last term is *plus* for *unity* and *lagging* power factors, and *minus* for *leading* power factors. Again, if the power factor is *unity*, the $\cos \theta_2$ term is unity and the $\sin \theta_2$ term is zero, so that the whole $V_2 \sin \theta_2$ term drops out. This condition forces Eq. (15-13) to become Eq. (15-10).

Be aware that the relationships in Eq. (15-13) and Eqs. (15-10), (15-11), and (15-12) can be performed by scaled diagrams without use of calculations.

The actual percent voltage relation is related to previous voltage regulation equations, but is modified by the transformation ratio or turns ratio α. This is because the output voltage of a transformer is not intended to match the input voltages, but rather the input voltage times the transformation ratio. The voltage regulation then becomes

$$\boxed{\frac{(V_1/\alpha) - V_2}{V_2} \, 100 = \text{percent regulation}}$$
Eq. (15-14)

Using this equation, we have

(1) V_2, the measured output terminal voltage under specified load.
(2) α, the actual turns ratio from known transformer design parameters.
(3) V_1, the calculated input voltage required to produce V_2.

An example with a few related sections will show the utility of this family of equations.

Example 15-4. Still following the same 2300/230 V step-down transformer as an example and repeating its important parameters, we have $R_1 = 0.093\ \Omega$, $X_{L1} = 0.280\ \Omega$, $R_2 = 0.000\ 93\ \Omega$, and $X_{L2} = 0.002\ 80\ \Omega$. The transformer's rated current at the secondary is 3261 A. Assume that $\alpha = 10$ from 2300/230 V and calculate the following:

(a) Equivalent winding resistance, inductive reactance, and impedance of both windings reflected to the secondary.
(b) The V_1/α and V_1 voltages for unity, 0.7 lagging and 0.7 leading PF.
(c) The percent voltage regulation for the three power factors.

Solution:
(a) Using Eq. (15-7).

$$R_{e2} = R_2 + \frac{R_1}{\alpha^2}$$

$$= 0.000\ 93 + \frac{0.093}{100} = 0.001\ 86\ \Omega$$

Using Eq. (15-8),

$$X_{e2} = X_{L2} + \frac{X_{L1}}{\alpha^2}$$

$$= 0.002\ 80 + \frac{0.280}{100} = 0.005\ 60\ \Omega$$

Using Eq. (15-9),

$$Z_{e2} = \sqrt{(R_{e2})^2 + (X_{e2})^2}$$

$$= \sqrt{(0.001\ 86)^2 + (0.005\ 60)^2} = 0.005\ 90\ \Omega$$

(b) Using the appropriate signs in Eq. (15-13),

$$\frac{V_1}{\alpha} = \sqrt{(V_2 \cos \theta_2 + I_2 R_{e2})^2 + (V_2 \sin \theta \pm I_2 X_{e2})^2}$$

With a power factor of 0.7, $\cos \theta = 0.7$, so $\theta = 45.572°$ and $\sin \theta = 0.714\,14$. V_2 = rated secondary voltage = 230 V. I_2 is given as 3261 A (from 750 000 VA/230 V = 3261 A). Then for *unity* power factor,

$$\frac{V_1}{10} = \sqrt{[230 + 3261(0.001\,86)]^2 + [3261(0.005\,60)]^2}$$

$$= \sqrt{55\,727 + 333} = \sqrt{56\,060} = 236.8 \text{ V}$$

The required V_1 is then

$$V_1 = 236.8(10) = 2368 \text{ V}$$

For 0.7 *lagging* power factor,

$$\frac{V_1}{10} = \sqrt{[230(0.7) + 3261)0.001\,86]^2 + [230(0.714) + 3261(0.005\,60)]^2}$$

$$= \sqrt{27\,911 + 33\,300} = \sqrt{61\,210} = 247.4 \text{ V}$$

$$V_1 = 247.4(10) = 2474 \text{ V}$$

For 0.7 *leading* power factor,

$$\frac{V_1}{10} = \sqrt{[230(0.7) + 3261(0.001\,86)]^2 + [230(0.714) - 3261(0.005\,60)]}$$

$$= \sqrt{27\,911 + 21\,303} = \sqrt{49\,215} = 221.8 \text{ V}$$

$$V_1 = 221.8(10) = 2218 \text{ V}$$

(c) For percent voltage regulation use Eq. (15-14):

$$\frac{V_1/\alpha - V_2}{V_2} 100 = \text{percent regulation}$$

$$\frac{236.8 - 230}{230} 100 = 2.9\% \text{ regulation at unity PF}$$

$$\frac{247.4 - 230}{230} 100 = 7.57\% \text{ regulation at 0.7 lagging PF}$$

$$\frac{221.8 - 230}{230} 100 = -3.57\% \text{ regulation at 0.7 leading PF}$$

The interpretation of these answers is that, to *deliver* a secondary voltage of 230 V, this particular transformer needs a primary voltage of 2368 V at unity power factor. $V_1 = 2474$ V at 0.7 lagging PF, and a $V_1 = 2218$ V when its load is 0.7 leading PF.

This is not backward as it might first seem to be. The delivered load voltage is desired to be a *constant* 230 V regardless of load conditions. A transformer with *tapped windings* would be used in this situation. This means that the actual turns ratio may be varied by selecting more or less turns on one of the windings. Typical tap selection points that vary α by 2.5 percent increments are available on standard distribution transformers. Since the load power factor does not vary greatly in any one installation, this transformer would be set at installation to match its situation. If the load were 0.7 lagging PF, the α would be adjusted up about plus 5 percent over the unity condition, or perhaps about 2.5 percent above a more normal 0.8 lagging power factor setting. This problem would find an identical answer if all data had been reflected to the primary rather than the secondary.

15-5 TRANSFORMER VOLTAGE REGULATION BY SHORT-CIRCUIT TEST

The determination of transformer voltage regulation by the use of Eq. (15-13) requires the knowledge of the winding resistances and inductive reactances. An actual test procedure that yields realistic values of X_{e1} and X_{e2} is the transformer *short-circuit test*.

From Fig. 15-3a, it can be seen that the simplified equivalent circuit of a transformer involves two voltages. The $I_1 Z_{e1}$ voltage is the total voltage drop across the *primary* and *reflected secondary* impedances. The αV_2 voltage is the *load terminal voltage reflected* back to the primary. This circuit is identical to Fig. 15-1c and was discussed in Sect. 15-2. It is repeated here for convenience. If a transformer secondary is *short circuited* deliberately so that there is essentially zero volts across the secondary terminals, the circuit in Fig. 15-3b results. Since the V_2 voltage is considered to be zero, the αV_2 voltage is also zero. As a result, the input voltage V_1 is the $I_1 Z_{e1}$ voltage drop.

If a transformer is deliberately short circuited, and also instrumented with a suitable voltmeter, ammeter, and wattmeter, the circuit shown in Fig. 15-3c then exists.

The transformer short circuit test using this Fig. 15-3c circuit proceeds as follows:

(1) The adjustable voltage source is set as low as possible and then shut off; then the secondary terminals are short circuited. Usually, the high-voltage coil is driven through the H_1 and H_2 terminals. Then the low-voltage coil is short circuited by connecting the X_1 and X_2 terminals. This is not an invariable rule since the test works the other way around. However, so much *less* than rated input voltage is needed that up to a 20 kVA or more, 2300 V, or even a 4600 V transformer coil can be driven by a normal adjustable transformer, such as a Variac or Powerstat. The limiting condition is that the I_1 current must be obtainable. The variable transformer itself can be powered by normal 115 to 120 V line voltage if the required current can be handled.

(2) The adjustable primary voltage is then cautiously raised until the *rated primary*

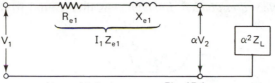

Note: this is the same as Fig. 15.1c

(a) Simplified equivalent circuit, loaded transformer

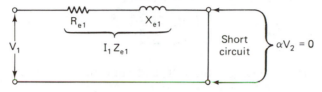

(b) Short circuited secondary, equivalent circuit

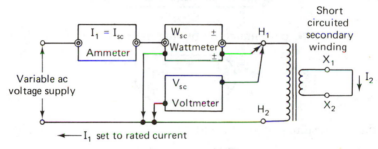

I_1 set to rated current

(c) Short circuit test circuit

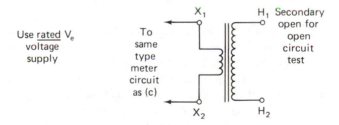

(d) Open circuit test changes

Figure 15-3. Transformer short-circuit test.

current is shown on the I_1 ammeter. This is the rated transformer kilovolt-amperes divided by the rated primary voltage, or

$$\frac{\text{rated kVA (1000 VA)}}{\text{kVA}(V_1)} = \text{rated } I_1 \text{ current}$$

(3) At this level of V_1 voltage, record the W_{sc} wattage, the V_{sc} voltage, and the $I_{sc} = I_1$ rated current.

(4) The *reflected equivalent impedance* is then found from

$$Z_{e1} = \frac{V_{sc}}{I_{sc}}$$

Eq. (15-15)

(5) The desired *reflected equivalent resistance* can then be determined from $I^2R = W$ or

$$R_{e1} = \frac{W_{sc}}{I_{sc}^2}$$

Eq. (15-16)

(6) With Z_{e1} and R_{e1}, the reflected equivalent inductive reactance can then be found:

$$X_{e1} = \sqrt{(Z_{e1})^2 - (R_{e1})^2}$$

Eq. (15-17)

This can also be determined by finding the power factor angle:

$$\cos \theta_1 = \frac{R_{e1}}{Z_{e1}}$$

Eq. (15-18)

From the resulting angle θ_1, find

$$X_{e1} = Z_{e1} \sin \theta$$

Eq. (15-19)

An example using a transformer that can be handled in a school laboratory will be used to clarify this procedure. However, it is good to first realize the assumptions that the short-circuit test involves.

Under short-circuit conditions, with no secondary voltage allowed or $V_2 = 0$, the αV_2 also is zero. This means that the only voltage needed in the secondary is that required to overcome the Z_2 impedance and cause rated I_2 current to flow. Since this impedance is very low, the voltage E_2 required to produce the current is small. This, in turn, means that a *very low magnetic flux* is required. A transformer with a high-side voltage of 2300 V and an efficient magnetic circuit will typically need only about 40 to 75 V to drive the rated V_1 current. Since the magnetic flux density varies directly with the induced voltages E_1 and E_2, the flux density is approximately $58/2300 = 0.025$ that of normal design conditions.

The causes of core losses in the magnetic circuit are the same as in dc machines. Thus, core losses are a function of the *square* of the flux density. Since the flux density is very nearly related to the V_1 voltage, the short-circuit core loss P_{sc} will be, in this case, only $P_{sc} = (58/2300)^2 P_{cl} = 0.000\ 64\ P_{cl}$. This short-circuit core loss then is less than one thousandth of the normal loss P_{cl}. Since the *normal* core loss is only a percent or so of the power handled, this tiny fraction can be safely *ignored* for normal transformers. Thus, the short-circuit test *power* is used only in overcoming the I^2R losses in the reflected total winding resistance.

Example 15-5. A 2300 to 230 V step-down transformer of 15 kVA rating is tested by the short-circuit test when connected as in Fig. 15-3c. The data taken on the high-side meters at rated I_1 current are $W_{sc} = 170$ W and $V_{sc} = 45.7$ V. Use $\alpha = 10$ because $V_1/V_2 = 10$ here again. Determine the following:

(a) $I_1 = I_{sc}$ test current level.

(b) Equivalent impedance, resistance, and inductive reactance from the test or the high-voltage side.

(c) Matching low-voltage side equivalent impedance, resistance, and inductive reactance.

(d) Voltage regulation at 0.773 lagging PF.

Note that this example can be solved by Eq. (15-13) or by a carefully scaled phasor diagram.

Solution:

(a) $I_1 = 15\ 000$ VA/2300 V $= I_{sc} = 6.52$ A.

(b) Equation (15-15): $Z_{e1} = V_{sc}/I_{sc} = 45.7/6.52 = 7.01\ \Omega$.

Equation (15-16): $R_{e1} = W_{sc}/I_{sc}^2 = 170/(6.52)^2 = 4.00\ \Omega$.

Equation (15-17): $X_{e1} = \sqrt{(Z_{e1})^2 - (R_{e1})^2} = \sqrt{(7.01)^2 - (4.00)^2}$

$$= 5.76\Omega$$

(c) $Z_{e2} = Z_{e1}/\alpha^2 = 7.01/100 = 0.0701\ \Omega$.

$R_{e2} = R_{e1}/\alpha^2 = 4.00/100 = 0.0400\ \Omega$.

$X_{e2} = X_{e1}/\alpha^2 = 5.76/100 = 0.0576\ \Omega$.

(d) Preliminary calculations for the determination of voltage regulation: $\cos\theta = 0.773$, $\theta = 39.38°$, $\sin\theta = 0.634$, rated $I_2 = 15\ 000$ VA/230 V $= I_2 = 65.2$ A. Using the lagging PF version of Eq. (15-13),

$$V_1/a = \sqrt{(V_2\cos\theta_2 + I_2R_{e2})^2 + (V_2\sin\theta_2 + I_2X_{e2})^2}$$

$$= \sqrt{[230(0.773) + 65.2(0.0400)]^2 + [230(0.634) + 65.2(0.0576)]^2}$$

$$= \sqrt{(180.40)^2 + (149.58)^2} = \sqrt{54\ 916} = 234.3\ \text{V}$$

$$V_1 = 234.3(10) = 2343\ \text{V}$$

Using Eq. (15-14).

$$\frac{V_1/\alpha - V_2}{V_2}100 = \frac{234.3 - 230}{230}100 = 1.87\%\ \text{regulations at 0.773 lag PF}$$

15-6 OPEN-CIRCUIT TEST FOR MAGNETIC LOSSES

It has been shown that essentially all the input power drawn during the short-circuit test is used in overcoming the total reflected primary copper loss. At the same time, little or no power is consumed by the magnetic losses during the short-circuit test.

For practical purposes, the *open-circuit transformer test* has the opposite effect. During the open-circuit test, there are virtually no copper losses in the primary winding and none in the secondary, because I is very small compared to rated I. As a result, the power consumed in the open-circuit test is all chargeable to the magnetic circuit losses. These losses include the same hysteresis and eddy current losses and the magnetization power losses discussed under dc motor and generator design and efficiencies. When both the copper losses and the magnetic circuit losses can be evaluated, the overall transformer efficiency can be readily determined. There is no transformer equivalent to the mechanical rotational losses of a motor or generator because there are no moving parts to consume energy.

The connections used in the open-circuit test are almost the same as those used in the short-circuit test, which were shown in Fig. 15-3c. The differences are that the open-circuit test is usually performed with the *low-voltage* winding excited, using the X_1 and X_2 terminals. Also, the secondary winding, now the H_1 and H_2 terminals, is *open circuit*. Hence the name of the test. See the transformer terminal circuit changes in Fig. 15-3d.

Since there is no secondary current during an open-circuit test with the terminals unconnected, there is no I_2 current and thus no I_1' current. The only current in the primary is that necessary to meet the I_m current requirement. Since this current is nearly 90° lagging, the power factor is very poor. Thus, very little power is drawn. All this power is used to overcome the various magnetic circuit losses. This is the power necessary to force the required magnetic flux through the magnetic circuit reluctance and the power to supply the hysteresis and eddy current losses, since this flux is cyclically varying.

Even though the open-circuit volt-amperes are substantially greater than the open-circuit wattage owing to the poor power factor, the current is still low. Typically, an open-circuit test on a transformer uses roughly one-quarter of the power that the same transformer draws on the short-circuit test because core loss roughly equals copper loss at half load. At the same time, the transformer connections are reversed so that the open-circuit test is supplying power to the normally low voltage, high current winding. The current drawn is of the order of 1 to 7 percent of the normal full-load current on that same winding. Since resistance power losses are I^2R losses, they are proportional to the *current squared*. If the current were 1 percent of normal, it then causes 0.0001 and at 7 percent no more than 0.005 times the normal copper losses in the transformer. Even in this day of pocket calculators with eight- or 10-significant-figure accuracy, the voltmeters, wattmeters, and ammeters used in a test of this type are rarely better than 1 percent accuracy. Thus, from a practical standpoint the copper losses in the open-circuit test are small compared to the test data accuracy and can prudently be *ignored*.

At the same time, by virtue of the fact that this test uses the *rated* voltage on the excited winding, the normal induced secondary voltage E_2 will exist. This implies that the full normal magnetic flux is present. As a result, the open-circuit test power measurement is a realistic measurement of the sum of the normal magnetic circuit losses.

Be very aware during this test that the normal *high voltage* exists at the open-circuit terminals. These terminals may reach a level of thousands of volts even though the test input voltage is only 120, 230, or 460 V. These terminals should then be treated with the same respect as a running circular saw or a live poisonous snake. Be aware of the hazards. Take due precautions by insulating the open terminals and no trouble will be encountered.

The open-circuit test proceeds as follows:

(1) The adjustable voltage source is carefully raised up to the normal rated voltage of the connected winding and its level is recorded as $V_{oc} = V_{rated}$.

(2) At the same voltage level, the open-circuit power is recorded as W_{oc} and the current as I_{oc}.

(3) The magnetic core loss is then W_{oc} or the total open-circuit watts.

15-7 TRANSFORMER EFFICIENCY

The efficiency of any device is its output power divided by its input power. When the efficiency is desired in percent, the answer is multiplied by 100. This basic relation was shown in Eq. (7-1):

$$\text{efficiency} = \frac{\text{output}}{\text{input}} \times 100 = \eta\%$$

A review of Eqs. (7-1) through (7-5) is entirely in order since the discussion in Chapter 7 up to Eq. (7-5) is very general in that its principles hold for *any* device.

It was shown in Sect. 15-5 that the short-circuit test will enable determination of the winding copper losses in a transformer. Section 15-6 shows that the open-circuit test will enable determination of the combined magnetic circuit losses in a transformer. Since these are all of the losses involved, we are then able to calculate the total losses or the Σ losses term in the efficiency equations.

$$\text{efficiency} = \frac{\text{input} - \Sigma \text{ losses}}{\text{input}} \times 100 = \eta\% \qquad \text{Eq. (7-4)}$$

or

$$\text{efficiency} = \frac{\text{output}}{\text{output} + \Sigma \text{ losses}} \times 100 = \eta\% \qquad \text{Eq. (7-5)}$$

In alternating current the wattage measurement is $EI \cos \theta$ or $VI \cos \theta$, depending on how the voltage term is labeled. Since the power in watts is measured by the wattmeter, and voltmeters and ammeters read their units directly, power is either $VI \cos \theta$ or watts. The $\cos \theta$ term normally requires *both* VI and watt readings, since cos θ = W/VA. Transformer power can then be measured at either the input or the output by appropriate meters. However, an efficiency measurement by direct meter readings while under load requires appropriately scaled and extremely accurate meters and adequate power. Input power in large enough quantity may simply not be available at the laboratory. Even if the input were adequate, the absorption of the output power in large quantities is a problem. The available meter errors may be larger than the transformer losses. As a result, *indirect methods* are developed and used for transformers just as they are for motors and generators. This is the major reason for the use of the short-circuit and open-circuit tests. The short-circuit test power required is usually a very small percent of rated transformer power. Open-circuit test power is usually a still lower percent. Using Eq. (7-5) as the most appropriate, since the output power is the rated value of most machines, we may derive the following variation.

$$\text{efficiency in percent} = \eta = \frac{V_2 I_2 \cos \theta_2 \,(100)}{V_2 I_2 \cos \theta_2 + \Sigma \text{ losses}}$$

This expands to the following useful form:

$$\eta\% = \frac{V_2 I_2 \cos \theta_2 \,(100)}{V_2 I_2 \cos \theta_2 + P_{\text{core}} + I_2^2 R_{e2}}$$

Eq. (15-20)

where the usable units are

V_2 = rated output voltage

I_2 = output current determined in logical steps from the rated output volt-amperes designation. Frequently steps of 1/4, 1/2, 3/4, full, and 1 1/4 of rated output current are chosen. The steps could well be by increments of 0.1 times rated current, if desired.

cos θ = decimal equivalent of the power factor chosen for the particular calculation; if no special value is chosen, 0.8 is representative.

P_{core} = core loss power in watts from the *open-circuit test*.

R_{e2} = equivalent total winding resistance reflected to the secondary as determined by the *short-circuit test*.

Example 15-6. The same transformer as Example 15-5 is tested by the open-circuit test when connected as in Fig. 15-3d. The data are taken on the low side at the rated 230 V. Note during the open-circuit test that the *low voltage* is V_1. The data taken for open circuit are V_1 = 230 V, P_{oc} = 45 W, and I_1 = 1.95 A. The short-circuit data are from Example 15-5. Calculate the following:

(a) The transformer efficiency at rated load for the following power factors: unity, 0.8 lag, 0.6 lag, 0.4 lag. Note from Eq. (15-20) that the equation does not differentiate between lagging and leading power factor.

(b) The transformer efficiency at 0.8 lagging PF and at 0, 0.1, 0.2, 0.3, 0.5, 0.75, 1.0, 1.25 times rated load and plot the results. Show part (a) data on the curve.

Solution: Preliminary data of R_{e2} from Example 15-5 yield $R_{e2} = 0.0400 \ \Omega$. From this example's data, $P_{core} = P_{oc} = 45$ W from open-circuit test. Rated I_2 current = 15 000 VA/230 V = 65.22 A.

(a) The losses may be tabulated for convenience (see Table 15-1).

$$I_2^2 R_{e2} = (65.22)^2(0.0400) = 170 \text{ W}$$

at *rated load*. This holds for all power factors.

(b) With a variable-load current, the $I_2^2 R_{e2}$ losses are now a variable. The P_{core} remains a constant loss. In this case, cos θ is specified as a constant. This was also tabulated as solution b for convenience (see Table 15-1).

The results are plotted in Fig. 15-4. Notice that the efficiency reaches a very high level at relatively low output power. The efficiency peaks very broadly at around one-half rated power. Chapter 7 showed that in *any device* the efficiency peaks when the fixed losses and variable losses are equal. Here the fixed losses are the P_{core} loss. The variable losses are the $I_2^2 R_{e2}$ loss. An inspection of the

TABLE 15-1
Solution a

PF	Core loss $= P_{oc}$ $= P_{core}$ (W)	Copper loss = $I_2^2 R_{e2}$ (W)	Total loss (W)	Output W = $V_2 I_2$ cos θ₂	Output + Σ loss (W)	Eff. η (%)
Unity	45	170	215	15 000	15 215	98.6
0.8	45	170	215	12 000	12 215	98.2
0.6	45	170	215	9 000	9 215	97.7
0.4	45	170	215	6 000	6 215	96.5

Solution b

Decimal load = L_{dec}	Core loss = P_{core} (W)	$I_2 =$ 65.22 × L_{dec}	$I_2^2 R_{e2}$ copper loss (W)	Σ loss (W)	Output watts = $V_2 I_2$ cos θ₂	Output + Σ loss (W)	Eff. η (%)
0	45	0	0	45.0	0	45	0
0.1	45	6.52	1.70	46.7	1 200	1 247	96.2
0.2	45	13.04	6.81	51.8	2 399	2 451	97.9
0.3	45	19.57	15.3	60.3	3 601	3 661	98.4
0.5	45	32.61	45.5	90.5	6 000	6 091	98.5
0.75	45	48.92	95.7	141	9 001	9 142	98.4
1.0	45	65.22	170	215	12 000	12 215	98.2
1.25	45	81.53	266	311	15 002	15 313	98.0

NOTE: $I_2^2 R_{e2}$ for rated load or $L_{dec} = 1.0$ is 170 W. The other entries in this column can be found from $L_{dec}^2 \times 170$ so that R_{e2} need not be found unless desired. The 170 W was direct experimental data from Example 15-5.

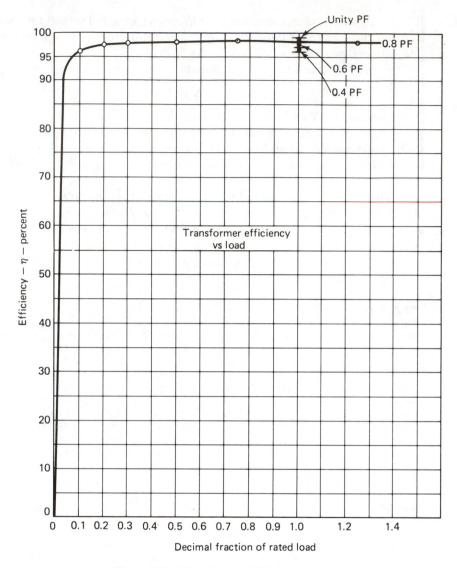

Figure 15-4. Transformer efficiency versus load.

calculation table shows that at 0.5 rated load $P_{\text{core}} = 45$ W and $I_2^2 R_{e2} = 45.5$ W, so $P_{\text{core}} \cong I_2^2 R_{e2}$ at this point.

A motor or generator usually has this peak efficiency point designed to fit the 0.75 to 0.85 load situation, since the highest efficiency is desired near the usual operating point. With large transformers, the overall efficiency is very good. By designing for best efficiency at 0.5 load, a very good efficiency is achieved over most of the range. The shape of the curve in Fig. 15-4 confirms this. The large dot symbols in Fig. 15-4 show the rated load efficiencies at the other specified power factors. If all curves were shown completely, the family of efficiency curves would show all efficiencies peaking at the 0.5 load, although at varying levels.

TRANSFORMER EQUIVALENT CIRCUITS   CHAP. 15

In rating the efficiency of distribution transformers used to supply a load district or a commercial area, a term called the *all-day efficiency* is used. This is simply the *total energy delivered* by the transformer in 24 hours divided by the *total energy input*. A determination of this efficiency soon shows the advantage of the fact that the best efficiency is placed somewhere near mid-load. The all-day efficiency takes into account periods of no load when the input is solely the P_{core} requirement.

Distribution transformers and voltage-dropping transformers for machine tools typically are never switched off. The no-load losses are not figured as enough to warrant the cost and complexity of a circuit to sense load demand. There is even an advantage in keeping the windings mildly warmed by the magnetizing current. This helps prevent insulation breakdown in humid climates since the windings tend to stay dry when they are warm.

QUESTIONS

15-1. What is meant by reflected impedance?

15-2. Why can the magnetization current component of the primary current be reasonably ignored in the simplified equivalent circuit of a loaded transformer?

15-3. Figure 15-2 shows that the reflected primary voltage and the secondary voltage of a transformer are not in phase. What is the major cause of this phase difference?

15-4. Why does the same transformer show different voltage regulation for different load power factors?

15-5. Why is a transformer short-circuit test not destructive to the transformer under test?

15-6. Why does the short-circuit test disclose the winding copper losses without being involved with the core losses?

15-7. Why does the open-circuit test disclose the transformer core losses without being involved with the winding copper losses?

15-8. Why are indirect methods used for transformer efficiency testing, especially in large sizes?

15-9. Why is a transformer efficiency normally at its highest at about one-half full load?

PROBLEMS

15-1. A transformer has a turns ratio of $\alpha = 2$. If its input voltage is 230 V and its output current is 8.70 A, what is (a) secondary voltage, (b) load impedance, (c) primary current, (d) primary input impedance?

15-2. The transformer in Problem 15-1 has a primary coil resistance of 0.293 Ω and a secondary coil resistance of 0.0733 Ω. What is its internal resistance reflected to the primary?

15-3. The transformer in Problem 15-1 has a primary inductive reactance of 1.15 Ω and a secondary reactance of 0.288 Ω. What is its internal reactance reflected to the primary?

15-4. With the internal resistance and reactance shown in Problems 15-2 and 15-3, what is the internal impedance of the transformer of Problem 15-1?

15-5. The transformer load impedance of Problem 15-1b is entirely due to a resistive load, and the internal resistances and reactances of the transformer are as in Problems 15-2 and 15-3. Using the transformer internal plus load impedance reflected to the primary, what primary current will be drawn?

15-6. Using the transformer coil resistances in Problem 15-2 and coil reactance in Problem 15-3, what will be the following:
(a) Transformer resistance reflected to the secondary?
(b) Transformer reactance reflected to the secondary?
(c) Transformer equivalent impedance reflected to the secondary?

15-7. Again using the transformer that has been developed in this sequence of problems, if it carries the load impedance developed in Problem 15-1b (13.22 Ω) and this is at unity power factor, what is its input voltage required to be if it is delivering 115 V to the load?

15-8. What input voltage would be required for this same transformer if the same load impedance and thus load current were to be carried at a 0.75 lagging load power factor?

15-9. What input voltage would be required for this same transformer if the same load were to be carried at 0.85 leading power factor?

15-10. Using the input required voltages from Problem 15-9, what would be the various transformer voltage regulations at (a) unity power factor, (b) 0.75 lagging power factor, and (c) 0.85 leading power factor?

15-11. The same 230 to 115 V step-down transformer shows the following results in a short-curcuit test: watts input with short-circuited secondary = 11.25 W at 8.70 A secondary current. The input primary voltage at this time was 10.51 V and the primary current was 4.36 A. Under these conditions, what are the following:
(a) Reflected equivalent impedance?
(b) Reflected equivalent resistance?
(c) Reflected equivalent inductive reactance?

15-12. Under the conditions of Problem 15-11:
(a) What power factor angle is shown under the input conditions?
(b) What reflected inductive reactance is shown?

15-13. If the transformer used in this series of problems is tested for its open-circuit input wattage and shows W_{oc} = 6.33 W, what is the probable full-load efficiency of the transformer at a 0.75 lagging power factor?

Answers. **15-1(a)** 115 V, **(b)** 13.22 Ω, **(c)** 4.35 A, **(d)** 52.9 Ω; **15-2**, 0.586 Ω; **15-3**, 2.30 Ω; **15-4**, 2.37 Ω; **15-5**, 4.30 A; **15-6(a)**, 0.147 Ω, **(b)** 0.576 Ω, **(c)** 0.594 Ω; **15-7**, 232.6 V; **15-8**, 238.6 V; **15-9**, 227.2 V; **15-10(a)**, 1.13%, **(b)** 3.74%, **(c)** −1.22%; **15-11(a)**, 2.41 Ω, **(b)** 0.592 Ω, **(c)** 2.34 Ω; **15-12(a)**, 75.8°, **(b)** 2.34 Ω; **15-13**, 97.7%.

16

The Polyphase
Induction Motor

The usual ac motor that has a substantial job to do is an induction motor. Most of the machine tools in a shop or factory are driven by induction motors. Even in a household, the larger motors are single-phase induction motors. These will be discussed in Chapter 19 because, strange as it may seem, the single-phase motor is more complicated and harder to understand than the two- or three-phase induction motor.

A typical factory will have almost all its machine tools individually driven by various sizes and classes of three-phase induction motors. The exception is when a speed higher than the synchronous speed of a two-pole motor is needed. Here a universal motor is usually found. These types along with special varieties of single-phase motors will be discussed in Chapter 20. The same factory will probably have its own transformer bank substation for reduction to 230 or 440 V, or whatever the local standard may be. It may also have one or more large synchronous motors driving the large loads, such as the ventilation system or the shop air system. These will be used for their power factor improvement, if present. Finally, the basic machines, presses, drills, grinders, and so on, will each have one or more induction motors.

The induction motor is chosen for its simplicity, reliability, and low cost. These features are combined with good efficiency, good overload capacity, and minimal or no service requirement. When the facts of very wide availability and simple installation by relatively little trained personnel are added, the choice of an induction motor seems well founded.

16-1 INDUCTION MOTOR ACTION

So far, all dc or ac dynamo types that have been discussed have been what is called *doubly excited motors*. This means that both the armature and the field, whether rotating or stationary, have had electrical connections and coil windings with excitation currents. An induction motor has a *single source of power* which powers the fixed stator coils. The excitation to the rotor is achieved by *induction* or transformer action.

The Biot–Savart law requires that motor action be achieved by a conductor that carries a current and that is immersed in a magnetic field. In a three-phase induction motor, the required magnetic field is produced by the stator as outlined in Sect. 16-1-1. This *rotating* field fulfills a double task, since it also supplies the energy to the rotor. An induction motor rotor usually has no discernible wires at all and is then known as a *squirrel-cage rotor*. This term is very graphic but a bit dated. In the early years of the century, when this type of winding first appeared, the common squirrel was a frequent house pet. The usual pen or cage that housed them contained a rotating wheel that the animal could enter. This wheel afforded exercise and amusement to the pet. The same device is more commonly found now in a cage for a mouse or a gerbil.

Returning to the motor winding, a squirrel-cage winding is a series of uninsulated bars that are inserted longitudinally in the exterior slots of the rotor. The bars are attached at each end of the magnetic structure to a solid conducting ring or rim. The windings certainly resemble an animal exercise cage when extracted from the laminated magnetic core. Since the voltages are low in these windings, the natural oxides are enough to form sufficient insulation. The most usual construction in small to medium sizes is to cast the winding bars, end rings, and ventilating fan paddles all in one solid unit. The variations from one maker to another, and one model to another, are achieved by the change in bar cross section and variation in the conducting alloy used.

16-1-1 Rotating Magnetic Field from Polyphase Fixed Stator Coils.
The simplest three-phase, two-pole stator will be taken as an example for the production of constant-strength uniformly rotating field. Any ac stator winding, except in the very simplest fractional-horsepower sizes (fractional kilowatts), has distributed coil winding such as described in Sect. 11-2 and shown in Figs. 11-2 to 11-4. Even so, a single pole phase group of coils can be said to have a magnetic centerline, which will normally be at the geometric center of the group. In addition, any pole–phase group will be followed, 180 electrical degrees in space, by the opposite pole–phase group. Thus, in a two-pole, three-phase stator, there are *six* pole–phase groups. The *A* phase will have a pole group that will be at maximum north pole strength at a particular instant. This will be directly opposed 180 electrical degrees away, and also 180 mechanical degrees away, in a two-pole structure by a maximum south pole. This south pole is also electrically an *A* phase group. The geometric center of the *B* phase group of coils is 120 electrical degrees around the stator. This *B* phase is also opposed by its own opposite pole 180 electrical degrees across the stator. In a similar fashion, the *C* phase has an

opposing pair of pole–phase coil groups, but this pair is 240 electrical degrees around the stator from the *A* phase. Picture individual coils that are formed as in Fig. 11-2. These coils are combined in pole–phase groups as in the unwrapped view of Fig. 11-3. These pole–phase groups are spaced around the stator, as is shown in *simplified* form for the *four-pole* stator in Fig. 11-4. A *two-pole*, three-phase stator is then shown in another simplified fashion in Fig. 16-1a.

The *currents* in these coils are shown in three-phase sinusoidal form in Fig. 16-1b. These coils are represented in concentrated mechanical form for simplicity in Fig. 16-1a. The coils are also represented as wye connected, which is the usual form. In Fig. 16-1a, if the current is flowing into the *A* group coils at the maximum rate or at the 90 electrical degree position of Fig. 16-1b, then the coil is producing a south pole as marked. At that same instant the bottom coil in Fig. 16-1a is producing a north pole. This may be verified by the right-hand rule, where, if the fingers represent the direction of current in a coil, the thumb points in the direction of magnetic flux from a north pole. This specific relation of current and pole polarity depends on the coil turn direction being wound as shown in Fig. 16-1a.

In this fashion, it can be seen that phase *A* has its maximum magnetic flux at 90 electrical degrees and because of its position in Fig. 16-1a it can be represented as a phasor pointing straight up. The phase *A* magnetic phasor will be maximum and straight down at 270 electrical degrees. The position of *electrical degrees* on Fig. 16-1a and 16-1c to f is zero at the *left horizontal centerline* and increasing *clockwise*. This is *arbitrarily chosen* for this illustration.

By similar reasoning, phase *B* has its maximum flux at 210 electrical degrees, which is 120° later than the *A* maximum, and has a negative maximum at 30 electrical degrees. Also, phase *C* is at positive maximum at 330° and negative maximum at 150°. It is worthwhile to take the time to understand the "ground rules" of the diagrams in Fig. 16-1.

To demonstrate the fact that a three-phase stator has a *fixed-strength magnetic field rotating at a constant speed*, the following step-by-step construction is offered. With the stated rules above and starting with Fig. 16-1c, the phasors are constructed as follows.

Selecting 30 electrical degrees as a first trial point, we find that phase *A* is at sin 30° = 0.5 times its maximum, which produces a magnetic phasor of *one-half* the radius of the diagram in Fig. 16-1c. At the same time, phase *B* is at full strength *negatively*. This calls for a full-length phasor pointing *away* from the position of the *B* pole. In the views of Fig. 16-1c to 16-1f, the *physical position* of the coils, and consequently the *orientation* of their individual magnetic flux, is the same as in Fig. 16-1a. This is represented by the letters *A*, *B*, and *C* on these four views. Phase *C* can be seen from Fig. 16-1b to have one-half its full positive current at this 30 electrical degree point. Since phase *C* is 240 electrical degrees lagging with respect to phase *A*, this is at the (360° − 240°) + 30° = 150° position of phase *C*, or sin 150° = 0.5. Thus, the phase *C* phasor is drawn at one-half radius at the *C* position.

We now have a 0.5-radius phasor pointing toward *A*, a full-length dotted phasor pointing *away* from *B*, and another half-length phasor, shown as dash–dot–

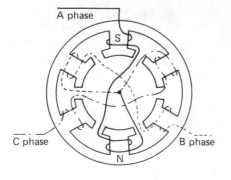

A phase

S

C phase   B phase

N

(a) Two pole, three phase,
Y connected stator

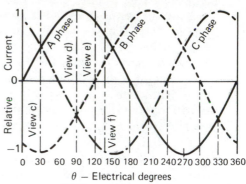

θ — Electrical degrees

(b) Current-phase relation

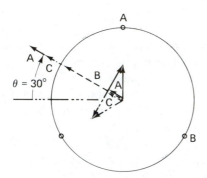

A

C

θ = 30°

B

A

C

B

(c) Phasor sum flux θ = 30°

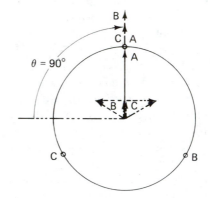

B

C A

A

θ = 90°

B C

C   B

(d) Phasor sum flux θ = 90°

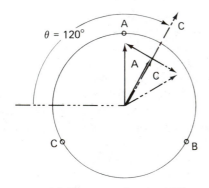

A   C

θ = 120°

A

C

C   B

(e) Phasor sum flux θ = 120°

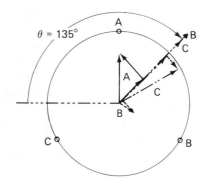

A   B

θ = 135°   C

A

C

B

C   B

(f) Phasor sum flux θ = 135°

Figure 16-1. Development of constant rotating field.

dot–dash, pointing toward C. As mentioned, the electrical degrees on this diagram start at the left and rotate clockwise. Therefore, phasor B is pointing along the desired 30° position, while phasors A and C have symmetrically opposite positions at 90° and 330°, respectively. These degree positions are the *mechanical position* of the phasors in relation to the coils at the 30 electrical degree *time*. Since phasor A is clockwise from B by 60°, and phasor C is counterclockwise by 60°, their *sine components* in relation to phasor B will cancel. Their cosine components then point in the same direction as B and can be considered to add to it. The small parts to add to B are *both* cos 60° times 0.5 = 0.25 the length of the B phasor. When added as shown in Fig. 16-1c, the B phasor *sum* is then 1.5 times its original length. This will be found to be a *general condition* in that the total magnetic phasor is 1.5 times the strength of the maximum of any one component. This will hold for *any position* around the circle, as will be shown.

Moving to Fig. 16-1d, which shows the conditions at 90 electrical degrees, we find a similar situation to Fig. 16-1c. This time phase A is at full strength positive and pointed toward point A on the circle of the figure. Both phases B and C are at one-half strength negative and thus oriented away from points B and C. The phasor resultant is a magnetic phasor in line with A and at 1.5 times the A maximum component. The magnetic phasor resultant has then held at 1.5 times the maximum of any single component and has moved from the 30° point to the 90° point as expected.

Now moving to Fig. 16-1e, we will investigate the 120 electrical degree point. At this point on Fig. 16-1b, we can see that phase B has a zero current strength and thus has *no contribution* to the phasor sum. Construction in a similar fashion to Fig. 16-1c and 16-1d shows that here we have a phasor of 1.5 length at the desired 120°. Therefore, the uniform phasor sum has moved in step with the specified electrical degrees and has kept its *constant* strength.

Another condition to investigate in order to feel confident in the generality of the uniform phasor sum at any position is where the sample time is taken such that all three phasor components are *different*. Consult Fig. 16-1f for this situation. We choose 135 electrical degrees as representative of *any random choice*. Here again the summaries of the sine components are $-0.5 + 0.25 + 0.25 = 0$, and they cancel as might be expected. The cosine components are $0.5 + 0.067 + 0.933 = 1.5$, so again they total 1.5!

This process is vitally important to the understanding of why the resultant magnetic field strength remains constant. The constant magnetic field rotates through 360 electrical degrees in position while the three different phases pulsate cyclically through 360 electrical degrees in time from their various starting points. This is the reason for the smooth rotational torque of any of the various types of three-phase motors. The field is as if it were caused by the mechanical rotation of a fixed magnetic field.

This uniform and rotating magnetic field is then present in synchronous motors and induction motors. The phenomenon is *very real* and not just an esoteric exercise.

16-1-2 Induction Motor Rotor Torque Development. An induction motor rotor is then swept by the moving magnetic flux from the stator. Any conducting windings in the rotor, whether of the simplified squirrel-cage type or of the more conventional wound-rotor type, are then traversed by the moving magnetic field. As long as there is a *speed difference* between the rotor and the moving or rotating magnetic field, there will be a voltage generated in the rotor windings. It will be shown that there must be some slip or difference between the speed of the magnetic field and the speed of the rotor windings. Remember that in the study of Chapter 2 it was shown that any motion of a conductor in a magnetic field generated a voltage. In Chapter 14, it was shown that the field could be moved mechanically, as in a motor or generator, or could produce generator action by a change of field strength in a cyclic fashion.

A normal induction motor runs at a *small percentage less speed* than its synchronous speed. For example, when a *four-pole* stator is excited with 60 Hz, the *synchronous speed* from Eq. (11-2$_E$) or (11-2$_{SI}$) modified algebraically becomes

$$S = 120 \frac{f}{P} \text{ rpm} \qquad \text{Eq. (16-1}_E\text{)}$$

or

$$\omega = 4\pi \frac{f}{P} \text{ rad/sec} \qquad \text{Eq. (16-1}_{SI}\text{)}$$

Using Table 11-1, the *synchronous speed* would be 1800 rpm (or 60π rad/sec, which is 188.50 rad/sec). A typical operating speed under normal load would then be around 1750 rpm (or 183.26 rad/sec).

This small difference from synchronous speed is the *slip* and it is normally expressed as a percent. As in any situation involving the artificial term of percent, the quantity is normally divided by 100 to convert percent to the decimal equivalent of percent. Slip is then defined as shown:

$$s = \frac{\text{synchronous speed} - \text{rotor speed}}{\text{synchronous speed}}$$

or

$$s = \frac{(S - S_r)100}{S} \text{ in percent} \qquad \text{Eq. (16-2}_E\text{)}$$

or in SI units,

$$s = \frac{(\omega - \omega_r)100}{\omega} \text{ in percent} \qquad \text{Eq. (16-2}_{SI}\text{)}$$

where

> s = percent slip or decimal slip without 100
>
> S or ω = *synchronous speed* of the rotating magnetic field in the appropriate units
>
> S_r or ω_r = *actual* rotor speed in operation

The four-pole motor running at 1750 rpm is then operating at the slip shown:

$$s = \frac{(1800 - 1750)100}{1800} = 2.778\% \text{ slip}$$

or 183.26 rad/sec operating speed is

$$s = \frac{(188.50 - 183.26)100}{188.50} = 2.780\% \text{ slip}$$

The difference between these two slip percentages is due to roundoff in the decimal speeds, which are necessary when operating in radians per second. Bear in mind that, if slip is stated as 0.027 78, the source probably means 2.778 percent, and vice versa. The practical world pays scant attention to the niceties of unit conversions when operating in a familiar field. Slip percent is usually *around 2 to 5 percent at full load*, sometimes higher. It is *much less at no load*. A typical number might be 1797 rpm or

$$\frac{(1800 - 1797)100}{1800} = 0.167 \text{ percent slip}$$

Note that this number is still larger than the 0.027 78 above.

If the operating slip of a motor is given, the operating speed may be determined by

$$\boxed{S_r = S(1 - s) = \left(\frac{120f}{P}\right)(1 - s)} \qquad \text{Eq. (16-3}_\text{E})$$

or

$$\omega_r = \omega(1 - s) = \left(\frac{4\pi f}{P}\right)(1 - s) \qquad \text{Eq. (16-3}_\text{SI})$$

where the units are the same as before and s is *decimal slip*.

It is the slip speed *difference* from the synchronous speed that the induction motor *rotor* sees. This small speed difference is sufficient to generate a low but real voltage in the squirrel-cage bars. This low voltage, and at the same time low frequency, is sufficient to cause large circulating currents to flow in the rotor. The generated voltage in each different rotor bar is related to the local strength of the moving magnetic field. Since the field strength varies in relation to the center of the moving magnetic flux, not all rotor bars see the same strength or the same polarity of magnetic flux. The result is that there are regions of circulating currents that correspond to the number of stator poles. There are few conductors in a squirrel cage, but the currents are correspondingly high and motor action is present and adequate. The frequency of the ac voltage in the rotor is a function of the basic frequency and the slip in *decimal notation*:

$$f_r = sf$$

Eq. (16-4)

where

f_r = rotor frequency in hertz
s = decimal slip
f = line frequency in hertz

Note that if percent slip is used here the *s* term must be divided by 100, and Eq. (16-4) is modified as shown:

$$f_r = \frac{sf}{100}$$

Eq. (16-5)

where this time *s* is percent slip.

The student is advised to *watch any calculation involving percent very carefully* to find if really percent or a decimal equivalent should be used.

Since the rotor frequency is normally low in operation, the rotor inductive reactance is also low. The current that flows is then largely limited by rotor resistance rather than reactance. Starting conditions, or decimal slip = 1, mean that the rotor sees the full-line frequency, and its inductive reactance is effective or even dominant compared to its resistance.

For special purposes of high-torque starting and/or a degree of variable speed control, induction motor rotors are sometimes made with conventional windings. Here the same magnetic flux will develop a higher voltage in a coil that has many turns. The windings will naturally have a higher resistance owing to their smaller cross section and greater conductor length. When the rotor coils are wound, they are brought out through slip rings to externally adjustable resistances. These resistances limit starting currents and serve to increase starting torque, as we shall see later.

It should also be realized that whether squirrel cage or wound rotors are used, slip is present. Thus, the regions on the rotor that have circulating currents, and thus magnetic poles, are also moving on the rotor surface at slip speed. The induced rotor magnetic poles move in respect to the rotor surface, but hold a fixed relation and linkage with the moving stator field.

A squirrel-cage induction motor is frequently identified by the abbreviation SCIM. The wound-rotor induction motor is identified as WRIM. These acronyms are not invariably used as identity but will save time and space, so their use will be progressively adapted here.

16-2 INDUCTION MOTOR PERFORMANCE PARAMETERS

To discuss induction motor starting, accelerating, running, and efficiency, for both SCIM and WRIM, some terms must be defined and means found to determine their values.   The tests are as follows:

(1) *No-load test* to determine the mechanical and magnetic losses.   This is analogous to a transformer open-circuit test and a dc rotational loss test.

(2) *Short-circuit* or *blocked-rotor* or *locked-rotor test* to determine the effective total resistance, the rotor resistance, and the electrical resistance losses or copper losses.

(3) *Stator-resistance test* to determine the stator resistance separately from the rotor resistance.

(4) *Load test* to determine load power, current, and power factor.

These useful tests enable the determination of the value of the individual parameters that will be used for calculating a number of useful motor relations.

16-2-1 No-Load Test or Rotational-Loss Test.
This test is performed by simply running the motor at rated voltage with the shaft running free.   Just as in the transformer open-circuit test, all the losses are charged to the primary or, here, to the stator circuit.   With instruments connected as in Fig. 16-2, readings are taken from all instruments under equilibrium conditions.   Note that an appreciable warm-up time is usually required to fluidize the grease in the bearings or reduce the viscosity of the oil in plain sleeve bearings.   The power reduction during warm up may be from 10 to 20 percent.   Since the slip is minimal during this test, the rotor currents, rotor magnetic losses, and rotor copper losses are considered to be negligibly small.   When the losses due to the ac resistance of the stator windings R_s are determined, the remaining power may be assigned to all the mechanical and magnetic power losses.

$$P_{\text{rot}} = \sqrt{3}\, V_l I_l \cos\theta - 3I_s^2 R_s$$

Eq. (16-6)

Note that $\sqrt{3}\, V_l I_l \cos\theta$ will be determined by the *difference* between two wattmeter readings, since the power factor is well below 50 percent at no load. Under these conditions, one wattmeter will read downscale and will need to be reversed if the two-wattmeter method is used.   From basic circuit theory then,

$$\sum W = \sqrt{3}\, V_l I_l \cos\theta$$

so that

$$P_{\text{rot}} = \sum W - 3I_l^2 R_s$$

Eq. (16-6)

where

$I_l = I_s$ for wye connection
$R_s = R_{\text{dc}} \times 1.25$ in ohms *per phase*

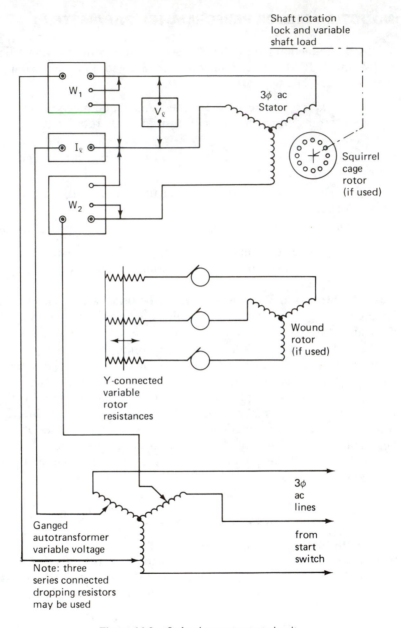

Figure 16-2. Induction-motor test circuit.

16-2-2 Blocked Rotor Test. This test is performed with the motor rotor mechanically locked against rotation. Some school laboratory machines have a ready provision for this blocking. Normal motors will require careful study to rig a *safe* means of blocking. The potential hazard is that if the block is inadequate and gets away from locked position the piece used may be *violently thrown* and cause damage or injury. The circuit in Fig. 16-2 is again used, but this time the voltage is set *very low*. By watching the ammeter (or ammeters), the voltage is carefully raised until the nameplate value of full-load current is reached. All readings are then taken quickly to avoid overheating. The motor cannot cool itself by its built-in fan action during this test.

Just as in the transformer short-circuit test, the relationships developed here include the *reflected equivalent* of the rotor parameters. In this case, the equivalent to the turns ratio α may not be known and is not needed, since no secondary readings are applicable on a squirrel-cage rotor.

With a *Y-connected stator*, we will then have the following short-circuit relations:

$$R_{es} = \frac{P_{br}}{3I_{br}^2}$$

Eq. (16-7)

$$Z_{es} = \frac{V_{br}/\sqrt{3}}{I_{br}}$$

Eq. (16-8)

$$X_{es} = \sqrt{Z_{es}^2 - R_{es}^2}$$

Eq. (16-9)

where

R_{es} = equivalent resistance reflected to the stator *per phase* in ohms
Z_{es} = equivalent impedance reflected to the stator *per phase* in ohms
X_{es} = equivalent inductive reactance reflected to the stator *per phase* in ohms
P_{br} = *total input watts* under blocked-rotor conditions
I_{br} = input line current under blocked-rotor conditions; this should approximate normal rated I_l amperes or $I_{br} \cong I_l$, when in wye
V_{br} = line-to-line *voltage* under blocked-rotor test conditions

From these equivalent values, we may then state that

$$R_{er} = R_{es} - R_s$$

Eq. (16-10)

and

$$X_{ebr} = X_s = \frac{X_{es}}{2}$$

Eq. (16-11)

where

R_{er} = resistance of the rotor per phase *reflected to the stator*
R_s = resistance of the *stator* in ac equivalent ohms from the stator resistance test of Sect. 16-2-3
X_{ebr} = inductive reactance of the rotor under blocked conditions, where slip $s = 1$ or 100 percent and thus the inductive reactance is that determined by the full-line frequency; remember that $X = 2\pi f L$ in any inductance.

X_s = inductive reactance of the stator; note that conventionally the X_s and X_{ebr} are *assumed* to be equal; hence the convenient relation of Eq. (16-11)

16-2-3 Stator Resistance Test. The stator resistance per phase is found from the voltmeter–ammeter procedures in Sect. 13-7 and used as in Eq. (16-6). For induction motors, the R_{dc} per phase is usually multiplied by 1.25 rather than 1.5 as in synchronous alternators.   Hence

$$R_s = R_{dc}(1.25)$$
Eq. (16-12)

With the synchronous alternator, this term was called R_a for resistance of the armature winding per phase.   Here the different designation R_s is used to recall the difference between the 1.25 and 1.5 arbitrary multiplication factors.

16-2-4 Induction Motor Load Test. The load test uses the same instrumentation again that is shown in Fig. 16-2.   The line-to-line voltage is returned to rated conditions.   The motor is connected to its normal load or to an adjustable laboratory load.   While under load, all readings are taken and recorded.   The load need not be a known or calibrated value.   However, frequently in a school laboratory situation the only convenient load is a dynamometer.   The dynamometer can give reasonably precise values of the load for checking purposes.   It will be found, as the methods are developed herein, that the use of these parameters, which have been measured in Sects. 16-2-1 through 16-2-3, will both enable the understanding of the equations that will be developed and enable realistic and practical motor testing as well.

16-2-5 Motor Parameters Summarized. The various terms developed and their relationships so far are summarized below.   Note that X_{er} is dependent on rotor speed because that determines the rotor excitation frequency.   Thus,

$$X_{er} = sX_{ebr}$$
Eq. (16-13)

Remember that $f_r = sf$, Eq. (16-4), since the rotor excitation frequency is determined by rotor speed.

Individual impedances can be constructed from the Pythagorean relation:

$$Z_s = \sqrt{R_s^2 + X_s^2}$$
Eq. (16-14)

$$Z_{er} = \sqrt{R_{er}^2 + X_{er}^2}$$
Eq. (16-15)

A summary of induction-motor terms and relationships follows.

STATOR

R_{es} = equivalent stator resistance, *ohms*, Eq. (16-7)

X_{es} = equivalent stator reactance, *ohms*, Eq. (16-9)

Z_{es} = equivalent stator impedance, *ohms*, Eq. (16-8)

R_s = stator res/ph, *ohms*, Eq. (16-12)

X_s = stator reactance, *ohms*, Eq. (16-11)

Z_s = stator impedance, *ohms*, Eq. (16-14)

f_s = *f* supplied by line, *hertz*

ROTOR

S = synchronous speed, *rpm*, Eq. (16-1$_E$)

ω = synchronous speed, *radians per second*, Eq. (16-1$_{SI}$)

s = slip, Eqs. (16-2$_E$), (16-2$_{SI}$)

S_r = rotor shaft speed, *rpm*, Eq. (16-3$_E$)

ω_r = rotor shaft speed, *radians per second*, Eq. (16-3$_{SI}$)

R_{er} = equivalent rotor resistance, *ohms*, Eq. (16-10)

X_{er} = assumed reflected rotor reactance, *ohms*, Eq. (16-13)

Z_{er} = equivalent rotor impedance, *ohms*, Eq. (16-15)

f_r = rotor frequency, *hertz*, Eq. (16-4)

Note that rotor resistance, reactance, and impedance are kept as *reflected quantities*.

16-3 INDUCTION MOTOR PERFORMANCE

Having shown that there are voltages induced in the rotor of an induction motor, and having shown that those voltages vary in direct relation to the slip, we are now ready to make use of these voltages and the relations in Sect. 16-2-5.

16-3-1 Rotor Current and Power. The inductively produced rotor voltage will cause a cyclic current to flow in the rotor at the slip frequency. This current will obey the same conditions as the current in a shorted secondary of a transformer. The current that flows will be developed into rotor power relationships by steps.

The rotor voltage may not be directly known in a squirrel-cage rotor, but it can be measured on an open-circuit basis on a wound rotor. This voltage will be the stator phase voltage times the slip times the turns ratio α on a per phase basis. Thus,

$$\boxed{E_r = \alpha s V_s}$$

Eq. (16-16)

The term α is readily determined on a wound rotor but not on a squirrel-cage rotor. However,

α = turns ratio stator to rotor

s = slip as defined before

V_s = stator applied voltage per phase

E_r = rotor induced voltage per phase

These quantities are readily determined in a *blocked-rotor open-circuit* test on a wound-rotor motor. They are as true in a squirrel-cage rotor, but the actual turns ratio is not apparent. We will subsequently find that it is possible to realistically consider rotor values as reflected equivalents at the stator level.

First, let us consider some illustrations at the rotor level using a wound-rotor motor:

Example 16-1. A three-phase, four-pole, 230 V wound-rotor induction motor has both its stator and rotor connected in Y. The rotor has half as many turns per phase as the stator. If the rotor is turning at 1725 rpm (180.64 rad/sec), determine the following:

(a) Slip in decimal and percent.

(b) The blocked-rotor voltage per phase.

(c) The rotor voltage per phase at the specified speed.

(d) The rotor voltage between slip ring terminals (open-circuit conditions assumed).

(e) The rotor frequency at operating speed.

Solution:

(a) Using Eq. (16-2$_E$),

$$s = \frac{(S - S_r)100}{S} = \frac{(1800 - 1725)100}{1800} = 4.167\%$$

In decimal form, this is

$$s = 0.041\ 67$$

When using or desiring decimal form, use Eq. (16-2$_E$) without the 100 factor. Using Eq. (16-2$_{SI}$),

$$s = \frac{(\omega - \omega_r)100}{\omega} = \frac{(188.50 - 180.69)100}{188.50} = 4.170\%$$

or, in decimal form,

$$s = 0.041\ 70$$

Consult Eq. (16-1$_E$) or (16-1$_{SI}$) to find the synchronous speeds, or find them in Tables 11-1 and 11-2. However, one should soon commit to memory the synchronous speeds of the common pole numbers and frequencies. This especially holds for two-, four-, and six-pole machines at 60 Hz. One of the *few* penalties of SI units is the unending decimals for synchronous speeds, 1800 rpm $= 60\pi$ rad/sec $= 188.495$, etc., rad/sec. It will pay to stay in 60π form for synchronous motors, but induction motors will *always* operate below synchronous speed. Customary usage is not yet established in SI unit measure.

(b) Using Eq. (16-16), we have

$$E_r = \alpha s V_s$$

where $\alpha = 0.5$, and in blocked condition $s = 1$. $V_s = 230/\sqrt{3}$ for wye connection.

$$E_r = \frac{0.5(1)230}{\sqrt{3}} = E_{br} = 66.4 \text{ V}$$

Note that the V_{br} in Eq. (16-8) is a *stator* voltage. The E rather than V implies an induced secondary voltage. The motor is simply a transformer under these conditions.

(c) The rotor is now moving at speed (either 1725 rpm or 180.69 rad/sec) and the slip s is as in part a. Again using Eq. (16-16),

$$E_r = \frac{0.5(0.0417)230}{\sqrt{3}} = 2.77 \text{ V}$$

(d) Since the rotor is in wye, we have

$$2.77(\sqrt{3}) = E_{rY} = 4.80 \text{ V}$$

(e) The rotor frequency is, from Eq. (16-4),

$$f_r = sf = 0.0417(60) = 2.50 \text{ Hz}$$

Note that the rotor voltage and frequency are very low under operating conditions.

Since the conversion of electrical power to mechanical power takes place in the rotor, it is necessary to consider rotor currents and wattage. This power will be found to be all the stator input that is not consumed by the stator in I^2R losses and magnetic losses. The rest is transferred to the rotor by induction or transformer action.

If we now identify the actual rotor voltage as sE_{br}, which is the slip times the voltage that would exist under blocked rotor conditions, a few current relations will evolve. Some thought will show that whatever this voltage is, even if it cannot be directly measured in a squirrel-cage rotor, it will obey Ohm's law for an ac circuit, or $I = E/Z$. Thus,

$$I_r = \frac{sE_{br}}{\sqrt{R_r^2 + (sX_{br})^2}}$$

This may be simplified by dividing through by s:

$$I_r = \frac{E_{br}}{\sqrt{(R_r/s)^2 + X_{br}^2}} \qquad \text{Eq. (16-17)}$$

If we still consider a wound-rotor induction motor, where the rotor conditions can be directly measured, we can develop a further example to show rotor currents.

Example 16-2. Using the motor in Example 16-1, calculate the rotor current per phase if $R_r = 0.075$ Ω and $X_{br} = 0.375$ Ω.

Solution: Using Eq. (16-17) and Example 16-1, we have

$$I_r = \frac{66.4}{\sqrt{(0.075/0.0417)^2 + (0.375)^2}} = 36.1 \text{ A}$$

Because of the way s has been used, this is really the current that 2.77 V forces through the operating rotor impedance. The rotor can then be represented as a series circuit of $R_r/s + X_{br}$, with E_{br} voltage forcing the current to flow. R_r/s is awkward, so this can be modified to

$$\boxed{\frac{R_r}{s} = R_r + R_r\left(\frac{1-s}{s}\right)} \qquad \text{Eq. (16-18)}$$

The rotor circuit is then visualized as a circuit containing actual R_r plus the $R_r[(1-s)/s]$, which is the *equivalent resistance of the mechanical load* per phase.

Power *delivered to the rotor* then can be broken into two parts: rotor power input = rotor copper loss + rotor power developed:

$$\boxed{\text{RPI} = \text{RCL} + \text{RPD}} \qquad \text{Eq. (16-19)}$$

The following then hold, using induced rotor current *per phase* in an I^2R power form:

$$\boxed{I_r^2\left(\frac{R_r}{s}\right) = I_r^2 R_r + I_r^2 R_r\left(\frac{1-s}{s}\right)} \qquad \text{Eq. (16-20)}$$

where

$$\boxed{\text{rotor power in } \textit{per phase}, \text{ RPI} = I_r^2\left(\frac{R_r}{s}\right)} \qquad \text{Eq. (16-21)}$$

$$\boxed{\text{rotor copper power loss } \textit{per phase}, \text{ RCL} = I_r^2 R_r} \qquad \text{Eq. (16-22)}$$

$$\boxed{\text{rotor power developed } \textit{per phase}, \text{ RPD} = I_r^2 R_r\left(\frac{1-s}{s}\right)} \qquad \text{Eq. (16-23)}$$

These hold for wound or squirrel-cage rotors.

These conveniently rearrange as *total powers*;

$$\boxed{\text{RPD} = \text{RPI}(1 - s)} \qquad\qquad \text{Eq. (16-24)}$$

and
$$\boxed{\text{RCL} = \text{RPI}(s)} \qquad\qquad \text{Eq. (16-25)}$$

The rotor power input, *either total power or per phase depending on usage*, is then input power to the motor minus stator losses.

The rotor power developed, RPD, can be further broken down into rotor output mechanical power and rotational losses. Remember that the rotational losses can be separately measured, so this is also a practical relation. We have

$$\boxed{\text{RPO} = \text{RPD} - P_{\text{rot}}} \qquad\qquad \text{Eq. (16-26)}$$

where

P_{rot} = rotational loss from Eq. (16-6)
RPO = rotor mechanical power output

Using the values developed in the two previous problems, we can evaluate rotor power relations in a practical way.

Example 16-3. Using the motor in Example 16-1 and the data developed in Examples 16-1 and 16-2, calculate the following as total powers:
  (a) Rotor power input, RPI.
  (b) Rotor copper losses, RCL.
  (c) Total rotor power developed, RPD.
  (d) The rotor power developed in horsepower.

Solution:
(a) From Eq. (16-21), allow for three phases:

$$\text{RPI} = I_r^2 \left(\frac{R_r}{s}\right)3 = (36.1)^2 \frac{0.075(3)}{0.0417} = 7032 \text{ W}$$

(b) From Eq. (16-22), use three phases:

$$\text{RCL} = I_r^2 R_r(3) = (36.1)^2 0.075(3) = 293 \text{ W}$$

(c) From Eq. (16-23), use three phases:

$$\text{RPD} = I_r^2 R_r \left(\frac{1 - s}{s}\right)3 = (36.1)^2 0.075\left(\frac{1 - 0.0417}{0.0417}\right)3 = 6738 \text{ W}$$

Alternatively, RPD = RPI − RCL, which can be seen from Eq. (16-19), so that RPD = 7032 − 293 = 6739 W as a check.

(d) The RPD in watts is the correct units for an SI calculation, but since 745.7 W = 1 hp,

$$\text{RPD} = 6739 \text{ W} \left(\frac{\text{hp}}{745.7 \text{ W}}\right) = 9.04 \text{ hp}$$

It will be seen that we never have to determine the actual rotor current unless some information is needed for an exterior rotor circuit in a wound-rotor motor. The remaining clue is in the following relation:

$$\text{Stator power input} - \text{stator copper loss} = \text{RPI}$$

This is exactly the same relation that is used for the rotational loss test, but in this case the *stator* power quantities are for a *loaded condition*. Thus, if SPI − SCL = RPI,

$$\boxed{\text{RPI} = \Sigma W - 3I_l^2 R_s} \qquad\qquad \text{Eq. (16-27)}$$

or $\qquad \boxed{\text{RPI} = \sqrt{3}\, V_l I_l \cos\theta - 3I_s^2 R_s} \qquad\qquad \text{Eq. (16-27)}$

Note that the term $\sqrt{3}\, V_l I_l \cos\theta$ fits the situation for either a wye or delta motor stator connection, but that the second term shows I_s rather than I_l. The currents are the *same* in wye and differ by $\sqrt{3}$ if the motor is delta connected. Be aware of the conditions.

With input measurements and Eq. (16-27) for both the load involved and for no-load conditions, all needed power calculations can be carried out if s is carefully determined under load.

16-3-2 Induction Motor Torque.
Returning to the basic equations that relate torque, speed, and power, we find from Chapter 6 that

$$\text{hp} = \frac{TS}{5252.1} \text{ in English units} \qquad\qquad \text{Eq. (6-1E)}$$

$$\text{kW} = t\omega \times 10^{-3} \quad \text{or} \quad W = t\omega \text{ in SI units} \qquad \text{Eq. (6-1SI)}$$

Multiplying each side by 745.70 to convert the English horsepower to watts,

$$W = \frac{TS(745.7)}{5252.1}$$

Rearranging, we have $T = 5252.1W/745.7S$ so that

$$\boxed{T = 7.0432\, \frac{W}{S}} \qquad\qquad \text{Eq. (16-28E)}$$

where

$\qquad T$ = torque in foot-pounds

$\qquad W$ = power in watts

$\qquad S$ = speed in revolutions/minute

This is done because all internal motor calculations are performed in watts rather than horsepower.   Rearranging in SI, we have

$$t = \frac{W}{\omega}$$

Eq. (16-28$_{SI}$)

where

t = torque in newton-meters

W = power in watts

ω = speed in radians/second

Putting these relations in the motor units used in this chapter, we have

$$T_g = \frac{7.04 \text{ RPD}}{S_r}$$

Eq. (16-29$_E$)

and

$$t_g = \frac{\text{RPD}}{\omega_r}$$

Eq. (16-29$_{SI}$)

where

T_g = gross developed torque in foot-pounds

t_g = gross developed torque in newton-meters

RPD = same rotor developed in watts in Eq. (16-24)

S_r = speed of the rotor in rpm

ω_r = speed of the rotor in radians/second

Now remembering that RPD = RPI$(1 - s)$ from Eq. (16-24), we can also state that

$$T_g = \frac{7.04 \text{ RPI}}{S}$$

Eq. (16-30$_E$)

and

$$t_g = \frac{\text{RPI}}{\omega}$$

Eq. (16-30$_{SI}$)

where

S = synchronous speed from Eq. (16-1$_E$) in rpm

ω = synchronous speed from Eq. (16-1$_{SI}$) in radians/second

As a result, if we know the RPD and the actual speed or the slip in appropriate units, we can find the developed torque. The RPI and the synchronous speed, in appropriate units, will produce the developed torque. If we know the rotational loss P_{rot}, we can relate the *output torque*, since, from Eq. (16-26), RPO = RPD $- P_{rot}$. Thus,

$$T_n = \frac{7.04\text{RPO}}{S_r}$$

Eq. (16-31$_E$)

and

$$t_n = \frac{\text{RPO}}{\omega_r}$$

Eq. (16-31$_{SI}$)

where

T_n = net output torque in foot-pounds

t_n = net output torque in newton-meters

With these families of equations, we can of course work both ways depending upon the data we have and the values that we will try to determine. If the basic stator power input SPI is measured, and if the rotational loss P_{rot} and the actual speed S_r or ω_r are determined along with the stator resistance R_s, all the torque or power determinations in Sects. 16-3-1 and 16-3-2 can be worked out. These torque relations are generalized operating conditions. However, sometimes *specific* torque conditions are needed.

16-3-3 Maximum Torque Developed.

There are really three different points where the actual value of the induction-motor torque is needed: the starting torque, the maximum torque, and the load torque or rated torque. The knowledge of how to determine these torque values enables one to fit a motor to its load in the most economical manner.

The highest torque that a squirrel-cage motor can develop is variously known as the *maximum torque*, the *breakdown torque*, or the *pull-out torque*. In a SCIM, this torque is ordinarily substantially higher than the starting torque. The maximum torque is usually developed at from half to three-quarters the synchronous speed and may be three to four times the rated full-load torque.

A wound-rotor motor has the ability to be adjusted so that its maximum torque is also its starting torque. This is done by adding resistance in the rotor circuit.

The maximum torque of a SCIM is well named as a *pull-out torque* because of the way the motor reacts to an overload. A typical situation, which many students will have experienced, is when a bench circular saw or a lathe will stall on a heavy cut. The machine will slow down as its cutting load is increased until suddenly it will stall and hum or growl loudly. The condition will persist until the load is relieved or a fuse or breaker blows. The motor has simply reached a point where it cannot continue to increase its torque. A further increase in load will cause a stall.

By means beyond the scope of this book, we find that

$$R_r = s_b X_{br}$$ Eq. (16-32)

where s_b is breakdown slip.

The significance of this is that, as the changing relation of torque to resistance reaches its peak, the slope of the curve reaches zero. At this point, which is the maximum torque, the relation shown in Eq. (16-32) emerges.

The term breakdown slip or s_b is used to show a difference from a *br* subscript, which we have taken to mean blocked rotor. This slip is the particular slip at maximum torque. The temptation to use s_m for slip at maximum torque is resisted because this is not the maximum slip possible. The greatest slip possible is at stall or blocked rotor, and would be $s_{br} = 1$ because there is no rotor speed under these conditions.

A wound-rotor motor can have external resistance, or R_{ex}, adjusted into its rotor circuit. When $R_r + R_{ex} = X_{br}$, then $s_b = 1 = s_{br}$. Under these conditions, the maximum torque is also the starting torque.

Another useful relation, whose mathematical background will not be shown, relates the torque at any speed to the maximum or breakdown torque.

$$T = T_b \left[\frac{2}{(s_b/s) + (s/s_b)} \right]$$ Eq. (16-33)

where

s = slip at any speed in question

s_b = slip at breakdown torque

Note that this equation can be used for SI units by substituting t for T and t_b for T_b, since the term in brackets is a dimensionless ratio.

Note also that the torques stated above are gross torques that have not been subjected to rotational losses.

Returning to a SCIM, if the motor is loaded to its stall point and the speed or the slip is carefully measured while the input power is measured, we can readily determine the torque developed at that point. Whether the result is developed torque or net delivered torque depends on whether the rotational losses have been measured by running light on the same motor.

Example 16-4. A squirrel-cage rotor induction motor is rated at 5 hp (3.725 kW) at 1745 rpm (182.74 rad/sec) and 15 A with 220 V line-to-line and a power factor of 72.9 percent lagging. The motor rotational loss is 208 W. If its maximum torque or stall point is found to be at 1260 rpm (131.95 rad/sec) determine (a) the maximum torque, and (b) the probable power input required at that maximum torque, neglecting copper losses.

Solution:

(a) First, the slip is needed at both stated speeds. Note that the number of poles was not specified, but operation so close to 1800 rpm (or to 188.5 rad/sec) seems to dictate that it *must be* a four-pole motor. Then from Eq. (16-2$_E$) or (16-2$_{SI}$), we must determine slips. At the operating speed, the *decimal slip* is

$$s = \frac{S - S_r}{S} = \frac{1800 - 1745}{1800} = 0.030\ 56$$

or

$$s = \frac{\omega - \omega_r}{\omega} = \frac{188.50 - 182.74}{188.50} = 0.030\ 56$$

Note that more significant figures are necessary in the speed measurement than are available in the slip. Also, speed determination to five significant figures is extremely difficult. Slip figures are rarely good to more than two significant figures in real practice.

At the maximum torque speed,

$$s_b = \frac{1800 - 1260}{1800} = 0.3000 \quad \text{or} \quad s_b = \frac{188.50 - 131.95}{188.50} = 0.3000$$

Next we need to know the value of the rated net torque. This should come from the rated mechanical output without regard to the electrical phenomena in the motor. Since rotational loss data are available in this problem, we can take advantage of Eq. (16-26) to convert the mechanical power specified in the problem to the developed or gross power.

$$\text{RPO} = \text{RPD} - P_{\text{rot}} \quad \text{or} \quad \text{RPD} = \text{RPO} + P_{\text{rot}}$$

$$3725 + 208\ \text{W} = \text{RPD} = 3933\ \text{W}$$

The gross torque under running conditions can then be found. From Eq. (16-29$_E$),

$$T_g = 7.04\ \frac{3933}{1745} = T_g = 15.87\ \text{ft-lb}$$

From Eq. (16-29$_{SI}$),

$$t_g = \frac{3933}{182.7} = 21.53\ \text{N·m}$$

A check shows that 1 ft-lb = 1.356 N·m, from Chapter 6. Hence 15.87(1.356) = 21.52 N·m. This checks.

Note that a torque taken from the specified power would be a net torque, T_n, but here we desired the gross torque, T_g. The specified speeds are S_r or ω_r rather than synchronous speeds S or ω.

Now applying Eq. (16-33), we can find the maximum torque. To find the breakdown torque, rearrange:

$$T_b = T\left[\frac{(s_b/s) + (s/s_b)}{2}\right]$$

$$= 15.87\left[\frac{0.3/0.030\ 56 + 0.030\ 56/0.3}{2}\right] = 78.9\ \text{ft-lb}$$

or in SI,

$$t_b = 21.53[\text{same}] = 107\ \text{N·m}$$

This very great increase in torque will not be achieved unless the magnetic iron is still relatively unsaturated at the higher current levels required. An increase by a factor of 4.5 or more is readily achieved in practice, so that the situation is not seriously misrepresented by the simplified mathematics used.

(b) A choice of a suitable equation suggests that since the T_b (or t_b) developed above is a gross torque or T_g (or t_g) in this case, and the power wanted is a power input, then Eq. (16-30$_E$) or (16-30$_{SI}$) would be the best choice. Rearranging, we have $T_g = 7.04\text{RPI}/S$, or

$$\text{RPI} = \frac{T_g S}{7.04} = 78.9\left(\frac{1800}{7.04}\right) = 20\ 170\ \text{W}$$

and $t_g = \text{RPI}/\omega$, so that

$$\text{RPI} = t_g\omega = 107(188.5) = 20\ 170\ \text{W}$$

The various copper losses were not accounted for in this last step, but very obviously they would be significantly increased due to the much higher currents. The rotational losses are implied to be the same at each speed and torque. The mechanical parts of the losses would reasonably decrease at the maximum torque speed. At the same time, the magnetic losses would increase due to the higher flux level and the greater frequency in the rotor. These various losses can be analytically studied, but the processes involve steps that are beyond the level of the text.

16-3-4 Starting Torque Relations. The starting torque of an induction motor can be determined by design over a reasonably wide range. The relationship in Eq. (16-32) and discussed in Sect. 16-3-3 shows that the maximum torque speed depends upon the *ratio* between the rotor resistance per phase R_r and the blocked-rotor reactance per phase. The reactance is determined by the magnetic structure of the rotor and by the number of turns per phase. This number of turns is not clearly seen in the peculiar all-short-circuit structure of a squirrel-cage rotor. Whatever it is though, it is fixed by the number of conductor bars per phase region on the rotor.

The rotor resistance is readily controlled in design by the resistivity of the alloy used in the squirrel-cage casting or by the use of external resistances in a wound-rotor motor. It follows from the relationship in Eq. (16-33) that the greater the slip at maximum torque or the lower the speed at maximum torque, the greater the starting torque. This could be stated by showing the starting conditions with the subscript *st*. Hence

$$T_{st} = T_b \left[\frac{2}{(s_b/s_{st}) + (s_{st}/s_b)} \right]$$ Eq. (16-34)

It can be readily seen that if $s_b = s_{st}$ the relation is $T_b(2/2) = T_{st}$. Thus, T_b becomes the starting torque T_s.

This is one of the major reasons for the wound-rotor motor. By adjusting the external rotor resistance so that the conditions of $R_r \cong X_{br}$, a starting torque can be achieved that is over four times the rated running torque. Then when the load is started the external resistance is reduced to as low a value as can be reasonably achieved. This has the effect of raising the speed of the maximum torque point. The final running condition then is that of a motor with very stiff speed-versus-load characteristics, or a relatively low percent speed regulation.

Squirrel-cage rotors can be designed to have a range of over four or five to one in their rotor resistance. The low-resistance rotor is achieved with large bars of low-resistance pure copper. A high resistance is achieved with bars of smaller cross section and of higher-resistivity aluminum alloy. With simple rotor bar design, the condition is fixed.

A very clever compromise is achieved by a *double-cage* rotor. In this type of design the rotor has two levels of bars built in or cast in. The outer layer has small cross section and relatively high resistance. The deeper or inner layer has large low-resistance bars. Recall that the inductance of a winding, the X_r factor, is a function of the rotor frequency. Also, the inductance of a particular part of the winding is a function of its local magnetic structure. A double-cage winding *outer layer* has relatively open slots and a poorer flux path around its bars, so that it has a relatively low inductance. At the same time, the inner-layer bars are thoroughly buried in a good magnetic structure and thus have relatively high inductance.

At start and at the very low speed range, the rotor frequency is the full-line frequency. Thus, the rotor induced current follows the low-inductance but high-resistance path in the outer layer. The inductance of the inner layer is such that the current is effectively blocked.

When nearing running speed, the rotor frequency is lower. Equation (16-4) shows that $f_r = sf$, so that when the slip is a small decimal the rotor frequency is low. This causes a large reduction in the inductive reactance of the inner layer, and thus the low-resistance current path is available.

Example 16-4 showed that, if a motor rating is known and if tests show its maximum torque or breakdown speed, the actual maximum torque can be calculated. By use of the relation in Eq. (16-33), the starting torque can then be calculated.

In the wound-rotor motor, where the actual rotor resistance R_r can be measured and controlled and where the X_{br} and E_{br} can be measured, the starting torque can be directly calculated. Equation (16-17) develops the rotor current per phase at any slip; if the slip is $s = 1$ at starting, the starting current in the rotor becomes

$$I_{rst} = \frac{E_{br}}{\sqrt{R_{rt}^2 + X_{br}^2}}$$

Eq. (16-35)

where

I_{rst} = current in the rotor at start in amperes *per phase*

R_{rt} = total rotor resistance per phase in ohms

Since this is current per phase, power input *per phase* is $I^2 R$, so that by squaring the I term and multiplying by R the rotor power input RPI is

$$RPI_{st} = \frac{E_{br}^2}{R_{rt}^2 + X_{br}^2}(R_{rt})$$

Eq. (16-36)

where RPI_{st} is rotor power input *per phase* at start.

This then enables the use of Eq. (16-30$_E$) or (16-30$_{SI}$) to calculate the torque, since these equations relate RPI and synchronous speed to torque.

Example 16-5. Using the wound-rotor motor of Examples 16-1 and 16-2, determine (a) the starting resistance to add per phase for maximum starting torque, and (b) the starting torque that results.

Solution:

(a) The rotor resistance per phase is $R_r = 0.075\ \Omega$. In order to match the inductive reactance of the rotor, $X_{br} = 0.375\ \Omega$, $0.375 - 0.075 = 0.300\ \Omega$ must be *added* to each phase externally.

(b) The RPI is found from Eq. (16-36), and the E_{br} from Example 16-1:

$$RPI_{st} = \frac{3(66.4)^2}{(0.375)^2 + (0.375)^2}(0.375) = RPI_{st} = 17\ 640\ W$$

The factor 3 is used because there are three phases in the rotor. Using Eq. (16-30$_E$), the torque is

$$T_g = \frac{7.04(17\ 640)}{1800} = 69.0\ \textit{ft-lb}$$

and using Eq. (16-30$_{SI}$), the torque is

$$t_g = \frac{17\ 640}{188.5} = 93.6\ \textit{N·m}$$

Again there are 1.356 N·m/ft-lb, so $69.0(1.356) = 93.6$ N·m. It checks.

QUESTIONS

16-1. What is meant by a doubly excited motor?

16-2. How does a uniform strength but rotating magnetic field induce voltages in an induction motor rotor?

16-3. How do voltages that are induced in the rotor produce a rotor magnetic field?

16-4. What is slip in an induction motor?

16-5. Why must some slip be present for motor action?

16-6. Why does running an induction motor unloaded enable the rotational losses to be determined?

16-7. What is the utility of the blocked-rotor test?

16-8. Why is it desirable to know the resistance of the stator windings?

16-9. Why is induction-motor rotor current related to slip?

16-10. What is the difference between gross developed torque and net output torque?

16-11. Why is maximum torque called breakdown torque?

16-12. How may the maximum torque be made to be the starting torque in a wound-rotor induction motor?

PROBLEMS

16-1. What is the synchronous speed of an induction motor with six poles operating on 60 Hz?   Calculate, and verify in Table 11-1.

16-2. What is the synchronous speed of an induction motor with four poles operating on 400 Hz?   Calculate in English and SI units, and verify in Tables 11-1 and 11-2.

16-3. If the motor in Problem 16-1 operates at 1142 rpm, what is its percent slip?

16-4. If the motor in Problem 16-2 operates at 11 200 rpm, what is its decimal slip?

16-5. If the motor in Problem 16-2 operates at 1200 rad/sec, what is its decimal slip?

16-6. An induction motor operates at 4.45 percent slip and has four poles.   What is its rpm on 60 Hz?

16-7. An induction motor operates at a slip of 0.0633 and has six poles.   What is its speed on 400 Hz when figured in SI?

16-8. What is the frequency in the rotor of the motor in Problem 16-6?

16-9. What is the frequency in the rotor of the motor in Problem 16-7?

16-10. A three-phase induction motor draws 4.5 A from its lines at 230 V line to line at a power factor 0.153 while running at no load.   Its dc resistance line to line between two phases of the stator is 1.863 Ω.   What is its rotational loss?

16-11. An induction motor is tested in the blocked-rotor test. Its rated line current of 8.5 A is drawn when the line voltage is 16.6 V and the total wattage is 48.8 W. Under these conditions, what is:
(a) The equivalent resistance reflected to the stator per phase?
(b) The equivalent impedance per phase?
(c) The equivalent inductive reactance per phase?

16-12. If in the motor in Problem 16-11 the ac resistance of the stator had been 0.127 Ω/phase, what would be:
(a) The resistance of the rotor per phase reflected to the stator?
(b) The inductive reactance of the rotor under blocked rotor condition?

16-13. An induction motor is running at a slip of 4.53 percent. If its equivalent blocked-rotor reactance is 0.555 Ω, what is its rotor reactance per phase when operating?

16-14. A wound-rotor motor has a turns ratio between its stator and rotor of 0.5 and operates on a 440 V, line-to-line, three-phase circuit. When operating at a slip of 14.5 percent, what is its rotor voltage per phase?

16-15. Calculate the rotor current per phase for a three-phase motor having the characteristics of that in Problem 16-14, but operating at a slip of 0.0482 and having $R_r = 0.13 \ \Omega$ and $X_{br} = 1.30 \ \Omega$.

16-16. What rotor power input (RPI) is developed in the motor of Problems 16-14 and 16-15?

16-17. What rotor copper losses may be expected for the motor in Problems 16-14 to 16-16 under the conditions stated?

16-18. Still following the motor in Problems 16-14 to 16-17, what rotor power will be developed?

16-19. An induction motor develops 48.7 hp when running at 1722 rpm. What is its output torque?

16-20. An induction motor develops 57.3 kW output when operating at 183.2 rad/sec. What is its torque in SI units?

16-21. A squirrel-cage induction motor is rated at 15 hp at 1745 rpm, and 44.1 A with 220 V line-to-line and a lagging power factor of 0.757. It is found to develop its maximum torque at 1365 rpm. Determine the maximum gross torque in foot-pounds. The rotational loss is found to be 335 W.

16-22. A wound-rotor induction motor with the characteristics found in Problems 16-14 to 16-18 is used where it is desired to develop the maximum starting torque. Determine the starting resistance that must be added to each rotor phase in an external starting circuit.

Answers. **16-1**, 1200 rpm; **16-2**, 12 000 rpm or 400π rad/sec; **16-3**, 4.83%; **16-4**, 0.0667; **16-5**, 0.0450; **16-6**, 1720 rpm; **16-7**, 785 rad/sec; **16-8**, 2.67 Hz; **16-9**, 25.3 Hz; **16-10**, 204 W; **16-11(a)**, 0.225 Ω, **(b)** 1.13 Ω, **(c)** 1.11 Ω; **16-12(a)**, 0.098 Ω, **(b)** 0.555 Ω; **16-13**, 0.025 Ω; **16-14**, 14.6 V; **16-15**, 42.4 A; **16-16**, 14 550 W; **16-17**, 701 W; **16-18**, 13 850 W; **16-19**, 149 ft-lb; **16-20**, 313 N·m; **16-21**, 167 ft-lb; **16-22**, 1.17 Ω.

17

Polyphase Induction Motor Characteristics

An induction motor, or any other motor for that matter, is rated at the load power at which it will operate continuously and efficiently. There are exceptions to this in order to get high mechanical power from a small physical size. In this special case, a motor nameplate will state the type of duty, such as intermittent duty, 15 minutes maximum. The rated power in horsepower (or kilowatts) and speed in rpm (or radians/second) will be displayed on the nameplate. The current at rated power, standard line voltage, and frequency are all a part of the rated conditions. All this information covers steady-state operation.

17-1 MOTOR CLASSES OF THE NATIONAL ELECTRICAL MANUFACTURERS ASSOCIATION

The transient operation of starting and accelerating a load are covered by the *class* designation. This class information is and has been standard in the United States and those countries that recognize the National Electrical Manufacturers Association (NEMA) code efforts. There is not yet a comparable recognized SI-based class specification. Such a type of rating will ultimately arise and be debated, modified, and adopted by the International Standards Organization (ISO). At this point we should bear in mind that some such type or class designation will have to be thrashed out and used.

Figure 17-1 shows a related set of squirrel-cage induction-motor speed versus torque curves. These curves are plotted on a per unit basis so that any power and number of poles may be accommodated.

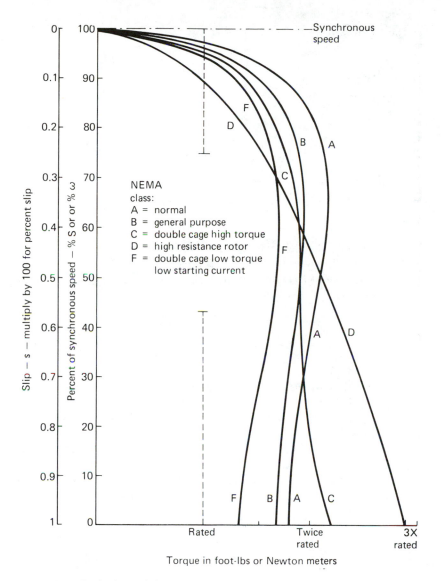

Figure 17-1. Squirrel-cage induction motor speed versus torque characteristics for NEMA classes.

Example 17-1. A certain industrial task requires about 15 hp (11.20 kW is also a NEMA guide recommended metric for comparably rated power; 15 hp × 0.746 kW/hp = 11.19 kW) and about 1200 rpm (125.7 rad/sec). A NEMA class A motor with six poles has a rated nameplate speed of 1160 rpm (121.5 rad/sec). Determine (a) the rated torque at full load, and (b) the approximate starting torque to be expected at full-line voltage.

Solution:
(a) Using Eq. (6-1$_E$) or (16-28$_E$) for English units and Eq. (16-28$_{SI}$) for

the SI situation, we have hp $= TS/5252$ or $T = 5252$ hp$/S$, so that

$$T = \frac{5252(15)}{1160} = 67.9 \text{ ft-lb}$$

Using Eq. (16-28$_E$), $T = 7.043$ W/S, where 15 hp (746 W/hp) $= 11\ 190$ W, so that

$$T = \frac{7.043(11\ 190)}{1160} = 67.9 \text{ ft-lb}$$

In SI, this is

$$t = \frac{W}{\omega} = \frac{11\ 200}{121.5} = 92.2 \text{ N·m}$$

This will not check exactly since the power is not quite the same, but 69.7 ft-lb (1.356 N·m/ft-lb) $= 92.1$ N·m. It checks.

(b) From Fig. 17-1, the starting torque of a class A squirrel-cage motor is very approximately 1.75 times rated torque:

$$T_{st} = 67.9 \times 1.75 = 119 \text{ ft-lb}$$

$$t_{st} = 92.2 \times 1.75 = 161 \text{ N·m}$$

Returning to the curve, this means that if the desired load has a starting requirement reasonably below these calculated starting torques, a class A motor can be used. If the starting torque requirement were as much as 2.5 times the normal rated torque, a class D motor would be needed.

The other side of the picture is the currents that are required and the efficiency and relative speed stability desired.

Using means that are not within the scope of this book, it is possible to show that the torque of an induction motor varies as the applied voltage squared.

$$\boxed{T_{st} = KV_s^2} \qquad \qquad \text{Eq. (17-1)}$$

where

$K =$ constant for a motor, which includes the winding configurations and the magnetic characteristics that are built in

$V_s =$ line-to-line stator voltage

The real utility of this equation is to determine what happens to the starting torque as the applied voltage is changed. The K term is very specific and is not ordinarily available without direct laboratory test, but it is a constant for a particular motor.

With the motor in Example 17-1 or any other induction motor, if the line voltage is reduced by a starting control to reduce the starting current inrush, the

TABLE 17-1 NATIONAL ELECTRICAL MANUFACTURERS ASSOCIATION CLASS SCIM MOTOR CHARACTERISTICS

NEMA Class	Full-load speed % regulation	Start torque times rated	Start current times rated	Characteristic name
A	2–5 <5	1.5–1.75	5–7	Normal
B	3–5 <5	1.4–1.6	4.5–5	General purpose
C	4–5 <5	2–2.5	3.5–5	High torque, double cage
D	5–8 and 8–13 ranges	Up to 3	3–8	High torque, high resistance
F	Over 5	1.25	2–4	Low torque, double cage, low starting current

starting torque drops still faster. With any specific motor, one-half voltage for starting will mean one-half the peak starting current, but only *one-quarter* the torque.

17-1-1 NEMA Class Motor Characteristics.

The various values assigned to NEMA motor classes are summarized in Table 17-1. For any specific and apparently critical situation, consult the manufacturer's specification sheets and performance curves. Remember that the values shown in Table 17-1 are for full-line voltage starting. Equation (17-1) always holds for reduced voltage starting. The problems of starting and controlling these motors are covered in Chapter 21.

17-2 WOUND-ROTOR MOTOR CHARACTERISTICS

The characteristics of a wound-rotor induction motor (WRIM) certainly resemble those of a SCIM. The advantage of the wound rotor is that the windings of the phases can be specifically terminated and brought out through slip rings. Then, by the use of external resistance, the total rotor resistance can be changed without a corresponding change in the winding inductive reactance. With normal proportions and windings, the typical inductive reactance under blocked-rotor or full-slip conditions is about three to five times the minimum rotor resistance or $X_{br} \cong 4R_r$. We have seen from the relation in Eq. (16-32) that the maximum torque is obtained when $R_r = s_b X_{br}$. Thus, to get the maximum torque to take place at the standstill, or $s = 1$, conditions of starting, it is required that $R_r = X_{br}$. This is quite simply accomplished with external resistance such that

$$R_r + R_{ex} = X_{br}$$ Eq. (17-2)

where R_{ex} is the external resistance per phase in ohms, and R_r and X_{br} are as before.

The ability to adjust the R_{ex} resistances grants four important advantages to a wound-rotor motor.

(1) The maximum torque and the starting torque may be made the same.

(2) The high variable torque allows reduced starting voltage to be used and still obtain a starting torque that will handle a difficult load. See Eq. (17-1) for this.

(3) The normal full-voltage starting current is moderate for an induction motor. It is around 2.5 times the rated running current.

(4) A degree of variable-speed control is obtained by adjusting the external resistance.

While wound-rotor motors are less important relatively than in the past, owing to the development of the double-cage squirrel-cage rotor induction motor, they are still used and should be understood. Figure 17-2 shows a typical set of wound-rotor slip versus torque curves, which may be compared to Fig. 17-1.

The curve labeled R_r alone is the speed versus torque curve for the rotor with its slip ring terminals short circuited. This characteristic is similar to a low-resistance squirrel-cage rotor. The starting torque is about 1.7 times the rated running torque. Maximum torque is reached at about three-quarters of synchronous speed. The speed droop at rated torque is about 3 percent. This then is the normal operating condition of a WRIM. Projecting directly up from its torque axis intercept to the intersection with the heavy dotted current curve, we find that the starting current would be roughly 3 to 3.2 times the normal rated current.

When an optimum external resistance is added to each rotor phase, the characteristic changes to the curve labeled $R_r + 0.6R_{ex}$. At this point, the $R_{ex} + R_r = X_{br}$, and the maximum torque is at the start condition. In this situation the starting current is found to be about 2.4 times the normal running current. Thus, the starting torque has *increased* from 1.75 to about 4 times normal running torque, but the starting current has *decreased* from over 3 times to less than 2.5 times the normal running current. At the same adjustment, the normal load is carried with around an 18 percent slip. This adjustment would not be left standing unless a lower operating speed were deliberately sought. More on this later.

Carried to a reasonable limit, the added resistance shown as the $R_r + R_{ex}$ curve results in the original starting torque, but only at about rated running current. This high rotor resistance results in a very soft speed regulation, where the operating speed is roughly only 60 percent of synchronous speed or a 40 percent slip.

Intermediate rotor resistance adjustments are shown whereby various levels of torque can be achieved. A frequent use of a WRIM is when the maximum starting torque is selected by the correct external resistance. Then as the speed increases, the high torque is roughly held by steady reduction of the external

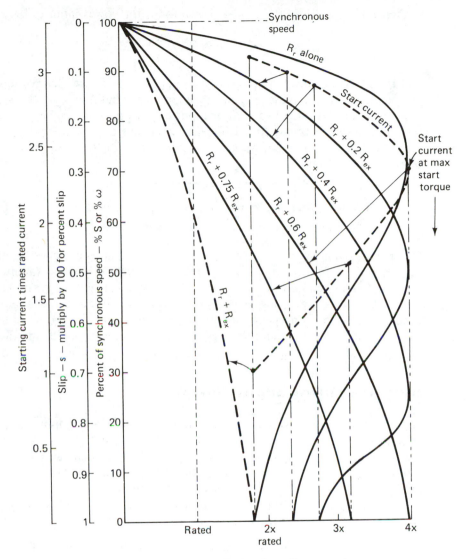

Figure 17-2. Wound-rotor induction motor, I and S (or ω) versus torque characteristics for various rotor circuit resistances.

resistance. This achieves the maximum load starting and acceleration that an ac electric motor can achieve. At the same time, running is very efficient.

Speed control will be covered in general terms later, but it should be recognized that when a wound rotor is operated steadily with an external resistance there are important power losses in the resistances. Equation (16-23) shows rotor power developed as follows:

$$RPD = I_r^2 R_r \left(\frac{1-s}{s} \right)$$

At the same time, Eq. (16-21) had shown the rotor power input to be

$$\text{RPI} = I_r^2 \frac{R_r}{s}$$

If we assume here that the R_r is the total of $R_r + R_{ex}$ at whatever adjustment is at hand, we then have a specific slip s for that condition. At the same time, we have simple expressions for rotor input power and rotor power developed, or input and output expressions. Efficiency is always output divided by input. If these equations are divided, we have an expression for the rotor circuit efficiency under these conditions.

$$\text{Rotor efficiency} = \frac{\text{RPD}}{\text{RPI}}(100) = (1 - s)100 \qquad \text{Eq. (17-3)}$$

This relation shows that the *rotor efficiency* alone can never be any better than $(1 - s)100$ *in percent*, where s is in decimal form. Thus, if the motor is set to run at a slip of 50 percent or $s = 0.5$, the rotor efficiency can be no better than 50 percent. This expression neglects any stator or rotational losses, but it is bad enough as it is. Summarizing, then, if low-speed operation is going to be used as a regular operation, the high rotor circuit losses must be expected, and the external resistances must be capable of dissipating the power that is lost.

17-3 INDUCTION MOTOR SPEED CONTROL

First, it is good to understand that fundamentally the induction motor is a relatively constant speed type. The induction-motor speed regulation is about comparable to a dc shunt motor when the dc motor has a fixed field adjustment. While the dc motor has no comparable feature to the synchronous speed of an ac motor, the slope of the speed versus load curve is roughly comparable. There are a few means of achieving variable speeds with an induction motor that allow considerable variation in speed. The three principal means are as follows:

(1) Frequency of the power supply determines synchronous speed. Changing the frequency changes the speed (see Chapter 21).

(2) The number of poles in the stator winding changes the speed of rotation of the moving magnetic field. Modifying the number of poles changes speed.

(3) Voltage changes affect the torque. By reducing voltage, the torque is reduced and thus the slip increases.

These methods of speed changing each have limitations and will be discussed individually. There is one more method that is adaptable to the wound-rotor motor:

(4) Resistance change in the rotor circuit can be accomplished in the WRIM. This changes the shape of the characteristic speed versus torque curve, and thus will reduce the speed of a *loaded* WRIM.

These are by no means the only methods of controlling ac motor speeds. However, most of the systems beyond those mentioned require so much auxiliary equipment that they become costly, large, and inefficient. The modern developments in solid-state circuit control are changing and will continue to change relative priorities of speed control methods. For example, the development of dc shunt motors with all-laminated field structures that allow the use of rectified alternating current in a solid-state field control has revolutionized the use of dc motors for variable-speed applications. This type of motor control is changing rapidly and must be assessed from manufacturers' literature. Modern solid-state devices are also changing the methods of varying an ac frequency. This must also be judged from specifications that are constantly being revised (see Chapter 21).

17-3-1 Motor Speed Change by Frequency Change. This means of speed change controls the f term in the synchronous speed equations, which we have seen are $S = 120f/P$ [Eq. (16-1$_E$)] or $\omega = 4\pi f/P$ [Eq. (16-1$_{SI}$)]. As an example, a four-pole motor has a synchronous speed of 1800 rpm (188.5 rad/sec) at $f = 60$ Hz. If the *frequency* were raised to 90 Hz, the *synchronous* speed would then be $S = 120(90)/4 = 2700$ rpm. The SI result would be $\omega = 282.7$ rad/sec. An induction motor, whether SCIM or WRIM, operates with a small slip. The slip percentage would be roughly the same so that a motor speed is changed in direct ratio to the frequency:

$$S_{f2} \cong S_{f1}\frac{f_2}{f_1}$$

Eq. (17-4$_E$)

or in SI units

$$\omega_{f2} \cong \omega_{f1}\frac{f_2}{f_1}$$

Eq. (17-4$_{SI}$)

where

S_{f1} or ω_{f1} = speed at the first frequency or reference frequency
S_{f2} or ω_{f2} = speed at the second frequency
f_1 = first or reference frequency such as 60 Hz
f_2 = any other reasonable frequency, say from 25 to 180 Hz

There is a serious effect with the supply voltage, however, in that it must change linearly with the frequency, as shown in Chapter 14. Small frequency changes can ignore the otherwise required voltage change. If the frequency change is substantial, the voltage *must* change in proportion:

$$V_2 = V_1\frac{f_2}{f_1}$$

Eq. (17-5)

The voltage change implies a power change as well. This is not an unreasonable situation since a driven load will usually require substantially more power as the speed is increased. Since the rotative speed increases, the internal cooling ability of the motor will also improve because the fan effectiveness will increase.

When the variable frequency is achieved by simply varying the rotational speed of an alternator, the variable voltage is automatically achieved by holding a constant field excitation. This begins to show the complexity required, since the variable-speed motor or motors now require a separate alternator with its own variable-speed prime mover. The problem of variable speed is transferred from one apparatus to another. Much progress is being made in solid-state electronic variable frequency power sources. They are expensive, however, in large sizes and bid fair to remain so. (Again, see Chapter 21.)

17-3-2 Motor Speed Change by Pole Change. The original discussion in Sect. (11-2) showed how poles are formed by grouping and connecting stator windings. Figure 11-4 shows the connections involved in a typical four-pole, three-phase winding. This winding diagram assumed a 36-slot stator with three coil groups per phase per pole. It would be equally practical to wind this stator with a six-pole winding having two coil groups per phase per pole. This would give a synchronous speed of two-thirds that of the four-pole winding. The practical two-speed combination results when *two separate winding groups* are placed in deep stator slots.

This type of two-speed motor is widely used for fan or blower drives and in some machine tool situations. The requirement for two complete windings is what limits the practicality of having each winding connected and available for series or parallel connections. This type of motor, in machine tool sizes, is almost always limited to the 208 to 230 V range. A two-winding motor can be quickly identified because its pole numbers are not at a 2-to-1 ratio or a 4-to-1 ratio to each other. The example above is a 3-to-2 ratio and is perhaps the most common.

When 2-to-1 pole numbers are wanted, a different approach is used. The *consequent pole* motor that results is achieved in the following manner, using Fig. 11-4 as a guide: tracing the wiring in Fig. 11-4a you will see that phase *A* has two separate groups of coils. One group from terminal 1 to terminal 4 covers the coils in poles 1 and 3. The other group from terminals 7 to 10 covers the coils in poles 2 and 4. Referring again to the diagram, you can see that if poles 1 and 3 are north poles at the moment, poles 2 and 4 must be south because their coil connections are reversed. The same conditions are true with phases *B* and *C*.

If terminals 7 and 10 were reversed so that 10 connected to 1 and 7 connected to the wye junction in place of 10, then in phase *A all poles* would be north. When this happens, the south poles that *must exist* develop automatically between existing north poles. Obviously, the process must take place in all phases. When this procedure is adopted, the coils *must be fractional pitch* so that they do not cover the iron area that is needed for the opposite polarity consequent poles. A consequent pole motor can then be shifted at will by external connection over a 2-to-1 range of pole numbers. Two poles can become four, four become eight, and six become twelve. Notice that if the coil connection in Fig. 11-4 had been such that poles 1 and 2 were one group and poles 3 and 4 the other, a consequent pole connection could not be achieved. In this case, the connected poles would have been wired in opposition. Swapping connections would still leave them opposite. This method has the shortest coil leaders and is usually adopted for reasons of cost and space. Consequent pole connections and their switching circuits will be explored in Sect. 21-10-1.

Another obvious method now becomes the combination of two windings and consequent poles. By this means, *four* different synchronous speeds are readily achieved. One method has a four-pole winding that can become a consequent eight-pole winding combined with a six-pole winding that can become a consequent twelve-pole winding. This will produce a motor that has *synchronous* speeds of 1800, 1200, 900, and 600 rpm (188.5, 125.7, 94.2, and 62.8 rad/sec). In this case, the required switch gear may well cost more than the motor and take more space.

17-3-3 PAM Motors.
Ever since the invention of the original induction motor in 1888 and the Dahlander or consequent pole motor in 1898, the only way to change the number of poles in an induction motor has been to use separate complete windings or to reverse the current direction in half of the coils and create a consequent pole relation.

Now, with the developments of Professor G. H. Rawcliffe of the University of Bristol in England, a new means is emerging. These PAM motors have been shown to be a satisfactory method and are covered by international patents and licensed for production in five different countries. Even so, they are virtually unknown in the United States.

PAM is an acronym for Pole Amplitude Modulation, and its method of operation is based upon selective switching of coil polarities so that pole ratios of other than two to one (2:1) can be achieved. Externally, the motors are connected and switched exactly as consequent pole motors. The same switching circuit that will handle a 2:1 speed ratio consequent pole motor will equally well serve PAM motors of 6:4 ratio, or 10:8, 10:6, and so on.

The related pole-phase group interconnections are such that when the external connection is switched from series delta to parallel wye, four out of six coils per phase, for example, will be reversed. This will produce two wide poles out of four original poles and leave two unchanged. If, originally, one of the six pole-phase groups had had an instantaneous magnetic polarity of N–S–N–S–N–S, or a standard six-pole relation, the switching would produce N–S–S–N–S–S. This produces a four-pole situation from the same number of pole-phase groups that had been six-pole. Obviously, the poles are not of equal face area; hence the name *Pole Amplitude Modulation*. However, the inequalities cancel out in the three phases, and the motor runs smoothly and draws balanced currents.

It is extremely timely to note that PAM motors are usually more efficient than two winding motors of the same power and speed ratios. They use the lamination steel and the winding copper as efficiently as normal single speed motors. As a result, it seems likely that their use will increase. The efficient low-cost motors that may be achieved can be used in many different situations because their initial cost is less than two winding two-speed motors and their efficiency on either speed is nearly as good as single-speed motors of the same speed.

Typical results that may be expected from both consequent pole and PAM motors may be gleaned from Fig. 17-3. The motor used in the test shown was a General Electric school laboratory unit built in a NEMA 213 frame size, but rated at only two horsepower. The advantage for this test was that this particular motor type had all 72 ends of its 36 individual winding coils brought out to a large circular terminal board. As a result, any reasonable number of coils may be made into pole-phase groups. The motor can then be run as a two-, four-, six-, eight-, or twelve-pole machine. The four-pole connection can be made as a consequent pole portion of a two-pole/four-pole two-speed unit or as a normal four-pole arrangement; or it can be used to simulate the unequal pole areas of a typical PAM motor. Further, any reasonable combination of series or parallel wye or series or parallel delta can be conveniently set up. Other, more unusual connections can be used on this machine. Here it was used to show the benefits or penalties of consequent pole and PAM winding connections.

If one takes the care to follow the relatively complicated curves on Fig. 17-3, the following comparisons can be made:

Test 1 shows a normal four-pole parallel wye connection where there were three coils in each pole-phase group and the winding connections and arrangements were perfectly symmetrical. Note that this machine was arranged for 220 V service on a series wye connection, but that here an unusual 110 V three-phase line was used so that the line voltage could be held the same for all of the tests. The resulting line currents are exactly double what they would be for 220 V connections, but this was chosen because 220 V service could not be used in some of the other connections. It was the *relative comparison* that was desired and achieved. Test 1 is completely normal for a two hp motor on three-phase service. The maximum efficiency was shown at 80% of rated load, or 1.6 hp. The power factor reached 84.5% at rated load. The line current ranged from about 6.0 A at no load to about 11.1 A at full rated load. The shaft speed showed a normal droop to about 1760 rpm at full load.

Test 2 shows the results when the unit was reconnected with two coils per pole-phase group instead of three, as before. In order to accommodate the six-pole connection and its lower synchronous speed of 1200 rpm, a series delta connection was used. This is because the same total magnetic flux could generate only about two thirds as much back emf. Actually, this factor is $(1/2) \div (1/\sqrt{3}) = 0.866$ in going from parallel wye to series delta, so the coils then operate at a bit more than the 0.667 factor of voltage that might have been desired. Since the coils in this specific machine span five slots in either connection, they have a very low pitch factor k_p when in four-pole configuration and a higher k_p when in six-pole, and this partially compensates for the voltage-per-coil discrepancy mentioned above. The low k_p when in four-pole connection also favors operation as a consequent pole motor. The operation was again in the normal range. The fact that the efficiency now peaks at almost exactly the rated 1.33 hp shows that the motor might well have operated slightly better on a bit lower voltage, but the effect is not serious. The Test 2 current is a bit high, which confirms this judgement that this particular configuration is better suited to a lower voltage. Test 4 ahead shows the same effect as will be discussed. However, both Test 1 and Test 2 can

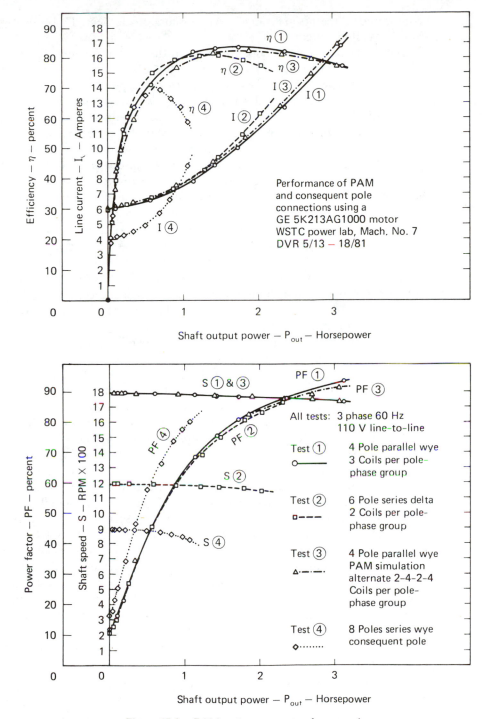

Performance of PAM
and consequent pole
connections using a
GE 5K213AG1000 motor
WSTC power lab, Mach. No. 7
DVR 5/13 – 18/81

All tests: 3 phase 60 Hz
110 V line-to-line

Test ①   4 Pole parallel wye
           3 Coils per pole-
           phase group

Test ②   6 Pole series delta
           2 Coils per pole-
           phase group

Test ③   4 Pole parallel wye
           PAM simulation
           alternate 2-4-2-4
           Coils per pole-
           phase group

Test ④   8 Poles series wye
           consequent pole

Figure 17-3. PAM and consequent pole comparison.

be considered in the normal range. The discussions in Sect. 17-4-3 below should clarify this voltage effect.

Test 3 shows the motor operation when the motor is connected to simulate a PAM motor operating on its *higher* speed, in this case a four-pole/six-pole machine where the six-pole configuration would *exactly* correspond to Test 2. Here there are six pole-phase groups of coils in each phase, or two coils per group. However, for four-pole operation, every other pole-phase group in a particular phase is connected with *two* of the two coil groups. Stated a different way, a four-pole phase was connected with a two-coil group, a four-coil group, another two-coil group, and again another four-coil group. Each phase was treated the same way, but rotated in physical position so that the overall result was symmetrical. Test 3 should be compared to Test 1 to see how alternate four-pole connections perform. Again, parallel wye was used. Notice that the efficiency curve has the same shape, but drops by 1.3% to 82.2%. The Test 3 curve peaks a little nearer to the two hp rating point. The speed curves are indistinguishable. Power factor and current curves have the same shape in Tests 1 and 3, but Test 3 shows 0.1 to 0.2 A higher current and about one percent lower power factor. These small discrepancies might even disappear on a slightly lower voltage or if the coils had all been built with some more turns. The point of all this is that a PAM four- to six-pole machine in this size would operate at least as well as Tests 3 and 2 respectively, or even fractionally better, since turns per coils or a lower line voltage would improve operation at either of the two conditions.

Test 4 was with the same three coil groups per pole-phase group as in Test 1, but in series wye connection as a consequent pole eight-pole configuration. In fact, Test 1 and Test 4 used the *same* coil groups, and the external connection corresponded to Fig. 21-8. Thus, parallel wye operation produced four poles and series wye operation produced eight. Here, as is usually the case in consequent pole operation, a significant loss of efficiency was shown. The efficiency curve peaks at only 69.2%. On the other hand, the line current is significantly lower and the power factor significantly higher. This means that under Test 4 conditions, the motor is operating *below* its best voltage. The drooping speed characteristic confirms this. Operation in series delta instead of series wye, corresponding to Fig. 21-6, had the opposite effect. The motor was noisy and it drew twice as much current, until the line voltage was reduced to 90 V. This was because, even with the low k_p pitch factor discussed before, the iron saturated before the required voltage per coil was generated as back emf. For best operation on series wye, the motor needs fewer turns per coil if the operation is to be on 110 V, or more turns per coil if the eight-pole operation is to be series delta on 110 V. Either way the four-pole operation would be affected somewhat adversely, since it is nearly optimum as shown. This is typical of the consequent pole circuit operation.

In summary, then:

(a) Consequent pole operation achieves a 2 : 1 speed range only, four-pole/eight-pole or two-pole/four-pole, etc. Consequent pole operation results in more or less degraded operation at one of its speeds, usually the low speed, where less power is normally required. If operation is to be at constant horsepower, with the connections discussed along with Fig. 21-6, then the compromise is usually taken at the higher speed. However, consequent pole construction and operation is simple and has been in use for over 80 years.

(b) PAM operation achieves any reasonable speed range except 3 : 1. The more usual speed ranges of 3 : 2, as in six-pole/four-pole discussed above, are easily achieved. In fact, the comparative performance is still more closely matched to a single-speed motor at each speed as the speed ratios more closely approach each other, such as 4 : 3 or 5 : 4. PAM operation has a minimum loss of efficiency and thus should become more widely used. The external connections required are the same as for consequent pole operation, so no installation or service penalties will be encountered. It is a pity that PAM motors are not made in the USA, but foreign manufacturers will supply the market until production is licensed and started in the States. The last eight or ten years have seen a widespread shift toward capacitor-run single-phase motors because of their better efficiency. The next few years may see a comparable shift toward PAM motors for multispeed service.

(c) Two-winding motors will continue to be built for years to come, but they are built on a larger frame size for the same power because space is needed for the deeper slots required to make room for adequate coil wire sizes. They have the advantage that they are widely available and well understood.

17-3-4 Motor Speed Change by Voltage Change.

A squirrel-cage or wound-rotor motor can be controlled by dropping the voltage. The control is crude and inefficient and therefore only used in very small sizes, particularly in small single-phase drives. A study of Fig. 17-1 shows that any induction motor has a slope to its speed versus torque curve within its normal operating range. Since the torque is proportional to the voltage squared, a reduction of the voltage seriously reduces the torque and causes increased slip. The result is an unstable speed and an overheating problem. The motor operation slides down its new speed versus torque curve and operates much nearer to its maximum torque. This method of speed control does not have a wide useful range and is not frequently used.

17-3-5 Motor Speed Change by Rotor Resistance.

The wound-rotor motor has the ability to adjust its rotor resistance. This effect can be seen in Fig. 17-2. Note that when lightly loaded the motor persists in operating at nearly its particular synchronous speed. When loaded at or near to its rated torque, the operating speed can be smoothly and continuously raised or lowered by changing the external rotor circuit resistance. A precarious speed reduction of as much as 50 percent can be achieved, but the slope of the curve is too steep for a steady control. Remember the comments about the energy transfer to the external resistances and the drastic loss of efficiency.

With special provision in the rotor winding configuration, a WRIM may be operated as a consequent pole if its stator windings are switched. This procedure has been used in railroad traction motors in the past.

17-4 INDUCTION MOTOR EFFICIENCY

The efficiency of any process or mechanism is assuming a greater importance as the relative cost of energy continues to climb. Induction motors obey the same laws as any other device; efficiency represents the ratio of output divided by input. The same question of formula choice that was discussed in Sect. 7-1 is applicable here, and this time the best choice is Eq. (7-4), because with a motor the input quantities are readily taken.

$$\text{Efficiency} = \frac{(\text{input} - \Sigma \text{ losses}) \ 100}{\text{input}} = \eta\%$$

Two principal methods are used in induction-motor efficiency determinations, depending upon whether the motor can be fully loaded or whether load simulation must be used.

17-4-1 Efficiency from No-Load and Blocked-Rotor Tests.
An induction motor is in many ways related to a transformer. As a result, the open-circuit test of Sect. 15-6 and the short-circuit test of Sect. 15-5 are quite applicable. One specific modification to these tests is described in Sect. 16-2-1 as the *no-load test*, which is the motor equivalent of an open-circuit test. The other modification is described in Sect. 16-2-2 as the *blocked-rotor test*.

The no-load test yields the rotational losses, which are the mechanical and the magnetic losses. As mentioned, the stator copper losses are not negligible owing to appreciable line currents even when running light or no load. The applicable equation for three-phase motors is repeated from Eq. (16-6) as modified in Sect. 16-2-1:

$$P_{\text{rot}} = \sqrt{3} \ V_l I_l \cos\theta - 3I_s^2 R_s$$

and also

$$\Sigma \text{ watts} = \sqrt{3} \ V_l I_l \cos\theta$$

These can be stated in words as rotational power loss = total input watts − stator copper loss. This situation is true because the unloaded rotor is operating at a very small slip and is thus seeing negligibly small voltage, current, and rotor copper loss.

The blocked-rotor test, as described in Sect. 16-2-1, yields values for the reflected rotor parameters. The total power drawn by the motor at the reduced voltage during this test is dissipated in the stator and rotor copper losses. The voltage is low and core losses are small, since only a small part of the basic magnetic flux flows.

The relationship shown in Eq. (16-7) produces the equivalent resistance *per phase* of the stator plus the rotor reflected to the stator:

$$R_{es} = \frac{P_{br}}{3(I_{br})^2}$$

Since this relationship is true, without regard to whether the stator is connected in wye or delta or whether the rotor is squirrel cage or wound, it can be used universally. Some references use this resistance as the resistance between lines, which would be *twice* this value. When this is done, it is usually shown as R_{el} or equivalent resistance between lines. By this fashion, $R_{el} = 2R_{es}$ in Y. Be aware of which value you are using. It makes a *2-to-1 difference!*

The stator resistance test in Sect. 16-2-3 will yield a value of the *stator resistance only* that is usable in the rotational loss test. Remember that $R_s = R_{dc}(1.25)$, as shown in Eq. (16-12). This value should be identified as to whether it is from line-to-line or of *one phase* only. Either way can lead to correct results, but be careful. A typical use of these measurements is shown next.

Example 17-2. A three-phase, 220 V, 60 Hz, two-pole induction motor has a rated current of 13.0 A at 0.885 PF at 3500 rpm (366.5 rad/sec) while carrying 5 hp (3.73 kW). A no-load test discloses that I_l = 6.5 A, V_l = 220 V, Σ watts = 350 W. A blocked-rotor test is recorded as I_l = 13.0 A, V_l = 51.2 V, Σ watts = 448 W. Determine the following:

(a) Equivalent total resistance per phase.
(b) Rotational loss.
(c) Equivalent copper loss at 0.5, 0.75, full, and 1.25 times rated load.
(d) Efficiency at each of these loads.

Solution:
(a) Using Eq. (16-7),

$$R_{es} = \frac{448}{3(13.0)^2} = 0.883 \ \Omega/\text{phase}$$

Note that this resistance assumes a working rotor circuit and is not quite correct for use in a rotational loss test, but it is used since it gives a pessimistic or safe result (lower efficiency than actual).

(b) Rotational loss from Eq. (16-6):

$$P_{rot} = 350 - 3(6.5)^2 0.883 = 238 \ \text{W}$$

(c) The full-load total copper loss from the data is 448 W. At 0.5 load,

$$448(0.5)^2 = P_{0.5} = 112 \ \text{W}$$

At 0.75 load,

$$448(0.75)^2 = P_{0.75} = 252 \ \text{W}$$

At full load,

$$P_{1.0} = 448 \ \text{W}$$

At 1.25 load,

$$448(1.25)^2 = P_{1.25} = 700 \ \text{W}$$

Remember that copper losses vary as the *square* of the load.

(d) Efficiency from Eq. (7-4):

$$\eta = \frac{\text{input} - \Sigma \text{ losses}}{\text{input}} (100)$$

The input is not supplied for all loads specified, so we must *approximate*. With this process, the input power is taken from the nameplate ratings at full load; $P_{\text{in}} = \sqrt{3} (V_l)I_l \cos \theta$:

$$P_{\text{in}} = \sqrt{3} (220)13.0(0.885) = 4384 \text{ W}$$

$$\eta_{0.5} = \frac{[0.5(4384) - (238 + 112)]100}{0.5(4384)} = 84.0\%$$

$$\eta_{0.75} = \frac{[0.75(4384) - (238 + 252)]100}{0.75(4385)} = 85.1\%$$

$$\eta_{1.0} = \frac{[4384 - (238 + 448)]100}{4384} = 84.3\%$$

$$\eta_{1.25} = \frac{[1.25(4384) - (238 + 700)]100}{1.25(4384)} = 82.9\%$$

This conventionalized method is most applicable on *very large motors* that cannot be fully loaded conveniently or safely.

17-4-2 Efficiency from the AIEE Load Slip Method.

This method is recommended as a standard means for testing induction motors that can be conveniently physically loaded. Quite obviously, a direct dynamometer method would still be preferable, but a dynamometer is not often available. The AIEE method can be used in place on an installation if voltmeters, ammeters, and wattmeters of the proper range are available.

The AIEE method depends upon some of the relations developed earlier in Chapter 16. Equation (16-19) states that RPI = RCL + RPD, or transposed that rotor power developed (RPD) equals rotor power input (RPI) minus rotor copper loss. Equation (16-25) states that RCL = RPI(s), or that rotor copper loss equals rotor power input times slip (in decimal form). This relation enables the following steps to be taken for this efficency test:

(1) Determine the stator resistance R_s from dc voltmeter-ammeter methods and reduce to a per phase equivalent from Eq. (16-12).

(2) Determine the rotational loss from the no-load test from Sect. 16-2-1 and as related in Eq. (16-6).

(3) Test and record input parameters and *actual speed* at each convenient load level. The load used may simply be the driven machine working in normal fashion at different loads. Note that the speed must be determined *very accurately* or large errors in the value of slip will be introduced.

(4) The stator copper loss is determined at each load level and at no load using recorded line currents and the value of R_s that has been found in part a. The loss for three phases is $3I_s^2 R_s$. It does not matter if the motor is connected in wye or in delta if the *per phase* current and *per phase* resistance are used.

(5) Total metered power input minus the stator copper loss at each current level is the rotor power input RPI for each level. At *no load*, the RPI is the rotational loss, since it is all consumed in turning the rotor.

(6) RPI(*s*) = RCL for each level.

(7) Rotor power out (RPO) is RPI − RCL − P_{rot}. RPO is the watt equivalent of the mechanical power developed.

(8) Efficiency is, as always, (output/input) × 100 = $\eta\%$. Thus, η = (RPO/Σ watts$_{in}$) 100.

Example 17-3 will illustrate the application of this method.

Example 17-3. The three-phase motor of Example 17-2 is tested at its full nameplate load level and at the *same* no-load conditions. The stator resistance test discloses an R_s = 0.525 Ω/phase. Calculate (a) rotational loss, and (b) efficiency at full rated load.

Solution:

(a)
$$350 \text{ W} - 3(6.5)^2 0.525 = P_{rot} = 283 \text{ W}$$

Note that this loss is probably *more realistic* than the P_{rot} = 238 W in Example 17-2 owing to the better value of R_s.

(b) The stator copper loss at the full-load point is $3(13.0)^2 0.525 = P_{scl} = 266$ W. The rotor power input is then

$$\text{RPI} = \sqrt{3} \ (13.0)220(0.885) - 266 = \text{RPI} = 4118 \text{ W}$$

The full-load slip at load from Eq. (16-2$_E$) is

$$\frac{3600 - 3500}{3600} = s = 0.0278$$

Note, use *decimal* slip in this situation. It may not be possible to get three-significant-figure accuracy.

Using whatever means of speed determination is available, remember that the no-load speed will be a *very small amount below* the synchronous speed for that motor. If testing a four-pole motor such as in this problem, the no-load speed on 60 Hz *cannot* be over 1800 rpm (or 188.5 rad/sec). The no-load speed is probably about 1794 to 1797 rpm (187.9 to 188.2 rad/sec). A hand-held tachometer is often in error by 5 to 10 percent. Thus, if 1825 rpm is disclosed at no load, 30 rpm should arbitrarily be subtracted from *all* the no-load and loaded speed readings. A no-load speed of 1725 rpm is equally unacceptable and should be arbitrarily adjusted. The arbitrary adjustment factor must be recorded on the data sheet and justified by appropriate notes.

The rotor copper loss is

$$\text{RCL} = \text{RPI}(s) = 4118(0.0278) = \text{RCL} = 114 \text{ W}$$

Rotor power output is then

$$\text{RPO} = \text{RPI} - \text{RCL} - P_{\text{rot}}$$

$$\text{RPO} = 4118 - 114 - 283 = \text{RPO} = 3721 \text{ W}$$

The efficiency is therefore

$$\frac{\text{output } (100)}{\text{input}} = \frac{3721(100)}{\sqrt{3}\,(13.0)220(0.885)} = \eta = 84.9\%$$

This efficiency is slightly higher than that shown in Example 17-2, which is typical of the two methods. Note that the RPO gives the mechanical power directly for an SI unit situation where it would be stated at 3.721 kW. This case is $3.721/0.746 = 4.99$ hp $\cong 5$ hp when working in English units.

17-4-3 Efficiency Improvement with Lower Voltage at Reduced Power.
The continuing and increasing necessity to reduce energy waste wherever possible makes it important to match the motor characteristics to the load requirement. Any induction motor will operate easily when not loaded to or near its rated power. However, there is a continuous need to generate a back emf that nearly matches the line voltage. This is required to limit the input line current which, in turn, requires an excitation current to develop the needed magnetic flux. The way to reduce this needed excitation current is to reduce the line voltage. As a result, the reduced level of needed back emf can be supplied by a lower magnetic flux.

As in a transformer, the induction motor is self-adjusting in this respect. Since the excitation current is all loss and is also nearly 90° lagging, it makes sense to reduce it as far as possible and yet have the motor be able to carry its load. The benefits of reduced line voltage, where possible, are improved efficiency and power factor.

The methods of achieving low voltage operation at part power and a discussion of the actual measured benefits that can be realized are shown in Section 21-5. This section discusses the control circuits needed and shows the benefits obtained by using lower voltage when operating at less than rated power.

QUESTIONS

17-1. What portion of the power that a motor is capable of developing is normally used as the rated power?

17-2. What is the utility of the various NEMA motor classes within the same power rating?

17-3. If a motor is required to develop a starting torque of a bit more than double its normal operating torque, what NEMA classes are applicable?

17-4. What is a double-cage squirrel-cage rotor?

17-5. What are the benefits of a double-cage rotor construction?

17-6. What added equipment is required to take full advantage of the torque range of a wound-rotor motor?

17-7. Why is the operating efficiency of a wound-rotor motor so seriously affected when external resistance is left in the rotor circuit?

17-8. Name three overall means of controlling the speed of induction motors.

17-9. What is a two-winding, two-speed induction motor?

17-10. What is a consequent pole two-speed induction motor?

17-11. How may a four-speed induction motor be constructed?

17-12. Why is a four-speed induction motor rarely used?

17-13. What limits speed control by voltage change?

17-14. What type of efficiency test should be used on an induction motor when it is not practical to operate the motor under load?

17-15. What information is gained in the no-load motor test?

17-16. What information is gained by the blocked rotor test?

17-17. What data are needed to perform the AIEE load slip efficiency test on an induction motor?

PROBLEMS

17-1. An induction motor is an NEMA class A and is rated at 5 hp on 440 V line-to-line. What ratio of reduction of starting torque is to be expected if it is started on 60 percent of normal line voltage?

17-2. A wound rotor is found from a blocked-rotor test to have a blocked-rotor inductive reactance of $0.412 \ \Omega$ and a rotor resistance of $0.103 \ \Omega$, each value being per phase. It is wished to set the external resistance for maximum starting torque. What value of external resistance per phase should be used?

17-3. The motor of Problem 17-2 is to have sufficient added resistance in its external rotor circuit so that its starting torque will be reduced to the amount it would have with no external resistance. What value of R_{ex} should be used per phase? Use Fig. 17-2.

17-4. What is the maximum rotor efficiency that can be expected in a wound-rotor induction motor that is operating at 0.25 slip?

17-5. A squirrel-cage induction motor is used for high rotative speeds when the frequency is increased. If a two-pole motor will turn at 3480 rpm when driving its load on a 60 Hz supply, what speed may be expected on 180 Hz?

17-6. Variable-speed high-speed spindles are needed for a high-production wood-working plant. If the motors will turn at 365 rad/sec on 60 Hz power, what speed will be obtained if the supply is increased (a) to 150 Hz, (b) to 240 Hz?

17-7. With the high-frequency situation in Problem 17-5, what supply voltage will be needed if the 60 Hz supply is 220 V line-to-line?

17-8. With the motor in Problem 17-6, what voltages will be needed if the original supply is 110 V line-to-line for (a) 150 Hz supply, (b) 240 Hz supply?

Answers. **17-1**, 0.36-to-1; **17-2**, 0.309 Ω; **17-3**, 0.515 Ω; **17-4**, 75%; **17-5**, 10 440 rpm; **17-6(a)**, 913 rad/sec, **(b)** 1460 rad/sec; **17-7**, 660 V; **17-8(a)**, 275 V, **(b)** 440 V.

18

The Synchronous Motor

As the name implies, the synchronous motor is related to the synchronous alternator. The name also implies synchronous motion. The real meaning is that a synchronous motor *moves in synchronism* with the rotating magnetic field that is created by its *stator* windings.

The speed–pole–frequency relationship that exists, as has been shown, is found by rearranging Eq. (11-2$_E$), which is $f = PS/120$, or Eq. (11-2$_{SI}$), which is $f = P\omega/4\pi$. This produces $S = 120f/P$ rpm, which is Eq. (16-1$_E$), and, in SI units, $\omega = 4\pi f/P$ rad/sec [Eq. (16-1$_{SI}$)].   Here the units are exactly the same as in Chapter 11.

f = frequency in hertz
P = number of poles per phase
S = shaft speed in revolutions per minute
ω = shaft speed in radians per second

The constants were explained in Chapter 11 and are the same here.   Table 11-1 lists some of the pole–frequency–speed relations in English units; Table 11-2 lists the same in SI units.   If Table 11-1 is entered from the left at the supply frequency and from above at the number of poles, the intersection is the rotating speed.   For example, if the supply is 60 Hz and there are eight poles in the machine, the *synchronous speed* is 900 rpm.   Table 11-2 shows that the same inputs produce 30π rad/sec or 99.248 rad/sec.

A synchronous motor will produce exactly the rotative speed dictated by its number of poles and supply frequency.   It is the average speed that is held exactly. The instantaneous angular speed varies a small amount as the load torque changes. However, the *overall speed regulation is zero percent in a synchronous motor.*

A synchronous motor of the larger sizes that are considered in this chapter requires a *dc excitation for its rotating field*. By varying the excitation, the operating power factor of the motor may be varied over a wide range. As a result, a synchronous motor can be adjusted to draw a *leading power factor* from the lines even while it is carrying a normal mechanical load. In large sizes this feature is more widely used than the absolute speed ability. Since many commercial and industrial customers are billed by kilovolt-amperes used instead of kilowatts, an improved power factor may mean large savings to the customer. In the medium range of roughly 50 to 500 hp (37.5 to 375 kW), a synchronous motor may actually cost less than other types and be used because of its first cost.

The smaller synchronous motors, which have permanent magnet or other nonadjustable rotating fields, are used for their precise speed, since their power factor cannot be controlled. These small types will be discussed later along with single-phase motors. This chapter is confined to polyphase, usually three-phase, *separately excited field* synchronous motors. The synchronous motor is then still a *dual excitation* motor in that both its rotor and its stator are excited from outside sources.

18-1 SIMILARITY TO SYNCHRONOUS ALTERNATORS

If, for a moment, we leave out the means of *starting* a polyphase synchronous motor, it then can be the same machine as a synchronous alternator. This is analogous to the dc shunt dynamo which, as we have seen, can be either a motor or a generator depending on its use at the moment. Motor and generator action is inseparable when the machine is loaded, as has also been discussed before.

The strong tendency for one synchronous alternator to pull into phase with another matched unit is due to synchronous motor action. In this situation, as soon as the generated emf of one alternator differs from the other parallel unit, there is a strong circulating current. This current produces a *motor action* in both units so that the one with the least dominant torque is pulled into phase position with the other.

The motor action involved is produced by the attraction between the *magnetic poles*, which are created by the *rotor field coils* and by the *stator coils*. The field coil magnetism is a function of the ampere turns in the field coils. These ampere turns produce a field that is in geometric relation to the coils and the coil core structure. Since the field rotates, its field magnetic flux rotates at the same speed. The stator coils do not rotate, but the individual phases produce a sinusoidally pulsating magnetic field in response to the sine-wave current form. Since the phase coil currents are excited in timed phase relation to each other, the result will be seen to be a uniformly rotating but constant-strength magnetic field. This was discussed in Sect. 16-1-1. There are then *two linked rotating fields*.

18-1-1 Synchronous Motor Action. The magnetic field strength in the stator has been identified as having 1.5 times the maximum strength of any one

component without identifying what the magnetic strength may be. Remember that in a dc motor the current in the armature is ultimately limited by the difference between the line voltage and the back emf. In the dc motor, the shunt field is relatively constant in strength.

The synchronous motor situation is somewhat similar. In this case the rotating magnetic field from the *rotor* is also relatively constant. The synchronous motor is then developing a back emf as it rotates. This is the E_{gp} voltage of a synchronous alternator. The phasor resultant of the line voltage and the back emf forces the current through the impedance of the stator winding. The $I \cos \theta$ component is automatically adjusted to match the load requirements. Whatever the load is, within reason, an appropriate line current is drawn from each of the three phases. It is the phasor combination of these three limited phase currents that produces the constant rotating magnetic field in the stator. The strength of this rotating field flux is then a function of the load. A small flux corresponds to a light load, and a large flux corresponds to a heavy load. The *motor action* is then the result of (1) the magnetic attraction between this rotating magnetic field in the fixed stator, and (2) the fixed strength but mechanically rotating field in the rotor.

The rotating field of the *stator*, which has whatever flux strength the situation requires, is linked with and pulls along the flux that emanates from the *rotor*. This magnetic linkage *does not slip* and causes *synchronous* rotational speed.

18-2 SYNCHRONOUS MOTOR STARTING

A synchronous motor is not inherently self-starting. One or the other of the various starting aids to be described below must be provided. The reason for the inability to start is that from the instant the line switch is closed the stator-produced magnetic flux is moving at full synchronous speed. Since the rotor is not yet moving, there is alternate forward and reverse torque as the stator flux passes the rotor flux. As the moving flux approaches an opposite pole of the now fixed rotor flux, a torque is developed, first in the wrong direction and then, as the rapidly moving flux passes the aligned position, finally in the correct direction. The rotor will move minutely, but cannot accelerate to synchronous speed before the correct moving flux gets away. As a result, a synchronous motor will growl loudly and draw a very high current. The situation is exactly like an attempt to synchronize two alternators with one standing still and the other at full speed. Therefore, synchronous motors must receive starting aid in one of the following ways:

(1) Mechanical drive is provided by what is normally the load. For example, if the load is a dc shunt generator, such as part of an urban electric railroad power supply, there will normally be one or more generators on the line. If not, there will be a standby battery bank. The dc motor is used to bring the ac synchronized motor up to synchronous speed. A normal synchronizing process is carried out to match the line frequency, phase sequence, and phase position. This requires that a dc source excite the synchronous rotor field so that it can generate the proper

E_{gp} voltage. When synchronized, the line switch is closed and the motor is then floating on the line and unloaded. Finally, the dc machine field current is increased so that it shifts from motor action to generator action. This loads the synchronous motor and the starting cycle is completed. Obviously, all these procedural steps are less than ideal and would be performed only to gain the advantage of better ac load power factor. If the load could not be self-driving it could not be performed at all.

(2) The synchronous motor requires a dc supply for its rotating field. This supply is frequently in the form of a small direct-connected *exciter* generator. By taking advantage of a dc motor's inherent overload ability for a short duty cycle, this small motor can be used to accelerate the large synchronous motor. After synchronizing, the exciter can return to its normal duty. This option obviously requires a substantial dc power source. Also, the driven load for the synchronous motor must be of such a character that either it can be accelerated by the exciter or it can be declutched.

(3) A third alternative provides a small ac induction motor to accelerate the synchronous motor. It has been shown that an induction motor runs at *less* than synchronous speed. If it is direct connected, the induction motor usually will have two fewer poles than the synchronous unit. This will allow a full synchronous speed to be reached even though the induction motor is "slipping" heavily. This alternative means that the character of the synchronous motor load must be such that an economically small induction motor can handle it briefly.

(4) The last and by far most attractive alternative is to build a degree of squirrel-cage windings directly into the synchronous motor rotor. These types of windings have been discussed in Chapter 16. However, by properly proportioning additional windings, the synchronous motor may be started even though directly connected to its full high torque load. One or the other variation of the induction-motor rotor windings is now the preferred means of starting synchronous motors. In fact, if the synchronous motor is to be coupled to a very high starting torque load, such as a jaw crusher or a large conveyor, this may well be the only reasonable way to start it. The dc field is normally shorted during the start to serve as an induction winding also.

When the induction-motor rotor windings are present, they are frequently known as *damper* or *amortisseur windings*. The other function of these windings is to damp out any tendency to hunt or oscillate ahead and behind the actual synchronous angular position.

A synchronous motor with a high-torque, wound-rotor, induction-motor-type starting winding is the only type of synchronous motor that will continue to run if the torque load exceeds the pull-out torque for synchronous operation. This type of synchronous motor with its *phase wound* or Simplex rotor may be immediately recognized by the presence of *five slip rings* on the rotor. There are three slip rings for the induction windings and two for the normal rotating field. The other types will stall with an overload.

18-3 SYNCHRONOUS MOTOR POWER FACTOR CONTROL

A synchronous motor will run and carry its load with a wide range of rotating field dc excitation. Since the relationship between the rotating field and the stator coils is the same as in a synchronous alternator, the resulting generated voltage per phase E_{gp} can also vary over a wide range. The E_{gp} voltage would exactly oppose the applied voltage per phase V_p if there were no requirement to supply torque, and if E_{gp} were adjusted to equal V_p. However, even if running light, there is some torque requirement. When running with a normal load, there is the full normal torque requirement and therefore the full requirement for the stator current required to develop the needed strength of the rotating stator magnetic field.

Figure 18-1 shows the various phasor relations for different dc field excitation values. When the field is of such strength that E_{gp} approximates V_p, the condition is called *normal field excitation*. The situation is shown in Fig. 18-1a. Under these conditions, E_{gp} swings around just enough so that the phasor sum of E_{gp} and V_p, which is the resultant voltage E_r, is sufficiently large to produce the required armature or stator current I_a. In a normal stator winding, the resistance R_a is held as low as possible in order to reduce the I^2R losses. The dominant part of the winding impedance is the inductive reactance. The angle between the E_r and the I_a phasors is nearly but not quite 90°. The whole winding impedance triangle fits into this corner of the phasors and is suggested by the dotted line in Fig. 18-1a.

By adjusting dc excitation to the correct amount, the armature current I_a may be made to be exactly in phase with the supplied voltage V_p. This is the situation for unity power factor shown in Fig. 18-1a.

Assuming the same motor load, which will mean the same required input power or the same $V_pI_a \cos \theta$, but with substantially less dc field excitation, the situation shown in Fig. 18-1b arises. Since less dc field excitation means less E_{gp}, the phasor sum of a normal V_p and a low E_{gp} will take a direction of E_r as represented in Fig. 18-1b. Since the requirement is for the same value of effective stator power, I_a must be larger in order that $I_a \cos \theta$ match the original value of I_a in Fig. 18-1a. This means that the E_r phasor must grow proportionately. For E_r to increase with the shorter E_{gp}, the angle α_2 must be larger than that of the original α_1. The only way that a smaller E_{gp} can contribute to a larger E_r is for this angular relation to change. The same phase angle between E_r and I_a will still hold, since the same winding impedance is effective.

Externally, under these conditions, the motor becomes a *lagging power factor load* on the power supply. The percent of lag, or the cosine of angle θ, depends on how short E_{gp} is allowed to become as the field excitation is reduced.

Internally, the phasor relation shown in Fig. 18-1b is created by the *increased* field magnetization that results from a lagging power factor stator current.

The opposite situation holds if the dc field excitation is *increased*. This results in an E_{gp} voltage greater than the V_p voltage, and creates the phasor relation shown in Fig. 18-1c. Under the same load condition, the stator power drawn from the

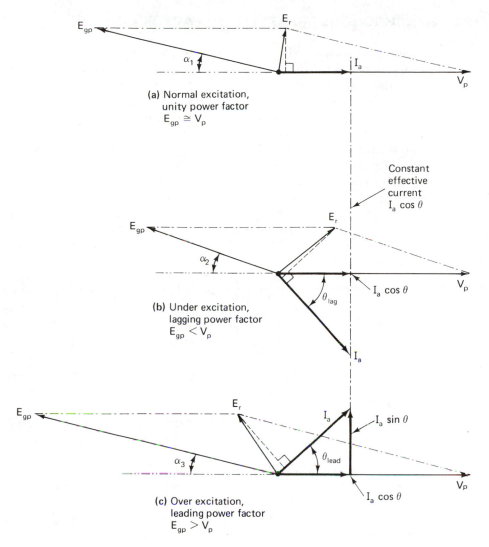

Figure 18-1. Effect of changing field excitation on synchronous motor power factor.

line must be the same, so that again $I_a \cos \theta$ must match the original I_a in Fig. 18-1a.

The phasor relation then becomes changed by the larger E_{gp}. The resultant stator phase voltage E_r is pulled around counterclockwise, as shown in Fig. 18-1c. Since the E_r phasor has rotated, so must the I_a phasor, as the angle between them is still set by the winding impedance triangle. The result is a strong *leading load power factor* with the *overexcited field adjustment*.

When the synchronous motor is deliberately adjusted this way, it draws a *leading I_a* current from the line by contributing an internal $I_a \sin \theta$ reactive current component. At comparable leading and lagging power factors the I_a stator current is the same. The benefit is that the created $I_a \sin \theta$ component is opposite in phase

to whatever $I \sin \theta$ components may exist in the rest of the installation that the synchronous motor serves. Substantial system power factor improvement is achieved commercially by overexciting large synchronous motors while they work.

When a synchronous motor is used solely to produce a large leading $I_a \sin \theta$ current component, it is called a *synchronous capacitor*. This name comes from the effect, which is the same as if a giant capacitor were placed across the line. When the motor carries a normal mechanical load, it is also normally used as a *synchronous power factor corrector* when it is deliberately overexcited.

18-4 SYNCHRONOUS MOTOR V CURVE

The power factor response of a synchronous motor under various dc field excitation currents while holding a constant power is clearly shown by the V-curve test. The curves are so named because of their distinctive shape when plotted. A synchronous motor that is to be tested is connected to instrumentation and a variable load as shown in Fig. 18-2. The two-wattmeter method is shown, but any of the acceptable wattmeter circuits as discussed in electric circuits courses are applicable, bearing in mind that the three phase loads are balanced. Three voltmeters are shown, but only one reading is needed if balance is shown. The motor is tested by applying a load and varying the dc field excitation in logical steps. Data for all meters are recorded at each step so that volt-amperes and watts power can be determined. This allows the power factor of the motor to be determined for each field current setting for each load.

Figure 18-3 shows a typical family of V curves. The curves may be taken

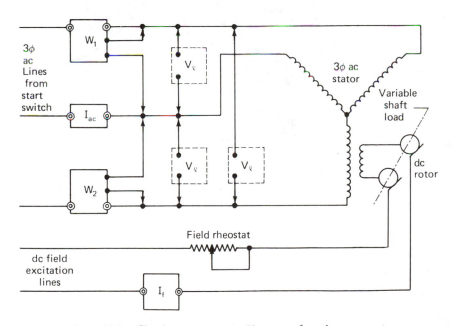

Figure 18-2. Circuit connections for V curves of synchronous motor.

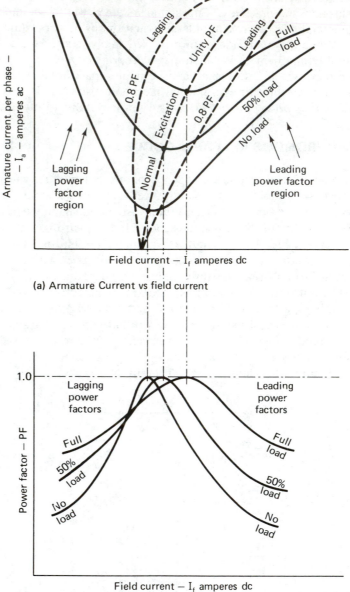

(a) Armature Current vs field current

(b) Power factor vs field current

Figure 18-3. Synchronous motor V curves.

over more closely spaced load increments to truly define the curve shapes, depending on the ease of load control and time available.

Notice that in Fig. 18-3a the no-load curve drops to a minimum, but not to zero. This minimum current can be related to a minimum power necessary to overcome the fixed internal losses, such as the rotational losses that are always present. The shape of the curves clearly shows that for each load there is a distinct minimum armature phase current I_a at a specific dc field current I_f. This specific field current is known as *normal excitation*. Unless otherwise specified, the labeled dc field current for a synchronous motor will be the current that produces minimum armature winding current in the region of 80 percent to full load.

Figure 18-3b shows the same data plotted as load power factor versus dc field current. These curves show that a synchronous motor can be overexcited and carry a substantial leading power factor. This process is limited by the maximum current rating of the stator windings. It is well to realize that, even though the increasing dc field current brings higher and higher leading power factors, the main stator winding current is increasing at the same time. The current-handling capability of the motor is fairly well taxed by full-load currents at 100 percent PF. If it is desired to carry a strong leading PF for load power factor improvement of a factory, and at the same time to power the factory air compressors, conveyor system, etc., the synchronous motor may well need to be a larger size. This is because a synchronous motor may be rated at unity PF or perhaps 80 percent PF at a given load. The nameplate usually states the load and power factor conditions. If a more leading PF is desired, it can be met by a motor of one or more frame sizes larger than the basic power requirement would ordinarily need.

The V-curve intersection with the normal excitation line in Fig. 18-3a illustrates the phasor relation in Fig. 18-1a. The V-curve intersection with the 0.8 PF lagging dotted line in Fig. 18-3a illustrates a phasor relation like that in Fig. 18-1b. The intersection of a load curve with the 0.8 PF leading line in Fig. 18-3a is illustrated in phasor language in Fig. 18-1c.

18-5 SYNCHRONOUS MOTOR POWER FACTOR CORRECTION

The potential for power factor improvement by the use of an overexcited synchronous motor is very useful. However, it is not often practical to attempt to carry the power factor improvement all the way to unity. Remember, the needed kilovolt-ampere reactive (kVAr) component of the armature current may be so large that it would require an impractically large frame size motor. A reasonable and significant improvement has been routinely obtained by using a synchronous motor or motors to power some substantial portion of an industrial plant load. Example 18-1 shows the steps involved.

Example 18-1. A factory has an average total electrical load of 41 300 kW at 0.810 PF lagging. Part of the load is incurred by a large three-phase induction motor of 6800 hp (5073 kW), which operates at 0.730 PF lagging and at 92 percent efficiency. The motor is in need of rewinding and requires extensive mechanical rebuilding so that replacement is scheduled. Two different synchronous motors are investigated, one to carry the same 6800 hp load at unity PF and the same efficiency. The other one is a larger frame unit and carries the same load at the same efficiency and at 0.780 PF leading. Calculate the following:

    (a) Overall system power factor using the unity PF motor.
    (b) Overall system power factor using the 0.780 leading PF motor.
    (c) The difference in required kilovolt-ampere rating of the two motors.

Solution:

(a) The original system kilowatt power will be expected to remain the same through these alternatives since the motor load and efficiency remain the same. Kilovolt-amperes and kilovolt-amperes reactive then are

$$\frac{41\ 300\ \text{kW}}{0.810\ \text{PF}} = 50\ 988\ \text{kVA originally}$$

$$\cos\theta = 0.810, \quad \theta = 35.9°, \quad \sin\theta = 0.586$$

$$50\ 988(0.586) = 29\ 900\ \text{kVAr originally}$$

The original motor kVA and kVAr are

$$\frac{6800\ \text{hp}\ (0.746\ \text{kW})}{0.92\ \text{eff (hp)}} = 5514\ \text{kW motor input}$$

Note: Do not neglect to account for efficiency.

$$\cos\theta = 0.730\ \text{(motor)}, \quad \theta = 43.1°, \quad \sin\theta = 0.683$$

$$\frac{5514}{0.730} = 7553\ \text{kVA motor input}$$

$$7553(0.683) = 5159\ \text{kVAr motor}$$

The original and still factory load less the motor is

$$41\ 300\ \text{kW} - 5514\ \text{kW} = 35\ 790\ \text{kW}$$

The kVAr component is

$$29\ 900\ \text{kVAr} - 5159\ \text{kVAr} = 24\ 740\ \text{kVAr}$$

The unity PF motor will then create a total factory kW the same as the original or 41 300 kW, but no more kVAr than the factory without the motor or 24 740 kVAr. The total factory power factor will then be

$$\text{arc tan}\ \frac{24\ 740}{41\ 300} = 0.599$$

which corresponds to $\theta = 30.92°$; thus

$$\cos\theta = 0.858 \quad \text{or} \quad 85.8\%\ \text{PF}$$

When the data are presented this way, the problem becomes quite an exercise in handling phasor relations of kW, kVA and kVAr. It is excellent practice and should be vigorously pursued by technical students. If the data had been available separately for separate plant load, it would have been much more simple, but plant loads are metered in total and the motor perhaps separately tested. The data on plant load less the motor may not be available.

(b) A synchronous motor with the same horsepower (or kW) and a 0.780 PF leading will have the same kW as the other motors, but will contribute a leading kVAr component.

$$\frac{5514 \text{ kW}}{0.780 \text{ PF}} = 7069 \text{ kVA}$$

$$\cos \theta = 0.780, \quad \theta = 38.74°, \quad \sin \theta = 0.626$$

$$7069(0.626) = 4424 \text{ kVAr leading}$$

The total plant kW is the same 41 300 kW, but the kVAr is less:

$$24\,740 \text{ kVAr} - 4424 \text{ kVAr} = 20\,316 \text{ kVAr}$$

The total plant power factor is then

$$\text{arc tan} \frac{20\,316}{41\,300} = \theta = 26.19°$$

$$\cos 26.19° = 0.897 \quad \text{or} \quad 89.7\% \text{ PF}$$

(c) Step (a) requires a kVA rating the same as its kW rating owing to unity power factor: 5514 kVA. The step (b) situation requires 7069 kVA; hence

$$7069 - 5514 = 1555 \text{ kVA}$$

The motor that produces the improvement in plant power factor from 85.8 to 89.7 percent requires an *additional* kVA rating of over 1500 kVA. This means a substantially larger frame size is needed to dissipate the added heat, since 100 − 92 or 8 percent of this difference is lost in heat in the stator windings. The economics of the choice between alternatives a and b is beyond the scope of this text. However, the engineering data needed for such an economic choice are well within the capability of this level.

18-6 SYNCHRONOUS CAPACITOR

A different circumstance arises when an industrial or public utility installation desires to improve its system power factor but does not have a suitably large shaft load to use an overexcited synchronous motor. In this case, a synchronous machine may be applied and yet perform no shaft work. This type of unit is properly called a *synchronous capacitor*. It can be designed to draw a very substantial kVA load. Here a *leading* power factor of around 0.1 or slightly lower can be expected. As a result, nearly all of the kVA drawn is available as a leading kVAr phasor. Only the rotational losses, the I^2R losses, and the rotating field losses are chargeable in kilowatt power. All the rest of the input can be directly applied to improving a system power factor by canceling a substantial part of the system lagging kVAr vector.

The mathematical procedure is very similar to Example 18-1 except that the kilowatt input to the synchronous capacitor must be added to the existing system kilowatt load. There is no substitution of motor load. Even so, with the very high leading power factor angle, a large power factor improvement will result in a *reduced* system kVA, which will reduce cost.

18-7 SYNCHRONOUS DYNAMO EFFICIENCY

The means of determining the efficiency of a synchronous alternator and a synchronous motor are so similar that they will be considered together. It is necessary to determine the rotational losses, the dc field losses, and the effective ac resistance of the stator windings so that the variable I^2R losses may be found.

Rotational losses may be found by the same running-light method as used in an induction motor. The field current is normally adjusted to the nameplate level. If not, a selected field current is chosen by the V-curve method when the desired motor power and power factor are selected. The test circuit in Fig. 18-2 is suitable. The rotational loss of a synchronous dynamo, modified from Eq. (16-6), is

$$\text{watts } P_{\text{rot}} = \sqrt{3}\, V_l I_l \cos\theta - 3I_a^2 R_a = \Sigma W - 3I_a^2 R_a \qquad \text{Eq. (18-1)}$$

where

> I_a = armature current per phase; this is line current if the dynamo is connected in Y.
>
> V_l = line-to-line voltage; if V_p or phase voltage is used, then $\sqrt{3}$ becomes 3; bear in mind that three-phase power is *always* $3V_p I_p \cos\theta = W$ regardless of whether Y or Δ
>
> $\sqrt{3}$ = factor of difference between line-to-line and phase volts
>
> $\cos\theta$ = motor power factor under this test condition
>
> R_a = stator effective ac resistance *per phase*, discussed under synchronous impedance test, Sect. 13-7 and in Eq. (16-12)

In a similar dc rotational loss test, the I^2R term can usually be ignored. However, in an ac dynamo under no load, the power factor is usually very poor unless the field excitation has been deliberately set for unity power factor. If a rotating field overexcitation current has been set for a strong leading power factor under load, the no-load PF may be approximately 30 percent. With a low power factor, leading or lagging, the I_a current at no load is significant.

The dc field losses are easily found by determining the dc field resistance, R_f, by use of voltmeter–ammeter methods. The field loss is then

$$\boxed{\text{watts } P_f = I_f^2 R_f}$$

Eq. (18-2)

where

I_f = adjusted field amperes used; note that this sets the field flux and thus causes part of the rotational loss

R_f = rotating field ohms resistance terminal to terminal, including slip rings

Since either a synchronous alternator or a synchronous motor is strictly a one-speed device, most of the rotational loss is a constant. The portion of the loss that is dependent on the eddy current and hysteresis losses in the core is a function of the rotating dc field current.

The rotational losses may also be determined during a synchronous impedance test. The prime mover that is used must be calibrated so that its delivered power is known in relation to its input volt, ampere, and watt readings. A review of Sects. 13-7 and 13-8 and Example 13-2 will refresh one on this subject. The heart of the matter is that the known mechanical power delivered to the synchronous rotor during the synchronous impedance *open-circuit* test is the complete rotational loss. Since driving the synchronous dynamo with a calibrated source enables synchronous impedance and rotational loss tests to be performed simultaneously, it has become an approved method of the IEEE. The steps in this procedure are as follows:

Step	Data
(1) The synchronous dynamo is driven at its rated synchronous speed (or as near as can be achieved with the test rig), but with no field excitation.	(1) Prime-mover input watts times known efficiency is the ac dynamo friction and windage power.
(2) Again drive to synchronous speed, but with desired normal dc field excitation. This will either be the value to produce the rated E_{gp} voltage if an alternator, or the desired PF excitation if the unit is to be used as a motor.	(2) The added prime-mover watts input, again times the prime-mover efficiency, represents the ac machine magnetic core losses. The dc field losses can be determined at this time by $V_f I_f = P_f$. These losses and step 1 are fixed losses.
(3) Perform the synchronous impedance short-circuit test per Sect. 13-8 with all due care.	(3) Under these conditions the magnetic core losses are negligible. The calibrated motor input now represents the full-load copper loss plus the friction and windage losses. Copper loss is then step 3 minus step 1.
(4) Complete the synchronous impedance test by measuring the open circuit E_{gp} voltage at the field excitation used in step 3.	(4) The armature phase synchronous impedance Z_s and reactance X_s are then determined by Sect. 13-8.

The choice of method or whether the IEEE test or the self-powered test is used depends upon available apparatus, time, and contractural specifications, if any. However the test is performed, the data are used as in the following example:

Example 18-2. A similar 1000 kVA synchronous alternator to that used in Examples 13-2 and 13-3 is run light to determine its rotational losses. The line-to-line voltage is 2300 V and the no-load line current is $I_l = 20.9$ A. The wattmeters show Σ watts $= 17.1$ kW. The field situation at the desired E_{gp} for full load at 0.8 PF is $I_f = 140$ A and $V_f = 105$ V. Owing to a large frame, the $R_a = 0.155$ Ω. Find the following:

    (a) The rotational loss.
    (b) The dc field copper loss.
    (c) The armature copper losses at 0.25, 0.5, 0.75, full load and 1.25 times full load.
    (d) The efficiencies at these loads at the same 0.8 PF as in Example 13-3.

Solution:
    (a) Directly from input data, Σ watts $= 17.1$ kW. Using a version of Eq. (18-1), where $\sqrt{3}\, V_l I_l \cos\theta$ is given as Σ watts, we have

$$P_{\text{rot}} = 17.1 \text{ kW} - 3I_a^2 R_a = 17\,100 - 3(20.9)^2 0.155 = 16.9 \text{ kW}$$

Be careful between watts and kilowatts!
    (b) The dc field losses are

$$140 \text{ A}(105 \text{ V}) = P_f = 14.7 \text{ kW}$$

Note that a field current was chosen which was found necessary to produce the needed E_{gp}, not just the $V_p = 2300/\sqrt{3} = 1328$ V. This is because of the poor voltage regulation of a synchronous alternator. Had this been an over-excited synchronous motor to be used for power factor improvement, the losses might well be taken at a different but known I_f current.
    (c) The armature copper losses are simply $I_a^2 R_a$ losses times 3 for three-phase coil groups. Rated current at 1000 kVA from Example 13-3 $= I_a = 251.0$ A. At 0.25 output,

$$I_a = 0.25(251.0) = 62.75 \text{ A}$$

$$3(62.75)^2 0.155 = P_{0.25} = 1.83 \text{ kW}$$

By a similar process, allowing for the appropriate fraction of I_a in each case, the armature copper losses in three phase are $3I_a^2 R_a$, so that

$$P_{0.5} = 7.32 \text{ kW}, \qquad P_{0.75} = 16.5 \text{ kW}$$

$$P_{1.0} = 29.3 \text{ kW}, \qquad P_{1.25} = 45.8 \text{ kW}$$

    (d) Efficiency is always available from

$$\frac{\text{output}}{\text{output} + \Sigma \text{ losses}} \, 100 = \eta\% \qquad\qquad \text{Eq. (7-5)}$$

which is appropriate for an *alternator*. If this were being tested for use as a *motor*, *input values* would more likely be available:

$$\frac{\text{input} - \Sigma \text{ losses}}{\text{input}} 100 = \eta\% \qquad \text{Eq. (7-4)}$$

Note: For a *motor*, P_f must also be supplied as an *input* so that $P_{in} = P_{stator} + P_f$. At 0.25 output or 250 kVA and 0.8 PF, using Eq. (7-5),

$$\frac{250(0.8)100}{250(0.8) + (16.9 + 14.7 + 1.83)} = \eta = 85.7\% \text{ eff}$$

At 0.5 output or 500 kVA and 0.8 PF,

$$\frac{500(0.8)100}{500(0.8) + (16.9 + 14.7 + 7.32)} = \eta = 91.1\% \text{ eff}$$

At 0.75 output or 750 kVA and 0.8 PF,

$$\frac{750(0.8)100}{750(0.8) + (16.9 + 14.7 + 16.5)} = \eta = 92.6\% \text{ eff}$$

At full output or 100 kVA and 0.8 PF,

$$\frac{1000(0.8)100}{1000(0.8) + (16.9 + 14.7 + 29.3)} = \eta = 92.9\% \text{ eff}$$

At 1.25 output or 1250 kVA and 0.8 PF,

$$\frac{1250(0.8)100}{1250(0.8) + (16.9 + 14.7 + 45.8)} = \eta = 92.8\% \text{ eff}$$

18-8 SYNCHRONOUS MOTOR CHARACTERISTIC CURVES

Once operating in synchronism, with the starting cycle completed, a synchronous motor has a flat speed versus torque curve. There is no change of speed until the torque load rises to the *synchronous pull-out torque*. This torque value is dependent upon the mechanical and electrical design features of the specific motor. The pull-out torque may be around 200 percent of the normal running torque.

During starting, the characteristic curve of speed versus torque depends upon the starting conditions that are present. Unless some form of damper windings are used, the starting torque characteristic is wholly that of the other machine that starts the synchronous motor. With amortisseur or damper windings, the motor then has starting characteristics similar to the type of induction motor that its auxiliary windings simulate. This was discussed in more detail in Chapters 16 and 17. Figure 18-4 shows a few of the types of curves that may be expected from synchronous motors. Note that the *synchronous pull-in torque* is a lower value than the synchronous pull-out torque. This pull-in torque value is related to the higher part of the speed torque curve of the starting windings. The nearer the

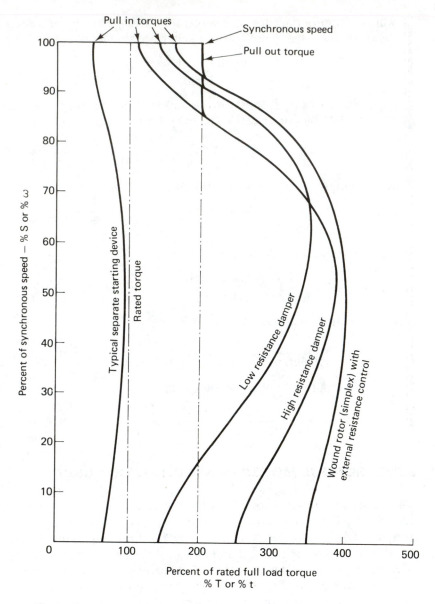

Figure 18-4. Synchronous motor starting and pull-out torque characteristics.

starting circuit can come to a true synchronous speed, the higher the pull-in torque is up to the pull-out torque level.

As a result, if the starting circuit can reach synchronous speed, the synchronous motor can grasp and pull into synchronism a large-torque load. If the starting circuit or starting machine falls below synchronous speed, the synchronous pull-in torque may be only a fraction of the normal running torque. Damper windings of various capabilities are available to match the specific load characteristics. It is best to actively consult the manufacturer's engineering representatives when a nonstandard problem is under study.

QUESTIONS

18-1. What does synchronous mean in regard to a synchronous motor?

18-2. What circumstances require the use of a synchronous motor where its exact rotational speed is not needed?

18-3. Describe some of the means for starting a synchronous motor.

18-4. Why is a synchronous motor not self-starting?

18-5. What action enables a synchronous motor to draw a leading power factor?

18-6. What is a synchronous capacitor?

18-7. Of what use is a synchronous motor V-curve relation?

18-8. What is the economic benefit of using a synchronous motor with a leading power factor in conjunction with a lagging power factor factory or business load?

18-9. What is the economic benefit to a public utility power plant in operating a synchronous capacitor?

18-10. What feature must a synchronous motor have in order to start a very high torque load?

PROBLEMS

18-1. What speed will a synchronous motor develop if it has 12 poles and runs on 60 Hz? Calculate in English and SI units; then verify in Tables 11-1 and 11-2.

18-2. What is the operating speed of a four-pole synchronous motor on 400 Hz? Calculate in English and SI units and verify in Tables 11-1 and 11-2.

18-3. If a synchronous motor is rated at 720 rpm and is to be started with a smaller 60 Hz induction motor, how many poles must the small motor have?

18-4. A factory has an average total electric load of 7850 kW at 0.793 PF lagging. Part of the load is a 1200 hp (or 900 kW) motor driving a low-voltage dc generator. This motor has a loaded PF of 0.831 lagging. The motor is replaced with a synchronous motor that will run at the same 92.5 percent efficiency and 0.800 PF leading. Following Example 18-1 determine:
(a) Original factory total kVA and kVAr.
(b) Original motor input kW, kVA, and kVAr.
(c) Synchronous motor input kW, kVA, and leading kVAr.
(d) Factory total kW, kVA, and kVAr with synchronous motor.
(e) Factory PF with synchronous motor.

18-5. A three-phase multipurpose school laboratory ac machine is run as a synchronous motor. At *no* load and a field excitation of 6.5 A, which represents unity power factor, its input wattmeters read 170 total. It is drawing a line current of 0.45 A on 220 V line-to-line. Its stator resistance per phase in this configuration is $R_a = 0.45$ Ω. What is its rotational loss?

18-6. Additional data are taken on the motor in Problem 18-5, but this time at a field current of 7.8 A. The two wattmeters now read −80 and 260 W, and the line current is 1.75 A. What is the rotational loss?

18-7. The school laboratory motor in Problem 18-5 is run with the field excitation of Problem 18-6, but under its rated load of 2 hp as determined by a dynamometer. The field resistance is 1.84 Ω. The line current is 4.70 A and the two wattmeters read 680 and 1010 W. Follow the procedure in Example 18-2 for a motor and find (a) the dc field copper loss, (b) the armature copper loss, (c) the motor efficiency at this load.

18-8. What is the loaded power factor of the motor under the conditions of Problem 18-7?

Answers. **18-1**, 600 rpm or 20π rad/sec; **18-2**, 12 000 rpm or 400π rad/sec; **18-3**, 12 poles; **18-4(a)**, 9900 kVA, 6031 kVAr, **(b)** 973 kW, 1171 kVA, 651 kVAr, **(c)** same 973 kW, 1216 kVA, 730 kVAr leading, **(d)** same total 7850 kW, 4650 kVAr, 9127 kVA, **(e)** 0.860 or 86.0%; **18-5**, 170 W; **18-6**, 176 W; **18-7(a)**, 112 W, **(b)** 29.8 W, **(c)** 82.4%; **18-8**, 0.944 or 94.4%.

19

The Single-Phase
Induction Motor

Many millions of single-phase induction motors are in use in homes, businesses, and small industries. It is probably safe to say that a single-phase induction motor is always a less efficient substitute than a three-phase motor, but three-phase power is normally not available except in the larger commercial and industrial establishments. Since electric power was originally generated and distributed for lighting only, millions of homes were connected to single phase. With this limitation, early motor-driven appliances depended upon the development of practical single-phase motors.

There are four basic classes of single-phase motors that are used in roughly equal quantities:

(1) Single-phase induction motors that perform the larger home and small business tasks. Furnace oil burner pumps and hot water or hot air circulators are one typical use. Refrigerator compressors and power tools such as lathes and bench-mounted circular saws are also powered with induction motors. The power outputs range from about $\frac{1}{6}$ hp (0.125 kW) up to $\frac{3}{4}$ hp (0.560 kW). Some larger sizes are used, but most will fall in this range. The majority are four-pole motors, with lesser quantities of two and six pole. The pole–speed relationship is the same as that for three-phase motors because the operation of the motors is essentially the same. Thus Eq. (16-1$_E$) or (16-1$_{SI}$) holds:

$$S = 120\,\frac{f}{P}\,\text{rpm} \quad \text{or} \quad \omega = 4\pi\,\frac{f}{P}\,\text{rad/sec}$$

(2) *Shaded-pole motors* are used in the smaller sizes for quiet, low-cost applications. Typical are small fans and blowers of less than the power ratings in part

1. There is, of course, an overlap in use. The shaded-pole motor is simple and reliable, but has low starting torque and poor efficiency relative to the induction motor. This type will be discussed in Chapter 20.

(3) *The universal motor* closely resembles a dc series motor and, as its name implies, will operate on any household ac frequency or on direct current. The real reasons for their widespread use are that they can develop very high speed while loaded and very high power for their size. Any service that requires a speed beyond that possible with a two-pole induction motor, where $S = 3600$ rpm (or $\omega = 377$ rad/sec), is a normal use for a universal motor.

A vacuum cleaner depends upon a suction-induced air flow. Since reciprocating pumps with the requisite air flow capability would be larger and more expensive, a domestic vacuum cleaner uses a high-speed centrifugal blower that is directly connected to a universal motor. Kitchen food mixers, portable electric drills, and portable circular saws take advantage of the high torque at low speeds, the speed control, and the high power relative to its size of a universal motor. This type is also discussed in Chapter 20.

(4) The last basic type of single-phase motor is the *synchronous motor*. The single-phase synchronous motor invariably has one of the various forms of magnetized but unwound rotors. As a result, they do not have the ability to control power factor. The *exact* speed relation of a synchronous motor is used in electric clocks and various cyclic timing devices. These motors are built in the very smallest sizes, where the power output is only a very few watts. Single-phase synchronous motors are also discussed in Chapter 20.

The relative quantities of these four basic types of single-phase motors are almost evenly distributed on a one-for-one basis in a modern home. The single-phase induction motor is usually of so much higher power rating than the other types mentioned that they collectively develop as much power as the aggregate of the other three put together. The special features of the single-phase induction motor are the subject of this chapter.

19-1 SINGLE-PHASE INDUCTION-MOTOR STARTING

Unlike the polyphase induction motor, single-phase induction motors are *not self starting*. The single-phase magnetic field pulsates in strength in a sinusoidal manner, as shown in Fig. 19-1b. The field polarity reverses each half-cycle, but it *does not rotate* by itself. On the other hand, if a single-phase induction-motor rotor is rotated rapidly by some mechanical means, it will continue to run and develop power with no other change.

Although the reasons are different, the result is the same as the three-phase synchronous motor. A single-phase induction motor will not start without modifying its circuit, but will run once it is brought up somewhere near its synchronous speed. Historically, the first single-phase induction motors were started by wrapping a rope or a strap around the shaft and pulling to spin the rotor. Fortunately, this difficulty can be overcome by various relatively simple methods.

The single-phase induction motor is started by *introducing a second phase*

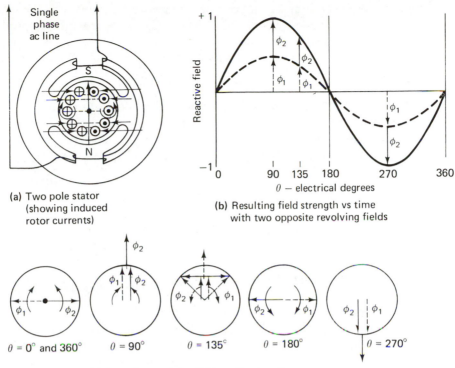

(a) Two pole stator
(showing induced
rotor currents)

(b) Resulting field strength vs time
with two opposite revolving fields

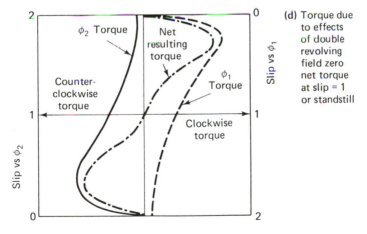

$\theta = 0°$ and $360°$ $\theta = 90°$ $\theta = 135°$ $\theta = 180°$ $\theta = 270°$

(c) Opposite revolving field resultants vs time
ϕ_1 Clockwise, ϕ_2 counterclockwise

(d) Torque due
to effects
of double
revolving
field zero
net torque
at slip = 1
or standstill

Figure 19-1. Torque at standstill in single-phase induction motor.

that is displaced in space and in time. Once started, however, the motor will self-generate the needed second phase internally. The rotation is produced by the rotating magnetic vector caused by the second phase. The single-phase motor then *operates as a two-phase motor* even though it is supplied a single phase from its power source.

19-1-1 No Torque at Standstill with Single Phase. To see why no torque is developed by a single-phase induction motor under start conditions, consult Fig. 19-1a, which shows a two-pole salient-pole stator and a simplified squirrel-cage rotor. The following facts may be observed:

(1) The magnetic field from the stator pulsates sinusoidally with time, but is always vertically up or down in the view. It does not rotate.

(2) Following transformer action, there are voltages induced in the squirrel-cage rotor. These voltages cause circulating currents to flow just as in a short-circuited transformer secondary. The current directions are as shown in Fig. 19-1a with the symbols $\oplus$, meaning into the page, and $\odot$, meaning out of the page. Dotted lines show the circulating currents as if they were in separate loops of conductor rather than a squirrel-cage rotor. The symmetry of the situation is such that the current flow is approximately as shown.

(3) These rotor currents then cause motor action forces orthogonally to the field flux. The forces act about their particular moment arm depending upon their position in the rotor, causing individual torque forces. The torques are, however, completely symmetrical and there is no net rotational torque. The facing pairs of horizontal arrows in Fig. 19-1a represent the symmetrically opposed forces.

19-1-2 Double Revolving Field Theory. The *double revolving field theory* is one of the two principal theories proposed to explain this dilemma of no torque at start and yet torque once rotated. Figure 19-1b shows a sinusoidal curve that represents the *phasor sum* of two smaller fields, which are presumed to *revolve in opposite directions* at the cyclic rate. It has been shown by Ferraris that any sinusoidally pulsating phasor or force can be represented mathematically by two phasors each of one-half the total strength that revolve in opposite directions. Figure 19-1c shows five different situations where constant strength fields ϕ_1 and ϕ_2 are combined by phasor methods. At the $\theta = 0$ or $360°$ position, each individual field is exactly canceled by the other and the phasor sum is zero. At $\theta = 90°$, the phasor sum is then twice the value of one individual phasor. At $\theta = 135°$, the components add to produce 1.414 times a single value. At the same time, their horizontal components cancel each other completely. The $\theta = 180°$ position reproduces the situation at $\theta = 0$ or $360°$, but the phasor components have changed places. The double revolving field components can then be seen to completely simulate the single large pulsating field.

Granting the plausibility of the double revolving field as a substitute for a single pulsating field, we will explore its effect when the rotor is moving. If the effect of only the ϕ_1 revolving component is considered, the *dotted curve* of Fig. 19-1d represents the torque on a squirrel-cage rotor. The curve is drawn from zero slip at the top to a slip of 2 or 200 percent at the bottom. The *first quadrant* of the figure represents a normal SCIM torque with its peak torque at about 0.25 slip. The fourth quadrant shows that the torque continues to drop off as the slip increases, because the rotor inductive reactance X_r increases as the slip increases. Remember that $X_{er} = sX_{ebr}$ from Eq. (16-13). A slip of 2 corresponds to full synchronous speed *backward*. The large increase in rotor reactance reduces rotor current and pulls that current far out of phase with the voltage that is induced in the rotor. Hence, the very low torque at $s = 2$.

The left side, or the second and third quadrants of Fig. 19-1d, shows the effect of revolving field ϕ_2 on the same rotor. As might be guessed, the effect is exactly opposite to that of field ϕ_1. The effect is the solid line in the figure. If the rotor is moving at nearly its synchronous speed clockwise, the ϕ_1 torque is large and the ϕ_2 is small. If the rotor is moving at a similar speed counterclockwise, the result is exactly opposite.

With this theory, if the slip is $s = 1$, it is then $s = 1$ for *both* ϕ_1 and ϕ_2, and the condition at the middle of the figure holds. The two different flux torques are opposite, so there is *no net torque*.

The double revolving field theory can then be seen to account for no torque at standstill or $s = 1$, and full normal torque once well started in either direction. In actuality, a single-phase induction motor, without its normal starting circuit, will run equally well *either way* once it is mechanically started.

19-1-3 Rotor Cross Field Theory. A second theory describes the operation of a single-phase induction motor in a very different fashion. This *cross field theory* is perhaps more realistic for the *operation* rather than the starting condition. The cross field theory also clearly shows why a simulated second phase serves to start a single-phase induction motor.

First, considering the situation of a single-phase field with no auxiliary circuits, the situation in Fig. 19-1a still holds. The motor will not self-start. If the rotor is mechanically started and then allowed to reach its equilibrium speed with a normal small slip, the situation in Fig. 19-2 exists.

(1) Figure 19-2a shows a two-pole field and squirrel-cage rotor as in Fig. 19-1a, but this time the rotor is moving as shown by the arrow. Voltages are induced in the secondary or the rotor conductor bars. The voltages cause currents to flow in the rotor. This time the phase lag in the current flow caused by the rotor inductive reactance results in the current peak being almost 90 electrical degrees lagging behind the induced voltage peak. These circulating currents cause a rotor magnetic field ϕ_r displaced from the stator field ϕ_s. This rotor cross field is not quite 90° displaced from the stator field, because the rotor power factor is not fully lagging since there is some resistance present.

(2) The view in Fig. 19-2b shows the *two fields* that are now present. The main or stator field ϕ_s is drawn in relation to the abcissa scale of electrical degrees. The rotor cross field ϕ_r is drawn as a dotted sine wave that lags the main field by *about* 80°. Since there are losses, the ϕ_r cross field is not quite as strong as the main field. Both the phase position and strength or amplitude effects are shown.

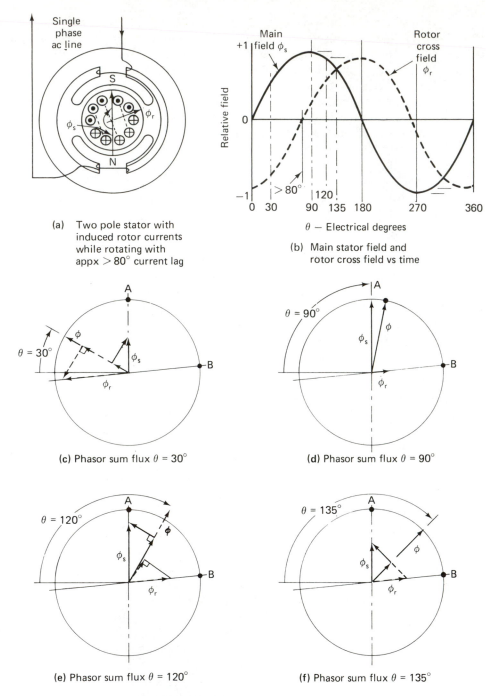

(a) Two pole stator with induced rotor currents while rotating with appx $> 80°$ current lag

(b) Main stator field and rotor cross field vs time

θ — Electrical degrees

(c) Phasor sum flux $\theta = 30°$

(d) Phasor sum flux $\theta = 90°$

(e) Phasor sum flux $\theta = 120°$

(f) Phasor sum flux $\theta = 135°$

Figure 19-2. Development of approximately steady rotating field flux from effect of rotor-induced cross field.

THE SINGLE-PHASE INDUCTION MOTOR CHAP. 19

(3) The result of two sinusoidally pulsating magnetic fields that are out of phase with each other is a rotating field. Since the fields are not quite at 90° to each other and are not quite the same strength, the resultant field is not of a constant flux intensity nor is its motion quite uniform. The phasor resultant field is identified as being *elliptical*. It also gallops and pulsates, to use graphic terms. However, the resultant rotating field is fully acceptable, and the motor performs *almost the same* as it would if it were supplied a true two-phase source.

The resulting rotating field produces torque in the rotor exactly as if it had been produced by a true two-phase field or, for that matter, any number of co-ordinated phases.

Since a single-phase rotor will rotate if it has a rotating magnetic field present, it remains to find a means of generating a rotating field *at the start*. There are a number of practical means for generating a rotating field. All the ones used for single-phase induction motors involve the *simulation of a second phase for a starting circuit*.

19-2 RESISTANCE SPLIT-PHASE STARTING

The problem of starting a single-phase induction motor resolves into the various means of producing a reasonable approximation of a rotating magnetic field. Then when the motor has started and is nearly up to speed, it produces its own rotating field from the cross field effect. The starting circuit becomes a detriment and is removed from the circuit by a centrifugally operated switch that is adjusted to open the starting circuit when the speed climbs to about three-quarters of normal, or s = 0.25.

The usual single-phase induction motor has a squirrel-cage rotor that is in-distinguishable from a two- or three-phase motor rotor of the same size except for the starting switch actuation mechanism. There are many varieties and variations of starting switches, and, good as they are, they are almost always the root cause of an electrical failure in a single-phase induction motor.

The resistance split-phase starting circuit has two meanings within its name. Split phase means that the phase relation of the starting circuit is split apart from the phase relation of the normal running circuit. The resistance term means that resistance is the added element used to create the phase split. Be aware that this type of motor is most often referred to as a *split-phase* motor with the resistance term usually ignored and dropped.

The circuit diagram in Fig. 19-3a shows the connections in a normal resistance split-phase motor. Note that the starting windings are connected in parallel with the running windings, and the starting switch can open the starting winding circuit. The starting winding is *physically displaced by 90 electrical degrees* from the running winding.

The phasor diagram in Fig. 19-3b shows the means of achieving the different or split-phase relation. The normal running winding has a relatively large number of turns and is wound with as large a wire diameter as can be conveniently fitted into the stator slots. This results in a relatively high inductance and low resistance.

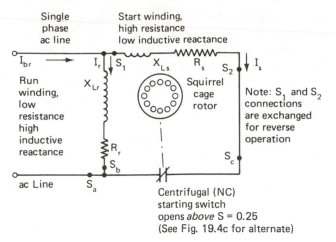

Single phase ac line

Start winding, high resistance low inductive reactance

Run winding, low resistance high inductive reactance

Squirrel cage rotor

Note: S_1 and S_2 connections are exchanged for reverse operation

ac Line

Centrifugal (NC) starting switch opens *above* S = 0.25 (See Fig. 19.4c for alternate)

(a) Complete motor circuit at start

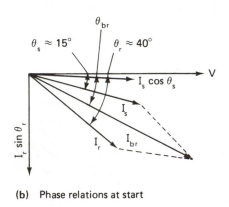

θ_{br}

$\theta_s \approx 15°$ $\theta_r \approx 40°$

$I_s \cos \theta_s$

V

$I_r \sin \theta_r$

I_s

I_r I_{br}

(b) Phase relations at start

Figure 19-3. Resistance split-phase induction motor relations.

Note here that when running normally the running winding resistance is a cause of substantial I^2R copper loss, so it is kept as low as feasible. The usual running winding has a power factor angle of $\theta_r \cong 40$ degrees under blocked-rotor conditions at start.

The associated starting winding has less turns, so its inductance is less than the running winding. It is also wound with a substantially smaller diameter wire, so its resistance is higher than the running winding. Under start conditions, the starting winding has a power factor angle of $\theta_s \cong 15°$. Note that the subscript r means running and s means starting, instead of r being rotor and s stator as in the three-phase circuit nomenclature.

Figure 19-3b shows the relationship between the normal running winding current phasor at a lagging power factor angle of about 40° and the normal starting winding current at 15°. Note that the cosine of the starting current, $I_s \cos \theta_s$, and the sine of the running winding, $I_r \sin \theta_r$, are about equal. These quadrature-related currents achieve the best rotating current effect when they are equal, so the windings are designed to achieve this effect.

The rotating magnetic field that results is very elliptical, but it serves the purpose and develops a starting torque of around 1.5 to 2 times the rated running torque.

19-2-1 Starting-Switch Problems. As mentioned, most of the trouble in these very simple motors is traceable to the starting switch. The low-cost construction that is usually adopted in good competitive business purposes usually results in plain sleeve bearings for the shaft. A mixture of surplus oil and bearing wear particles eventually finds its way to the starting switch contacts. The contacts will then arc and burn owing to the presence of this conductive material. On the other hand, normal use causes some arcing and pitting, and the contact gap changes. There is also an effect of the end play adjustment of the rotor shaft, which may cause a worn switch to malfunction. The spring material that is used to create a snap action in the switch may lose its spring temper owing to overheating in the presence of any contact arcing.

There are two end results to these progressive starting-switch deteriorations:

(1) The most common failure is that the switch will ultimately fail to open when needed, particularly if the motor is heavily loaded and does not pass the switch actuation speed by sufficient margin. If the motor duty cycle is long, the starting winding will almost always overheat and burn out. The *next* attempt to start the motor will result in a growling sound and no rotation. This can cause failure of the main winding unless a fuse or breaker blows.

(2) The start switch may fail to close, whereupon the motor will overheat its main winding without any failure to the starting winding.

These characteristic failures have resulted in the destruction of millions of single-phase motors. If the motor installation is periodically observed, this type of failure can be caught and only the starting switch will need replacement. There is a readily measurable current difference in a running motor with the starting circuit still active. In addition, the motor will hum more loudly than normal because the running cross field produced rotating magnetic field is distorted by the starting circuit.

It is safe to say that 90 percent of the single-phase motor failures are traceable to one or more of the problems above. When a failed motor is disassembled and the smaller starting winding circuit is black from overheating, one may be sure that this has happened. The usual single-phase appliance motor is lubricated for life and buried into its installation where it is forgotten. When the motor won't start and the room is full of acrid smoke, it is too late and replacement is needed.

Example 19-1. A $\frac{1}{2}$ hp (0.375 kW), 115 V resistance split-phase motor has experimentally measured starting currents in the different windings of $I_r = 12.0$ A at $\theta_r = 40°$ lagging and $I_s = 8.0$ A at $\theta_s = 15°$ lagging. At the moment the line switch is closed, what are the following quantities?

(a) The components of each winding current in phase with the line voltage.

(b) The quadrature components of each winding current that lag the supply voltage by 90°.

(c) The total blocked-rotor current and its power factor and power factor angle.

(d) Compare the in-phase component of the starting winding current with the lagging component of the running winding current.

Solution:

(a) $\cos 40° = 0.766$, $\cos 15° = 0.966$.

$$12.0(0.766) = I_r \cos \theta_r = 9.19 \text{ A}$$

$$8.0(0.966) = I_s \cos \theta_s = 7.73 \text{ A}$$

(b) $\sin -40° = -0.643$; $\sin -15° = -0.259$.

$$12.0(-0.643) = I_r \sin \theta_r = -7.72 \text{ A}$$

$$8.0(-0.259) = I_s \sin \theta_s = -2.07 \text{ A}$$

(c) From Fig. 19-3b, the tangent of θ_{br} is $-(7.72 + 2.07)/(9.19 + 7.73)$ = $\tan \theta_{br} = -0.579$.

$$\text{arc} \tan -0.579 = \theta_{br} = -30.1°$$

But $9.19 + 7.73 = I_{br} \cos \theta_{br}$, so that

$$I_{br} = \frac{16.92}{\cos \theta_{br}} = \frac{16.92}{0.865} = 19.6 \text{ A}$$

$$\cos \theta_{br} = \text{PF} = 0.865 \text{ lagging}$$

This is also $19.6 \angle -30.1°$ A in complex notation if this method is preferred.

(d) $I_s \cos \theta_s = 7.73$ A, and $I_r \sin \theta_r = 7.72$ A, so these quadrature components are approximately the same. These values represent a resistance split-phase winding pair that is about optimized for the rotating field component that can be generated by this method.

When the starting cycle is completed, the resistance split-phase motor operates with characteristics that are very similar to those of a three-phase induction motor with the same type of squirrel-cage rotor. The characteristics will be studied with the other single-phase induction motors.

19-2-2 Reversing Resistance Split-Phase Motors. The direction of rotation of a resistance split-phase motor is determined by the starting winding connections. See terminals s_1 and s_2 in Fig. 19-3a for connections that may be interchanged for a reverse direction start. Remember that the motor will *continue to run* in the direction that it is started, as was discussed with Fig. 19-1. As a result, the motor cannot be plugged or reversed while running by simply changing the starting circuit connection, because at running speeds the cross field torque exceeds the starting winding torque. The resistance split-phase motor is then a *reversing* motor, but a *nonreversible* motor in that a reverse may be made from standstill but not while running.

19-3 CAPACITOR-START INDUCTION MOTORS

When a split-phase induction motor is coupled to a load that requires a high starting torque, a resistance split-phase circuit does not supply adequate starting torque. Rather than simply use a motor with a higher power rating, which would then be inefficient with the operating load, a different starting circuit is adopted. The *capacitor-start split-phase* induction motor can develop starting torques of more than 4.5 times the rated running torque when using a moderately high resistance rotor. Note that this type of motor is most commonly identified as a *capacitor-start* motor, and the term split-phase is usually dropped.

Comparing Figs. 19-4 and 19-3, it can be seen that a capacitor is added in the starting winding circuit. This capacitor is chosen so that its capacitive reactance is sufficient to more than overcome the inductive reactance of the starting winding. The power factor angle of the starting winding is usually adjusted until $\theta_s \cong 42°$ *leading*.

A substantial θ_s leading angle is possible if the start winding is identical to that used in the resistance split-phase circuit. However, to get the full benefit of a circular field rotating magnetic field, the start circuit blocked-rotor current I_s should be approximately the same as the running winding current I_r. At the same time, the leading power factor angle θ_s should be as large in magnitude or larger than the lagging power factor angle of the running winding θ_r. The practical optimum that still uses a reasonably sized capacitor achieves starting current phasor relations about as shown in Fig. 19-4b. With $\theta_s \cong 42°$ leading and $\theta_r = 40°$ lagging, a total phase angle between the start and run windings of around 82° is achieved. This compares to a typical resistance split-phase shift of only about 25°.

The ratio of the starting torque of a capacitor-start to a resistance split-phase start motor is the ratio between the sine of the phase split angles. This is not precisely true unless the currents are all the same amperage, but it is approximately true in practice. The result is that the starting torque of a capacitor-start motor, with the same rotor and the same running windings, is about sin 82°/sin 25° = 0.990/0.423 $\cong$ 2.34 times that of a comparable resistance split-phase motor. The starting current, I_{br} in Fig. 19-4b, is also slightly reduced at the same time in comparison to the resistance split phase.

19-3-1 Capacitor Requirements.
The capacitors involved in this service are special dry electrolytic-type units that are rated for the peak ac voltage involved. As a result of careful cost balancing and their very compact size, these units typically are rated for a very short duty cycle of a total of 1 min of service per 1 hour of time. If the load is difficult to accelerate or if the starting switch is consistently slow in opening, the capacitor will fail. Usually, this means that the capacitor and the starting switch need to be replaced. The starting winding is not quite so likely to be "cooked" at the same time. Another advantage is that, because of the generally higher cost and more difficult service, the capacitor-start motor is more

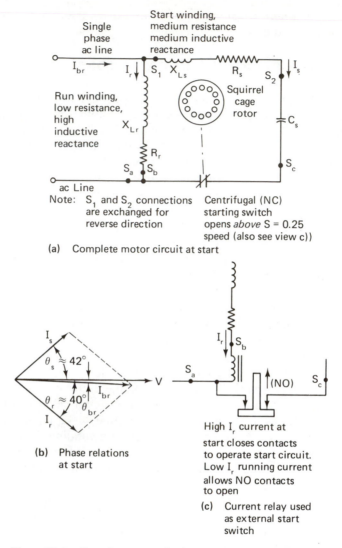

Figure 19-4. Capacitor-start split-phase induction motor relations.

likely to have sealed ball bearings. As a result, the starting switch is much less likely to become contaminated in service.

19-3-2 Starting-Switch Problems. There are some motor applications where a centrifugal starting-switch failure may be particularly damaging to the equipment and to the manufacturer's reputation. Customers have learned to expect that a domestic refrigerator may operate for 20 years or so with little or no service. Since these units are usually built with the motor and compressor in a single hermetically sealed unit, a centrifugal start switch failure would be intolerable.

The answer to this problem is to move the starting switch outside the hermetically sealed unit. Figure 19-4c shows the usual solution to this dilemma, which is a current-sensitive relay. Visualize this circuit interchanged for the centrifugal start switch at points s_a, s_b, and s_c in either Fig. 19-3a or 19-4a. At blocked-rotor conditions, the inrush current is from five to seven times the normal running current in a resistance split-phase motor and from three to five times normal in a capacitor-start motor. As a result, the relay armature snaps in and closes the *normally open* contacts to operate the starting winding circuit. The normal current reduction, as the motor approaches the point where the start circuit should be opened, is enough to allow the relay to *drop out* and open the starting circuit.

By control of the winding turns, the spring loading, and the armature air gap of the relay when both open and closed, the actuation and release points of the relay may be closely held. When the drop-out point is tailored to $s = 0.25$, the operation is indistinguishable from a centrifugal switch. The current relay is easily replaced when required, thus saving the motor.

A related starting relay is a *voltage*-sensitive relay connected across the starting winding and equipped with *normally closed* contacts. This relay is adjusted to recognize the increase in E_s voltage as the slip decreases.

The modern capacitor-start motor is usually built for services requiring $\frac{1}{2}$ hp (0.375 kW) up to $7\frac{1}{2}$ hp (5.60 kW). Sizes beyond this are much more apt to use a three-phase source.

The capacitor-start motor is both reversing and reversible in that its starting circuit torque can overcome its cross field torque and reverse while running.

Example 19-2. A $\frac{1}{2}$ hp (0.375 kW), 115 V capacitor-start split-phase motor with the same running winding as the motor in Example 19-1 has a starting winding and capacitor such that its current is 0.9 times the running winding current and at a θ_s power factor angle of 42° leading. At the start condition, determine the following:

(a) The in-phase and quadrature components of the starting winding current.

(b) The total blocked-rotor starting current and its power factor.

(c) Compare probable starting torque relation between this and the motor in Example 19-1.

Solution:
(a) $\cos \theta_s = 0.743$; $\sin \theta_s = 0.669$.

$$I_s = 0.9 \times I_r = 0.9 \times 12 = I_s = 10.8 \text{ A}$$

$$10.8(0.743) = I_s \cos \theta_s = 8.02 \text{ A in phase}$$

$$10.8(0.669) = I_s \sin \theta_s = 7.23 \text{ A quadrature}$$

(b) From Fig. 19-4b, the tangent of θ_{br} is $(-7.72 + 7.23)/(9.19 + 8.02)$ $= \tan \theta_{br} = -0.0285$.

$$\text{arc tan } -0.0285 = \theta_{br} = -1.63°$$

But $9.19 + 8.02 = I_{br} \cos \theta_{br}$, so that

$$I_{br} = \frac{17.2}{\cos \theta_{br}} = \frac{17.2}{0.999} = 17.2 \text{ A}$$

$$\cos \theta_{br} = 0.999 \quad \text{or} \quad \text{PF} \cong \text{unity}$$

Note that the starting current is *reduced* from that of Example 19-1, or 19.6 A down to 17.2 A, a 2.4-A reduction.

(c) The starting torque will be proportional to the sine of the phase angles between the start winding and the run winding, or sin 82°/sin 25° = 0.990/0.423 = 2.34 to 1.

The capacitor-start motor will have at least 2.34 to 1 *more* starting torque than the resistance split-phase motor. It will probably have even more since the start winding current is *larger* than the motor in Example 19-1. In actual service, the line voltage will not be depressed as far in this case owing to a lower total starting current, so the capacitor-start motor has real starting superiority.

19-4 CAPACITOR-START CAPACITOR-RUN MOTORS

The resistance split-phase and the capacitor-start motors share one undesirable quality in that their operating power factor is lower than that of a corresponding polyphase motor. As a result of a lower power factor requiring a higher current for the same wattage, their stator copper losses are higher than desirable. This reduces their operating efficiencies to a percentage in the high 60s or low 70s.

The *capacitor-start capacitor-run* or *two-value capacitor motor* overcomes these difficulties. One capacitor is chosen for the optimum running power factor and is left in the circuit to the starting winding. This winding is now more properly identified as the *auxiliary winding* since it remains in the circuit. The best running condition is when the auxiliary winding is 90 electrical degrees ahead of the running winding, which is now the *main winding*, while the load current is normal. This means that the motor carries its load as a *two-phase motor* with balanced pulsating fields. The result is a constant field flux that rotates steadily. Under these conditions the same motor will see an improvement of efficiency of from 10 to 15 percent. The operating power factor is more favorable, the rated current is reduced, and the motor is quieter in operation.

The running capacitor is augmented in parallel by a starting capacitor, which supplies the added capacitance needed to produce the auxiliary winding phase shift with the starting current level. Figure 19-5a shows a two-value capacitor motor circuit. The start capacitor is a short-duty-cycle dry electrolytic unit with a capacitance value of from 10 to 15 times that of the run capacitor. The run capacitor, on the other hand, must be of a constant duty type. It is then an oil-filled paper type or one of the proprietary fluid-filled types, such as Pyranol or Inerteen.

19-4-1 Autotransformer and Capacitor Circuits. The circuit shown in Fig. 19-5b produces exactly the same two-value capacitor effect. When a capacitor is connected to the secondary terminals of a step-up transformer, it appears to the

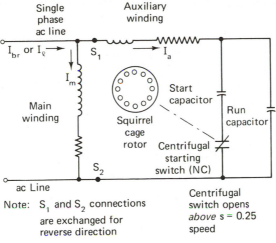

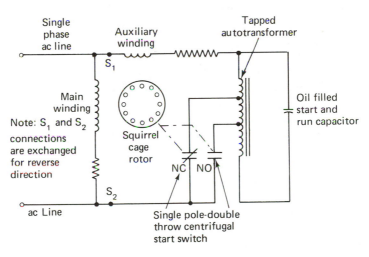

(b) Two value capacitor motor curcuit using
autotransformer to effect change

Figure 19-5. Capacitor-start capacitor-run or two-value capacitor motor.

primary side to be multiplied by the turns ratio squared. This actually uses the effect that was covered in transformer equivalent circuit reflected values. In this manner, a low-capacitance but high-voltage oil-filled capacitor can do both tasks.

If the autotransformer has, for example, a total of 150 turns and is tapped at the 25-turn point its turns ratio is $\alpha = 150/25 = 6$. An 8-microfarad (μF) running capacitor would then appear as $C_{es} = \alpha^2 C_r$, or an equivalent starting capacitance of $6^2(8) = 288$ μF. Normally, the running capacitance wanted is from one-tenth to one-fifteenth of the start value. This could be achieved by shifting the transformer primary tap from the 25-turn point to the 90-turn point. The turns ratio is now $\alpha = 150/90 = 1.67$, and the reflected capacitance is $(1.67)^2(8) = 23.3 \ \mu F$. This is a capacitance ratio of $288/23.3 = 12.9$. Other values can of course be achieved by varying the autotransformer taps.

The point of the circuit is that the autotransformer may well cost less than a suitable start capacitor of the full required capacitance. The other side of the picture is that the capacitor will see the line voltage times the turns ratio. Typically, a 1000 V ac rating capacitor is required.

19-5 PERMANENT-SPLIT CAPACITOR MOTOR

A specific form of split-phase capacitor-type motor is the *permanent-split capacitor motor* or *single-value capacitor motor*. This type of motor has certain advantages for loads that are relatively easy to start.

(1) No centrifugal switch is required.

(2) Easy to reverse while in service.

(3) Effective speed control is possible.

These motors are most often encountered in fan or blower service because of some further advantages:

(4) Low cost and compact size.

(5) Silent operation with no television or radio interference.

(6) Can be specifically matched to a particular load.

The single-value capacitor motor is ordinarily built in small sizes of up to $\frac{1}{6}$ hp (0.125 kW). Usually, the desired speeds are such that four- or six-pole units are found.

The usual single-value capacitor motor has *identical* main and auxiliary windings. The direct winding is proportioned for about 40° of phase lag under load. Auxiliary winding circuits and capacitor are proportioned for about 40° of leading phase angle. The motor can be operated with identical characteristics in either direction. A typical circuit connection is shown in Fig. 19-6a. Note that with the use of the single-pole multiposition reversing switch the capacitor is transferred from one winding to the other. If single-direction service is desired, as in a kitchen vent fan, the reversing switch is omitted. An adjustable autotransformer can give a continuously adjustable voltage for smooth speed control. Where cost is most important, tapped fixed resistances are used for two or three speeds under load. In Fig. 19-6a, the typical fan service switch circuit shown combines the functions of reverse and speed control. In each direction the first switch position from the center, or off, is full speed. Further switch motion introduces resistance which reduces the effective voltage across the motor.

The characteristic curves shown in Fig. 19-6b show that this sequence is necessary. Typically, the lowest speed setting reduces the voltage until there is virtually no available torque to start the load. If the slow speed point were selected first, the motor might not start and would then overheat.

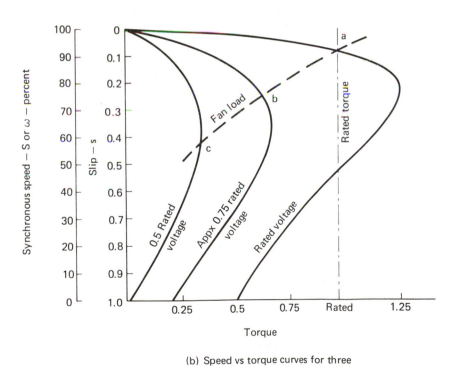

(a) Three speed reversing single value
capacitor motor circuit

(b) Speed vs torque curves for three
speed single value capacitor motor

Figure 19-6. Permanent split capacitor or single-value capacitor motor.

The rated load position is far out toward the maximum torque point. Operation this far out is permissible in fan service where the air blast is available for motor cooling. The usual open frame construction aids ventilation. Since torque is proportional to the square of the applied voltage in any induction motor, the half-voltage point represents about one-quarter of rated torque. A fan or blower is a nonlinear load, and one-quarter rated torque will produce around 60 percent of full speed. This is about the natural response desired for a fan load because very low speeds are almost useless. The speed drop shown in Fig. 19-6b from points *a* to *b* and *c* results in very quiet operation and a gentle but perceptible air motion. Note further that, since separate resistances are used for reverse motion, the reverse speeds can be different if desired. With all these beneficial design qualities, the permanent-split or single-value capacitor motor is fairly efficient under load because the load power factor is very nearly unity.

Another typical use for this type of motor is in dishwasher service. Here reversing the motor changes the direction of water flow without the requirement for electrically operated valving.

19-6 SINGLE-PHASE INDUCTION MOTOR LOAD CHARACTERISTICS

The first three types of single-phase motors listed have performance characteristics that are sufficiently related to be considered together. The permanent-split capacitor motor characteristics were shown separately in the last section because they are ordinarily tailored to be substantially as shown in Fig. 19-6b. The larger single-phase induction-motor types exhibit speed versus load characteristic curves that are very much like polyphase induction motors.

Figure 19-7 is a composite series of characteristic curves in per unit fashion so that comparative judgment may be made. With the same rotor and stator, a resistance split-phase motor and a capacitor-start motor have identical characteristics above the speed where the centrifugal starting switch opens. The large difference in the starting torques available is the real reason for the capacitor-start motor. The two-value capacitor motor is very nearly similar to the capacitor-start motor except for its lower noise level and greater efficiency under load.

Most single-phase induction motors are made with the running winding in two winding groups. This enables series operation on 220 to 230 V lines. Parallel operation then uses the normal 110 to 115 V lines. Usually, the starting winding is not subdivided, so fewer connections are involved in changing service voltage. This can be done because of the very short duty cycle of the starting winding. As a result, the starting characteristics are not identical on the two different voltages. A two-value capacitor or capacitor-start capacitor-run motor is a bit more complicated, since the now auxiliary winding is in service all the time. Where dual-voltage versions of these motors are made, the starting or auxiliary winding must also be duplicated. Individual treatment varies, and the manufacturer's characteristic curves and connection requirements should be followed.

Single-phase motors for operation on 440 to 460 V are only used for the larger

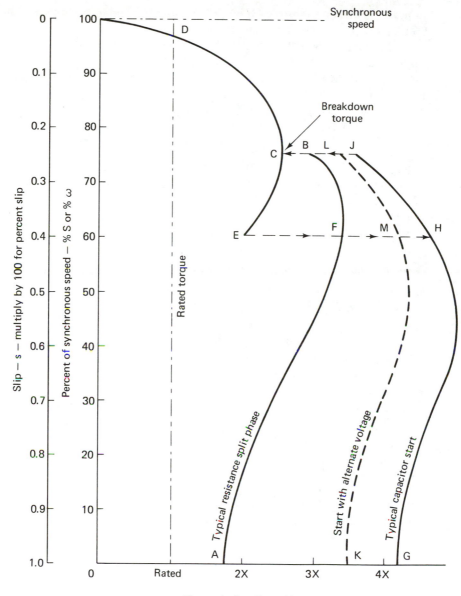

Figure 19-7. Single-phase induction motor speed versus torque characteristics.

sizes, since single-phase lines are not usually available at that voltage level. In modern usage, 440 to 460 V is almost always associated with three-phase systems.

19-6-1 Resistance Split-Phase Motors. The composite performance curves of Fig. 19-7 do not refer to any particular motor. Rather, they are intended to show how the various starting circuits affect the speed versus torque curves of motors. The representative motors have identical main or running windings and identical rotors and stator magnetic structure.

Starting at point A in Fig. 19-7, this represents the starting torque of a typical resistance split-phase motor. As the typical motor accelerates, the motor speed moves up the curve to about $s = 0.25$, where the centrifugal starting switch opens. At this point, B in Fig. 19-7, the motor characteristic shifts to C on the curve of rotor torque under the main or run winding only. The starting circuit cutoff point is set in this particular speed region because of the following:

(1) This is the maximum torque or breakdown torque point and is therefore the highest torque point for the main or running circuit.

(2) The torque with the starting circuit combined with the main circuit is past its peak. A further speed increase would bring the combined torque below the main circuit torque.

(3) This speed is below any normal operating point so that the switch actuation is decisive rather than marginal.

(4) The normal production tolerances of the centrifugal switch affect the cutoff speed and this region has sufficient tolerance.

After the transition from one type of circuit to the other between B and C, the motor continues to accelerate to the operating point at D on the diagram. Part-load operation will cause the motor to ride farther to the left of point D. An overload that is within the acceptable range, say 1.15 times normal rated load, will result in the load point moving to the right of D.

A heavy overload will force the motor speed to drop down past the point C to a point where the centrifugal switch cuts in and reactivates the starting circuit at point E. This will increase the torque available, since the motor is now operating on the starting circuit curve at F. If the torque F is sufficient, the motor will accelerate to point B. When the starting switch cuts out at B, the motor may cycle around points B, C, E, and F until the load is relieved or a protective circuit blows. Operation in this overload region may be easily simulated with a laboratory dynamometer. If a single-phase motor operates this way in service, either the load has temporarily changed, the line voltage is low, or the motor is inadequate for the task. A serious overload, such as a jammed part in the driven load, will entirely stall the motor if the torque at point F cannot hold the load.

A normal shutdown allows the speed to drop all the way to zero and the starting switch to reset for another operation.

19-6-2 Capacitor-Start Motors. A capacitor-start single-phase motor will ordinarily have a large reserve of starting torque, such as shown in Fig. 19-7, point G. This enables a tough load, such as a reciprocating compressor, to be driven past the first compression without the benefit of the momentum that will be stored in a moving flywheel.

The operation of the starting switch is identical except that the decrease in torque from J to C is more drastic. Normal acceleration from point C to D is exactly similar to the resistance split-phase motor situation.

A severe overload situation that drops the motor speed down past C to E will then cause the starting switch to cut in, as in the case of the split-phase motor. In this case, the available torque will roughly double from E to H, and the load can usually be kept running for awhile.

It has been mentioned that the starting winding circuit is not often changed when going from 115 V service to 230 V service. As a result, the compromise start winding circuit may result in less torque on one voltage, such as is seen in the dotted curve from K to L. The characteristics on the main winding are usually identical. If the voltage increases by a factor of 2, the current drops in the same ratio. The fact that the rated torque operating point is so far to the left at point D is based upon the ability to cool the motor in service.

19-6-3 Capacitor-Start Capacitor-Run Motors. The operation of a specific capacitor-start capacitor-run or two-value capacitor motor can be expected to parallel very closely the capacitor-start motor in Sect. 19-3. The real difference will show in characteristic curves of current, power factor, and efficiency, for which see Fig. 19-8. The line current is less, due to the better power factor achieved with the leading auxiliary winding in parallel with the lagging main winding. Less current means lower kilovolt-amperes, thus reducing the cost of the installation. The motors are more efficient in their use of watts. This is a further saving.

Single-phase induction motors are available with various squirrel-cage resistances. As a result, the operating point D on Fig. 19-7 may vary a small amount up or down as the running slip changes. By using a high resistance rotor, a substantial starting torque may be handled with a simple resistance split-phase starting circuit. The tradeoff between a simple low-cost motor that can start a particular load and high efficiency while operating is an economic study. At present, the most efficient long-term energy use is becoming more and more vital.

Various design changes whose need became evident during the 1973 oil embargo have produced a major shift toward capacitor run motors in various forms. Many appliances are now advertised and sold on the basis of their reduction in energy use. In many cases the improvement has come by changing to capacitor run motors. General Electric, for example, markets a line of "Watt Saver" motors, and Dayton makes a line called "Wattrimmer." These are capacitor-run in single-phase units. Some of the motors that are designed for relatively easy starting applications, such as fans and blowers, are a new but logical combination of split-phase start combined with capacitor run.

The very large improvements that have been achieved can be seen in Fig. 19-8. This compares two modern NEMA 56 frame size 0.5 hp four-pole single-phase motors. Both are cataloged and available. The GE 5KC 37 NN 6X is a conventional capacitor start induction run unit. The GE 5KHC43MG5 is a "Wattsaver" that is split-phase start, but capacitor run.

These units were compared for use in a solar heating system air handler blower which was expected to run whenever the sun shines, 12 months a year. The motors are not exactly comparable, but probably represent the improvements possible in appliance service. Notice from Fig. 19-8 that a capacitor-run unit draws less line current at full load than the conventional unit does at no load! The power

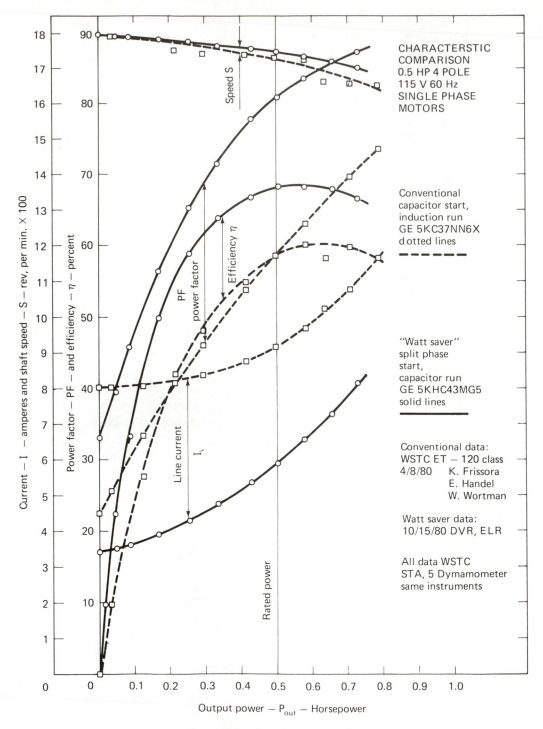

Figure 19-8. Capacitor run benefits.

factor curves diverge widely as a result. The "Wattsaver" motor operates at about 80.7% PF at load, which is in the range of a three-phase unit. Even so, the capacitor-run motor can still be improved, since its efficiency peaks at about 68.5%.

The improvement shown here is readily attainable with conventional motors and existing tooling. Still greater improvement is reached in General Electric's "Supersaver" line of motors. These represent what can be achieved with a complete redesign having more lamination iron, more copper, and a generally optimized design. All major manufacturers have, or shortly will have, comparable improvements because there are no new developments involved, just application of proven but slightly more expensive principles which are now economically justified and even demanded.

Further study of Fig. 19-8 will show that the conventional motor actually operates *below* the point of best efficiency at its rated load. This was probably done to give the motor some overload capacity, since general purpose replacement motors are frequently misapplied and thus overloaded. On the other hand, the capacitor-run motor is more nearly matched to its rating. It can be expected to handle an overload easily because its internal heating is mostly caused by its winding current. This current is distributed between both windings, and its heating effect finally reaches that of the conventional motor at no load when the capacitor-run motor is at nearly 50% overload at around 0.72 hp!

Qualitatively, the new design runs barely warm to the touch at full load, while the conventional motor is too hot to touch comfortably at no load. *Both* motors met their nameplate currents as tested, and both were new out-of-the-box and clean. The same dynamometer and the same instruments were used in the two tests.

19-7 SINGLE-PHASE INDUCTION-MOTOR EFFICIENCY

Except for the fact that a single-phase induction motor is usually substantially less efficient than its polyphase counterpart, the efficiency is determined in the same fashion. The blocked-rotor method of simulating full-load conditions is less preferable than the AIEE method, but may still be needed to test a large single-phase motor.

The differences show in the formula handling of single phase versus three phase, where $\sqrt{3}$ is no longer needed. They also show where the line-to-line resistance is the desired value rather than one-half line-to-line resistance, as used in three-phase circuits. These differences show as follows.

19-7-1 No-Load or Rotational-Loss Tests.
The rotational losses when running light are now found from a further modification of Eq. (18-6).

$$P_{rot} = VI \cos \theta - I^2 R_r = W - I^2 R_r$$
Eq. (19-1)

where

$VI \cos \theta$ = watts input

I = total current in a resistance split-phase or capacitor-start motor

$R_r = R_{dc} (1.25)$ = effective ac resistance of the running winding

Note that if the motor were a capacitor-start capacitor-run type, the $I^2 R$ losses in *both* stator windings would be needed.

$$P_{rot} = VI \cos \theta - I_m^2 R_m - I_a^2 R_a$$

or

$$\boxed{P_{rot} = W - I_m^2 R_m - I_a^2 R_a}$$

Eq. (19-2)

where

I_m = current in main winding

I_a = current in auxiliary winding; I_a and I_m may or may not be the same

$R_m = R_{mdc}(1.25)$ = effective ac resistance of the main winding

$R_a = R_{adc}(1.25)$ = effective ac resistance of the auxiliary winding

19-7-2 Blocked-Rotor Test. Here the situation in Eq. (16-7) modifies a bit:

$$\boxed{R_{es} = \frac{P_{br}}{I_{br}^2}}$$

Eq. (19-3)

where

R_{es} = equivalent stator winding resistance

P_{br} = watts input at stall

I_{br} = stator winding current

Note that the start winding *may be disconnected* for a simple blocked-rotor test, since it is not effective when running at speed. This value is not needed in this test unless the motor is a capacitor-start capacitor-run type, where this type test is not usually used.

19-7-3 Stator Resistance Test. The procedures in Sect. 16-2-3 hold for determining the stator dc resistance by voltmeter–ammeter methods, then multiplying by 1.25. Therefore, Eq. (16-12) holds here. Both windings should be tested for a two-value capacitor motor.

19-7-4 Efficiency from the AIEE Method. The procedures in Sect. 17-4-2 hold for single-phase induction motors. Be aware that the single-phase situation is more simple because the square root of three and the multiplier three are not needed. The procedural steps are identical. The efficiency of an integral horsepower (kilowatt) motor will typically reach 70 to 75 percent. Fractional-horsepower (kilowatt) units will operate in the 60 percent range.

19-7-5 Improved Efficiency by Using Lower Voltage at Part Power Loads. Many motor uses involve a range of loads from full rated or even an overload down to only a small fraction of full load. In these cases the motor runs very inefficiently while at part load. This is because the constant excitation portion of the stator current is automatically maintained even when the motor is unloaded mechanically. This low efficiency coupled with low power use can be significantly improved by reducing the input voltage. Since the change of loading may not be anticipated, a rapid and coordinated adjustment would be required. Automatic controls have been developed to control the motor input voltage as the load changes.

A growing group of licensed manufacturers now supply this need by making available controls that are licensed to the National Aeronautics and Space Administration (NASA). These types of controls, as invented by Frank Nola of NASA, are effective on single-phase or three-phase motors, depending upon their circuit design. The controls sense the motor input power factor and serve to reduce the input voltage when the PF is lower than desired. The control response time is sufficiently short so that a sudden transient load increase is easily met by a rapid rise of voltage.

The effect of a typical control of this type is shown in Fig. 19-9. The specific control tested was a Nordic ES-1, which is one of a family that is NASA licensed. The specific motor, a GE 5KC 37 NN 6X of 0.5 hp, was a normal capacitor-start, four-pole, 56 frame size, induction motor. This was the *same* motor tested as the conventional motor in Fig. 19-8 and discussed in Sect. 19-6-3.

Study of Fig. 19-9 shows that when the NASA control was compared to a conventional motor operation, it was essentially identical at full load efficiency and current. Reducing the load to the 0.2 to 0.3 hp range shows a large gain in efficiency for the control. On the other hand, an increase in load showed detrimental effects of the control. When control-equipped, the motor reached its maximum torque when the load was increased to 0.7 hp. The motor then cycled in and out of its starting winding connection because at maximum torque the speed dropped suddenly to the point of the starting switch operation. This undesirable effect was probably because the control limited the motor voltage to 103 V maximum at its tested adjustment.

In summary, then, this specific NASA-related control had the following results and effects:

(a) No significant change of performance at rated power, except a slight reduction of shaft speed.

(b) Very much improved performance at part power, which continued all the way down to no-load conditions.

(c) Some reduction in starting torque because of lower maximum voltage with the control.

(d) Inability to carry a load beyond 0.7 hp or a 40% overload.

This type of control should be beneficial in cyclical processes in manufacturing.

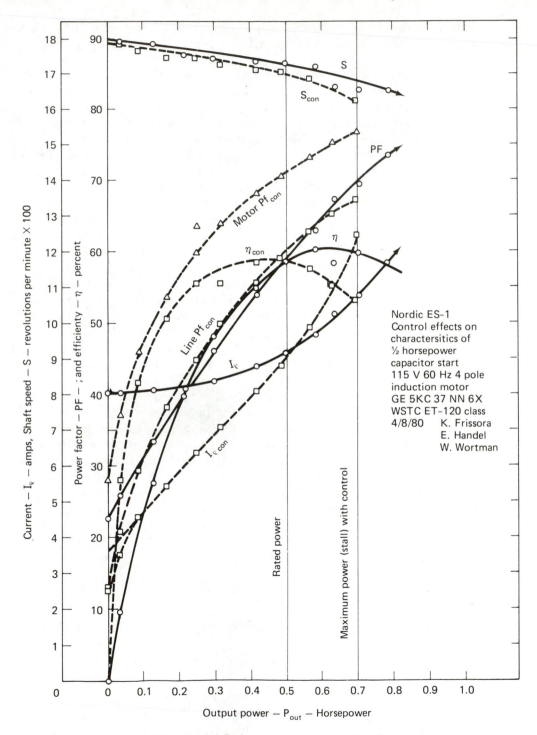

Figure 19-9.   Reduced voltage operation benefits.

On the other hand, it would show no improvement in situations where a motor is either off or fully loaded. Examples of this type of operation are in refrigerators, freezers, and air conditioners, where the motor should be proportioned for maximum efficiency at its load. This constant load situation is better served by a capacitor-run motor in single-phase service rather than by use of an add-on control to help a mismatched motor. The capacitor-run motor in Fig. 19-8 shows better performance than the NASA-controlled motor in Fig. 19-9 on down to about 0.2 hp. However, the NASA control applied to the already efficient capacitor-run motor would probably show further improvement at part power conditions. Lastly, this type of NASA control is now being largely supplied as an after-market device to correct existing installations. Its real benefits will come with mass production, when it can be built into the motor originally, at very low cost.

QUESTIONS

19-1. What prevents a single-phase induction motor from being self-starting unless it has special starting circuit provisions?

19-2. Describe the basis of the double revolving field theory.

19-3. Describe the basis of the rotor cross field theory.

19-4. How does the creation of a second artificial phase enable a single-phase induction motor to develop starting torque?

19-5. How is the required phase shift accomplished in a resistance split-phase motor?

19-6. What is the function of the centrifugal starting switch in a single-phase motor?

19-7. What happens when the centrifugal starting switch fails closed?

19-8. What happens when the centrifugal starting switch fails open?

19-9. How is the required phase shift accomplished in a capacitor-start induction motor?

19-10. What circuit change enables a resistance split-phase or capacitor-start induction motor to be reversed?

19-11. What advantage does a capacitor-start capacitor-run motor have over a capacitor-start motor?

19-12. How does an autotransformer enable a capacitor to perform as two different values of capacitor in a capacitor-start capacitor-run motor?

19-13. What feature limits the utility of a permanent-split capacitor motor?

19-14. What combination of features makes the permanent-split capacitor motor particularly useful in fan service?

19-15. How does a resistance split-phase or capacitor-start induction motor respond to a brief overload torque that is a bit beyond its breakdown torque level?

19-16. Why is a single-phase induction motor less efficient than a comparable-power three-phase induction motor?

PROBLEMS

19-1. If a single-phase induction motor runs normally at 1125 rpm on 60 Hz power, how many poles must it have?

19-2. If a single-phase induction motor runs normally at 183 rad/sec on 60 Hz power, how many poles must it have?

19-3. A $\frac{1}{3}$ hp (0.250 kW) resistance split-phase motor has the following currents in the various windings at the moment of starting: running winding current $I_r = 9.0$ A at $\theta_r = 40.5°$ lagging, and starting winding current $I_s = 6.0$ A at $\theta_s = 15.5°$ lagging. What are the following values:
(a) The components of each winding current in phase with the line voltage?
(b) The quadrature components of each winding current that are lagging the voltage?

19-4. With the motor of Problem 19-3, what are the following:
(a) The total blocked-rotor current?
(b) The blocked-rotor power factor?
(c) The blocked-rotor power factor angle?

19-5. With the motor of Problems 19-3 and 19-4, compare the in-phase component of the starting winding current with the lagging component of the running winding current.

19-6. A $\frac{3}{4}$ hp (0.563 kW), 230 V capacitor-start single-phase motor has the following currents in the various windings at the moment of starting: $I_r = 6.75$ A at $\theta_r = 40.2°$ lagging, and $I_s = 6.08$ A at $\theta_s = 41.5°$ leading. What are the following values:
(a) In-phase components of each winding current?
(b) Quadrature components of each winding current?

19-7. With the motor in Problem 19-6, what are the following:
(a) The total blocked-rotor current?
(b) The blocked-rotor power factor?
(c) The blocked-rotor power factor angle?

19-8. If the motor in Problems 19-6 and 19-7 has a running winding resistance of $R_r = 4.1\ \Omega$ and the motor draws 120 W and 3.91 A running light, what is its rotational loss?

19-9. If the motor in Problems 19-6 to 19-8 runs at 1725 rpm (180.6 rad/sec) and 6.04 A at 800 W, what is its efficiency by the AIEE load slip method? Follow the procedure in Sect. 17-4-2, but be aware of the differences between the three-phase and single-phase situations.

Answers. **19-1**, 6 poles; **19-2**, 4 poles; **19-3(a)**, $I_r \cos \theta_r = 6.844$ A and $I_s \cos \theta_s = 5.782$ A, **(b)** $I_r \sin \theta_r = -5.845$ A and $I_s \sin \theta_s = -1.603$ A; **19-4(a)**, 14.66 A, **(b)** 0.861 or 86.1%, **(c)** 30.54° lagging; **19-5**, 5.782 A ≈ 5.845 A; **19-6(a)**, $I_r \cos \theta_r = 5.156$ A, $I_s \cos \theta_s = 4.554$ A, **(b)** $I_r \sin \theta_r = -4.357$ A, $I_s \sin \theta_s = 4.029$ A; **19-7(a)**, 9.716 A, **(b)** 0.999 or 99.9% **(c)** 1.935° lagging; **19-8**, 57.3 W; **19-9**, 70.8%.

20

Shaded-Pole, Synchronous, Universal, and Other Single-Phase ac Motors

There are many types and subtypes of single-phase and polyphase ac motors. The grouping into families and the resulting chapter coverage in a textbook becomes a bit arbitrary. Chapter 19 was confined to single-phase induction types that create their starting torque from various varieties of phase splitting. In this chapter the induction-motor types create some degree of rotating magnetic field flux by various ways of selectively delaying the increase and decrease of the main magnetic field. These types are the *reluctance-start induction motor* and the *shaded-pole motor*.

The reluctance-start and shaded-pole motors may have perfectly ordinary squirrel-cage rotors and operate subsynchronously. With special rotors, some of these motors can then pull into step and operate synchronously. These are the *reluctance motors* or *reluctance synchronous motors*. Further subdivisions are the *hysteresis* or *hysteresis synchronous* motors. The various subtypes will be discussed in order.

Another major type of single-phase motor is the *universal motor*. These motors most closely resemble dc series motors and will operate on 60, 50, or 25 Hz or on direct current without modification or adjustment. The type was originally developed to take advantage of the ability to operate on various frequencies and/or direct current. However, the United States is supplied exclusively almost with 60 Hz power at this time. Universal motors have proliferated because they are well adapted for uses where their special features can predominate. These are as follows:

(1) High speed from above 3600 rpm (377.0 rad/sec) to around 25 000 rpm (2618.0 rad/sec).

(2) High power output in small physical sizes for use in portable tools.

(3) High torque at low and intermediate speeds to carry a particularly severe load.

(4) Variable speed by adjustable governor, by line voltage, or especially by modern pulse techniques.

All these advantages are not gained without some disadvantages, which should also be recognized:

(5) Increased service requirement due to use of brushes and commutators. The life of these parts is limited in severe service.

(6) Relatively high noise level at high speeds.

(7) Moderate to severe radio and television interference due to brush sparking.

(8) Requirement for careful balancing to avoid vibration.

(9) Requirement for reduction gearing in most portable tools.

The true *ac series motor* is almost extinct but had some advantages in the past. This motor type was built as a variable-speed traction motor in the early days of electric railroads. Its use has been almost entirely superseded by dc series motors owing to their smaller size and more flexible control. It is good to avoid the tendency to disregard obsolescent types of mechanisms in any field. The development of some new material or design modification may well bring a type back into prominence, perhaps for a totally new service.

Another obsolete type of motor is the various *repulsion* and *repulsion induction* motors. Their high cost of manufacture and the high state of development of capacitors suitable for capacitor-start motors have forced them out. Neither the ac series nor the various repulsion types will be discussed here due to space limitations. References (4), (13), and (14) give good treatments on these types.

The relative uses of the various single-phase motors will be discussed after their technical principles have been presented.

20-1 SHADED-POLE INDUCTION MOTORS

A shaded-pole motor develops the required rotating magnetic flux in a different manner from the two coil motor types in Chapter 19. A shaded pole is a magnetic pole that is physically divided and has its smaller segment surrounded with a short-circuited *shading coil*. Figure 20-1a shows the construction of a typical two-pole, open-frame, shaded-pole motor. The action of the shading coils may be understood by following Fig. 20-1b. As the magnetic field intensity begins to increase in the field lamination structure, a voltage is induced in the shading coil. The resulting current that flows in the low-resistance shading coil develops an opposing field in accordance with Lenz's law. The bulk of the magnetism that does flow then passes through the unshaded side of the pole shoe.

As the total flux reaches its maximum, its rate of change of flux has dropped to zero. At this point, there is no voltage induced in the shading coil and its effect

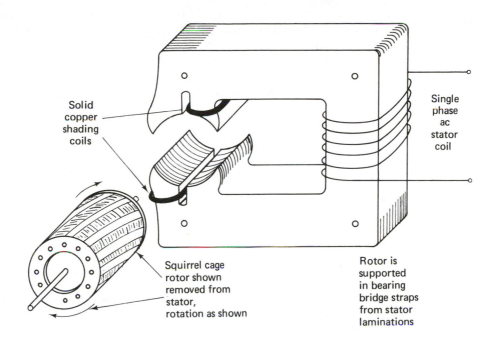

Solid
copper
shading
coils

Single
phase
ac
stator
coil

Squirrel cage
rotor shown
removed from
stator,
rotation as shown

Rotor is
supported
in bearing
bridge straps
from stator
laminations

(a) General appearance of typical open
frame two pole shaded pole motor

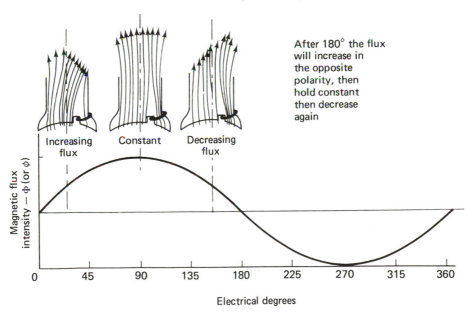

After 180° the flux
will increase in
the opposite
polarity, then
hold constant
then decrease
again

Increasing
flux

Constant

Decreasing
flux

Magnetic flux
intensity — ϕ (or ϕ)

0   45   90   135   180   225   270   315   360

Electrical degrees

(b) Sweeping magnetic flux with shaded poles

Figure 20-1. Shaded-pole construction and field flux motion.

has disappeared. The effective center of the total flux has then moved from the left of the pole to the center.

After 90 electrical degrees, the total flux begins to decrease. Since the flux is decreasing instead of increasing, the effect of the shading coil is opposite. Since the coil tends to resist the dropping of the field, it holds the flux at a high level at the shaded side of the pole. The result is to move the center of the effective flux from the center of the pole to the shaded side.

When the flux drops to zero and starts to increase in the opposite polarity, the shading coil still has the effect of delaying the buildup and decrease of the total flux. The field pole flux is then said to *sweep* from the unshaded side to the shaded side of the pole.

In this salient pole construction, the rotating flux effect is neither uniform nor continuous. There is sufficient rotational effect, though, to start the rotor and bring it up to such a speed that the cross field effect can dominate and maintain nearly uniform rotation.

As might be guessed from the nonuniformity of the effect, a shaded-pole motor is low in both torque and efficiency. These motors are used by the millions in devices where the power requirements are so small that the inefficiency is outweighed by the very low cost. This seems to be the least costly way to build an electric motor that has so far been devised. Efficiencies vary from as low as 5 percent in the very smallest sizes to as high as 35 percent in the larger-sized distributed pole construction motors.

Along with poor efficiency, these motors have relatively high slip. Typically, a slip of 7 to 10 percent at full load is expected. In common with other high-slip induction motors, some speed control is available by changing the applied voltage. This can be done by the use of series resistances or reactance coils. Some installations use tapped autotransformers to efficiently drop the voltage and consequently the speed of the motor. One very attractive simplification is where the stator coil itself is tapped and used as an autotransformer. In this case, the *low speed* connection uses the *full* stator coil. When a lesser number of turns is used, the volts per turn of the coil increases, and so does the field flux.

Shaded-pole motors have poor starting torque, usually not over one-half the rated full-load torque. Small open-frame motors such as the shaded-pole unit shown in Fig. 20-1a are usually rated in inch-ounces (or ounce-inches, which is the same) of torque. There does not seem to be an equivalent SI unit for small torque values. The probable and logical use is the millinewton meter or mN·m or N·m $\times\ 10^{-3}$. This text will not use the older metric measures of gram-centimeter or milligram-millimeter because in SI parlance grams, milligrams, or kilograms are used for mass, not force. The true force unit is the newton with its various prefix multipliers by 10^3 increments, kilo, milli, and so on.

Torque in foot-pounds is from Eq. (6-1$_E$); $T = 5252.1(\text{hp})/S$, where the speed S is the rotor speed S_r instead of synchronous speed. Since there are 16 oz in 1 lb and 12 in. in 1 ft,

$$ t' = \frac{5252.1(16)12(\text{hp})}{S} \quad \text{or} \quad t' = \frac{1\ 008\ 400\ \text{hp}}{S} $$

From Eq. (16-28), where $T = 7.0432 \ W/S$, treated similarly

$$t' = \frac{7.0432 \ W(16)12}{S} \quad \text{or} \quad t' = \frac{1352.3 \ W}{S}$$

But since it is extremely unlikely to be able to get data to any better than three significant figures in very small units, these should be rounded off to:

$$\boxed{t' = \frac{hp}{S} \times 10^6} \qquad \text{Eq. (20-1}_E)$$

and

$$\boxed{t' = \frac{W}{S} \times 1.35 \times 10^3} \qquad \text{Eq. (20-2}_E)$$

or

$$\boxed{hp = t'S(10^{-6})} \qquad \text{Eq. (20-3}_E)$$

and

$$\boxed{W = t'S(0.739 \times 10^{-3})} \qquad \text{Eq. (20-4}_E)$$

where t' = torque in *ounce-inches* instead of newton meters or foot-pounds. However, in SI units, $t = W/\omega$ and $W = t\omega$ from Eq. (6-1$_{SI}$) and the same transposed. The units are t = newton meters, as before. Note that any of the speed–power–slip–torque relations of Chapter 16 may be modified to use the lower torque range if proper care is used. In very accurate transducer or gyro torque work, it may be desirable to use a higher number of significant figures if the data justify the increase. The full constant in English units would then become 1 008 405.7.

When shaded-pole motors are built in their larger range of sizes, or up to about $\frac{1}{6}$ hp (or 0.125 kW), the construction usually becomes more conventional. Nonsalient pole units with distributed stator windings and continuous toothed stator lamination stacks are commercially built. These units are usually of a semiopen frame structure and frequently have skeletal die-cast end bells. They appear much like a resistance split-phase motor except that the shading coils can be found in the stator pole faces. Typically, six-pole motors of this construction are used in larger fans or in dishwashers.

20-1-1 Shaded-Pole Motor Reversing. The rotation direction of a motor such as is shown in Fig. 20-1a cannot be changed unless the motor is mechanically disassembled and the stator flopped over one half-turn. This places the shading coils in the opposite relation and the motor will then turn the other way. Obviously, the rotation direction must be specified when purchasing in quantity.

If reversible operation is desired, two differently wound shading coils are provided on each pole face, one at the one-third point and one at the two-thirds point. The motor will always rotate in the direction such that the shading coils are in the *trailing position*, as can be seen from Fig. 20-1b. With multiple coils, the trailing set on each pole face is short circuited through an external switch and the leading set is left as an open circuit. In order to reverse, the formerly shorted coils are opened and the other set then shorted. Rapid reversing operations are then possible with a single-pole double-throw switch.

A larger shaded-pole motor with the distributed windings will usually be reversed by actually changing the portions of the stator coils that are used. A full closed set of windings such as are used in three-phase stators is not required in single-phase service. However, by providing full sets of windings and dividing them into segments, the coil in use can be shifted in relation to fixed shading coils. These motors then can be as readily reversed as the simple single-value capacitor motors.

The shaded-pole motor costs slightly less, but is also much less efficient. Again, the cost of energy will probably change the fields of utility of various motor types in the near future. Already the advertising world has begun to pay attention to low power drain in small household appliances. Here and in other applications a trend away from cheap but inefficient motors can be seen.

20-2 RELUCTANCE-START INDUCTION MOTORS

Another form of single-phase induction motor with only the basic running winding and no special starting winding is the *reluctance-start* motor. In this case, the stator pole tips are modified in a still different fashion to achieve a degree of field flux rotation. Figure 20-2a shows a four-pole reluctance-start stator. Notice that the leading one-third portion of each pole shoe is cut away to create a substantially larger air gap.

Recall that the inductance of a coil with a magnetic core is dependent upon the magnetic reluctance of the flux path. The leading part of the magnetic flux path then has a higher reluctance, and the coil sees less inductance in regard to that part of the magnetic path. When the voltage in the coil starts to build up and create a coil current, the flux in the leading portion builds up relatively rapidly. This is shown in Fig. 20-2b. As the total current reaches a peak, the other portions of the core begin to acquire substantial flux. At last the low reluctance and the high inductance part of the path begins to build up its magnetic flux, as shown in Fig. 20-2b. The high reluctance leading portion shows a relatively small flux phase lag. Conversely, the low reluctance trailing portion lags the voltage by a phase angle of nearly 90 electrical degrees. This results in a sweeping flux that apparently moves across the pole face very much like the shaded-pole motor flux.

Once started by this sweeping flux action, the reluctance-start motor does not maintain a cross field action quite as well, owing to the mutilated flux path. As a result, the reluctance-start motor is also an indifferent performing unit. Its starting torque is not quite 50 percent of its rated running torque. The maximum torque is not much more than the rated torque, and the operating slip is high.

Due to the high slip, the speed can be controlled by reducing the applied voltage, much as in the shaded-pole motor. In common with other high-slip motors, the efficiency is poor, a bit lower than a shaded-pole motor of the same rating.

The reluctance-start motor *cannot* be electrically reversed. The flux always

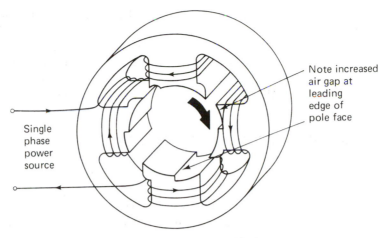

Single
phase
power
source

Note increased
air gap at
leading
edge of
pole face

(a)   Reluctance start stator pole shape

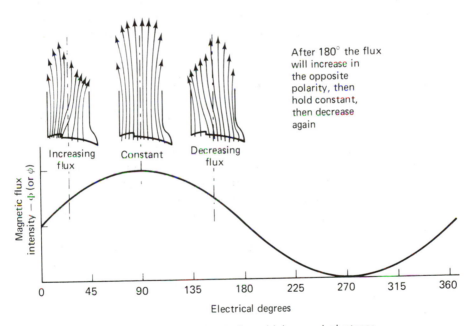

After 180° the flux
will increase in
the opposite
polarity, then
hold constant,
then decrease
again

Increasing
flux

Constant

Decreasing
flux

Magnetic flux
intensity — φ (or ϕ)

0   45   90   135   180   225   270   315   360

Electrical degrees

(b)   Sweeping magnetic flux with increased reluctance

Figure 20-2.   Reluctance-start pole construction and field flux motion.

sweeps towards the trailing or small air gap portion of the stator.   As with the
fixed shading coil in shaded-pole motors, the reluctance-start motor can be reversed
in direction by physical disassembly and exchange of stator position.

It is hard to justify the use of the reluctance-start motor since it is an even
poorer performer than the shaded-pole motor.   Even more than in the case of the
shaded-pole type, the reluctance-start motor will be less and less evident as the

energy cost of a motor use becomes either a factor of public awareness or of legislative attention.

On the favorable side, these motors are quiet and interference free. They are about equally low in first cost in comparison to the shaded-pole motor. It would seem that the permanent-split capacitor motor is a better candidate for any task with a substantial duty cycle because of better energy use.

20-3 SINGLE-PHASE SYNCHRONOUS MOTORS

No less than six different methods have been discussed so far that enable a rotating magnetic field to be simulated. As a result, a single-phase squirrel-cage induction motor can be started in various ways. We now will investigate the rotor modifications that will enable some of these types of motors to pull into step and run at their true synchronous speed. These synchronous motor types are known as *reluctance motors*, *hysteresis motors*, and *subsynchronous motors*. There is a related polyphase type of synchronous motor known as the *synchronous induction motor*. This last type will be explored when the concepts are explained. It is *not the same* as the polyphase synchronous motors discussed in Chapter 18, even though it uses precisely the same rotating-stator-caused field flux as discussed in Sect. 16-1-1.

20-3-1 Reluctance Motors. The *reluctance motor* is a contraction of reluctance synchronous motor. This type has a squirrel-cage rotor that has been mechanically modified so that its rotor magnetic lamination surface is cut away between what then become salient rotor poles. There may or may not be a continuous squirrel-cage bar structure through the cutaway magnetic structure. This depends upon the starting torque and breakdown torque requirements.

A normal type and even the modified type of squirrel-cage rotor will accelerate to near its synchronous speed. When the slip is reduced to near the normal running value, the salient poles pass slowly through the rotating synchronous speed field. At each passage there is a strong *increase* in circulating flux due to the reduced reluctance in this region. The rotor establishes a position where its salient-pole centers ride a few degrees behind the rotating magnetic field and move in synchronism.

At synchronous speed, there is no further generation of voltage in the squirrel-cage conductors since this requires slip. The magnetic pull of the moving field on the distinct salient rotor poles is sufficient to pull the rotor into step at torque values of up to about 1.2 times normal rated torque. Once in synchronism, it will normally require about double the rated torque to cause the motor to drop out of synchronism. The holding of synchronous speed is then due to the different magnetic path reluctance at the salient pole versus the between pole positions on the rotor. This *is not* the same use of reluctance as in the reluctance-start motor.

There is no rotor-produced magnetic field to generate a large E_{gp} voltage as in a large synchronous motor. As a result, the reluctance motor really operates with a severely underexcited field, and so draws a substantially lagging power factor.

A typical reluctance-motor performance curve is shown in Fig. 20-3. Three features are different from a normal resistance split-phase characteristic curve:

(1) The flat top of the curve shows that the motor will run at synchronous speed similar to Fig. 18-4. Note that the motor will pull into synchronism at around 1.2 times its rated torque. This is one of the parameters that sets the rated load for

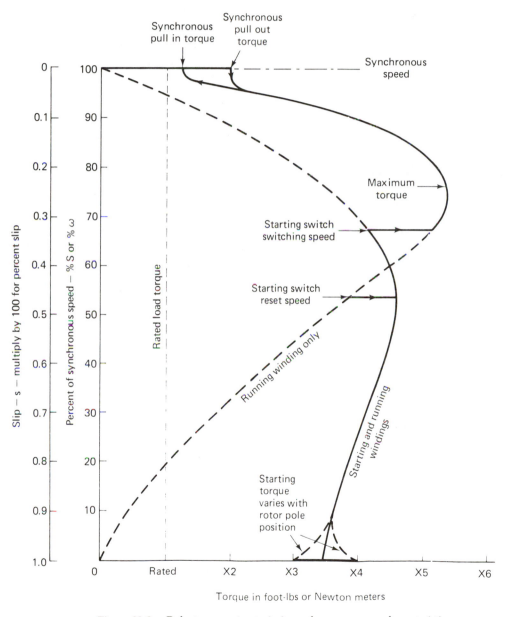

Figure 20-3. Reluctance motor typical speed versus torque characteristics.

a given motor. It must be able to show a small margin over its rated torque so that it might pull in with cold lubricants or low voltage. The pull-out torque is usually about twice the rated torque.

(2) The maximum torque is as much as five or six times the rated torque. Note that this does not indicate any especially high potency to this type of motor. The contrary is true because a larger than normal frame size is needed for the rated torque as a synchronous motor. As a result, the motor is apparently unusually powerful when acting as an induction motor during acceleration.

(3) The starting torque varies over a significant range depending upon where the rotor salient poles happen to stop after the last operation cycle. The structure of the stator in relation to the rotor causes this difference.

Other than these points, the characteristic curve resembles that of a normal high-resistance rotor squirrel-cage motor with as much as twice the rated torque. The centrifugal starting switch (or current relay if used) must be set to match the bulge of the torque curve on any motor. The efficiency will be poorer than a standard type of motor, depending on size.

20-3-2 Hysteresis Motors. The very common synchronous clock motor is usually a *hysteresis motor* with shaded-pole starting. This type of motor has a very different rotor structure. The rotor magnetic structure is a permanent magnetic material whose magnetic centerline can only be moved with difficulty. It is the hysteresis loss power required to move the retained magnetic vector that supplies the torque force.

The stator may be any of the constructions that supply a rotating magnetic field vector for starting purposes. This moving vector links with and attempts to drag around the magnetic flux in the motor. In medium sizes, the magnetism slips and moves in the rotor, only to come finally to a fixed position as the rotor locks into synchronism. Very small hysteresis motors will accelerate to synchronous speed in a very few cycles. These types have a hardened ring-shaped rotor with a cross bar structure to aid the alignment of the retained magnetic force. Originally developed by Telechron for high-quality clock drives, this type has proliferated. They are available with a very wide selection of sealed and lifetime lubricated reduction gear sets.

Slightly larger units are assembled with the rotor of hardened and laminated rings. The smooth exterior of the rotor combined with its stepless or any-position magnetic retention makes the type extremely quiet and smooth running. They are very well adapted for record player or tape deck capstan drives of the highest quality.

An extremely important use of this type of motor is for the rotation of gyroscope rotors in inertial navigation and control systems. Here the requirement is for as near absolute accuracy as can be achieved. One major component of the instrument accuracy that contains the gyroscope is that the gyroscopic moment be absolutely constant. This constancy requires a synchronous motor that is driven by a regulated constant-frequency source.

The higher-accuracy units are usually two or three phase to achieve a truly constant rotating flux vector in the stator. A capacitive phase shift may be used, but at a sacrifice of accuracy. In this type of service, all the input power is used in overcoming the rotational losses. There is no mechanical shaft output. The rotors are hermetically sealed in a low-density inert gas to achieve a low windage loss and to prevent lubricant deterioration. Since the hysteresis ring structure has appreciable mass, and rotational mass is desired, the units are usually turned inside out. The fixed stator is inside the cup-shaped rotor, and the smooth inner face of the rotor is the hysteresis ring. The small airborne units operate on 400 Hz power and rotate at 12 000 rpm (1256.7 rad/sec) with four-pole rotors. The very small units are two pole and rotate at 24 000 rpm (2513.3 rad/sec).

20-3-3 Subsynchronous Motors. When the motor has a rotor that has an overall cylindrical outline and yet is toothed as a many-pole salient-pole rotor, it is a *subsynchronous motor*. A typical rotor may have 16 teeth or poles, and in conjunction with a 16-pole stator will normally rotate at a synchronous 450 rpm (47.1 rad/sec) when operated on 60 Hz. If this motor were temporarily overloaded, it would drop out of synchronism. Then the speed drops down toward the maximum torque point, and the motor will again lock into synchronism at a submultiple speed of 225 rpm (23.6 rad/sec). Hence the name of the subsynchronous motor.

This type of motor starts and accelerates with hysteresis torque just as the hysteresis synchronous motor does. There is no equivalent of induction-motor torque as in the reluctance motors.

This type of motor in any given size will develop a higher starting torque but a lesser synchronous speed torque than a reluctance motor.

20-4 POLYPHASE SYNCHRONOUS INDUCTION MOTORS

Now that the operation of some of the varieties of synchronous motors that do not have dc-excited rotors has been explained, it is appropriate to discuss the *synchronous induction* motor. This type of motor is closely related to the reluctance motor of Sect. 20-3-1 but is normally built as a polyphase unit. The synchronous induction motor starts as a high-resistance squirrel-cage induction motor. The pull into synchronous speed is accomplished by the reluctance torque generated in its modified rotor. The rotor will externally resemble the rotor of a reluctance motor. Part of the surface of the magnetic lamination structure is relieved or cut away to form the proper number of salient poles.

There are also internal differences in the rotor laminations of a synchronous induction motor. The interiors of the laminations are punched out in such fashion that low reluctance paths exist from the side of one salient pole to the adjacent side of the next pole. This cutaway makes the reluctance of the magnetic path still greater when the rotor is away from its locked-in synchronism position.

The performance of the synchronous inductance motor is inflated in the same manner as the reluctance motor of Fig. 20-3. Since the reluctance torque that locks the motor in synchronism is limited, the rated torque of the motor is relatively small for its frame size. Then when acting as an induction motor during acceleration or during overload, the motor has an exaggeratedly large torque in relation to its rated torque.

The characteristic curves more closely resemble the right hand curve in Fig. 18-4. The starting torque is about four times the rated torque, and the maximum torque may be as high as seven times the rated torque.

The synchronous inductance motor finds its market and use where synchronous or *synchronized* operation is desired and where the load is difficult to start. It is also used where the power factor improvement available in a dc-excited synchronized motor is not deemed worth the extra cost. The usual motor of this size is not over 50 hp (37.5 kW).

20-5 INDUCTION-TYPE MOTOR EFFICIENCIES

In the various small motor types discussed in this chapter, the actual efficiency is usually low. The efficiency itself may not be particularly important in the economics of the product device. However, efficiency determines the power consumed and the heat that is wasted to the surroundings. Each of these factors may be of vital importance to a vehicle-borne, airborne, or space-exploration device. Here the power consumption budget and internal heat balance are frequently very critical. At the very least, the motor input requirements must be matched by the output capability of the power supply that drives the motor.

20-5-1 Nonsynchronous Motors. A direct output mechanical power measurement can be performed using the two-scale Prony brake, as mentioned in Sect. 6-3 and shown in Fig. 6-2. This test is relatively easy to set up and perform if small scales of the correct force sensitivity are available. It is usually easier to fabricate a special drum pulley of a suitable radius so that available scales can be used. The alternative is to borrow or purchase the appropriate force scales. The adjustable support arm in Fig. 6-2 is important. Holding by hand is totally unsatisfactory. The other requirement is a suitable means of measuring rotating speed. A calibrated stroboscopic device, such as a Strobotac or a Strobolux, is desirable. However, there is always the possibility of interpreting the speed incorrectly, either high or low, by flashing twice per revolution or every other revolution. Remember that the no-load speed will be close to or below the synchronous speed. The speed will depend on the supply frequency and number of motor poles.

Example 20-1. A small single-phase two-pole shaded-pole motor is operated under load, where it draws 359 mA at 115 V, 60 Hz. The power drawn under load is 12.4 W. Its loaded speed is 3050 rpm (319.4 rad/sec). A two-scale Prony brake rig shows scale forces of 38.0 and 42.2 oz (1077 and 1196 g). Note the now incorrect use of grams as force, which requires translation to newtons. Older metric force scales are still in evidence. The scale between the cords is wrapped around a convenient pulley, which is 0.3937 in. (10.00 mm) in diameter. The no-load speed is found to be 3520 rpm (368.6 rad/sec), and the no-load input power is 6.77 W. The stator R_{dc} resistance is 31.04 Ω. The current is 241 mA at no load. Find the following:

    (a) The motor power output at the measured load.
    (b) The rotational loss.
    (c) The stator copper loss under load.
    (d) The motor efficiency using the AIEE method.

Solution:

(a) From Eq. (20-3$_E$), hp $= tS(10^{-6})$. To find the torque in ounce-inches, use the *difference* between the scale readings times the pulley radius:

$$t = (42.2 - 38.0)\left(\frac{0.3937}{2}\right) = 0.827 \text{ oz-in.}$$

$$\text{hp} = 0.827(3050)10^{-6} = 0.002\ 52 \text{ hp}$$

The SI system approach will use the basic Eq. (6-1$_{SI}$) modified for watts rather than kilowatts or watts $= t\omega$. To find torque in newton meters, we must translate the obsolete force measurement from grams force to newtons.

Note that just because laboratory equipment is calibrated in metric units it is not necessarily usable in SI work! Laboratory force scales have a useful life of 20 years or so. Useful units that are calibrated in grams, kilograms, or dynes will not be junked. Take readings and *convert* to newtons. Refer to Table 6-4 and find that gram force $= 0.009\ 806\ 6$ N. The scale forces are then

$$1077(0.009\ 807) = 10.56 \text{ N}$$

$$1196(0.009\ 807) = 11.73 \text{ N}$$

The torque is $(11.73 - 10.56)0.005 = t = 0.005\ 85$ N·m.

$$\text{watts} = 0.005\ 85(319.4) = 1.87 \text{ W}$$

A check shows 0.002 52 hp(746 W/hp) $= 1.88$ W. Check. Note that the roundoff of the constant in Eq. (20-3$_E$) prevents a closer check agreement.

(b) For single-phase work, rotational loss from Eq. (19-1) is $P_{rot} = W - I^2 R_s$ since only one wattmeter is involved and the line current is the stator current.

$$P_{rot} = 6.77 - (0.241)^2 31.04(1.25) = P_{rot} = 4.51 \text{ W}$$

The rotational loss is *greater* than the output power!

(c) The stator copper loss is $I^2 R_s = P_s$:

$$P_s = (0.359)^2 31.04(1.25) = 5.00 \text{ W}$$

Copper loss in the stator is also greater than the output power!

(d) The efficiency should be found directly with actual torque measurements. Thus, $\eta = $ (output/input) $\times 100 = \eta\%$ efficiency.

$$\eta = \frac{0.002\ 52 \text{ hp}(746 \text{ W})100}{\text{hp}(12.4 \text{ W})} = 15.2\% \text{ eff}$$

Using the SI data,

$$\eta = \frac{1.87(100)}{12.4} = 15.1\% \text{ eff}$$

Since the actual measurement of output torque may have been very unsatisfactory with vibrating force scales and unknown scale calibration, the AIEE method should be used as a check.

$$\text{Input power} = 12.4 \text{ W}$$

$$\text{Stator copper loss} = 5.00 \text{ W}$$

$$\text{Rotational loss} = 4.51 \text{ W}$$

$$\text{Slip} = \frac{3600 - 3050}{3600} = s = 0.153$$

$$\text{Rotor input power} = 12.4 - 5.00 = 7.4 \text{ W}$$

$$\text{Rotor copper loss} = 7.4(0.153) = 1.13 \text{ W}$$

Thus, mechanical output power is

$$7.4 - 1.1 - 4.5 = 1.8 \text{ W}$$

Note that, because of subtraction, only two significant figures are available.

$$\eta = \frac{1.8(100)}{12.4} \cong 14.5\% \text{ eff}$$

Take your choice; one value is probably as good as the other. An exact agreement cannot be expected.

This rather lengthy procedure is applicable to any of the small induction-motor types. If the motor operates synchronously, the portion of the AIEE test that corrects for the rotor copper loss is not applicable. In synchronous operation, there is no slip.

A final note on small motor testing by the two-scale Prony brake is that the shaft conditions must be satisfactory. A very small motor usually has a smoothly ground shaft with no flats, keyways, or splines. Intermediate types generally have some provision for locking their driven element, so a pulley must be used.

Torque measurement units in small motor specifications have been a chaotic area. Dyne-centimeters is sometimes used and is readily translated to newton-meters. Milligram-millimeters needs to be treated with care to get the force unit to newtons. Frequently, the torque measurements are presented in a now obsolete metric measure combined with speed in revolutions per minute or even in revolutions per second. See Table 6-4 for help in this area.

20-6 UNIVERSAL MOTOR

A dc series motor will operate on a low-frequency ac line without distress. Normal ac frequency of 60 Hz is too high for good operation owing to the severe voltage drop caused by the series field inductive reactance. A shunt motor is virtually useless on alternating current since the normal shunt field has a very high inductance. As a result, the field current is low and at a very lagging power factor in relation to the armature current.

The major change that makes a satisfactory commutator-type motor for ac operation is the reduction of field turns. The resulting *universal motor* has a relatively weak field that requires more armature winding turns to produce the same torque force. Universal motors then require very careful design to avoid commutation troubles. This means that brush area, brush material composition, and brush pressure are quite critical.

Universal motors are reversible by interchanging the field connection or armature connection, but not both. Due to relatively strong armature reaction, the brushes may be offset to favor one direction or rotation.

Since the usual modern universal motor is packaged integrally with its related appliance, it can be carefully customized to have the best commutation under its normal load conditions and rotation direction. The usual duty cycle strongly favors one direction even if reverse operation is used.

Another problem specific to ac operation is that the field core is subject to hysteresis and eddy current losses, as is the armature. The universal motor then has a laminated field structure to prevent field core overheating. These changes are not detrimental to dc operation, so the motor retains its universal operation feature.

The ability to operate on any normal ac frequency or on direct current was a major reason for the original development of universal motors. This is of much less importance than it was two generations ago. Most overseas commercial markets outlets now have 50 or 60 Hz power, and the long-term trend to 60 Hz continues.

The immense proliferation of universal motor uses since World War II has come about because of the high power output for a given size or because of the high available speed, or both. A modern carpenter–builder's saw or Skillsaw could operate with a two-pole induction motor, but the weight and size of even a $\frac{1}{2}$ hp unit (0.375 kW) would prevent user acceptance. The builder's saw is a one-hand tool. These saws use universal motors that may have a peak output of over 1 hp (0.75 kW) and yet are around 4 in. in diameter (approximately 100 mm). In this case the saw blade is reduction geared to get appropriate blade torque under load. Since the duty cycle is low, even in steady professional use, the life is satisfactory. This tool has revolutionized the carpenter trade and is now considered almost indispensable. The prosperity of the last generation has seen literally millions of these tools purchased for home workshop hobby use.

Another example is the woodworking router. Here a no-load spindle speed of around 20 000 rpm or more (2094 rad/sec) is used. A substantial fraction of this speed is used during normal cutting and produces a smooth cut with a small-diameter two-lipped cutter bit.

The major long-term household use has been the vacuum cleaner where the power of the universal motor is still being increased. Household units are available with well over 1 hp (0.75 kW) under load.

Many small home appliances have started with universal motors and have been developed into cordless battery-operated units. These units are usually permanent magnet field dc motors when operated on batteries. However, a small motor-driven appliance with a line cord connection and a high power requirement is almost always universal motor powered.

Almost all universal motor services are with a short duty cycle because of marginal brush life under heavy load.

The universal motor does have the ability for a wide speed control range, which enables it to be used when other types would not be successful. An example of this is the domestic sewing machine. The vast majority of home sewing is performed at slow but variable machine speeds. The home dressmaker is not comfortable with a machine that will operate at the furious speed of a commercial machine.

Speed control is by series resistance or occasionally voltage reduction by variable transformer or autotransformer. Modern developments in solid-state silicon-controlled rectifiers (SCR) have enabled wide speed control with relatively constant torque. This feature is further adding to the uses of the universal motor.

A detriment to the long-term use of these motors is the radio and television interference caused by the brush sparking. This effect can be minimized by filtering, but the operation of a universal motor is usually evident on any television picture that is in the same building. This is not to say that the motors are intolerable, but some small lines of "snow" are usually evident on a television screen. An older unit without adequate filtering and with worn commutator and brushes will produce severe interference even though the appliance apparently functions well.

A universal motor is readily identified by the insulated brush holder caps that either protrude through the case or are accessible by removing an outer shroud.

20-6-1 Universal Motor Characteristics.
The output characteristic curve of a universal motor will show the same shape as that of a series dc motor (see Fig. 6-5). The higher torque is developed at low speeds. The speeds at no load rise to high levels but are not destructive. This is because the attached gear box has enough drag to keep the motor from overspeeding, or because the load is directly attached. Smaller units are simply structurally arranged to resist the bursting forces of their maximum free speeds. Usually, the combined brush, bearing, windage, and magnetic losses (the rotational losses) will limit the no-load speed to 25 000 or 30 000 rpm (2618 or 3142 rad/sec).

20-6-2 Universal Motor Efficiency.
Universal-motor efficiency determination is related to the dc series motor procedures of Chapter 7. The output power measurement is not convenient with a larger motor, such as that on a builder's saw or a vacuum cleaner, owing to the high speed. If the reduced speed saw arbor shaft were connected to a typical small 1 hp (0.750 kW) dynamometer, the actual armature speed could be used. The intervening reduction gear losses must then be accounted for. A dynamometer that will handle over 10 000 rpm (1047 rad/sec) is a very special item and is not normally available outside of the motor

manufacturer's testing laboratory. On the other hand, a two-scale Prony brake is applicable in principle, but would take careful design and use for high power.

The best school laboratory approach would be to measure winding resistances and then multiply to get the effective ac resistance, $R_{ac} = R_{dc}(1.25)$.

Rotational losses are found by measuring the armature voltage from brush to brush while running unloaded at the desired speed. This will require the adjustment of the line voltage supply. The field current and the armature current are usually the same, but the individual motor circuit must be traced. A common connection is where the two field coils of the usual two-pole motor are separated, with the armature between them. This means that there are two field coil losses although they can be combined conveniently.

Once the field and armature resistances are determined, and the rotational loss for the proper speed or family of speeds is found, the universal motor efficiency can be handled the same as a series dc motor. See Chapter 7 for many suggestions that are applicable.

20-7 SINGLE-PHASE MOTOR APPLICATION

Perhaps the most effective way to illustrate the very widespread use of single-phase motors is to count those that are in use in the average household. The author has used this procedure with many class groups. The results are surprisingly consistent. Each student is asked to count and identify the single-phase motors in his household. The reporting is listed under the categories of induction, shaded pole, synchronous, and universal. The usual class average shows between 30 and 40 single-phase motors per household. The four major categories are very evenly divided, even though many of the uses vary widely.

It may not be possible to identify whether a motor is shaded pole, reluctance start, or permanent split capacitor without some disassembly, so these types are usually lumped together. There may not be a clear identity between a resistance split-phase and a capacitor-start induction motor unless the capacitor mounting can be seen. Since the usual home appliance motor is fairly well or even completely concealed, this choice is bypassed and made an item for class discussion.

Having tried this investigation with a few hundred students covering a wide economic background range, the consistent range of quantities has convinced the author that his own list is quite representative. The *quantity* in use seems to be more consistent than the *type* of appliance. What one household uses another ignores, but substitutes a totally different appliance. There are certain staple uses, such as refrigerators, vacuum cleaners, and clocks. Table 20-1 shows the author's single-phase motor list.

There are other perfectly normal types of appliance and tool uses that are present in millions of households. The author's list is heavy with workshop tools and light with the basic synchronous motor clock use. Some other representative uses in the home are shown in Table 20-2.

When the typical office and business uses are added, the list continues to grow. The motor type adopted by one manufacturer for a particular service may not be adopted by a competitor since there are many marginal choices.

TABLE 20-1 SPECIFIC SINGLE-PHASE AC MOTOR USES

Induction	Shaded pole	Synchronous	Universal
Basement			
Oil burner	Drafting eraser	Off-peak electric	Small drill
Warm-air	Lathe coolant	meter time	Large drill
blower	pump		Sander–grinder
Metal lathe	Solar heat		Vacuum cleaner
Drillpress	hot water		High-speed hand
Combination	pump		grinder
woodworking	Solar heat		Lathe tool post
tool	panel vent		grinder
Vertical miller			Hedge clipper
Solar heat			Snow blower
air handler			
Garage			
Garage door			
opener			
Kitchen			
Dryer blower	Dishwasher	Dishwasher	Can opener
Refrigerator	pump	timer	Hand mixer
Washing	Vent fan	Dryer timer	Vacuum cleaner
machine		Stove clock	Carving knife
		Washing	Blender
		machine timer	Food processor
Living Areas			
Air conditioner	Hassock fan	Day–night light	Slide projector
compressor	Record player	timer	focus
Air conditioner	Slide projector	Tape deck	
blower	blower	capstan	
	Tape deck		
	reels		
	Typewriter		
Bedrooms			
Air conditioner	Fan	Alarm clock	Sewing machine
compressor	Record player		
Air conditioner	Hair dryer		
blower			
Attic			
Exhaust fan	Cross attic fan		
Subtotals by type			
16	15	9	16
Overall total: 56			

TABLE 20-2   OTHER FREQUENT USES OF SINGLE-PHASE MOTORS

Induction	Shaded pole	Synchronous	Universal
Swimming pool filter pump	Movie projector	Day–night thermostat	Builder's saw
Rotary lawn mower	Fish tank circulator	Clock–radio	Belt sander
Deep freeze	Remote TV tuner	Refrigerator defrost timer	Router
Jig saw		Record player (high quality)	
Band saw			
Planer			
Shaper			
Water pump			
Basement sump pump			
Furnace hot water circulator			

Office uses include all kinds of paper-handling machines, duplicators, print makers, shredders, calculating machines, computer card handlers, etc.

The important point to recognize is that the *new types of motor uses are constantly growing.* Many commercially successful motor uses of today would have been laughed at by serious product designers a generation ago. There are many motor uses in the near future that are not imagined today. Each new use is fair ground for a technician in product development work. In the future, each will have to be efficiently applied.

QUESTIONS

20-1. How does a shaded-pole motor produce the necessary rotating magnetic field?

20-2. Why does a shaded-pole motor have relatively high slip?

20-3. Why does a shaded-pole motor have low efficiency?

20-4. In view of the disadvantages, why are shaded-pole motors produced in such high quantity?

20-5. How may a shaded-pole motor be visually identified?

20-6. How is a shaded-pole motor reversed?

20-7. How are larger distributed winding shaded poles reversed when only one set of shading coils is provided?

20-8. How may a reluctance-start motor be identified?

20-9. Why is a reluctance-start motor less desirable than a shaded-pole motor for the same service?

20-10. How does a reluctance synchronous motor start?

20-11. How does a reluctance synchronous motor run synchronously?

20-12. How may a reluctance synchronous motor be visually identified?

20-13. How does a hysteresis synchronous motor start?

20-14. How does a hysteresis synchronous motor run synchronously?

20-15. What is a synchronous induction motor?

20-16. What is a subsynchronous motor?

20-17. Why is a subsynchronous motor so named?

20-18. What is a universal motor?

20-19. How may a universal motor be visually identified?

20-20. What are some advantages of a universal motor?

20-21. What are some disadvantages of a universal motor?

20-22. Prepare a list of single-phase motors subdivided as in Table 20-1 or 20-2 that are presently in your school or your home.

PROBLEMS

20-1. How many inch-ounces of torque are developed by a small motor that is rated at 0.035 hp at 2650 rpm?

20-2. How many inch-ounces of torque are developed by a small motor that is rated at 15 W output at 2575 rpm?

20-3. How many newton meters of torque are developed by a small motor that is rated at 12 W output at 150 rad/sec?

20-4. What is the horsepower of a small motor that is rated at 15 in.-oz at 3050 rpm?

20-5. What output power in watts is developed by a small motor that is rated at 20 in.-oz at 2850 rpm?

20-6. What output power in watts is developed by a small motor that is rated at 0.100 N·m torque at 165 rad/sec?

20-7. A two-scale Prony brake shows 43.7 and 29.4 oz when used on a 1 in. diameter pulley at 2775 rpm.
(a) What torque is shown?
(b) What power is developed?
Use English units.

20-8. A two-scale Prony brake shows 1205 and 1075 grams over a 1.27 cm diameter pulley, while a stroboscope shows 2915 rpm when a small motor is tested.
(a) What is its torque in newton meters?
(b) What is its output in watts?

20-9. A small shaded-pole motor is operated on a two-scale Prony brake rig. The scale readings are 17.7 and 21.9 oz on a 0.250 in. diameter pulley at 2785 rpm.
(a) What is its torque?
(b) What is its power?

20-10. The motor in Problem 20-9 operates on the same conditions and draws 8.9 W at 205 mA on 115 V. It runs light on 4.8 W at 145 mA. Its stator resistance $R_{dc} = 63.1\ \Omega$.

(a) What is its rotational loss?

(b) What is its stator copper loss under load?

(c) What is its rotor copper loss under load?

(d) What is its efficiency using the AIEE method?

20-11. Compare the output power from Problems 20-9 and 20-10. What is the percent difference?

20-12. The motor in Problems 20-9 to 20-11 is tested on a different two-scale Prony brake rig. This time the scales read 6.1 N and 4.9 N. The pulley diameter is 6.35 mm. The speed is 290 rad/sec.

(a) What is the torque?

(b) What is the power?

(c) Compare with result in Problem 20-9.

Answers. **20-1**, 13.2 in. oz; **20-2**, 7.86 in. oz; **20-3**, 0.08 N·m; **20-4**, 0.0458 hp; **20-5**, 42.1 W; **20-6**, 16.5 W; **20-7(a)**, 7.15 oz in., **(b)** 0.0198 hp; **20-8(a)**, 8.13 $\times$ 10^{-3} N·m, **(b)** 2.48 W; **20-9(a)**, 0.525 oz in., **(b)** 1.46 $\times$ 10^{-3} hp; **20-10(a)**, 3.14 W, **(b)** 3.32 W, **(c)** 1.26 W, **(d)** 13.3%; **20-11**, Problem 20-10 is 7.63% larger; **20-12(a)**, 0.003 81 N·m, **(b)** 1.10 W, **(c)** Problem 20-12 is 0.909% larger.

21

Alternating-Current Motor Control and Operation

The control of ac motors has traditionally followed the same concepts as that of dc control in terms of conventional control. Electromechanical relays are used to mechanically switch contacts which alter the circuit in control of the motor for various reasons, i.e., reverse polarity, switch resistors in or out, etc.

With the development and continual improvement of solid state power devices and the availability of complex, precise triggering sources for these devices, solid state control has shown to be superior to the conventional approach in many respects. As was seen in Chapters 8 and 9, such important factors as high reliability, low maintenance, greatly improved efficiency and the absence of moving parts make solid state control the clear choice of today when considering new installations or retrofitting and updating older systems.

Conventional ac motor control systems are, of course, still very much present today and will probably continue to be for many more years. In many "in-place" installations, it is just not economically possible to replace conventional control systems with their modern solid state replacements.

This chapter briefly addresses both conventional control and solid state control of ac motors (the induction motor in particular) so the reader has a general idea of the concepts involved. For detailed study of specific circuits of interest, the reader is referred to the references.

There are more similarities than differences between dc and ac motor control. As a result of this similarity, the vast majority of the items discussed in the first part of Chapter 9 are directly applicable to ac motors.

Alternating-current motors of substantial size are normally started by simply connecting them across the power lines. Their winding inductive reactances prevent the extreme starting currents of large dc motors.

21-1 ALTERNATING-CURRENT MOTOR LOAD CHARACTERISTIC LIST

This discussion will follow the form of Sect. 9-1 on dc control.

21-1-1 Power Supply. The brief discussion in Sect. 9-1-1 is applicable, but with ac, motor voltage rating can frequently be changed by either series or parallel stator winding connections. This holds for polyphase or single-phase induction motors in the vast majority of cases. Further voltage adjustments can be made by the use of transformers. It would rarely pay to try to adapt a 115 V motor to a 460 V line although a transformer can do the job.

Here also a three-phase motor cannot directly be operated from a single phase line. Some form of phase converter is required. A single-phase motor can always be operated from any two lines of a three-phase system providing the voltage level is made correct.

The final power supply consideration is that the frequency of the supply must match or nearly match the motor rating. A universal motor is relatively independent of the line frequency, except that a motor rated for 60 Hz or less will not usually operate well on 400 Hz.

The ac power supply versus matching situation is much more flexible than the dc situation. The supply must have adequate current and wattage capability.

21-1-2 Starting Requirement Considerations. See Sect. 9-1-2.

21-1-3 Stopping Requirement Considerations. See Sect. 9-1-3.

21-1-4 Reversing Requirement Considerations. See Sect. 9-1-4.

21-1-5 Normal Running Requirement Considerations. See Sect. 9-1-5.

21-1-6 Speed Control Requirement Considerations. See Sect. 9-1-6.

21-1-7 Safety and Environmental Requirement Considerations. See Section 9-1-7.

This repetitive listing above is adopted to save much greater repetition since the basic considerations that influence motor control design decisions are largely the same between ac and dc motors. Again, references (6) to (10) are all valuable. Note that references (6) and (9) have motor control texts that are separate books from their excellent motor principle texts.

21-2 COMPONENTS AND RELATED DESIGN SYMBOLS

Again, Sect. 9-1 should be reviewed because it is all applicable.

The transformer is an added device used for control circuits in ac machines that has no direct counterpart in dc machines.

21-2-1 Line Voltage Transformers.

Transformers are used for main line voltage control, for both reduced voltage starting, and for voltage matching. The difference here is that a starting voltage reduction will usually be an open delta or V–V circuit for economy on three-phase circuits.

When line-to-load voltage matching is used, such as to adopt a 220 V two-speed motor to a 440 V line, the transformers may be used in a bank of three. An example would be a delta–delta connection.

21-2-2 Pilot Control Circuit Transformers.

Under modern safety regulations, manually operated pilot devices, such as pushbutton controls, may not have more than 115 V in their connected circuits. To a lesser degree, a machine-mounted pilot device, such as an overtravel limit switch, should not be connected with more than 115 V. This results in the need to control 220 V, 440 V, or higher-voltage motors from 115 V pilot devices. Note that the control circuit is always single phase.

The *control circuit transformer* is used to meet this requirement so that all high voltages are reduced to 115 V for control purposes. The 115 V control lines operate the various sequence and control relays and devices. The main line contactors are operated by 115 V electromagnets. Only the primary line contacts, and overload heater circuits actually see the higher motor voltages.

The reason for all these precautions is that any device carrying the lower 115 V level is much less likely to break down. Even if there is a breakdown, the 115 V lines are not so likely to cause severe shock hazard to an operator.

The usual control circuit transformer has, in miniature, the same two primary winding coils and two secondary winding coils as a power distribution transformer. A 440 to 460 V power level will use the two primary coils in series. If 208 V to 230 V lines are used, the primary coils are connected in parallel. Normally, the 115 V secondaries are connected in parallel to yield the full current capacity. They may be used separately so that part of the control may be more thoroughly isolated. If some older control elements that have 220 V or 440 V magnet coils are being used, a second control circuit transformer may be used as a step-up device so that 110 V control levels are maintained, but a 440 V device may be actuated.

Control circuit transformers are widely available in various standard but low volt-ampere ratings. These transformers are also available for any normal primary voltage range, such as 2300 or 4600 V. These transformers are packaged for mounting in or on a motor control component box.

21-3 NONREVERSING FULL-VOLTAGE MOTOR STARTER

The usual single-phase motor, of almost any type, is started and operated by simply connecting it across the line. This is done by any of the simple manual switch types or by cyclically operated timer contacts. A better-grade home appliance will have a built-in thermal overload protective device. Typically, this may be a Klixon circuit breaker that can be reset with a push button.

Single-phase induction motors of up to about $1\frac{1}{2}$ hp (1.125 kW) when in a home, or up to about 5 hp (3.75 kW) or more in commercial use, are started by use of full voltage. These larger motors are usually permanently connected and provided with substantial switches. Builders' saws or household vacuum cleaners are about the upper limit for a device that is normally connected through a line cord and a plug connection. In either case, the actual switching is performed by a suitable switch rather than the plug.

Three-phase induction motors can be started with full line voltage without damage even in extremely large sizes. However, to avoid depressing the voltage of the line and thus affecting the surrounding equipment, motors of around $7\frac{1}{2}$ hp (5.625 kW) or larger usually use some form of reduced-voltage starting.

The usual situation for a *nonreversible three-phase* induction motor with *full line voltage starting* is shown in Fig. 21-1. Note that the complete array of controls and safety devices is provided. These include bus switches, protective fuses, magnetically operated contactors, thermal overload protection, and reduced voltage in the control circuit for pushbutton safety. In a commercial or industrial situation, *every machine* operating across 208 to 230 V or 440 to 460 V three phase will have at least these elements in its control circuit. Older installations may well have only two of the three lines protected with thermal overload elements. Many older installations will still have 220 V or even 440 V at the pushbuttons. Both of these conditions are rapidly being reworked and no longer appear in properly prepared new installations.

If the situation shown in Fig. 21-1 had used a 208 V to 230 V supply, the *primaries* of the control circuit transformer would have been connected in parallel. Under this lower voltage situation, the motor would be connected in parallel wye as in Fig. 11-4b. The *same equipment* would be used throughout in either the high- or low-voltage situation.

There are a number of items to note in Fig. 21-1 that will not be shown so completely in some of the later motor control circuit diagrams. These are as follows:

(1) Wiring color codes will follow those shown when flexible multiconductor cable is used. If rigid or flexible conduit is used, the same colors should be used. Many service departments, however, stock only black wire. The *terminal numbers* that appear at wire connection points should be permanently marked *on the wire* so that repairs may be made conveniently.

(2) Note that in the control circuit most conductors are connected to the *same*

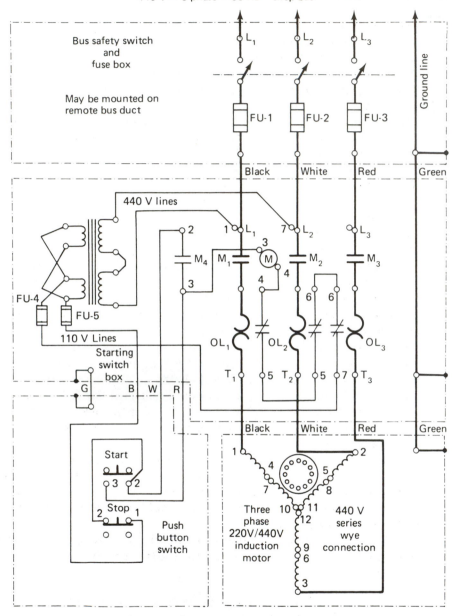

Figure 21-1. Three-phase nonreversible line voltage motor control.

terminal number at each end. Thus, 2 to 2, 3 to 3, 4 to 4, and so on. Terminal 1 on the starting switch would go to terminal 1 on the pushbutton control if the transformer did not intervene. Terminal 7 to terminal 7 is the same situation.

(3) The main line input L_1 follows through to T_1 and to terminal 1 on the motor. The same situation holds for L_2, T_2, and 2, and for number 3.

(4) When checking an installation, if the motor *rotation* is wrong, the lines to *any two* motor terminals should be *interchanged*. Usually, lines 2 and 3 are the ones to be corrected.

(5) Note that the cables from the bus to the safety switch, from the safety switch to the starting switch, from the starting switch to the motor, and from the starting switch to the pushbutton control *all* carry a *fourth ground line*. This is both a code requirement and simple prudence.

Note also that the entire control circuit is composed of two simple loops. One loop goes from high-voltage terminal L_1 through the control circuit transformer primary group and back to high-voltage line L_2. The other low-voltage loop is a series connection from the transformer secondary winding group out and return. This simple series loop has one parallel branch, where the start button and the main contactor maintaining contacts M_4 are in parallel with each other. The remainder of the series loop places the three overload release contacts OL_1, OL_2, and OL_3 and the main contactor solenoid coil M all in series with the *stop button*. *Any break* in this loop will *stop* the motor by allowing the main contactor to drop out. There might also be one or more overtravel limit switches, other off buttons, or emergency switches placed in this series loop. Any additional *start* buttons are placed in *parallel* at contacts 2 and 3.

21-4 REVERSING FULL-VOLTAGE MOTOR STARTERS

When the line voltage is low, as in household 115 V single-phase lines, and the motor is less than 1 hp (0.75 kW), a *reversing motor control* frequently uses a small drum rotary switch. In the past, *reversible* three-phase motor-driven tools such as a drill press with 1 hp (0.75 kW) or less used the drum rotary switch.

21-4-1 Three-Phase Integral Horsepower Controls. The control is more elaborate when the horsepower or kilowatts required becomes substantial or when higher voltages are used. Figure 21-2 shows a typical three-phase reversing line voltage control for 208 V or higher, and for up to *about* $7\frac{1}{2}$ hp (6.25 kW).

The two electromagnetic line contactors are usually mounted in a prepared frame so that the action of one can mechanically block the simultaneous action of the other. This is called *mechanical interlocking*. Electrical interlocking may be used instead or in addition, but it is much more liable to be found on reduced voltage starters. In that case, the extra electrical contacts required are already present on relays used for other parts of the circuit.

The control in Fig. 21-2 is widely used in industrial lathes, milling machines, and other machine tools. This circuit diagram is also shown in full detail rather than partial form because of its widespread application.

The pushbutton control shown may be traced to the maintaining contacts on the forward contactor and on the reverse contactor. When *forward* is pressed, the push button makes contact between terminals 3 and 2. When the forward contactor pulls in, the normally open maintaining contacts 3 and 2 are closed to *hold* the forward function. The same relation exists between 5 and 4 on both the

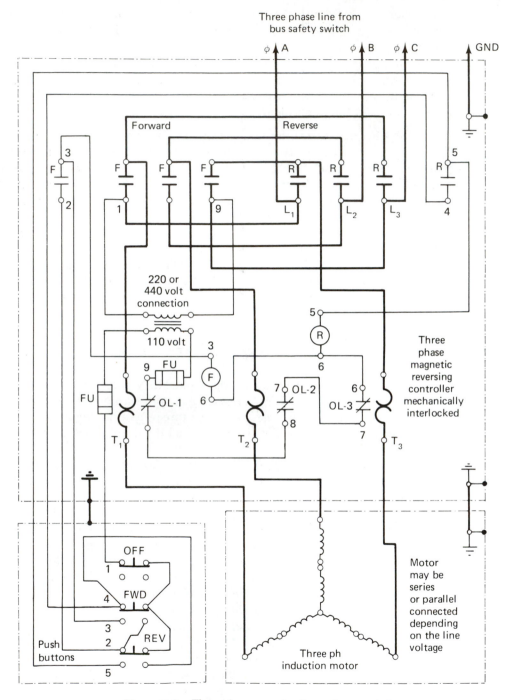

Figure 21-2. Three-phase reversing line voltage control.

reverse pushbutton and the reverse maintaining contacts. This circuit is very commonly applied, but until recently the thermal overload circuit in the T_2 line was not present. The use of the control circuit transformer to reduce the voltage on the pushbuttons is more apt to be present on an older installation. Both features should be incorporated in any new installation and retrofitted into older controls.

21-5 REDUCED-VOLTAGE THREE-PHASE MOTOR STARTING CONTROLS

When polyphase ac motors become large, their normal starting current will briefly reduce the line voltage and thus disturb the operation of other apparatus on the same distribution circuit. If the motor is large in relation to the total factory or business location, its starting current may be the single highest peak current during the metering period. If so, and if the utility uses peak current metering, as many do, then every time this particular motor is used, it will keep the kilowatt-hour power metering basis at a higher cost rate for the *entire metered installation*. Obviously, there is sound economic basis for a cost-reducing starting control for a motor of this character. Either for voltage regulation, or for cost reduction, one of the various methods of *reduced-voltage starting* is used. There are four basic ways to accomplish reduced voltage starting:

(1) *Line resistance starting* that uses suitable high-current, low-ohmic-value resistances in each line. After a suitable time delay the resistance is removed in steps or all at once.

(2) *Line reactance starting* uses suitable iron-core reactances in place of the resistances in (1) to accomplish the same result.

(3) *Autotransformer starting* uses tapped autotransformers in open delta or V-V connection to reduce the motor voltage. This is quite an efficient means of dropping voltage, and would therefore be chosen where the time delay during acceleration was substantial.

(4) *Wye–delta starting* is used when the motor is designed for *delta* operation at its rated voltage. The motor phase windings are reconnected by contactors for a wye circuit at starting. As a result, each phase will see the normal line voltage divided by $\sqrt{3}$.

$$V_l/\sqrt{3} = V_{st} = 0.577V_l$$

Eq. (21-1)

where

V_{st} = starting voltage per phase
V_l = line-to-line voltage

A variation of this method is *series delta–parallel wye starting*. Here advantage is taken of a normal dual winding motor operating on its *lower voltage*. Since the windings are in *parallel wye* for normal line voltage service, they may be placed in *series delta* for the same 0.577 or 58 percent reduction. It is necessary that the windings are not inadvertently reversed during the changeover. If a mild start where only one-quarter of the maximum starting torque is acceptable, the winding shift can be from parallel wye to series wye. The same effect is achieved by dropping from parallel delta to series delta.

All these schemes require a time delay between the first stage of starting and the final full-voltage situation. This time-delay circuit may be single stage, two stage, or more. Since this type of control was described in detail in Chapter 9, the student is referred to the sections on positive time-delay control circuits and current-limited time-delay circuits in Chapter 9.

The control circuit for ac motor controls is usually operated by the control circuit transformer secondary voltage. Since this voltage is most often 115 V, 60 Hz ac, the control relays and contactors must be suitable for that service. Frequently, a dc motor control will use ac-operated pilot devices. When it does, the control circuit must be fed from the local ac lines. Conversely, some ac motor control systems require direct current in some of the control elements. This usually is handled by ac-to-dc rectifier circuits in the control system. To comply with modern safety regulations, many dc motor control systems that had formerly operated on 250 V dc are rebuilt to use 115 V ac pilot devices so that the operator controls do not contain over 115 V. This also allows a factory or business service department to standardize its inventory of motor control system spares.

The control system shown in Fig. 9-1 is a generalized three-stage time-delay starting circuit when considered from a point *below* the control fuses FU-3 and FU-4. If this system contained all 115 V ac pilot devices, and if its control power came from the low-voltage secondary of a control circuit transformer, it would be entirely suitable for ac motor control.

The addition or subtraction of time-delay stages is relatively simple. To reduce the system in Fig. 9-1 to a *single time delay*, the control levels that are labeled 3, 4, 5, and 6 would be removed. The final levels 7 and 8 would then be grafted onto levels 1 and 2, where the present 3 and 4 levels are. The *M* contactor is used for the first stage or *starting* connections of a three-phase circuit. The 1*A*, 2*A*, and 3*A*, or however many stages there are, contactors are used to close the intermediate and final circuits. These contactors are sometimes labeled as *M* for main or *S* for start. The last stage of delayed contact, or the only delay if there is just one, may be labeled as *R* for run. Nomenclature is not fully standardized nor does it need to be if a technician understands the implications of the names and letter identities.

21-5-1 Specific Starting Circuits. The four different methods of reduced-voltage starting are shown in a related fashion in Fig. 21-3. The time-delay control circuit is not shown but is related to Fig. 9-1 as described above.

In Fig. 21-3a and 21-3b the main contactor *M* will normally carry a fourth or maintaining contact set, and will have the overload heater circuits and contacts. The run contactor *R* can be a simpler unit or a contactor instead of a starting switch. The autotransformer starting circuit in Fig. 21-3c has a *six-contact* starting contactor. A normal three main contact *run contactor* will usually carry the over-

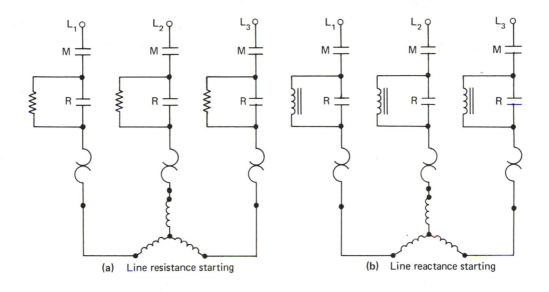

(a) Line resistance starting

(b) Line reactance starting

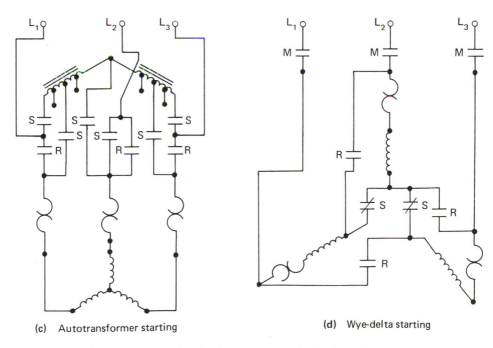

(c) Autotransformer starting

(d) Wye-delta starting

Figure 21-3. Reduced-voltage starting methods, three-phase motors.

load contact equipment. Notice that the circuit is such that the overload heater elements are always active, as are their contacts, which are not shown.

Figure 21-3d for the Y–Δ starting circuit shows three contact identities, M, S, and R. The M contacts are on a normal starting switch that carries the overload heaters and contacts. The start contacts S and the run contacts R are on a *double-throw* contactor.

This unit is normally closed on the S contacts until the control circuit finishes its time delay and causes its S contacts to open and its R contacts to close. Notice that all motor windings are protected in either circuit configuration. Other variations, such as series delta–parallel wye, are not shown, but use a similar school of thought even though more contacts are required. See the various references for specific circuits.

Notice that when shifting from Y to Δ the motor circuit is temporarily open. This feature may cause *brief* high-current transients when the circuit is reclosed. See reference (9) for further discussion on this transient feature.

21-5-2 Other Reduced-Current Starting Circuits.

The real goal of any starting device is to reduce the current at starting below that which would occur if the motor were placed across the line. When a motor is normally operated with its windings in parallel, an obvious reduction in current can be achieved by using *one-half* of the windings in a *part-winding circuit*. When the motor is moving at a large percentage of its running speed, the other half of the winding can be cut in to achieve full running torque. This scheme will reduce the inrush current at start to about 60 percent of the normal across the line value. At the same time, the torque will be slightly less than 50 percent of the full-winding value. If these current and torque levels are satisfactory, the method is attractive because the motor is a standard unit and no additional high-current primary control device beyond a second main contactor is needed. See Fig. 21-4a for a typical part-winding circuit. The contactor designations are M_1 for first main contactor and M_2 for second-stage main contactor.

It is good to investigate whether the motor windings involved will still include all poles when part windings are used. For example, the winding shown in Fig. 11-4, if operated according to Fig. 21-5a, would only include two live poles on part winding. When wired according to Fig. 11-4, its operation on part winding would still be as a four-pole unit, but two of the poles would be consequent poles. An even more common situation is where the first half of the winding is the first and second pole and the second half includes the third and fourth pole. Here consequent pole operation is still less efficient. Fortunately, larger motors, for which this form of control is mostly used, frequently have up to four parallel paths instead of the two shown in Fig. 11-4. This will allow part winding operation still to include all poles and thus gain the reduced current benefit without serious performance penalty.

21-5-3 Wound-Rotor Induction-Motor Starting.

The wound-rotor induction motor has the feature that its starting torque and its starting current can be controlled by varying the external resistance in the rotor circuit. See Sect.

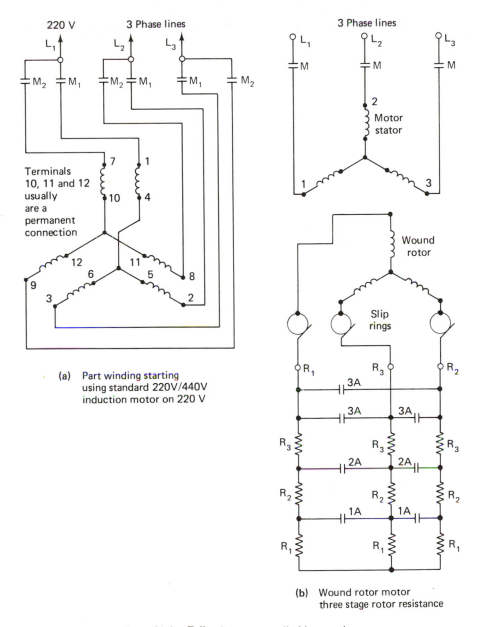

(a) Part winding starting
using standard 220V/440V
induction motor on 220 V

(b) Wound rotor motor
three stage rotor resistance

Figure 21-4. Full-voltage current-limiting starting.

17-2 and Fig. 17-2 for a review of these characteristics. Obviously, the external resistance can be controlled manually by the use of suitable potentiometers or tapped resistances. Frequently, wound-rotor motors are equipped with large face plate rotary switches and suitable high-wattage resistors. These are arranged so that the motor speed may be manually controlled while under load.

When it is desired to start a wound-rotor motor with a single pushbutton operation, the required change of resistance must be automatic. A definite time-delay control circuit having three delayed stages is suitable. This is shown in Fig. 9-15. With such a control the external resistance can be changed in three stages. See Fig. 21-4b for the high-current portion of a wound-rotor-motor starting control. The contactor designations of M, $1A$, $2A$, and $3A$ enable its use to be visualized with the control circuit portion of Fig. 9-15.

Depending upon the actual resistances of the resistance stages in the rotor circuit, the torque may be maximized or the current minimized. With resistances set for maximum torque values, the motor can be made to accelerate approximately up the right side of the speed versus torque curves in Fig. 17-2. If lower torques are acceptable, the starting and accelerating currents can be held to minimum values. The actual resistance values in either case can be developed from the per unit curves in Fig. 17-2 if the basic rotor resistance R_r and blocked-rotor reactance X_{br} are known. An actual set of performance curves for the specific motor is preferable if it can be obtained.

From Fig. 21-4b it can be seen that, when the M contactor closes, the resistance in the rotor per phase is $R_r + R_1 + R_2 + R_3 = \Sigma R$. The resistances are usually connected in wye as shown. When the first delay stage times out, the $1A$ contactor closes and shorts out the three R_1 resistors. As the second stage times out, the $2A$ contactor closes and removes the R_2 resistances. The action of the third stage is similar, except that frequently three sets of $3A$ contacts are used to produce the lowest available rotor resistance. The timing circuit shown in Fig. 9-15 then opens the circuits of the previous stages so that lower duty cycle contactors may be used and to reduce the control circuit power. Notice how similar this control action is in the dc motor and the ac wound-rotor situations.

21-6 REVERSING REDUCED-CURRENT CONTROLS

When reversing requirements are combined with the need to reduce current in starting, the circuits of Sects. 21-3 and 21-4 are combined. It is necessary for the current control mechanism to be on the motor side of the reversing control so that it does not need to be duplicated. In the case of a wound-rotor motor, the two control functions are entirely separate. The reversing control is entirely in the stator circuit, and the current controlling is wholly in the rotor circuit. The required additional circuits in the low-voltage control level are minimal. Once each function is separately understood, the combined circuit is relatively easy to follow. Detailed circuits will not be shown here, but a selection is shown in the references.

21-7 PLUGGING REVERSE OF ALTERNATING-CURRENT MOTORS

When an ac-powered machine or device requires a rapid reverse or pull down of speed, the motors may be *plugged* in similar fashion to that described for dc motors in Chapter 9. The ac induction motor has an advantage here because its line current will not seriously increase over its normal starting current.

If a motor does not require current reduction during a normal start, it will usually not need special current reduction during a plugging reverse. In this case, the *main contactors* need to be the next larger size so that their contacts can resist the longer duration or more frequent use of the high-current level.

If a motor is of sufficient size that it requires current reduction for normal starting, it will usually use the same current-reduction circuit for a plugging reverse. These circuits will need investigation for the *duty cycle* in the same sense that the motor does. If normal use of start and stop, reversing, or plugging reverse is sufficiently frequent, the motor itself may need to be a larger size than the power requirement would indicate. The line resistors or the wound-rotor-motor rotor circuit resistors will need to be analyzed for their probable temperature rise and will need to have their wattage rating adjusted accordingly.

With the circuit of Fig. 21-2, a plugging reverse may be performed by first pressing the off button to allow the forward contactor to drop out. Then the reverse button can be pressed, even though the motor is still moving forward at nearly its full normal speed.

Obviously, this is not a convenient plugging reverse cycle. By using forward and reverse pushbuttons that are connected as shown in Fig. 9-20, a plugging reverse may be achieved. With this circuit, merely pressing the reverse button allows the forward contactor to drop out and the reverse contactor to pick up. The time-delay starting circuit may or may not be present, depending upon the need discussed before.

If plugging is desired to rapidly reduce speed, but not actually change rotation direction, a *plug-stop* circuit is used. Just as were used in the dc motor plug stop, a *plug-stop switch* (PS) or a *centrifugal switch* (CS) needs to be used. These allow the motor to be cut off the line when its motion just reverses or nearly ceases. The control circuit technique used in Fig. 9-20 is applicable.

21-8 RETARDING AND STOPPING ALTERNATING-CURRENT MOTORS

Again, the similarity to dc motor control situations is dictated by the *type of service* rather than whether direct or alternating current is used. If the load is of an *overhauling* nature, as in a descending crane or elevator, there is a need for the motor to develop a torque that is in the opposite direction to the rotational motion. Here the characteristics of an induction motor are considerably different, so a retarding control will be different than for direct current.

An induction motor can serve as an *induction generator* if it is driven to a speed slightly *above* its synchronous speed. As a first approximation, if a motor develops its rated power at a slip of s *below* its synchronous speed S, it will *absorb* its rated power at a slip s above the synchronous speed S. The difference of $S - s$ to $S + s$ depends upon the rotor type and the normal slip value. If $s = 0.05$, then $S - s = 1.00 - 0.05 = 0.95$, and $S + s = 1.00 + 0.05 = 1.05$. A motor that has a normal loaded slip of 5 percent *below* the synchronous speed will rise to 5 percent *above* synchronous speed when *absorbing* its rated power as an induction generator. The motor *requires* its normal stator excitation to function in this manner. If its main contactor is opened either by a deliberate act or by the action of its overload breaker, it cannot self-generate because it has no self-excited mode of operation, such as a dc generator.

When an induction motor is used in hoist service, it has the speed range that its number of poles and its working slip range dictate. When descending with a hoist load, an induction motor will rotate just perceptibly faster than normal as long as it is across the lines. Changing wound-rotor resistance would *increase* the slip and thus increase the rate of load descent. The speed can be reduced by a plugging action, but this cannot be maintained for a slow descent. The usual small industrial hoist has a control with an up and a down button. When the buttons are released, the motor is plug stopped and held with a spring-operated magnetically released brake. Small motions are produced by stabbing the buttons to get a jog action. There is no holding circuit for safety. The hoist will stop unless the controls are held.

This induction generator action has been used in a few ac electric-powered mountain railroads. The Virginian, the Norfolk and Western, and part of the former Pennsylvania Railroad all used three-phase wound-rotor consequent pole motors in some of their locomotives. All these services used 25 Hz, single-phase overhead wire power supply. All converted the single-phase power to three-phase power by induction phase converters. While the gear ratios and the specific operating speeds varied, all these locomotives had only two normal operating speeds. One speed was for four-pole motor operation and another for eight-pole operation. The long periods of acceleration for starting and between speeds were handled by using very large blower-cooled resistances in the wound-rotor motor circuits.

When descending grades, these locomotives would automatically transform into induction generator mode. They were therefore fully capable of descending a long grade at about 10 percent greater speed than they would climb the same grade. During the descent the power was pumped back into the lines. The two-fixed-speed feature was not very flexible but was considered an advantage in single-track freight service because of the ability to keep trains spaced automatically. These locomotives served for 40 to 50 years with satisfaction but are now retired and scrapped. They were simply worn out from millions of miles of service.

This is another apparently obsolete type of mechanism that might very well be reconsidered and redesigned for high-speed commuting service in rapid-transit railroads. The shortage of energy will force reconsideration and modernization of many things. Time changes ground rules.

21-9 ALTERNATING-CURRENT MOTOR JOGGING CONTROLS

The problem of producing a small incremental motion or jogging in an ac-motor-driven mechanism is the same as a dc motor control. Jogging is produced by modifying the pushbutton and control circuit. When a jog button is used, the *M* contacts are closed, but the normal maintaining circuit is bypassed deliberately.

This is then a function of the control circuit rather than the motor. Again, review the section on dc jogging. The principles are the same.

21-10 ALTERNATING-CURRENT MOTOR SPEED CONTROL

Each ac motor type has its own variable-speed characteristics, which have been discussed in the appropriate sections. Squirrel-cage motors are not capable of a wide speed range unless either the number of poles or the line frequency is changed.

Frequency-changing circuits are being developed at a continuously increasing rate. This is due to the rapidly improving state of solid state circuit elements that can handle high voltage and/or high current.

Pole changing is handled by switching circuits that resemble some of those already shown.

21-10-1 Consequent Pole Speed Controls.
There are three classes of consequent pole motors depending on their intended service. The NEMA has standardized the terms as follows:

(1) *Constant torque* motors with about the same *rated torque* at both high and low speed. These motors are connected in parallel wye for high speed and series delta for low speed (see Fig. 21-5 for typical connections).

(2) *Constant horsepower* motors with about the same *rated power* at both high and low speed. These motors are connected in series delta for high speed and parallel wye for low speed (see Fig. 21-6 for typical connections).

(3) *Variable torque* motors with the rated torque *directly proportional to speed* at either high or low speed. These motors are connected in parallel wye for high speed and series wye for low speed (see Fig. 21-7 for typical connections).

The motor connections shown in Figs. 21-5, 21-6, and 21-7 are NEMA-standardized terminal numbers connections for single-winding two-speed motors. It is then obvious that it is necessary to know which type is being wired. The comparative speed versus torque curves on a per unit basis are shown in Fig. 21-8. In this figure the *high-speed* connection is assumed to produce the same characteristic curve in each case.

The choice of which motor type to use depends upon the speed versus torque requirements of the driven load. The actual circuit used will require a three-terminal contactor for one speed and a five-terminal contactor for the other. Figures 21-5 through 21-7 show the contacts required as *H* for high speed and *L* for low speed.

21-10-2 PAM Motor Speed Controls.
The circuits shown in Figs. 21-5, 21-6, and 21-7 are equally suitable for PAM motors. As presently manufactured in Europe, these motors have the same six terminal points that are NEMA-standardized for consequent pole motors. Even if not labeled the same, the manufacturer's literature should enable translation of terminal numbers. At this point

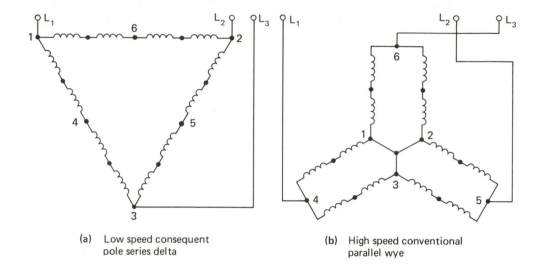

(a) Low speed consequent pole series delta

(b) High speed conventional parallel wye

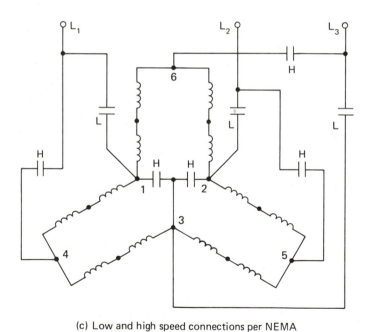

(c) Low and high speed connections per NEMA

Figure 21-5. Constant torque, consequent pole, single-winding motor circuits.

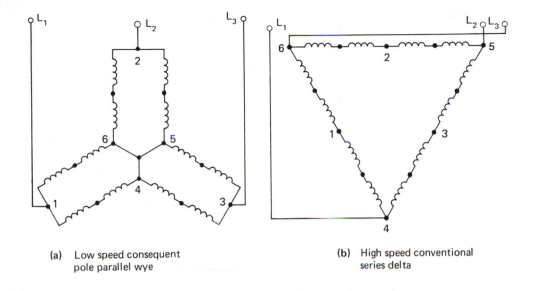

(a) Low speed consequent pole parallel wye

(b) High speed conventional series delta

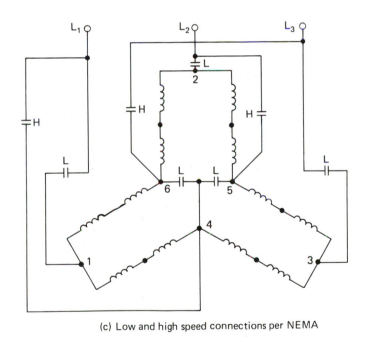

(c) Low and high speed connections per NEMA

Figure 21-6. Constant power, consequent pole, single-winding motor circuits.

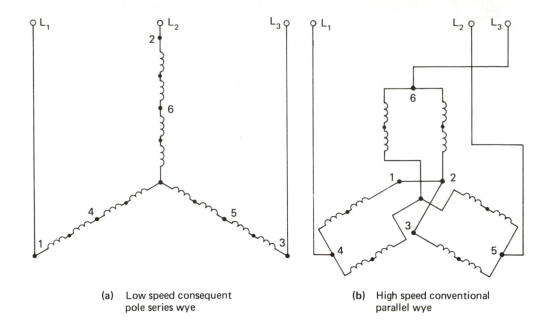

(a) Low speed consequent
pole series wye

(b) High speed conventional
parallel wye

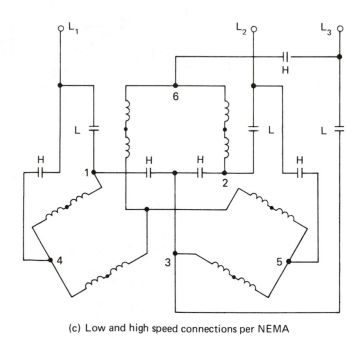

(c) Low and high speed connections per NEMA

Figure 21-7. Variable torque, consequent pole, single-winding motor circuits.

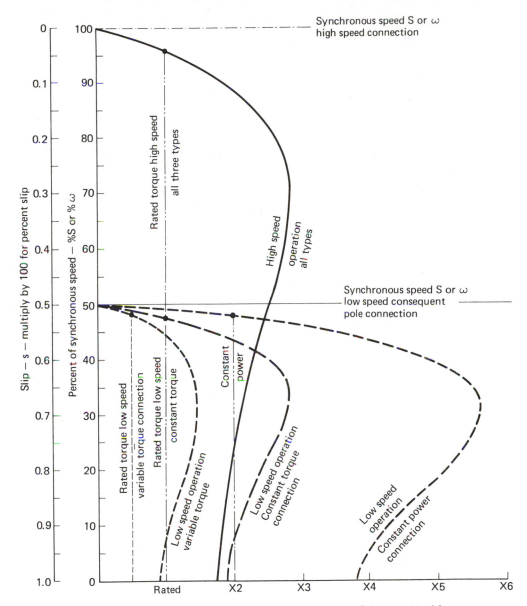

Figure 21-8. Speed versus torque for consequent pole motors.

then, consider a PAM motor installation to require the same controls as a similar size and voltage consequent pole control.

21-10-3 Two-Winding Speed Controls. When a three-phase motor has two distinct windings, such as a four-pole winding and a six-pole winding, the switching required is very similar to Fig. 21-2. Here one contactor is for high speed, *H* instead of *F*; the other is for low speed, *L* instead of *R*. The circuit and interlocking requirements are identical.

21-10-4 Two Speed with Reverse Controls. To have both two-speed service and reversing service for either two-winding motors or for consequent pole motors requires that the control circuit be cascaded. The reversing circuit that sets the *phase sequence* at the motor will precede the two-speed feature, but both are required. These circuits frequently require extensive interlocks to prevent certain undesirable sequences of operation. Consult the references for a specific version.

21-11 REDUCED VOLTAGE MOTOR OPERATION EFFICIENCY BENEFITS

The recent and still increasing rise in the cost of energy will force much study and change in motor installations and operation. It has long been known that it is uneconomical to install a higher power motor than a situation demands—not just because of the first cost of a motor and its associated switching controls, but because a motor that is operated at, say, one half its rated power will draw substantially more power from the lines than a unit that is rated at that lower power.

On the other hand, a great many factory motor installations work at a low percentage of their rated power for a substantial portion of their duty cycle. It will be shown that by appropriate switching to a lower supply voltage, much of the wasted energy can be saved. Overall factory reductions in power consumption of around 20 to 30% seem entirely within reason. This situation is rapidly moving from an interesting curiosity to a vital need.

The reduced voltage can be supplied by versions of the NASA-licensed Frank Nola controls as mentioned in Sec. 17-4-3, or by electromechanical controls from limit switch sensors, or even by manual programming. The paragraphs ahead will show the gross benefits to be obtained. Methods of obtaining them will be discussed after that.

Figure 21-9 shows two complete sets of characteristic curves of a typical three hp, four-pole 208–220 V, 60 Hz induction motor. The efficiency and current curves are separated from the power factor and speed curves to avoid clutter and confusion. These curves are typical of a number of different tests performed to illustrate the effects of line voltage, and may be considered as indicative of the results on a large range of motor sizes.

The solid line curves in both views are the normal full voltage characteristics as provided by the motor manufacturer. The efficiency is seen to reach a maximum at around 75 to 80% of the rated power. Again, reasonably typically, the efficiency at 50% of rated power is near that at full rated power. However, below half rated power the efficiency falls off rapidly. At about 10% of rated power the efficiency is only about $\frac{2}{3}$ that of full power.

This rapid fall in efficiency at low power is inescapable at full line voltage. The excitation current that is required to produce the magnetic flux is essentially

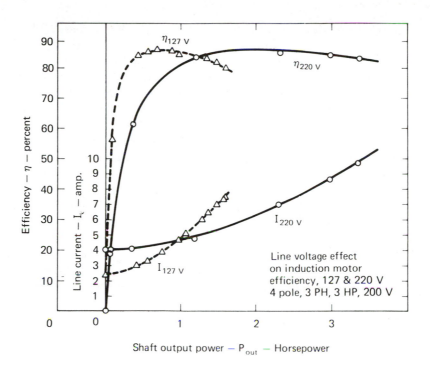

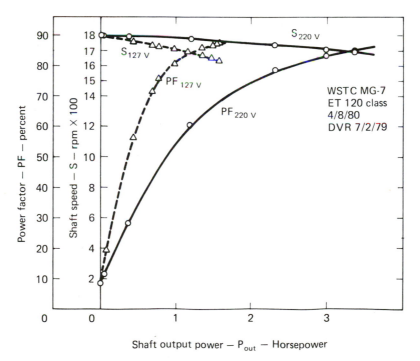

Figure 21-9. Reduced voltage effects.

constant and also lags the line voltage by nearly 90 degrees. All this is acceptable if the motor is loaded in its upper range most of the time. However, a turret lathe, for example, may be substantially loaded on only one or two of its turret positions. The other turret positions are frequently involved with tooling that only requires a very small portion of the rated power. Worse yet, many machines serve for months at a time performing a repetitive production task which never loads its installed motor as high as 50 or 60% of rated power.

The first step, and at the same time the greatest relief of this situation, is to reduce the voltage seen by the motor winding coils by a convenient fraction. Any motor which is rated in parallel wye can be easily reconnected to operate in series delta. This reduces the coil voltages to $(\frac{1}{2}) \div (1\sqrt{3})$, or 0.866. A more convenient reduction and also perhaps the most widely beneficial at the same time, is seen by going from parallel delta to parallel wye, as well as from series delta to series wye. These changes reduce voltage by 0.577.

The dotted line curves in Fig. 21-9 shows the characteristics of the same motor when it is operated either on 127 V instead of 220 V or by staying on 220 V and changing from parallel delta to parallel wye. In the region around one-third power, in this case 1 hp, the operation is essentially the same on either voltage. The motor will easily carry one-half rated power on the 0.577 times rated voltage per coil. The real benefit shows at say 10% of rated power. Here at 10%, or in this specific case 0.3 hp, the efficiency, which is dropped to 57% on rated voltage, is *improved* to 82%. At the same time the current is *reduced* from 4.2 A to about 2.8 A. As a result, the power factor climbs from 25% to 45%. The speed shows reasonable slip up to 50% of rated power. These benefits are too great to be lightly dismissed.

This type of motor operation can be achieved by the circuits used in wye delta starting, as in Fig. 21-3d. Autotransformer circuits, such as that of Fig. 21-3c, can be used to achieve any reasonable voltage reduction other than 0.577 to 1. For example, a machine that idles completely unloaded for any appreciable part of its cycle may profitably use 20 or 25% of its rated voltage. The motor shown in Fig. 21-9, when running light, drew 125 W, 4.05 A at 0.085 PF. When the voltage was drastically reduced to as low as 30 V, the motor drew only 25 W at 0.8 A and 0.6 PF. At this very low voltage, the motor accelerated impractically slowly and could not carry even an idling machine load. However, at 55 V, or 25% of the rated line, the motor still drew only 30 W at 1.15 A and 0.28 PF. At this point it could easily idle a reasonable machine spindle and gearing load.

Inspection of Fig. 21-9 again shows that suitable control points might well be the following:

a) A decreasing load from full power would drop the line current from around 8.8 A to, say, around 5 A, to as low as 4.5 A at a desired switch-over point. Current sensitive relays could then drop out and control the switching into the low voltage regime.

b) The same relay might well be able to pull back in at 5.5 A, with adjustment. This differential range could serve to prevent control oscillation, and yet not miss the major benefits. Even a sudden increase of load with a large machine would not cause motor stalling because the motor inertia is ample enough to allow the switching time transient to take place.

Where the load varies over a continuous range without previously known repeating conditions, a NASA-type control seems ideal. It seems reasonable that this type of control can be essentially built into a motor, which would result in the best of both worlds. Much change in factory operating procedures and resulting significant energy economics is cheerfully foreseen.

21-12 SOLID STATE CONTROL CONSIDERATIONS

As previously noted, solid state control of ac motors, and in particular the induction motor, is now commonplace and in some cases can be achieved more economically than ever before. The basic concepts of solid state devices as presented in Chapter 8 serve to enable the reader to understand the general concept of ac motor control and for this reason a brief review of that material is suggested.

21-12-1 Single Phase ac Motor Control. The speed of an induction motor can be controlled to a certain degree by varying the voltage applied to the motor input. With the use of a control circuit, as shown in Fig. 21-10, utilizing a diac and triac, the speed of a motor can be safely varied approximately $\pm 10\%$ around some chosen base speed of the particular motor. Note that the motor cannot quite reach its synchronous speed. The basic operation of Fig. 21-10 begins with the R-C portion of the circuit. As the voltage of the source swings between positive and negative, the capacitor charges to a positive voltage during the first half-cycle and then to a negative voltage during the second half-cycle. When the voltage across the capacitor reaches a value of breakover voltage, V_{BO}, of the diac in each direction, the diac conducts and triggers the triac to conduction. (The triac, then, receives a positive gate pulse and then a negative gate pulse.) Once the triac is triggered in either the positive or negative half-cycle, it will act like a short circuit for the rest of that half-cycle, and the motor receives that remaining line voltage. The firing delay angle to the triac can be varied so as to deliver more or less line voltage to the motor by varying the rheostat, R_B, so that the capacitor takes more or less time during each half-cycle to reach the appropriate breakover voltage of the diac. Due to the relatively high inductance of the induction motor, a snubber

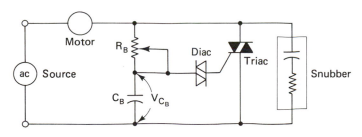

Figure 21-10. Single-phase induction motor speed control.

circuit is usually connected across the triac to insure that the triac shuts off as the line voltage changes polarity during each half-cycle.

The speed variation of the induction motor, as shown in Fig. 21-10, results from a voltage variation. Since the torque is directly proportional to the square of the applied voltage in the induction motor, it should be noted that the torque is also varying. As the torque varies, so does the efficiency of the unit, and due care must be exercised to insure that a reasonable balance between the two are maintained.

The capacitor-start split phase induction motor of Fig. 19-4 and the capacitor-start, capacitor-run motor of Fig. 19-5a each rely on a centrifugal switch. This switch engages a starting capacitor for phase shift during the initial starting of the motor at zero speed and until the shaft speed reaches approximately 75% of running speed. In essence, this action causes the motor to act as a two-phase induction motor during start-up. The switch then opens to run as a pure single-phase motor, as in the capacitor-start motor of Fig. 19-4 or as in the capacitor-start, capacitor-run motor of Fig. 19-5a. The capacitor-run motor continues to run as a two-phase motor because of the unswitched running capacitor in series with the starting winding. In either case, the mechanical centrifugal switch is subject to failure and is the major cause for motor downtime and repair. The centrifugal switch with its moving parts and mechanical problems can be replaced with a triac circuit as shown in Fig. 21-11. By placing a low resistance sensing resistor in series with the stator (running) circuit, a relatively high triggering voltage is developed across this sensing resistor due to the high starting current of the stator winding. As this high starting current flows, the voltage across the sensing resistor changes polarity with each half-cycle and is of a value sufficient to trigger the triac so that it acts as a closed switch to both positive and negative half-cycles. The sensing resistor is sized so that it delivers a sufficient triggering voltage to the triac based on the *peak* stator current. Thus, as the motor speed reaches about 75% of full speed with a

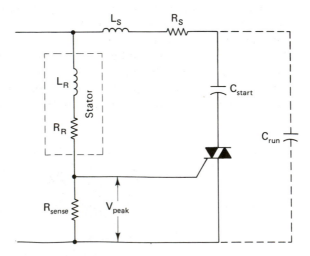

Figure 21-11. Solid-state control for capacitor-start or capacitor-start, capacitor-run induction motor.

corresponding decrease in peak stator current, the peak voltage across the sensing resistor is of a low value and not sufficient to trigger the triac. Since the triac is not being triggered on, it acts as an open switch and the starting capacitor is effectively removed from the circuit. As before, a snubber circuit is connected across the triac to maintain the switching integrity of the triac and to minimize transient disturbances.

21-12-2 Three-Phase ac Motor Control.

The synchronous speed of an induction motor depends on the number of poles and the supply frequency, as noted in Eq. (16-1), $S = 120\frac{f}{P}$ rpm (or $\omega = 4\pi\frac{f}{P}$ rad per sec). Since the number of poles is fixed for a given motor, the speed is really dependent on the frequency supplied to the motor and is directly proportional to that frequency, $S \sim f$.

On the other hand, the torque of an induction motor varies when the input voltage is varied. As the voltage is varied, the torque varies as the square of the applied voltage, as noted in Eq. (17-1), $T = KV_s^2$. The torque, then, is directly proportional to the square of the applied voltage, $T \sim V^2$.

A convenient way to illustrate the relationship between the torque and speed of the induction motor for speed control purposes is shown in Fig. 21-12a. In this figure, the torque vs. speed characteristic of a NEMA class A induction motor is shown in a plot of torque versus speed. It is important to note that this plot is related to the plot that was presented in Fig. 17-1 except that the axes have been switched. Torque is now shown on the vertical axis. The horizontal axis, instead of being labeled in terms of slip or percent slip, is represented as speed in rpm. The discussion involved here will center on the class A induction motor but the concepts apply to the other classes of polyphase induction motors.

As can be seen from Fig. 21-12a, there are points along the curve which are of particular interest and importance.

Imagine that the 15hp, 1160 rpm (11.20 kW, 125.7 rad per sec) polyphase induction motor of Ex. 17-1, which is a NEMA class A, is sized and chosen to drive a certain mechanical load. In order for the motor to bring the load from rest, 0 rpm (0 rad per sec), to the rated speed, 1160 rpm (125.7 rad per sec), it must supply a sufficient torque to accelerate the load up to rated speed. At the instant that this motor is energized (at rated voltage and frequency) by connecting it across the line, it will produce a "starting torque," point a. This point is easily recognized in Fig. 21-12a because it is the torque produced when the shaft is at standstill or 0 rpm (0 rad per sec). As the shaft is accelerated and the speed increases, the value of torque *increases* up to a "maximum torque" as indicated at point b on the curve. Note that this is not true for NEMA class C and D whose maximum torque is at standstill. As the speed continues to increase, the torque follows the curve and *decreases* until the operating point is reached at point c. While operating at point c, the motor supplies just enough torque to keep the load turning at the rated speed, in this case 67.9 ft-lb. (92.2 Nm). If the mechanical load is suddenly removed, the torque will automatically follow the curve and decrease toward zero foot-pounds (0 Nm), and the speed will increase and approach synchronous speed at point d, with 0 percent slip. Unless the load is overhauling

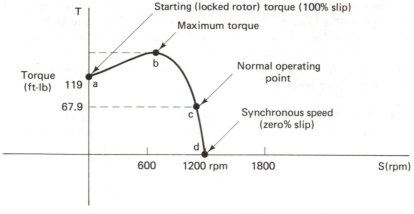

(a) Torque vs. speed characteristic (ABC rotation) of
Example 17.1 (15 hp, 6-pole induction motor)

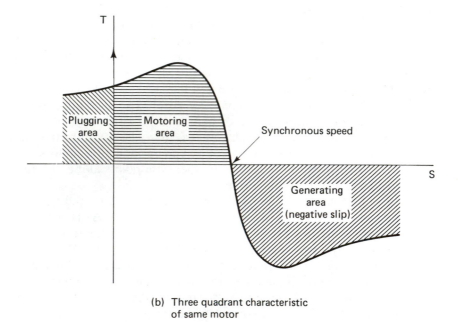

(b) Three quadrant characteristic
of same motor

Figure 21-12. Torque speed characteristics.

by nature, point d, of course, will never be reached because the unloaded motor must still supply enough torque to overcome its own counter torque resulting from friction, windage loss, etc.

Although the torque vs. speed curve, as shown in Fig. 21-12a, is very useful for a motor application that requires simple on–off operation, it really does not include enough information for an application that requires speed control. Figure 21-12b shows the curve of Fig. 21-12a continued into the second and fourth quadrant and represents the full torque vs. speed characteristic for the 15 hp (11.20 kW) induction motor of Ex. 17-1. This curve actually represents the three separate areas of possible operation for the same motor.

The motoring area, as indicated in Fig. 21-12b, is located in quadrant I. In this area, the induction motor delivers positive torque while the shaft is rotating in a positive direction; i.e., the motor delivers power to the mechanical load and the slip is between 100% and 0%. Under normal operation, the motor operates in this area.

The induction motor can actually generate power if it is operated with negative slip. Under this condition it is called an asynchronous generator. When the induction motor is running at a positive shaft speed greater than synchronous speed (negative slip), a negative torque results, as shown in quadrant IV of Fig. 21-12b. Under the proper conditions, this area of operation is utilized for regenerative braking and will return power to the supply.

Sometimes it is necessary to plug the induction motor. Plugging, it will be recalled, is a method that is frequently used to slow and/or stop a rotating mechanical load by applying a torque in a direction opposite to the direction of its speed. In Fig. 21-12b, it is observed that the plugging area is in quadrant II and occurs when a positive torque is applied to a load that is rotating in the negative direction in relation to its rotating magnetic field. As noted in Section 21-7, a plug-stop or plugging reverse condition is accomplished by changing the phase sequence of the supply.

Figure 21-13 illustrates graphically how this plug stop or plugging reverse action occurs. Note that this figure is actually comprised of two curves which are seen to be mirror images of each other. The solid line in quadrants I and II represents the torque versus speed characteristic of the 15 hp (11.20 kW) motor of Fig. 21-12a with a three-phase stator supply voltage of ABC rotation. If the supply voltage's phase sequence is reversed to an ACB rotation, the rotating magnetic field of the motor will reverse direction. When the direction of rotation of the field reverses, the torque reverses also, and the motor operates along the dashed curve in quadrants III and IV of Fig. 21-13.

To understand the plugging process, consider that the 15 hp (11.20 kW) motor is being operated from a supply of ABC rotation and is operating at its normal operating point in quadrant I at point Ⓐ of Fig. 21-13. When it is desired to stop the motor, a plug stop sequence is initiated either by physically pushing a reverse button or by an electronic control command. The supply phase sequence is then reversed via reversing contactors as noted in Fig. 21-2 or by electronic solid state switches. At the instant that the supply phase sequence is reversed, the motor will operate at point Ⓑ of Fig. 21-13 because the speed cannot change instantaneously due to the inertia of the rotor and mechanical load. The speed will then begin to decrease due to the negative torque applied to the load as a result of operating at point Ⓑ. As the speed decreases, the value of negative torque increases as the motor follows the dashed curve to point Ⓒ on the curve. At point Ⓒ the motor comes to a stop, and the voltage supply can be shut off either electromechanically or electronically and a brake applied if necessary to complete the plug stop process. It should be noted that the zero speed condition

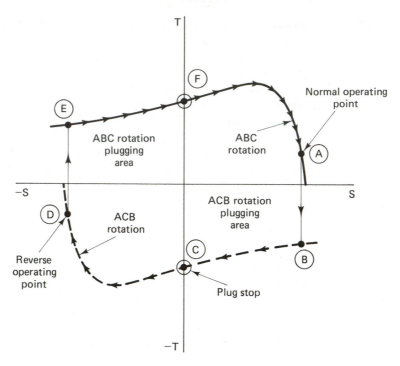

Figure 21-13. Phase reversal for three-phase induction motor plugging.

at point Ⓒ must be sensed somehow so that the voltage supply is turned off at the proper time. Zero speed can be sensed visually by the operator, mechanically by a zero speed centrifugal switch, or electronically by an incremental shaft encoder, among others.

A plugging reverse process follows the same sequence as that of the plug stop operation. If the 15 hp (11.20 kW) motor under discussion is running at point Ⓐ and a plugging reverse sequence is initiated, the voltage supply phase sequence is reversed as before and the motor's operating point jumps from point Ⓐ to point Ⓑ. The speed decreases along the dashed curve until zero speed is reached at point Ⓒ. At this point, however, the supply is not shut off as before but remains connected and the shaft begins to rotate in the opposite direction. As the shaft's speed increases along the curve between point Ⓒ to point Ⓓ, the torque increases to a maximum and then decreases to its normal operating value at point Ⓓ. Note that points Ⓐ and Ⓓ have the same values of torque and speed and that they represent the same operating point of the motor with opposite directions of rotation. The motor will continue to run quite nicely at point Ⓓ with ACB rotation until either a plug stop or plugging reverse is again initiated. When initiated, the phase sequence is again reversed back to an ABC rotation.

The motor's operating point immediately jumps from point Ⓓ to Ⓔ, and the motor operates along the solid characteristic line, where at point Ⓕ a plug stop condition is realized, or, if the supply is not disconnected, the motor's speed increases along the curve to point Ⓐ, where it settles at steady state operation at the normal operating point.

Plugging reverse or plug stop operation of the three-phase induction motor is a simple and straightforward control technique but is of no use when continuous, controlled variations of speed are required. Remember also that plugging is generally accomplished by "across the line" switching, and high switching currents (or slightly higher than normal starting currents) usually result, which must be considered when sizing switching components and evaluating peak demand costs.

Some motor applications require a continuous, smooth, stepless control of speed between zero and rated speed. Generally, in the past this task was assigned to the dc motor because of the ease with which speed control is achieved. It has always been relatively easy to vary the dc voltage applied to the armature of the dc motor, and thus its speed. On the other hand, the dc machine has a commutator subject to sparking (a hostile condition in some environments) and requires regular periodic maintenance. Furthermore, the dc motor generally costs more and weighs more than a comparably rated induction motor. Continuous speed control of the induction motor, however, requires that the supply voltage and frequency be varied, as will be seen. Historically then, the speed control tasks were handled by dc machines where possible.

When an environmental condition disallowed the use of the dc machine due to sparking, an induction motor was used but required a special motor-generator set for its power supply. This MG set consisted of a dc motor with variable speed control driving an ac generator, as shown in Fig. 21-17a. For obvious reasons, the induction motor was not generally applied to tasks requiring variable speed.

With the continuing progress and development of high power solid state devices, sophisticated sensing transducers, and microprocessor-based triggering sources, variable speed control of the induction motor is now a practical reality. The techniques discussed in Chapter 8 can be handily applied to the task. Before looking at some of the techniques in supplying variable voltage and frequency to the induction motor, it is important to consider exactly how these variables affect motor behavior.

Figure 21-14 is a graphical illustration of how the RMS current of the induction motor varies in relation to the general torque versus speed characteristics previously discussed. Note that the current is minimum at synchronous speed because no torque is developed, and the only current that flows is a small amount necessary to maintain a constant flux at the poles. This minimum value of current is called the excitation current. As noted in Fig. 21-14, however, the current drawn by the motor varies according to how much torque the motor is developing at any given instant of time. Notice that when the motor is started at rated voltage and frequency, a relatively high starting current is drawn by the motor. Then, as the torque rises to its maximum, the current decreases to a lesser value. Note that the current continues to decrease until the operating point is reached at rated torque and speed (normal operating point). As discussed previously, if a load is overhauling in nature and brings the motor's speed above the synchronous speed, the motor will behave as an asynchronous generator, and current will be fed back

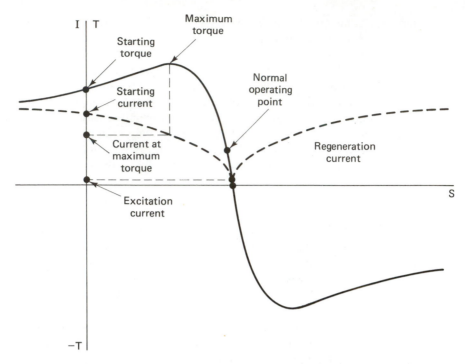

Figure 21-14. Induction motor current-speed relationship.

into the supply line. This regeneration current will increase as the speed increases above synchronous, as shown in Fig. 21-14.

The operating characteristics discussed above can be applied to any class of induction motor, and it should be emphasized that the supply voltage and frequency to this class A motor under discussion have been fixed at the rated values, in this case 600 volts and 60 hertz.

Basic across the line on/off operation of the induction motor at rated voltage and frequency, as described above, has been the most commonly used method of operation, but now variable speed induction motor drives have become common-place and are used extensively to conserve energy and provide a high degree of versatility for a given application. When controlled speed variations are required it becomes necessary to vary the frequency and voltage supplied to the motor.

Although it is possible to vary the speed of the induction motor solely by varying the applied voltage, it is obvious that the torque characteristics of the motor will suffer ($T \sim V^2$). A simple three-phase autotransformer is sometimes used to manually vary the speed of the motor, or a solid state switch composed of two parallel back-to-back SCRs per phase. The voltage applied to the motor is varied by the manual setting of the autotransformer or by varying the firing delay angle, α, sequentially to the three sets of SCRs via an electronic triggering device with a variable speed, set-point control. This type of variable speed solid state control is limited in its use because large firing delay angles cause intermittent and distorted voltage waveforms and result in higher operating temperatures.

Simultaneously varying the applied voltage and frequency to the induction motor results in the ability to smoothly vary the speed of the motor from zero speed to rated speed and vice versa, with added benefits. By varying the voltage and frequency in direct proportion to each other, the general torque versus speed curve of Fig. 21-12 is simply shifted sideways without any significant change in shape.

Figure 21-15 represents the general curve of an induction motor. Consider that the curve farthest to the right marked "rated" represents the basic torque versus speed curve of a certain motor being supplied by rated line voltage and frequency, say 600 volts and 60 hertz. By reducing the supply frequency by one-third of rated to 40 hertz and also reducing the supply voltage by one-third to 400 volts, the motor will operate according to the middle curve. Note that the curve retains approximately the same shape as the rated curve and crosses the horizontal axis at 800 rpm (83.8 rad per sec), indicating that the synchronous speed has also been reduced by the same ratio: one-third. If, on the other hand, the supply frequency is reduced by two-thirds of rated to 20 hertz and simultaneously the supply voltage is reduced by two-thirds of rated to 200 volts, the motor will operate along the curve on the left of Fig. 21-15. Notice again that this 200 volt/20 hertz curve retains nearly the same shape as the rated curve, but the synchronous speed has been reduced by the same two-thirds ratio so that it is now 400 rpm (41.9 rad per sec). It is obvious that the rated curve can be shifted anywhere along the horizontal axis when the applied voltage and frequency are varied in essentially the same proportion.

With this ability to shift the rated curve come a few distinct advantages. If the motor of Fig. 21-15 is started simply across the line at rated voltage and frequency (600 volts/60 hertz), the starting torque will be significantly less than the

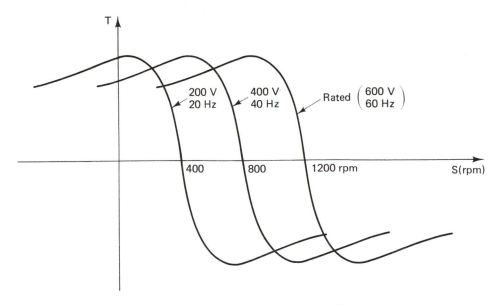

Figure 21-15. Induction motor curve shifting.

maximum torque available at the knee of the curve. The associated starting current for this rated starting torque is significantly higher than the current that is drawn at the point of maximum torque. If, on the other hand, the motor is started at the reduced voltage and frequency of 200 volts and 20 hertz, as shown in Fig. 21-15, the starting torque will essentially be equal to the maximum torque due to the shift, and the associated starting current will be significantly less than the starting current at rated voltage and current. The supply voltage and frequency can then be gradually increased electronically with actual feedback via transducers so that as the shaft speed increases, the maximum torque is applied throughout the acceleration of the load right up to the rated value and at essentially constant current. This is accomplished via the trigger source that monitors actual current, speed, and torque, and compares these to set point settings of these variables. The result is that the triggering source changes the delay angles to the various SCRs accordingly, and applied voltage and frequency are changed in direct proportion. The shift of the curve is a direct result of this change.

Note that during the time of acceleration from zero speed up to rated speed with constant maximum torque (and its associated constant current) that the horsepower (kW) of the machine is actually changing. At the lower speeds the horsepower (kW) delivered is lower than the rated horsepower. Note also that for sustained lower-than-rated speed operation, forced cooling of the machine must be supplied or the motor must be derated due to the lower rpm. Since the current is the same, however, the I^2R heating effect will be essentially the same and it may be necessary to augment the cooling flow somehow.

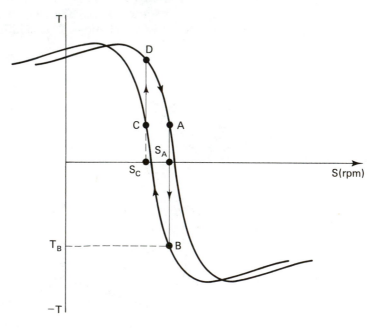

Figure 21-16. Speed control using curve shifting.

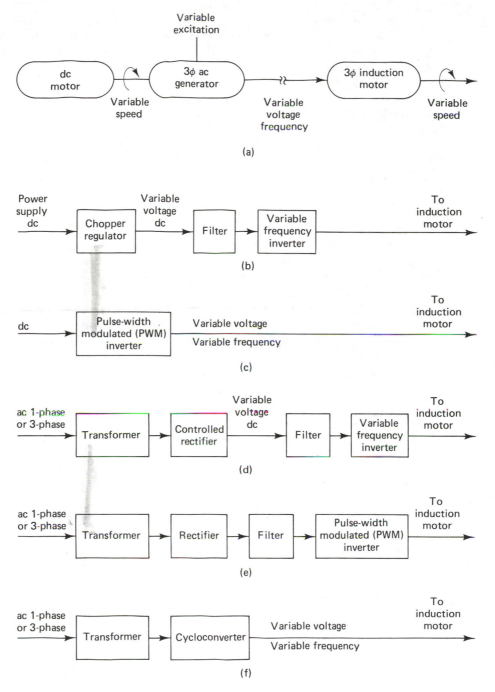

Figure 21-17. Schemes for induction motor speed control. (Samir K. Datta, *Power Electronics and Controls*, © 1985, p. 159. (A Reston Publication) Reprinted by permission of Prentice-Hall, Englewood Cliffs, New Jersey.

After the machine has accelerated to a desired operating point, it will continually be monitored by the sensing transducers that continually feed back the running information to the triggering source. The triggering source will change triggering pulses to the SCRs as necessary to maintain operation at the desired operating point.

Once the initial acceleration is completed and the motor is operating at a desired operating point, it may be necessary to change the speed. In order to understand just how this change occurs, consider that the three-phase, 15 hp, six-pole motor previously discussed is operating at point A of Fig. 21-16, and it is desired to run the motor at some lower speed. When the speed set point is lowered in the triggering source, the first delay angles are electronically changed so that the frequency and voltage are lowered in direct proportion. This action results in the shifting of the characteristic curve to the left, as shown in Fig. 21-16. However, the operating point does not move horizontally with the shift because the shaft speed of the motor cannot change instantaneously due to the inertia of the spinning mechanical load. For this reason, the operating point of the machine immediately drops down to point B on the shifted curve. It is obvious that while at point B, the motor immediately begins to regenerate due to the negative slip, and a negative torque, T_B, is administered to the load, which quickly slows it down. The speed, then, drops from S_A to S_C and follows the curve between points B and C during slowdown. Note that anytime the motor operates in quadrant IV, power is *returned* to the source.

If, on the other hand, the motor is operating at point C and the speed must be increased, a set point change is initiated and the supply frequency and voltage are increased in direct proportion, with the resulting shift of the curve to the right. Again, because the speed cannot change instantaneously, the operating point jumps from point C to D and then the machine accelerates along the shifted curve to point A, where it again operates at steady state, running at speed S_A.

By using the technique of voltage/frequency variation for speed control of the three-phase induction motor, the ability to control speed to a fine degree with high efficiency is possible. The hardware itself is complex and costly but will probably be well worth the investment when an application demands continuous speed control.

There are various component configurations which are currently being used to control the speed of the induction motor with variable voltage/frequency supply. Figure 21-17 depicts those in general use today, although other component configurations are sure to evolve.

QUESTIONS

21-1. Name some power supply condition differences that favorably affect an ac motor control in comparison to a dc control.

21-2. Where might line voltage transformers be used in an ac motor control?

21-3. What is the function of a pilot control circuit transformer?

21-4. Why are pilot control circuit transformers almost invariably used in factory ac motor controls in modern practice?

21-5. What is a full-voltage starter control?

21-6. How is reversing accomplished in a typical three-phase ac motor control?

21-7. What is the benefit of a reduced-voltage motor control in ac practice?

21-8. What conditions might make an autotransformer reduced-voltage motor starting control desirable?

21-9. What circuit parameter is varied in a wound-rotor induction-motor control?

21-10. How is part-winding starting accomplished?

21-11. What advantage may be taken of induction generator action in an ac motor control?

21-12. In general, what control circuit provision is used in a jogging control?

21-13. What is a consequent pole motor speed control?

21-14. How are two-winding speed controls accomplished?

21-15. Explain how the centrifugal starting switch in a single phase motor may be eliminated by use of a solid state device.

21-16. Using generalized curves explain how induction motor plugging is accomplished.

21-17. Explain how variable voltage/variable frequency is used in continuously variable speed control.

21-18. Illustrate three ways in which components can be arranged to achieve variable voltage/variable frequency motor control.

22

Alternating-Current
Motor Selection

Again, as in Chapter 10, it should be emphasized that motors and generators are normally rated by their *output power*. This power rating has been listed for many years as *horsepower* for motors and *kilovolt-amperes* or *kilowatts* at a particular *power factor* for ac generators. The widespread use of System International (SI) units in both equipment identity and design calculation will lead to motor specification in kilowatts (kW) of mechanical output power.

The factors specified for dc motor selection in the first part of Chapter 10 are again valid for ac machinery, with some additions. Without disturbing the listed order in Chapter 10, we will *add* the following:

(9a) Frequency of input or output power.

(9b) Number of phases in power supply, whether single, two, or three phase. Single and three phase prevail except for servocontrol motors, which are usually two phase.

A careful study of these various factors will then enable the basic motor type to be chosen. If a type is not specifically chosen, usually related types can be considered, such as shaded pole or permanent split capacitor. Various types that can perform a particular task may be further subjected to consideration for the following:

(12) Efficiency in service, which may be considerably different while doing the same task. Unfortunately, the more efficient motor is frequently more expensive. If the most efficient type were the least costly, all competing types would disappear from the market.

(13) Radio frequency interference (RFI) and television interference (TVI). Here some motor types are at a disadvantage, particularly the universal motor. All dc types are roughly similar.

22-1 SHAFT POWER

See Sect. 10-1. Note that the tentatively standardized SI power ratings for dc motors in Table 10-1 were really chosen for ac motors and were suggested for dc motors because of their logical basis. These ratings are more sure of application in alternating current. The very small power ratings used in various small ac appliances are not currently standardized, nor have tentative SI ratings been offered.

One reason for this state of affairs is that small-sized motors are very frequently so built into their associated product that they are virtually custom fitted. Some small sizes have been tentatively standardized by the first manufacturer in a particular field. The sizes were dictated by the single manufacturer for the duration of his patent control. Many of these proprietary sizes were so logical that competitive makes have later freely built motors that would interchange in the same sizes.

In very large motor sizes, in the range of thousands of horsepower or thousands of kilowatts, the motor design is usually very specific or custom tailored to the task. This does not imply "clean sheet of paper" approach. This probably means an existing frame design will be modified by changing the depth of the lamination stack and by use of a very specifically tailored winding. Almost every specific motor is then a scale up or a scale down of an existing design. The new machine will make use of as many existing design detail parts and as much existing manufacturing tooling as is reasonably possible. One manufacturer will frequently offer a competitive design that will interchange with another make that may have previously fit the service.

22-2 CHARACTERISTICS OF THE DRIVEN LOAD

See Sect. 10-2. A hard-starting load for ac service would require a universal motor, wound-rotor polyphase motor, or a capacitor-start single-phase motor depending upon the size, the needed speed range, and the type of power supply available.

In ac work, constant speed can be absolutely constant if needed by using one of the various types of synchronous motors. Relatively constant speed needs can be met by most types of induction motors. Variable speeds that cover a wide range are best met by a universal motor or by the various pole changing types of induction motors. Rather than chart the extremely wide ranges of motors that are available, the student is urged to become familiar with the performance characteristics of each type that has been presented in past chapters.

22-3 SPEED RATING

See Sect. 10-3. Alternating-current motor speed ranges are determined by the basic synchronous speed relations for all types except the universal motor. Equation $(16\text{-}1_E)$ states that $S = 120f/P$ rpm, and Eq. $(16\text{-}1_{SI})$ states that $\omega = 4\pi f/P$ rad/sec. Various *synchronous* speed relations for common frequencies and basic pole combinations are given Table 11-1 for English units and Table 11-2 for SI units.

An induction motor will not run precisely at these speeds, but will operate *below* a synchronous speed by its characteristic slip value. The synchronous speed will only be exceeded if the load is of an overhauling nature, as discussed in Sect. 21-8. Most of the standardized dc motor speeds are related to equally normal ac induction motor speeds. For example, 1750 rpm (183.3 rad/sec) corresponds to a four-pole induction motor running at a normal slip value.

Some of the standardized dc speeds, such as 2500 rpm (261.8 rad/sec), are not matched by a corresponding standardized ac motor speed. There is not a normal pole and frequency combination that would result in this speed for an induction motor. A universal motor could well be designed to operate at this speed, but this is on the low side for a modern unit.

22-4 FRAME SIZE

See Sect. 10-4. The motor frame sizes used for ac service are closely related to dc frame sizes. A 215 frame is the *same* for both dc and ac machines with the possible exception of the shaft size. There has been a modification to ac frame size numbers that has become widespread in the last decade. This is the T series or frame sizes with a T suffix.

The T series motors are designed to operate at higher temperatures in order to take advantage of more resistant insulation materials. As a result, the available power in a given frame size has increased. A T series motor, within a given frame number, has the same standardized dimensions except for an *increased shaft size* to accommodate the higher output power.

In ac machines, for example, an older 215 A frame would have a 1.125 in. shaft. A new 215 T frame has a 1.375 in. shaft. Study Table 10-3 for ac machine frame sizes, as they exist with the present NEMA standards that have evolved over the last generation. The full NEMA standards include many inactive frame sizes. For ac machines, there are generally only two used lengths within the standardized six lengths for a given diameter series. Thus, 213 and 215 or 213 T and 215 T are the only 210 series frame sizes used to any extent.

A range of available frame sizes for industrial polyphase squirrel-cage induction motors for both older frame sizes and the later T series sizes is shown in Table 22-1. The motors are NEMA class A and B, 60 Hz, totally enclosed, fan cooled (TEFC). This is typical of modern practice since they can be used in most environments. The information is specifically related to Westinghouse Life-Line T motors, but is representative of high-quality construction of a number of man-

TABLE 22-1 HORSEPOWER RATINGS AND FRAME SIZES
Polyphase Squirrel Cage NEMA Classes A and B TEFC

Horse-power	Approx. kW	3600 (120π)		1800 (60π)		1200 (40π)		900 (30π)	
0.5	0.375							182	H143T
0.75	0.56					182	H143T	184	H145T
1	0.75			182	H143T	184	H145T	213	182T
1.5	1.12	182	H143T	184	H145T	184	182T	213	184T
2	1.5	184	K145T	184	K145T	213	184T	215	213T
3	2.24	184	182T	213	182T	215	213T	254U	215T
5	3.75	213	184T	215	184T	254U	215T	256U	254T
7.5	5.6	215	213T	254U	213T	256U	254T	284U	256T
10	7.5	254U	215T	256U	215T	284U	256T	286U	284T
15	11.2	256U	254T	284U	254T	324U	284T	326U	286T
20	15	286U	256T	286U	256T	326U	286T	364U	324T
25	19	324U	284TS	324U	284T	364U	324T	365U	326T
30	22.4	326S	286TS	326U	286T	365U	326T	404U	364T
40	31.5	364US	324TS	364U	324T	404U	364T	405U	365T
50	37.5	365US	326TS	365US	326T	405U	365T	444U	404T
60	45	405US	364TS	405US	364TS	444U	404T	445U	405T
75	56	444US	365TS	444US	365TS	445U	405T		444T
100	75	445US	405TS	445US	405TS		444T		445T
125	90		444TS		444TS		445T		
150	112		445TS		445TS				

Shaft power — Synchronous speed in RPM (rad/sec) — Frame Sizes

NOTE: The suffix S signifies a smaller shaft size. It is representative of the high-speed and therefore low-torque uses of a frame size.

ufacturers. Notice that a given power and speed of motor is not required to have a particular frame size. Rather, the motor should conform to one or the other specific size so that its mounting conditions may be standardized.

Compare Table 22-1 with Table 10-2 to see that frequently an ac motor of a particular power and speed is found to be one frame size *shorter* than its dc counterpart. When T frames are used, the ac motor tends to be one whole diameter series *smaller*. The polyphase ac motor is efficient and can normally run hotter than its dc-power counterpart.

These standardized frame sizes have a number of subvarieties for their mounting characteristics. Table 10-3 refers to foot-mounted sizes. There are also standardized dimensions for various other mounting schemes. There are a few different standardized end flange or end face mounting types. These are for use where the motor in question will be directly bolted to a surface such that its shaft will protrude through the surface.

The NEMA type C flange is flat with a circular male rabbet that is concentric with the shaft. This type is designed to be attached by mounting bolts from within the device to which it is mounted. The NEMA type D flange mounting also has a concentric rabbet or spud, but is larger than the type C for the same motor frame size. This type of motor is designed to bolt on from the motor side of the flange. These flange-mounted motors have fully standardized mounting dimensions in NEMA-approved series.

Type P flange-mounted motors are usually mounted vertically, as they are especially tailored for pump service. Type PH motors are both flange mounted and have hollow shafts. These are also for specialized pump service.

Oil burner motors are especially standardized for use on furnace oil burners. *Sump pump* motors have an enlarged and elongated hub for mounting in a supporting tube.

The new NEMA-suggested guide contains recommended, but not yet approved, metric-dimensioned flange-mounted frame sizes. Some related standard will ultimately be adopted.

Returning to the basic NEMA foot-mounted motors, there are seen to be a few different standardized shaft diameters for the standard frame sizes. There are also from two to three standard shaft lengths corresponding to a particular diameter. Table 22-2 shows how these shaft sizes correspond with the various frames.

TABLE 22-2 NEMA AC MOTOR FRAME STANDARD SHAFT SIZES

Old NEMA series			T series			Dim. (see Table 10-3)		
Standard	Short S series		Standard	Short S series		U shaft diam.	N–W shaft length	Square key size
183 184			H143T H145T K143T K145T			0.875	2.25	0.187
			182T 184T			1.125	2.75	0.250
213 215						1.125	3.00	0.250
			213T 215T			1.375	3.37	0.312
254U 256U						1.375	3.75	0.312
	324S 326S			284TS 286TS		1.625	3.25	0.375
			254T 256T			1.625	4.00	0.375
284U 286U						1.625	4.87	0.375
	364US 365US			324TS 326TS 364TS 365TS		1.875	3.75	0.500
			284T 286T			1.875	4.62	0.500
324U 326U						1.875	5.62	0.500
	444US 445US 404US 405US			404TS 405TS		2.125	4.25	0.500
			324T 326T			2.125	5.25	0.500
364U 365U						2.125	6.37	0.500
				444TS 445TS		2.375	4.75	0.625
			364T 365T			2.375	5.87	0.625
404U 405U						2.375	7.12	0.625
			404T 405T			2.875	7.25	0.750
444U 445U						2.875	8.62	0.750
			444T 445T			3.375	8.50	0.875

The frames with an S suffix to their designation have shafts that are both shorter and smaller in diameter than is usual for the frame size. This type of designation is used on both the older NEMA frame sizes and the more recent NEMA T frame motors.

A study of Table 22-1 shows that motors using the S shafts are either 1800 or 3600 rpm types, where the torque does not warrant the massive shafts used on lower-speed motors.

The tentative SI-dimensioned frame sizes can be studied from Tables 10-4 and 10-6. Since it is logical to do so, the SI frame sizes that ultimately appear will probably recognize some such alternative shaft sizes as the NEMA S series. These sizes will probably drop from one preferred metric shaft size to one or two preferred sizes smaller. It is good to remember that SI-dimensioned parts of *all kinds* will avoid arbitrary sizes and use logical principles as far as possible. The discussion on probable SI-dimensioned International Electrotechnical Commission (IEC) approved sizes in Sect. 10-4 is perhaps even more applicable for ac motors.

An excellent reference for both ac and dc motor selection is *Machine Design* magazine's "Electric Motors Reference Issue," Penton Publishing Company. This compendium of information is periodically updated so that the latest issue is very reliable. Its detailed lists of manufacturers by specific types are especially useful.

22-5 SPEED CLASSIFICATION

See Sect. 10-5. The speed classifications that are recognized by the NEMA were excerpted in Sect. 10-5 for those terms that were applicable to dc motors. For ac motors, there are a few additional classifications worth noting:

(1) *Constant-speed* motors that stay within an arbitrary 20 percent variation from no load to full load include *most* induction motors. A slip of over 20 percent might be found in a small shaded-pole or reluctance-start motor. Synchronous motors of all types are constant speed.

(2) *Varying-speed* motors are represented by universal motors. Here the speed change from no load to full load is much more than 20 percent.

(3) *Adjustable-speed* motors are typically represented by pole changing induction motors, but they fit the multispeed classification better.

(4) *Adjustable varying-speed* motors are represented by wound-rotor induction types when there is substantial external resistance in the rotor circuit.

(5) *Multispeed motors* are specifically the pole changing motors where the speed can be specifically changed by pole changing. At the same time, the speed is relatively constant within any one of the specific pole configurations.

(6) *Nonreversible motors* are motors that cannot be electrically reversed. Reluctance-start motors and the shaded-pole motors with *fixed shading* coils and single stator winding cannot be reversed.

(7) *Reversible motors* include types that may be reversed by changing external connections, but not necessarily while running. All three-phase induction motors are reversible, but certain single-phase types are not reversible when running. A

resistance split-phase motor will not have enough torque supplied by its starting winding to overcome its cross field supplied running torque. As a result, the resistance split field cannot be reversed while running, even if a switching circuit were provided that would bypass the centrifugal starting switch. The motor is then reversible but not reversing.

(8) *Reversing motors* are motors that may be reversed while running under any circumstances. Again, all three-phase induction types are reversing motors. A single-phase capacitor-start motor is reversing since its starting circuit torque is dominant. Note that, when connected according to Fig. 21-2a, a capacitor-start motor will *not* reverse while running, since its starting winding is inactive once the centrifugal switch has opened. Special wiring provision is required so that the starting winding is accessible electrically while running. The resistance split-phase motor will not meet the reversing classification. All universal motors are both reversing and reversible.

Note that synchronous motors are not reversing unless their starting circuit has a dominant torque, and then they may be plug reversed. The sequence required is to open the dc field circuit and plug reverse the induction motor with its starting circuit, providing it is of a dominant torque character in a single-phase unit. Since almost all dc field synchronous motors are three phase, the direction of main stator magnetic field is easily reversed by interchanging any two of the three phases.

Small single-phase synchronous types, such as hysteresis synchronous motors, can be reversed if the starting winding can be reversed and is then dominant.

All ac motors except those mentioned in (6) above can be electrically reversed, but may or may not meet the *reversing motor* special requirements.

22-6 EFFECT OF DUTY CYCLE

See Sect. 10-6, where the information is exactly applicable to ac motors and should be reviewed. However, there are some considerations that should also be mentioned that seem particularly applicable to ac motor types.

Some motors are installed in locations where condensed moisture is a continual problem. Remote pump locations or spillway gate actuators are good examples. By deliberately applying a low ac voltage to one or two phases, the motor is kept warm enough so that moisture does not collect on the windings. By omitting one phase, no torque is developed, so the motor does not attempt to rotate when rotation is not desired. The low voltage is adjusted so that *just enough* wattage is drawn to keep the windings perceptibly warm. Since the windings in a three-phase motor are distributed around the stator, the whole stator winding is kept dry. This can be more effective than special heater windings and allows the use of a standard, totally enclosed motor.

Some gear and linkage mechanisms have considerable play or backlash when built of low-cost components. Normal antibacklash springs are difficult or impossible to put into a mechanism that has more than about one turn of rotation. In this case, a shaded-pole motor can be geared into one end of the total mechanism.

It is supplied just enough voltage to develop the torque required to keep all gears and links against one side of their allowed backlash. In the author's experience, this use has required about 20 to 30 V with a normal 115 V, single-phase, open-frame, shaded-pole motor. The motor windings remain just perceptibly warm in continuous service with voltages of this level. At the same time, the motor can be easily forced to run backward against its torque direction without problems. When the mechanism is driven the other way, this *antibacklash* motor will automatically follow along and keep the whole mechanism free of backlash. This type of torquer service can be applied in any situation where the line voltage is kept low enough so that no back emf is needed to control the current.

22-7 AMBIENT TEMPERATURE EFFECTS

See Sect. 10-7, where the comments on dc machinery are exactly applicable. The ac motor has some advantage here because there is no very localized high temperature spot such as a commutator, except in universal motors.

Alternating-current motors can be used in some very unusual environments. A typical example is the submerged deep well pump motor. Here the windings are encapsulated in a material that is impervious to the fluid, which might be either petroleum or water. Since there are no connections or electrical elements in an induction motor that need to be exposed, the whole motor is immersed in the pumped fluid. Very high power can be used in a small-sized motor because the ambient fluid then becomes a motor coolant. There are substantial rotational losses due to the fluid drag, but smooth rotor construction minimizes this effect. Even if the fluid were fairly hot, the windings will operate with a very small temperature rise.

22-8 ALLOWABLE TEMPERATURE RISE

See Sect. 10-8 and Table 10-6, which are applicable for ac or dc motors.

22-9 VOLTAGE AND CURRENT RATINGS

In this case, the main power line voltage levels are not the same for ac as they were for dc motors.

22-9-1 Alternating Current Voltage-Level Ratings.

The NEMA has recommended voltage levels as standard for ac motor ratings as shown in Table 22-3.

Notice that a three-phase motor may be rated as 220 V line-to-line and be fed from distribution service that is rated as 230 V line-to-line. When an alternator is used for this service, it is rated at 240 V. The differences are presumed to be normal line and switching voltage drops.

Obviously, if an alternator is driven locally and supplies a motor that is close to it, the motor and alternator will operate at essentially the same voltage.

TABLE 22-3 NEMA STANDARD VOLTAGE RATINGS FOR
AC MOTORS

	Line voltages
Single-phase motors	115, 230, 460
	Line-to-line voltages
Polyphase motors	110, 208, 220, 440, 550 2300, 4000, 4600, 6600

TABLE 22-4 NEMA STANDARD VOLTAGE RATINGS FOR AC
ALTERNATORS

Line-to-line voltages					
120*	240	480*	600	2 400*	4 160*
4 800	6 900*	12 000	13 200	13 800*	23 000

*Preferred for new work

TABLE 22-5 TYPICAL AIRCRAFT VOLTAGE
RATINGS FOR 400 Hz SERVICE

	Line voltage
Single phase	26.5, 120
	Line-to-line voltage
Three phase	208 with 120-V line to neutral for single phase

Table 22-4 shows the NEMA-standardized ac generator voltages. The reasons for these relatively few voltage levels over such a vast range of equipment sizes is to that standardized and competing equipment is possible. To use a voltage that is different from standard would require custom-built equipment, which is foolish. Voltages higher than 23 000 V are confined to cross-country distribution lines, the "high-tension" lines that are so evident almost everywhere.

It is not known what action will be taken, if any, on internationally standardized voltage levels; 150 V, 280 V, and 360 V have some use in overseas areas.

Table 22-5 shows that aircraft use much the same voltage. Lighting equipment in cabins uses 120 V service just as in a home. Voltages higher than 208 V line-to-line are unnecessary so far owing to relatively short line lengths in even the largest-sized aircraft. This may well change if substantially higher power levels are needed. If changes are made, probably a previously standardized voltage level will be chosen.

22-9-2 Alternating Current Frequency.
A motor must be matched to its line frequency if it is to operate satisfactorily. Even the small changes from 60 to 50 Hz affect the operating speed and current levels of all motors except the universal

motor. Where ac power was used for railroad service, 25 Hz power was formerly standard in the United States. The trend now in railroad service is to 60 Hz lines using commercial power. The voltage is dropped by transformer and rectified in the locomotive for use in dc series traction motors. Four-hundred-hertz service is nearly universal for alternating current on aircraft, missile, and space work; 60, 50, 25, and $16\frac{2}{3}$ Hz may be recognized internationally.

22-9-3 Alternating Current Motor Current Levels. Every motor type, size, and load circumstance leads to a different operating current level. However, to standardize the services, controls, and circuit protective devices, the National Electric Code has standardized current levels that are normally not exceeded in service. Table 22-6 shows full-load current levels for *single-phase motors*. This table is based upon NEC 430-148, but has equivalent power levels in kilowatts added.

A specific example of a single-phase-motor current range is shown in Fig. 22-1. This is a motor calibration curve taken at the Waterbury State Technical College during the course of normal laboratory work. This specific curve was taken with the school single-phase supply line at the 242 V level. The shapes of the curves are characteristic of most induction motors. The efficiency and power factor curve values are typical of single-phase squirrel-cage motors. Three-phase equipment will show better efficiency and lower current levels.

Note from these curves and a specific data point that the output power is specifically determined. If this motor is drawing 14 A at the same 242 V, it must be producing about 2.41 hp (1.80 kW).

Table 22-6 shows that a single-phase 230 V induction motor of 3 hp (2.24

TABLE 22-6 FULL-LOAD CURRENTS FOR SINGLE-PHASE AC MOTORS
(From NEC 430-148)

Motor (hp)	Power (kW)	Line currents in amperes		
		115 V	230 V	440 V
$\frac{1}{6}$	0.125	4.4	2.2	
$\frac{1}{4}$	0.190	5.8	2.9	
$\frac{1}{3}$	0.250	7.2	3.6	
$\frac{1}{2}$	0.375	9.8	4.9	
$\frac{3}{4}$	0.560	13.8	6.9	
1	0.750	16	8	
$1\frac{1}{2}$	1.12	20	10	
2	1.50	24	12	
3	2.24	34	17	
5	3.75	56	28	
$7\frac{1}{2}$	5.60	80	40	21
10	7.50	100	50	26

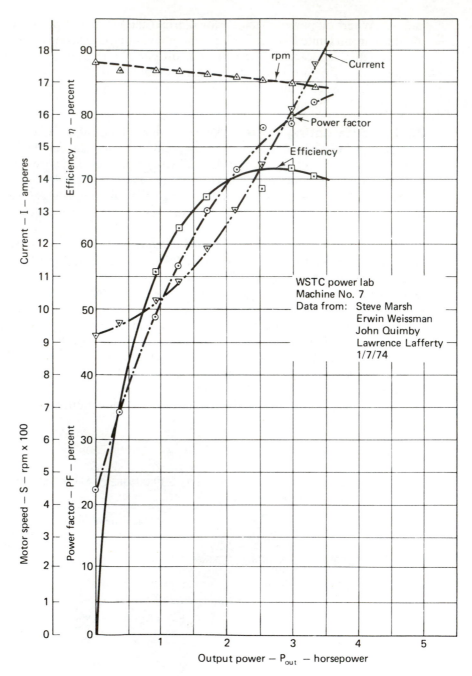

Figure 22-1. Single-phase, 242 V, capacitor start, four-pole, 3 hp induction motor characteristics.

kW) draws *not over* 17 A at full load. Experiments show that when *near* rated voltage, the current varies approximately directly with voltage:

$$I_{lb} \cong \text{operating current} = \left(\frac{V_b}{V_r}\right)I_{lr} \qquad \text{Eq. (22-1)}$$

where *r* is rated condition and *b* is specific condition.   Thus, we expect

$$I_{242} = \left(\frac{242}{230}\right)I_{230} \cong \left(\frac{242}{230}\right)17 \cong 17.9 \text{ A}$$

Figure 22-1 shows that 16.2 A was drawn at 242 V, so this particular motor is significantly *below* the NEC 430-148 expected current level.

Most manufacturers will supply characteristic curves of motor performance that may be interpreted as in Fig. 22-1.   However, the manufacturer's data are usually a guaranteed performance level.   The motor will usually perform perceptibly better than the characteristic curves.   A specific calibration curve such as Fig. 22-1 is very useful in product development work.   With a voltmeter and ammeter, the motor output can be very closely determined.

A similar table of full-load motor currents for three-phase motors is shown in Table 22-7.   This is based upon NEC 430-150.   These values are for normal motors and loads.   Motors built for especially difficult conditions may not conform.

When motors are held at locked-rotor condition or are stalled, they will draw line currents that are determined by their rotor construction.   Since this current affects their control circuit protective device settings, a coded situation has been developed that identifies their kilovolt-amperes per horsepower (or kilowatts) with a locked rotor.   Many modern motors carry this locked-rotor-indicating code letter on their nameplates.   This system has been standardized by the National Electrical Code as NEC 430-76.   It has been adopted as a standard by the NEMA.   No known international standard has been adopted yet nor can it be until the motor sizes themselves are standardized (see Appendix B).

When motors are *not marked* by code letters, the *starting currents* may be expected to correspond to Table 22-8.   This table is shown for three-phase, 220 V line-to-line conditions.   The values of current in amperes that are shown may be scaled for other line voltages within the range of the *same winding connection* as shown below:

$$I_{stb} = \text{starting current} = \left(\frac{V_b}{V_r}\right)I_{str} \qquad \text{Eq. (22-2)}$$

where

$$V_b = \text{new line-to-line voltage (or condition } b)$$

$$I_{str} = \text{rated line current at start from Table 22-8}$$

Note that this is the *same effect* as that of Eq. (22-1).   If a *different voltage* is used

TABLE 22-7 FULL LOAD CURRENTS FOR THREE PHASE AC MOTORS
(From NEC 430-150)

Motor output power		Squirrel-cage and wound-rotor motors: Line amperes, I_l						Synchronous at unity PF: I_l			
		(line-to-line volts)						(volts)			
hp	kW	110	208	220	440	550	2300	220	440	550	2300
$\frac{1}{2}$	0.37	4	2.1	2.0	1	0.8					
$\frac{3}{4}$	0.56	5.6	3.0	2.8	1.4	1.1					
1	0.75	7	3.7	3.5	1.8	1.4					
$1\frac{1}{2}$	1.12	10	5.3	5.0	2.5	2.0					
2	1.50	13	6.9	6.5	3.3	2.6					
3	2.24		9.5	9.0	4.5	4					
5	3.75		16	15	7.5	6					
$7\frac{1}{2}$	5.60		23	22	11	9					
10	7.50		29	27	14	11					
15	11.2		42	40	20	16					
20	15		55	52	26	21					
25	19		68	64	32	26	7	54	27	22	5.4
30	22.4		83	78	39	31	8.5	65	33	26	6.5
40	31.5		110	104	52	41	10.5	86	43	35	8
50	37.5		133	125	63	50	13	108	54	44	10
60	45		159	150	75	60	16	128	64	51	12
75	56		196	185	93	74	19	161	81	65	15
100	75		261	246	123	98	25	211	106	85	20
125	90			310	155	124	31	264	132	106	25
150	112			360	180	144	37		158	127	30
200	150			480	240	192	48		210	168	40

NOTE: If synchronous motor is excited for 90% PF or 80% PF, multiply currents by 1.1 and 1.25, respectively.

on a 220 V rated motor the *starting torque* will vary in proportion to the *line voltages squared*, within the range of the *same winding connection*, as

$$T_{stb} = \text{starting torque} = \left(\frac{V_b}{V_r}\right)^2 T_{str}$$ Eq. (22-3)

where

V_b = new line-to-line voltage

T_{str} = torque in percent of rated torque from Table 22-8

If this torque is wanted in either foot-pounds or newton meters, it can be determined from Eq. (16-28$_E$) or (16-28$_{SI}$) by using the actual rated speed in rpm (or radians per second) from the nameplate data.

When a different voltage motor is used on its *proper rated voltage*, its starting torque is assumed to be the *same* as in Table 22-8. The different motor on its *proper voltage* will draw a current that is now *inversely* proportional to the voltage, or

$$I_{stb} = \text{starting current} = \left(\frac{V_r}{V_b}\right)I_{str} \qquad\qquad \text{Eq. (22-4)}$$

Note the difference from Eq. (22-1). A 440 V motor will draw one-half of the starting current of the same power motor on 220 V.

TABLE 22-8 RATED LOCKED ROTOR STARTING CURRENT AND TORQUE FOR 220-V THREE-PHASE INDUCTION MOTORS OF VARIOUS CLASSES

Motor output power		Rated run	Starting amperes class		Start torque % of rated:					
					Class A and B motor poles			Class C motor poles		
hp	kW	A	B,C,D	F	4	6	8	4	6	8
$\frac{1}{2}$	0.37	2.0	12				150			
$\frac{3}{4}$	0.56	2.8	18				150			
1	0.75	3.5	24		275	175	150			
$1\frac{1}{2}$	1.12	5.0	35		265	175	150			
2	1.50	6.5	45		250	175	150			
3	2.24	9.0	60		250	175	150		250	225
5	3.75	15	90		185	160	130	250	250	225
$7\frac{1}{2}$	5.6	22	120		175	150	125	250	225	200
10	7.5	27	150		175	150	125	250	225	200
15	11.2	40	220		165	140	125	225	200	200
20	15	52	290		150	135	125	200	200	200
25	19	64	365		150	135	125	200	200	200
30	22.4	78	435	270	150	135	125	200	200	200
40	31.5	104	580	360	150	135	125	200	200	200
50	37.5	125	725	450	150	135	125	200	200	200
60	45	150	870	540	150	135	125	200	200	200
75	56	185	1085	675	150	135	125	200	200	200
100	75	246	1450	900	125	125	125	200	200	200
125	90	310	1815	1125	125	125	125	200	200	200
150	112	360	2170	1350	125	125	125	200	200	200
200	150	480	2900	1800	125	125	125	200	200	200

NOTE: Starting *current* of class A motor usually *higher* than B, C, D. Starting *torque* of class D motor usually *higher* than A, B, C. Starting *torque* of class F motor usually *lower* than A, B, C.

22-9-4 Motor Protective Devices. With the probable starting current and running current of the particular motor available from Tables 22-6, 22-7, and 22-8, the proper circuit protection elements can be assigned. Some of the types of motors considered are listed in Table 22-9, where fuse current ratings and overload heater ratings are shown in percent of *full-load* current. There are many variations to the few listings in this table, but they are shown as representative of available NEC code information. This information is part of NEC 430-153.

Many different ratings of fuses and other protective devices are available in various physical family groups. Consult manufacturer's literature to find the specific ratings available. The individual current-handling rating steps are sufficiently small for any degree of protection desired.

Whole families of circuit pilot devices are available to operate the motor control contactors that actually handle the currents that the motors draw. These contactor sizes have been standardized in numbered sizes by the NEMA. The contactor size ratings are listed in Table 22-10. Note that a given size contactor will have provision for adequate rating heater units to cover the range of service currents a motor might require.

For starting and protective equipment, consult the same references for Sect. 21-1. *Machine Design* magazine's "Electric Controls Reference" issue is particularly valuable in that it is periodically updated and has complete manufacturers' lists.

22-9-5 Motor Power Factor. Any ac motor, other than a specifically adjusted large synchronous motor, will operate at a lagging power factor. The power

TABLE 22-9 PROTECTIVE DEVICE RATING OR SETTING FOR MOTORS NOT MARKED WITH CODE LETTER INDICATING LOCKED ROTOR KILOVOLT-AMPERES

	Percent of full-load current	
Motor type	Fuse	Time-limit circuit breaker
Single phase, all types	300	250
Squirrel cage and synchronous with full voltage, line resistor or line reactor starting	300	250
Squirrel cage and synchronous with autotransformer start:		
If not over 30 A	250	200
More than 30 A	200	200
High-reactance squirrel cage		
If not over 30 A	250	250
More than 30 A	200	200
Wound rotor	150	150

TABLE 22-10 STANDARD NEMA RATINGS FOR AC CONTACTOR SIZES

Contactor size	8-hour open rating (A)	Motor power in horsepower (multiply by 0.75 for equiv. kW)					
		Single phase			Three phase		
		115 V	230 V	440 V / 550 V	110 V	208 V / 220 V	440 V / 550 V
00	10	$\frac{1}{3}$	1		$\frac{3}{4}$	$1\frac{1}{2}$	2
0	20	1	2	3	2	3	5
1	30	2	3	5	3	$7\frac{1}{2}$	10
2	50	3	$7\frac{1}{2}$	10	$7\frac{1}{2}$	15	25
3	100	$7\frac{1}{2}$	15	25	15	30	50
4	150				25	50	100
5	300					100	200
6	600					200	400
7	900					300	600
8	1350					450	900
9	2500					800	1600

factor will vary over the load range of the motor. Figure 22-1 shows a power factor curve for a specific 3 hp (2.24 kW), single-phase motor while operating on 242 V. This curve is typical in shape to all types of induction motors. Figure 22-1 clearly shows the very poor power factor displayed by a typical single-phase induction motor, especially during part-load conditions. This particular motor was a capacitor-start, split-phase unit. If it had been a capacitor-start, capacitor-run motor, the full-load power factor might be very near to unity. This type of curve is ordinarily part of a motor manufacturer's characteristic performance curve that is available upon inquiry.

Worked examples of motor selection problems will illustrate the use of these principles.

Example 22-1. An experimental packaging machine is being developed by a small custom-machinery builder. The machine has been driven during its development process by a 3 hp (2.24 kW), single-phase, four-pole, capacitor-start motor. Motor data during the development runs have shown occasional short period current peaks of 15.2 A, with most of the time at 12.5 A. The voltage was, respectively, 240 and 242 V. The starting or breakaway torque was measured at 10.6 ft-lb (14.3 N·m). The customer wishes a 220/440 V three-phase induction motor. Use Fig. 22-1 and the tables in this chapter and determine the following:

(a) What power rating motor to use.
(b) A suitable class motor to meet the starting torque requirement.
(c) The probable starting current of the chosen motor in both 220 and 440 V three phase.

(d) Select a starting contactor size that will be suitable for either service. Choose tentative line fuse and heater current ratings for both voltage levels.

Solution:

(a) From Fig. 22-1, 15.2 A at 240 V would closely correspond to 2.75 hp (2.05 kW), and 12.5 A at 242 V corresponds to 1.93 hp (1.44 kW). It appears that the requirement can be safely met by a motor of 2 hp or 1.5 kW. The 2.75 hp overload is stated as brief and is within the short-time capability of a 2 hp (1.5 kW) motor.

(b) A four-pole three-phase class A or class B induction motor of a nominal 2 hp (1.5 kW) will have a starting torque of 250 percent of its normal rated torque according to Table 22-8. Equation (6-1$_E$) rearranged or Eq. (16-28$_{SI}$), which is Eq. (6-1$_{SI}$) rearranged, shows the normal rated torque of a 2 hp (1.5 kW) motor with an assumed reasonable load speed of 1740 rpm (182.2 rad/sec) to be approximately

$$T = \frac{\text{hp}(5252)}{S} = \frac{2(5252)}{1740} = 6.04 \text{ ft-lb}$$

$$t = W/\omega = \frac{1500}{182.2} = 8.23 \text{ N·m}$$

Check: 1 ft-lb = 1.356 N·m, so 6.04(1.356) = 8.2 N·m. Note that 1500 W or 1.5 kW is *not* exactly 2 hp. The torque that must be met or exceeded from the data is 10.6 ft-lb (14.3 N·m). This is (10.6/6.04)100 = 175 percent of rated or (14.3/8.23)100 = 174 percent of rated. This torque is safely exceeded by a class B motor, which is a desirable general-purpose type. Thus, we can use a 2 hp, four-pole class B motor. The metric equivalent and *probable* SI standard is 1.5 kW. It is not known whether this will be identified as class B or if some entirely different classification will be used.

(c) Table 22-8 shows that a 2 hp (1.5 kW) class B motor will draw a starting current of 45 A on 220 V. There will be an inverse relation between rated volts and starting current, thus, 22.5 A on 440 V.

(d) The contactor sizes are chosen from Table 22-10: a number 0 contactor will adequately handle 3 hp (2.24 kW) on 220 V and 5 hp (3.75 kW) on 440 V. A number 00 contactor could handle the load on 440 V, but not on 220 V. Use size 0. Protective device selection from Table 22-9 is based on full-load current rather than starting current. The *full-load* rated current is found in Table 22-7 to be 6.5 A on 220 V, and 3.3 A on 440 V. Table 22-9 requires squirrel-cage motors that are started across the line to be protected by fuses rated at 300 percent and time-limit circuit breakers (or contactor heaters) to be rated at 250 percent of full-load current.

$$\begin{aligned} \text{Fuses:} \quad & 3 \times 6.5 & = 19.5 \text{ A at 220 V} \\ & 3 \times 3.3 & = 9.9 \text{ A at 440 V} \\ \text{Heaters:} \quad & 2.5 \times 6.5 & = 16.3 \text{ A at 220 V} \\ & 2.5 \times 3.3 & = 8.3 \text{ A at 440 V} \end{aligned}$$

22-10   MACHINE ENCLOSURE TYPES

See Sect. 10-10.   The information for motor enclosure ratings is applicable for ac or dc service.   There are, of course, ac motors fit for services that cannot be accommodated by dc motors.   Fluid-immersed pump motors are in this category, but these are usually very specific designs.

22-11   MAINTENANCE AND ACCESSIBILITY

See Sect. 10-11, where the information presented is again exactly applicable.   Unfortunately, many small single-phase motors are so designed into their particular appliance that the cost of servicing them exceeds the cost of a new replacement.

Single-phase resistance split-phase motors can be serviced as far as bearing replacement or starting switch replacement is concerned.   A burned-out motor of any type and of smaller than the integral horsepower sizes is usually junked and replaced.   The hand-labor cost of rewinding usually exceeds the cost of a new motor.

There are flourishing motor rewinding shops in most large cities in the United States.   These shops owe their existence to the fact that many industrial motors are poorly serviced or not serviced at all.   Integral-horsepower induction motors of all sizes may have new stator windings and/or bearings installed and are ready for another 20 years of hard service.

22-12   EFFICIENCY

Much has been covered in each motor chapter and section on the efficiency of various specific types.   Obviously, the efficiency of a motor affects its cooling problem and its input current requirements.   This in turn affects its starting circuit requirements.   Much tabulated information is available from industry sources.   Even so, a specific and different motor installation problem may require an efficiency calculation for its specific situation.

22-13   RADIO AND TELEVISION INTERFERENCE

The technician studying this material should be aware that any sparking device is a potential source of radio frequency interference.   Three-phase motors do not create any interference unless they have slip rings, as in the larger synchronous motor or in wound-rotor motors.   Slip-ring interference is usually not serious.

The real interference offender is the universal motor and all the dc motors with their commutators.   An arc at a commutator does not have to be visible to be an interference source.

The actual circuit protections that minimize radio and television interference are really the subject of a communications course.   However, if a potentially interfering motor has a completely enclosed and grounded case and has filter capacitors from its lines to ground, the interference problem is usually acceptable.

The filter capacitors are sufficiently small in capacitance so that their capacitive reactance is high at line frequency. As a result, they will drain off a very minimum of line current. Since the RFI or TVI is very high frequency, the same capacitor will have *low* capacitive reactance at those frequencies, and the interference frequency current is led to ground. This is in accordance with the *basic* equation

$$X_c = \frac{10^6}{2\pi fC}$$

where

$$X_c = \text{capacitive reactance in ohms}$$

$$C = \text{capacitance in } microfarads$$

In a similar fashion, series line inductances will have a much higher inductive reactance to the high frequency of the interference source since $X_L = 2\pi fL$. The series field coils may be sufficient for this purpose.

QUESTIONS

22-1. If the desired output speed must be held absolutely constant, what type of motor must be used?

22-2. What type of ac motor may be expected to produce television and/or radio interference?

22-3. What size ranges of motors may be expected to be custom built or especially tailored for the task?

22-4. What determines the number of poles specified in an ac induction or synchronous motor?

22-5. What circumstances would allow an induction motor to run at around 2500 rpm (262 rad/sec)? Consult Table 11-1 or 11-2 for help.

22-6. Within a given frame size, what dimension will change if the motor frame number has a T suffix?

22-7. What does an S suffix letter mean on a motor frame number?

22-8. What is a flange-mounted motor?

22-9. What is the difference between reversible and reversing motors?

22-10. Why are specific voltage levels standardized for motor design?

22-11. What information is provided on a motor characteristic curve?

22-12. What does a locked-rotor code letter mean on a motor nameplate?

22-13. What information is provided by a motor class letter, such as A or B?

22-14. What type of ac motor is capable of operating at unity or even leading power factor?

22-15. Why are single-phase fractional-horsepower induction motors not normally rebuilt when the windings are burned out?

22-16. Of what use are line filter capacitors on a universal motor application?

PROBLEMS

22-1. A machine is to be direct driven by a three-phase 60 Hz motor at just below 1200 rpm, and the load will require about 7 hp. What NEMA frame size is required?

22-2. A machine is to be direct driven by a three-phase 60 Hz motor at about 180 rad/sec and will require about 2.2 kW of shaft power. What NEMA frame size should be chosen?

22-3. What shaft diameter will result from the frame sizes in Problem 22-1?

22-4. What shaft diameter will result from the frame sizes in Problem 22-2?

22-5. What current will a single-phase, $\frac{3}{4}$ hp, 230 V ac induction motor draw when carrying a normal load?

22-6. How much current would a normally loaded 115 V, 0.375 kW ac induction motor be expected to draw?

22-7. If the motor in Fig. 22-1 were developing its full rated 3 hp, but the line voltage had dropped to 208 V, what line current might be expected? In calculating, consider the volts and amperes in Fig. 22-1 to be rated conditions.

22-8. A $7\frac{1}{2}$ hp, three-phase induction motor is wound in the usual fashion so that it may be connected in series wye or parallel wye.
(a) What line current will it draw at full load in parallel on 208 V lines?
(b) What current will it draw on 220 V lines?
(c) What current will it draw in series on 440 V lines?

22-9. A three-phase induction motor is rated at 1.12 kW and can be connected in parallel for 110 V or series for 208 or 220 V.
(a) What current will it draw at full load on 110 V?
(b) What current at 208 V in series at full load?
(c) What current at 220 V in series at full load?

22-10. What current will the motor in Problem 22-8 draw at start when operated at 220 V?

22-11. What current will the motor in Problems 22-8 and 22-10 draw at start if it is run on 208 V?

22-12. What current will the motor in Problem 22-9 draw at start when operated at 220 V?

22-13. If the motor in Problems 22-9 and 22-12 is reconnected in parallel and operated on 110 V, what starting current will it now require?

22-14. A 10 hp four-pole, class B 220 V motor is to be used on a machine that will require 135 percent of rated torque for starting. Determine the following:
(a) Its starting torque in percent of rated torque on its rated voltage.
(b) Its starting torque in percent of rated torque on 208 V. *Hint:* consider percent rated torque as if it were rated torque.
(c) Will it start the load on 208 V?

22-15. If the motor in Problem 22-14 is to be operated on 220 V, what protective fuse rating will it need?

22-16. What contactor size will the motor in Problems 22-14 and 22-15 require (a) if run in parallel on 220 V lines, or (b) if run in series on 440 V lines?

Answers. **22-1**, 256 U or 254 T; **22-2**, 213 or 182 T; **22-3**, 256 U is 1.375 in. dia., 254 T is 1.625 in. dia.; **22-4**, 1.125 in. dia.; **22.5**, 6.9 A; **22-6**, 9.8 A; **22-7**, 18.8 A; **22-8(a)**, 23 A, **(b)** 22 A, **(c)** 11 A; **22-9(a)**, 10 A, **(b)** 5.3 A, **(c)** 5.0 A; **22-10**, 120 A; **22-11**, 113 A; **22-12**, 35 A; **22-13**, 70 A; **22-14(a)**, 175%, **(b)** 156%, **(c)** yes; **22-15**, 81 A; **22-16(a)**, size 2, **(b)** size 1.

Appendix A

Unit Conversions

UNIT CONVERSIONS USEFUL IN MACHINE CALCULATION

Unit	cgs	English	English	RMKS or SI
Length	l'	L	L	l
1 cm =		0.393 70 in. =	0.032 808 ft =	10^{-2} m
2.54* cm =		1 in. =	0.083 333 ft =	0.0254 m
30.48 cm =		12 in. =	1 ft =	0.3048 m
10^2 cm =		39.370 in. =	3.2808 ft =	1 m
Area	a'	A	A	a
1 cm^2 =		0.155 00 in.^{2} =	0.001 076 4 ft^2 =	10^{-4} m^2
6.4516 cm^2 =		1 in.^{2} =	0.006 944 4 ft^2 =	6.4516×10^{-4} m^2
929.03 cm^2 =		144 in.^{2} =	1 ft^2 =	0.092 903 m^2
10^4 cm^2 =		1550.0 in.^{2} =	10.764 ft^2 =	1 m^2
Volume	u'	U	U	u
1 cm^3 =		0.061 024 in.^{3} =	$0.353\ 15 \times 10^{-4}$ ft^3 =	10^{-6} m^3
16.387 cm^3 =		1 in.^{3} =	5.7870×10^{-4} ft^3 =	16.387×10^{-6} m^3
283 17 cm^3 =		1728 in.^{3} =	1 ft^3 =	0.028 317 m^3
10^6 cm^3 =		610 24 in.^{3} =	35.315 ft^3 =	1 m^3
Force	f'	F	f	Obsolete metric f''
1 dyne =		$0.224\ 81 \times 10^{-5}$ lb =	10^{-5} newton =	$0.101\ 97 \times 10^{-5}$ kgf
4.4482×10^5 dyne =		1 lb =	4.4482 newton =	0.453 59 kgf
10^{5*} dyne =		0.224 81 lb =	1 newton =	0.101 97 kgf
9.8066×10^5 dyne =		2.2046 lb =	9.8066 newton =	1 kilogram force

Flux	ϕ'	Φ		ϕ
1 line	$=$	1 line	$=$	1 weber $\times 10^{-8}$
10^{8*} line	$=$	10^8 line	$=$	1 weber

Flux Density	β'	B		β
$\dfrac{1 \text{ line}}{\text{cm}^2}$	$=$	$\dfrac{6.4516 \text{ line}}{\text{in.}^2}$	$=$	$\dfrac{10^{-4} \text{ weber}}{\text{m}^2}$
$\dfrac{0.155\,00 \text{ line}}{\text{cm}^2}$	$=$	$\dfrac{1 \text{ line}}{\text{in.}^2}$	$=$	$\dfrac{0.155\,00 \times 10^{-4} \text{ weber}}{\text{m}^2}$
$\dfrac{10^4 \text{ line}}{\text{cm}^2}$	$=$	$\dfrac{6.4516 \times 10^4 \text{ line}}{\text{in.}^2}$	$=$	$\dfrac{1 \text{ weber}}{\text{m}^2}$

*Definitions:
$2.540\,000\,0$ cm $= 1$ in.
10^5 dyne $= 1$ newton
10^8 lines $= 1$ weber $= \dfrac{1 \text{ newton meter}}{\text{ampere}}$

$\dfrac{1 \text{ weber}}{\text{meter}^2} = \dfrac{1 \text{ newton}}{\text{meter ampere}}$

NOTE: 1 line = 1 maxwell.
$1 \text{ line/cm}^2 = 1$ gauss.

FORCE UNIT CONVERSIONS USEFUL WITH MOTORS

Unit:	cgs	English		SI	Obsolete metric
Quantity:	f'	F	F'		f
1 dyne $= 1.0197 \times 10^{-3}$ g		$= 3.5969 \times 10^{-5}$ oz $= 0.224\,81 \times 10^{-5}$ lb		$= 10^{-5}$ N $= 0.101\,97 \times 10^{-5}$ kgf	
980.66 dyne $= 1$ g		$= 35.274 \times 10^{-3}$ oz $= 2.2046 \times 10^{-3}$ lb		$= 9.8066 \times 10^{-3}$ N $= 10^{-3}$ kgf	
27.802×10^3 dyne $= 28.350$ g		$= 1$ oz $= 0.0625$ lb		$= 0.278\,02$ N $= 28.350 \times 10^{-3}$ kgf	
4.4482×10^5 dyne $= 453.59$ g		$= 16$ oz $= 1$ lb		$= 4.4482$ N $= 0.453\,59$ kgf	
10^5 dyne $= 101.97$ g		$= 3.5969$ oz $= 0.224\,81$ lb		$= 1$ N $= 0.101\,97$ kgf	
9.8066×10^5 dyne $= 10^3$ g		$= 35.274$ oz $= 2.2046$ lb		$= 9.8066$ N $= 1$ kgf	

NOTE: Force includes relationship to obsolete gram and kilogram force, which are still used in instruments.

WORK OR TORQUE UNIT CONVERSIONS USEFUL WITH MOTORS

Unit:	cgs	English	SI	Obsolete metric
Quantity:	$f'd'$ or t'	FD, $F'D'$, or T, T'		fd or t
1 dyne cm $= 1.0197 \times 10^{-3}$ g cm		$= 1.4161 \times 10^{-5}$ oz in. $= 7.3755 \times 10^{-8}$ lb ft	$= 10^{-7}$ N·m $= 1.0197 \times 10^{-8}$ kgf m	
980.66 dyne cm $= 1$ g cm		$= 0.013\,887$ oz in. $= 7.2329 \times 10^{-5}$ lb ft	$= 9.8066 \times 10^{-5}$ N·m $= 10^{-5}$ kgf m	
7.0615×10^4 dyne cm $= 72.008$ g cm		$= 1$ oz in. $= 5.2083 \times 10^{-3}$ lb ft	$= 7.0615 \times 10^{-3}$ N·m $= 0.720\,08 \times 10^{-3}$ kgf m	
1.3558×10^7 dyne cm $= 1.3826 \times 10^4$ g cm		$= 192$ oz in. $= 1$ lb ft	$= 1.3558$ N·m $= 0.138\,26$ kgf m	
10^7 dyne cm $= 1.0197 \times 10^4$ g cm		$= 141.61$ oz in. $= 0.737\,55$ lb ft	$= 1$ N·m $= 0.101\,97$ kgf m	
9.8066×10^7 dyne cm $= 10^5$ g cm		$= 1.3887 \times 10^3$ oz in. $= 7.2329$ lb ft	$= 9.8066$ N·m $= 1$ kgf m	

ANGULAR VELOCITY OR SPEED UNIT CONVERSIONS

Unit:	cgs	English	SI
Quantity:	s or rps	S or rpm	ω or rad/sec
1 rev/sec		= 60 rev/min	= 2π or 6.2832 rad/sec
0.016 667 rev/sec		= 1 rev/min	= 0.104 72 rad/sec
0.159 15 rev/sec		= 9.5493 rev/min	= 1 rad/sec

Appendix B

Induction Motor Locked Rotor Indicating Code Letters

Code letter	Kilovolt-amperes per horsepower with locked rotor	Kilovolt-amperes per kilowatt with locked rotor
A	0–3.14	0–4.21
B	3.15–3.54	4.22–4.75
C	3.55–3.99	4.76–5.35
D	4.0–4.49	5.36–6.02
E	4.5–4.99	6.03–6.69
F	5.0–5.59	6.70–7.50
G	5.6–6.29	7.51–8.44
H	6.3–7.09	8.45–9.51
J	7.1–7.99	9.52–10.72
K	8.0–8.99	10.73–12.06
L	9.0–9.99	12.07–13.40
M	10.0–11.19	13.41–15.01
N	11.2–12.49	15.02–16.75
P	12.5–13.99	16.76–18.76
R	14.0–15.99	18.77–21.44
S	16.0–17.99	21.45–24.13
T	18.0–19.99	24.14–26.81
U	20.0–22.39	26.82–30.03
V	22.4–and up	30.04–and up

Derived from NEC 430-76 and as adopted by NEMA modified for kilowatt rating.

Bibliography

(1) Liwschitz-Garik, M. M., and C. C. Whipple, *Direct Current Machines*, 2nd Ed. New York: Van Nostrand Reinhold Company, Inc., 1956.

(2) Kuhlmann, J. H., *Design of Electrical Apparatus*, 2nd Ed. New York: John Wiley & Sons, Inc., 1940.

(3) Siskind, C. S., *Direct-Current Machinery*. New York: McGraw-Hill Book Company, 1952.

(4) Kosow, I. L., *Electric Machinery and Transformers*. Englewood Cliffs, N.J.: Prentice-Hall, Inc., 1972.

(5) Dawes, C. L., *Electrical Engineering, Vol. I, Direct Currents*, 4th Ed. New York: McGraw-Hill Book Company, 1952.

(6) Kosow, I. L., *Control of Electric Machines*. Englewood Cliffs, N.J.: Prentice-Hall, Inc., 1973.

(7) McIntyre, R. L., *Electric Motor Control Fundamentals*, 3rd Ed. New York: McGraw-Hill, Inc., 1974.

(8) Millermaster, R. A., *Harwood's Control of Electric Motors*, 4th Ed. New York: John Wiley & Sons, Inc. (Interscience Division), 1970.

(9) Siskind, C. S., *Electrical Control Systems In Industry*. New York: McGraw-Hill, Inc., 1963.

(10) Alerich, W. N., *Electric Motor Control*. Albany, N.Y.: Delmar Publishers, 1965.

(11) James, H. D., and L. E. Markle, *Controllers for Electric Motors*, 2nd Ed. New York: McGraw-Hill, Inc., 1952.

(12) Liwschitz-Garik, M. M., and C. C. Whipple, *Alternating Current Machines*, 2nd Ed. New York: Van Nostrand Reinhold Company, Inc., 1961.

(13) Dawes, C. L., *Electrical Engineering, Vol. II, Alternating Currents*, 4th Ed. New York: McGraw-Hill, Inc., 1947.

(14) Siskind, C. S., *Electrical Machines, Direct and Alternating Current*, 2nd Ed. New York: McGraw-Hill Book Company, 1959.

(15) Johnson, E. R., *Servomechanisms*. Englewood Cliffs, N.J.: Prentice-Hall, Inc., 1963.

(16) Charkey, E. S., *Electromechanical System Components*. New York: John Wiley & Sons, Inc. (Interscience Division), 1972.

(17) Datta, S. K., *Power Electronics and Controls*. Englewood Cliffs, N.J.: Prentice-Hall, Inc., 1985.

(18) Wildi, T., *Electrical Power Technology*. New York: John Wiley and Sons, 1981.

(19) Gottlieb, I. M., *Power Control with Solid State Devices*. Englewood Cliffs, N.J.: Prentice-Hall, Inc., 1985.

(20) General Electric Co., *SCR Manual*. Syracuse, N.Y. Sixth Edition, 1979.

(21) Emanuel, P. J., *Motors, Generators, Transformers and Energy*. Englewood Cliffs, N.J.: Prentice-Hall, Inc., 1985.

Index

Armature (*cont'd*)
 frog leg, 57
 lap, 56
 multielement, 57
 multiplex, 57
 pitch factor, 357
 simplex lap, 58
 simplex wave, 58
 triplex, 58
 types of, 55
 wave, 56
Asynchronous induction dynamo, 553
Autotransformer
 polyphase motor starting, 547, 549
 single-phase speed control, 506
 starting, dual-capacitor motor, 504

Back emf. *See* Counter emf
Bearings, 52, 339
Biasing, 232, 237, 239, 240
Biot-Savart's law, 16, 36, 38, 39
Brush, 51
 drop voltage, 199, 201
 rigging, 53
 shifting, 62
Brushless or commutatorless motor, 65
Buildup, conditions for, 89
Bus safety switch, 283

Can Motor, 79
Capacitor
 motor, permanent split, 506
 start (*see* Motor, ac, single phase)
 synchronous, 483
Centrifugal switch, 497, 499
Chopper, 308
Circuital law of magnetics, 15, 69
Circuit breaker, 225
 rating, 590
Coil, 51, 53
 balance (*see* Balance coil)
 reactance, 64, 497
 resistance, 187
 span, 357
Commutating pole, 54, 62
Commutation process, 28
 brush neutral position, 60
Commutator
 brush spacing, 60
 segment, 51
 sparking of, 62

Compensating winding, 64
Compound motor. *See* Motor, dc, compound
Compounding, degree of, 104
Conductor, 15
 emf induced in, 17
 force on, 36
 per turn, 18, 31
Consequent pole, 460
Contactor
 definition, 222
 size, 591
Control, 218, 540
 ac, 540
  characteristic list, 541
  closed loop, 312, 314
  components, 221, 542
  environment, 280, 541
  nonreversing starter, 543
  normal running, 279, 541
  reduced current circuits, 550
  reduced voltage circuits, 547
  reversing full voltage starter, 545
  reversing requirement, 279, 541
  safety, 280, 541
  speed control, 555, 559, 569
  starting requirement, 278, 541
  stopping requirement, 278, 541
  transformers, 541
  transformers, control circuit, 541
  wound rotor starting, 550
  Y-Δ starting, 547
 dc, 277
  characteristic list, 278
  components, 218, 221
  component symbols, 221, 230
  current limit starter, 285
  definite time starter, 281
  direction of rotation, 160, 279, 288
  dynamic braking, 296
  electric brake, 298
  environmental requirements, 280
  field, 301
  jogging, 299
  normal running, 279
  pilot control device, 227, 231
  plugging reverse, 291
  power supply, 278
  primary control device, 221, 227
  pushbutton reverse, 288